CW01164034

A CASEBOOK FOR SPATIAL STATISTICAL DATA ANALYSIS

A CASEBOOK FOR SPATIAL STATISTICAL DATA ANALYSIS

A Compilation of Analyses of Different Thematic Data Sets

DANIEL A. GRIFFITH
LARRY J. LAYNE
with contributions by J. K. Ord and Akio Sone

New York   Oxford
Oxford University Press
1999

Oxford University Press

Oxford   New York
Athens   Auckland   Bangkok   Bogotá   Buenos Aires   Calcutta
Cape Town   Chennai   Dar es Salaam   Delhi   Florence   Hong Kong   Istanbul
Karachi   Kuala Lumpur   Madrid   Melbourne   Mexico City   Mumbai
Nairobi   Paris   São Paulo   Singapore   Taipei   Tokyo   Toronto   Warsaw

and associated companies in
Berlin   Ibadan

Copyright © 1999 by Oxford University Press

Published by Oxford University Press, Inc.
198 Madison Avenue, New York, New York 10016

Oxford is a registered trademark of Oxford University Press

All rights reserved. No part of this publication may be reproduced,
stored in a retrieval system, or transmitted, in any form or by any means,
electronic, mechanical, photocopying, recording, or otherwise,
without the prior permission of Oxford University Press.

Library of Congress Cataloging-in-Publication Data
Griffith, Daniel A.
A casebook for spatial statistical data analysis : a compilation of analyses
of different thematic data sets / Daniel A. Griffith and Larry J. Layne
with contributions by J. K. Ord and Akio Sone.
 p.   cm.
Includes bibliographical references and index.
ISBN 0-19-510958-9
1. Geography—Statistical methods.   2. Spatial analysis (Statistics).
I. Layne, Larry J.   II. Ord, J. K.   III. Sone, Akio.   IV. Title.
G70.3.G744   1999
910'.72—dc21   9853829

9 8 7 6 5 4 3 2 1
Printed in the United States of America
on acid-free paper

This book is dedicated to R. Dale and Mary E. Swartz.

Daniel A. Griffith

✪

This book is dedicated to Dr. Francesco Lagona and Ludmila Coromoto Ortegano Sanchez.

Larry J. Layne

PREFACE

If a lunatic scribbles a jumble of mathematical symbols it does not follow that the writing means anything merely because to the inexpert eye it is indistinguishable from higher mathematics.
                    Eric Temple Bell (Newman 1956, 308)

If a naive researcher completes a standard statistical analysis of georeferenced data, it does not follow that the data analytic results have turned data into meaningful information merely because to the inexpert eye they are indistinguishable from conventional statistics results! Empirical work in many scientific fields is based on data for which the location of observations is an important feature. These type of data, especially when they are accompanied by a locational tag (e.g., latitude and longitude, Cartesian or Universal Transverse Mercator [UTM] coordinates), can be referred to as georeferenced or spatial data. The observations often consist of a single cross-section of spatial units or sometimes of a time series of cross-sections. The distinctive characteristic of the statistical analysis of spatial data is that the spatial pattern of locations (point patterns), the spatial association between attribute values observed at different locations (spatial dependence), and the systematic variation of phenomena by location (spatial heterogeneity) become the major foci of inquiry. In addition to being of interest in and of itself (from a geographer's perspective), any spatial pattern embedded in data causes a number of measurement problems, referred to as spatial effects (spatial autocorrelation, spatial heterogeneity), that affect the validity and robustness of traditional statistical description and inference methods when applied to this category of data. Recognition of this complication has given rise to an increasingly sophisticated body of specialized techniques developed in the fields of geostatistics, spatial statistics, and spatial econometrics. These techniques are not just relevant in geography but are applicable to a wide range of scientific areas, such as agriculture, archaeology, criminology, demography, ecology, environmental studies, epidemiology, forestry, geology, international relations, natural resources, regional science, remote sensing, sociology, statistics, urban planning, and urban and regional economics.

With the increasing availability of geographic information systems (GIS), which has helped to pique interest in spatial data analysis in a wide range of fields (Wilson, 1996), the need has arisen to furnish many more empirical applications of good spatial analysis. This very goal is echoed by contributors to both *Spatial Statistics: Past, Present, and Future* (Griffith 1990) and *Geostatistical Case Studies* (Matheron and Armstrong 1987), and with the aim of helping to avoid spatial statistical malpractice, it was a principal motivation for compiling the *CRC Practical Handbook of Spatial Statistics* (Arlinghaus 1996). In a cross-disciplinary context this goal also is promoted as a fruitful direction for research by the editorial panel of and contributors to the National Research Council's *Spatial Statistics and Digital Image Analysis* (1991).

As one response to this literature gap and this clamor for more empirical applications, we
- discuss, in summary format, the theoretical basis of prominent spatial statistical techniques, in an applications format with implementation guidelines;
- provide additional numerical evidence establishing links between geo- and spatial statistics; and
- present a wide variety of data examples taken from a disparate set of disciplines.

Principal templates for this type of volume are provided by Hand et al. (1994), *Handbook of Small Data Sets*, and Andrews and Herzberg (1985), *Data: A Collection of Problems from Many Fields for the Student and Research Worker*. The data sets analyzed in this book comprise readily available ones (35), some of our own (3), and some unpublished ones from colleagues (4). But published georeferenced data sets are not always complete, frequently lacking locational tags. In some cases, we have secured locational tags from other (especially Internet) sources, or we have performed the digitizing task ourselves. Additional valuable georeferenced data sets are noted throughout, too, especially from such popular sources as Hamilton (1992), Rasmussen (1992), Carr (1995), and Chatterjee et al. (1995); and, government data sets are everywhere online, with many linked to the National Spatial Data Clearinghouse Web site being well worth retrieving. Analyzing such a broad collection of data sets, covering a wide range of situations, illustrates many types of spatial statistical analyses and helps point out gaps in current spatial statistical methodology.

Georeferenced data sets increasingly are massive; historically, one major source of such very large data sets has been remote sensing. Special concerns affiliated with this size of data set resulted in the National Research Council publishing *Massive Data Sets* (1996). To date, sampling often has been used to reduce such data sets to a manageable form, but sampling may well obscure latent spatial dependency. In §5.7, we illustrate the ability to analyze such voluminous data using implementations outlined in this text.

Software for implementation purposes discussed in this book includes SAS and SPSS routines (which, together with Dr. Akio Sone of Tsukuba University, we have developed), GEO-EAS, Heuvelink's semivariogram model fitting module, and VARIOWIN. Accordingly, the recent release of PROC GIS in SAS makes this book a timely addition to the literature. Many other dedicated software packages are available, too, and sometimes are mentioned in this book, including Deutsch and Journel's GISLIB, specially developed MINITAB macro code, and Anselin's GAUSS-based package for the analysis of spatial data (SpaceStat).

In addition to *Advanced Spatial Statistics* (Griffith, 1988), a number of books have appeared during the past decade that deal with spatial statistics, such as Anselin's *Spatial Econometrics* (1988), Arbia's *Spatial Data Configuration in Statistical Analysis of Regional Economics and Related Problems* (1989), Christakos's *Random Field Models in Earth Sciences* (1992), Christensen's *Linear Models for Multivariate, Time Series, and Spatial Data* (1991), Cressie's *Statistics for Spatial Data* (1991), Gregoire et al.'s *Modelling Longitudinal and Spatially Correlated Data* (1997), Isaaks and Srivastava's *An Introduction to Applied Geostatistics* (1989), Ripley's *Statistical Inference for Spatial Processes* (1988), Guyon's *Random Fields on a Network* (1995), Haining's *Spatial Data Analysis in the Social and Environmental Sciences* (1990), and Upton and Fingleton's *Spatial Data Analysis by Example* (1985, 1989). None of these publications, however, specifically addresses the goals that we have outlined. Furthermore, this volume differs from these others in that we deal with spatial statistics by

- pursuing unification of the perspectives advanced in geostatistics, spatial statistics, and spatial econometrics (although Cressie's treatise verges on being an exception), and
- combining a rigorous methodological treatment with a comparative applications focus, while including numerous new or freshly analyzed, illustrative data sets and worked examples.

In achieving these goals and pursuits in this volume, we hope to have furnished invaluable insights to professional researchers, academics, graduate students, and analysts in a range of disciplines that deal with spatial data and who are increasingly using the tool of GIS in their work.

Acknowledgments

The assistance of many people made this book possible. Both of us enjoyed the complete support of the Department of Geography, Syracuse University, with funding from the National Science Foundation (#SBR-9507855), over two years, enabling us to complete the research upon which this book is based. Parts of this volume were presented to the Spring 1997 students of GEO 686: Advanced Spatial Statistics; their feedback is most appreciated. We especially appreciate the efforts of Mr. Peter L. Fellows, a student in this class, who completed the task of double-checking much of the SAS code. Dr. J. Keith Ord, of the Pennsylvania State University, contributed a commentary on our paper appearing in the 1996

Proceedings of the American Statistical Association; it appears here in §3.3.3 in its entirety, and we thank him for this invaluable contribution. Dr. Akio Sone, of Tsukuba University, contributed the SPSS code that appears in Appendices 3-B, 3-D and 3-F; we thank him for his invaluable contribution, too. We are grateful to Dr. Gerard Heuvelink, of the University of Amsterdam, who made the latest version of his WLSFIT software available to us for use in our research. We also are grateful to Dr. Susanna McMaster, of Macalester College, for allowing us to analyze her forest stand data, to Dr. William T. Elberty, of St. Lawrence University, for allowing us to analyze his Adirondack data—and to Mr. Peter Fellows for arranging for us to secure these data—and to Dr. W. Scott Overton, of Oregon State University, for arranging for us to obtain the two EMAP data sets. We are indebted to Mr. Michael Kirchoff, of the Cartography Laboratory at Syracuse University, for providing a very high level of professional assistance with the figures and maps. We wish to thank Ms. Kay Steinmetz, at Syracuse University, for furnishing us with expert copy-editing advice. We also wish to thank Ms. Diane Griffith for helping us cope with several serious wordprocessing glitches, and for unselfishly giving of her time to help size the considerable number of graphics. And last, but certainly not least, we would like to express considerable thanks to Ms. Michele Griffith for helping us retrieve a number of data sets from Internet sources and for helping us compile the bibliography; she saved us from having to engage in an enormous amount of tedious and time-consuming work!

Needless to say, of course, we alone assume full responsibility for any errors in writing, logic, calculations, and content.

Syracuse, New York
March 1998

Daniel A. Griffith
(griffith@maxwell.syr.edu)
Larry J. Layne
(lortegan@reacciun.ve)

CONTENTS

List of Data Sets Analyzed — xv

Abbreviations — xvi

PART I: THEORETICAL BACKGROUND

1. Introduction — 3
   1.1. Parallels between Spatial Autoregression and Geostatistics — 6
   1.2. The Many Faces of Spatial Autocorrelation — 9
   1.3. The Moran Coefficient Scatterplot Tool — 15
   1.4. Multivariate Spatial Association — 18
   1.5. Heterogeneity and Locational Information — 25
   1.6. The Semivariogram Plot Tool — 34
   1.7. Computer Code for Implementing Spatial Statistical Analyses — 39
   Appendix 1-A: SAS Code for Computing MC Using Standard Regression Techniques. — 62
   Appendix 1-B: Directions for Using ArcInfo to Construct a Thiessen Polygon Surface Partitioning for a Set of Georeferenced Points. — 64
   Appendix 1-C: SAS Macros Used for Converting among Degrees, Decimal Degrees and Radians — 67

2. Important Modeling Assumptions — 69
   2.1. The Relative Importance of the Principal Assumptions — 71

	2.2.	Variable Transformations	73
	2.3.	Transforming in Search of Normality	75
	2.4.	Transforming in Search of Constant Variance	79
	2.5.	Linearity: Exploitation of Linear Relationships by Linear Statistical Models	83
	2.6.	An Absence of Independence: The Presence of Spatial Autocorrelation in Georeferenced Data	86
	2.7.	Exploring Residuals in Spatial Analysis	105
	2.8.	Other Statistical Frequency Distribution Assumptions	111
3.		Popular Spatial Autoregressive and Geostatistical Models	120
	3.1.	Spatial Autoregressive Models	121
	3.2.	Geostatistical Models	133
	3.3.	Articulating Relationships Between Spatial Autoregressive and Geostatistical Models	142
	3.4.	Computer Code for Spatial Autoregressive and Semivariogram Modeling	153
		Appendix 3-A: SAS Code for Estimating Equations (3.7a-c)—the CAR Model	165
		Appendix 3-B: SPSS Code for Estimating Equations (3.7a-c)—the CAR Model	169
		Appendix 3-C: SAS Code for Estimating Equations (3.8) & (3.9)—the SAR & AR Models	172
		Appendix 3-D: SPSS Code for Estimating Equations (3.8) & (3.9)—the SAR & AR Models	175
		Appendix 3-E: SAS Code for Selected Semivariogram Models	178
		Appendix 3-F: SPSS Code for Selected Semivariogram Models	194

PART II: GEOREFERENCED DATA SET CASE STUDIES 205

4.		Analysis of Georeferenced Socioeconomic Attribute Variables	207
	4.1.	The Cliff-Ord Eire Population Data	211
	4.2.	Urban Population Density	214
	4.3.	Residential Insurance Coverage in Chicago	218
	4.4.	Urban Crime in Columbus, Ohio	222
	4.5.	Geographic Distribution of Minorities across Syracuse, New York	227
	4.6.	Concluding Comments: Spatial Autocorrelation and Socioeconomic Attribute Variables	232
		Appendix 4-A: Centroids Derived from a Digitized Version of Cliff and Ord's Map of Eire (1981, 207) Using ArcInfo	234

5. Analysis of Georeferenced Natural Resources Attribute Variables ... 235
   5.1. Kansas Oil Wells Data ... 241
   5.2. Natural Resources Inventory Data ... 245
   5.3. Island Biogeography: Plant Species Data ... 265
   5.4. Weather Station Rainfall Data ... 272
   5.5. Drainage Basin Runoff Data ... 277
   5.6. Digital Elevation Data ... 284
   5.7. Preclassified Remotely Sensed Image Reflectance Data ... 292
   5.8. Concluding Comments: Spatial Autocorrelation and Natural Resources Attribute Variables ... 302

6. Analysis of Georeferenced Agricultural Yield Variables ... 305
   6.1. The Mercer-Hall Straw Yield Data ... 311
   6.2. The Wiebe Wheat Yield Data ... 316
   6.3. The Broadbalk Wheat and Straw Yield Data ... 320
   6.4. Sugar Cane Production in Puerto Rico ... 330
   6.5. Milk Production in Puerto Rico ... 337
   6.6. Concluding Comments: Spatial Autocorrelation and Agricultural Yield Variables ... 350

7. Analysis of Georeferenced Pollution Variables ... 353
   7.1. Southwestern Pennsylvania Coal Ash ... 360
   7.2. EMAP Indicators of Ecological Condition ... 365
   7.3. Great Smoky Mountains Water pH ... 373
   7.4. Chemical Elements in Northwest Texas Groundwater ... 378
   7.5. Hazardous Waste Contamination of Soil: Dioxin ... 390
   7.6. Concluding Comments: Spatial Autocorrelation and Pollution Variables ... 393

8. Analysis of Georeferenced Epidemiological Variables ... 396
   8.1. Glasgow Standardized Mortality Rates ... 405
   8.2 Pediatric Lead Poisoning in Syracuse, New York ... 410
   8.3. Fox Rabies in Germany ... 417
   8.4. Concluding Comments: Spatial Autocorrelation and Epidemiological Variables ... 423

PART III: VISUALIZING WHAT IS NOT OBSERVED ... 427

9. Exploding Georeferenced Data When Maps Have Holes or Gaps: Estimating Missing Data Values and Kriging ... 428
   9.1. An Introduction to EM Estimation ... 430

	9.2.	Estimating Missing Values: Two Simplified Georeferenced Data Illustrations	432
	9.3.	Estimating a Conspicuous Missing Data Value for the Coal-Ash Data Set	438
	9.4.	Estimating Conspicuous Missing Data Values for an Agricultural Experiment	441
	9.5.	Estimating Missing Median Family Income Data for Ottawa-Hull	445
	9.6.	Generalizing a Map Surface with Kriging	447
	9.7.	A Cross-Validation Example	451
	9.8.	Concluding Comments: Exploding Georeferenced Data	454
10.	Concluding Comments		457
	10.1.	More about the Nature of Georeferenced Data	459
	10.2.	Reflections on Spatial Data Model Specifications	460
	10.3.	Implications regarding Relations between Spatial Autoregressive and Geostatistical Models	469
	10.4.	Reflections on Kriging	473
	10.5.	Spatial Statistics and GIS	475
	10.6.	Now Is the Time for All Good Spatial Scientists to ...	476
	10.7.	Some Questions Yet Unanswered: Future Research	479

Epilogue	484
References	488
Subject Index	503

LIST OF DATA SETS ANALYZED

Anselin's Columbus, Ohio, Urban Crime Data
Anselin's Southwestern Ohio County Wage Data
Bailey and Gatrell's LANDSAT 5 Remotely Sensed Data: High Peak, England
Broadbalk Wheat and Straw Yield Data for Selected Time Periods
California-Mexico Coastal Localities Species Data
Chemical Elements in Northwest Texas Groundwater
Cliff-Ord's Eire Population Data
Cressie's North Carolina SIDs Data
Cressie's Wolfcamp Aquifer Piezometric-head Water Pressure Data
Cressie's Coal-Ash Data for Southwestern Pennsylvania
Cressie and Clark's Coal-Burning Particulate Emissions Data
Dioxin Contamination of Soil
EMAP Great Northern Forest Data
EMAP Near-Coastal Waters Data
Fox Rabies in Germany
Galapagos Islands Species Data
Great Smoky Mountains Water pH Data
Griffith and Amrhein's Ontario Drainage Basin Data
Haining's Glasgow Cancer Data
Haining et al.'s Weather Station Rainfall Data: Kansas-Nebraska
IDRISI Remotely Sensed Image (Worcester, Massachusetts) Classification Tutorial Data
Isaaks and Srivastava's Walker Lake Region Digital Elevation Data
Jones and Wrigley's Kansas Oil Wells Data
LANDSAT 5 "Thematic Mapper" Remotely Sensed Data: Adirondack Park, New York
Matui's Japanese Settlement Data
McMaster's Minnesota Township Forest Stands Data
Mercer-Hall Wheat and Straw Yield Data
Milk Production in Puerto Rico for Selected Time Periods: USDA Data
Ottawa-Hull, Ontario, Median Family Income
Ottawa-Hull, Ontario, Population Density
Pediatric Lead Poisoning in Syracuse, New York
Puerto Rican Agricultural Productivity Data
Puerto Rican Urban Population Data for 1990
Rayner's Vandalized Turnips Field Plot Experiment Data
Residential Insurance Coverage in Chicago
Simulated Data from Theoretical Frequency Distributions
Sugar Cane Production for Selected Time Periods: Commonwealth of Puerto Rico
Syracuse, New York, Minority Populations
Syracuse, New York, Population Density Data
Wayne County, Michigan, Air Pollution Data
Weather Station Rainfall Data: New England
Wiebe Wheat Yield Data

ABBREVIATIONS

A-D	Anderson-Darling normality test statistic
AAR	agricultural administrative regions (of Puerto Rico)
ANCOVA	analysis of covariance
ANOVA	analysis of variance
AR	autoregressive response (model specification)
ARA	arterial road network index (by Cliff and Ord)
AREA	surface area of an areal unit
C	the covariogram of geostatistics (scalar function); the geographic connectivity/ weights matrix (matrix); the EMAP acronym for Chesapeake Bay (Roman letter)
CAR	conditional autoregressive (model specification)
CBD	central business district
CMA	community medicine area
COS	trigonometric cosine function
COV(✡)	the covariance of two variables denoted by ✡
CPR	Commonwealth of Puerto Rico (denoting its Department of Agriculture)
CPRSC	Commonwealth of Puerto Rico sugar cane production figures
CT	census tract for Syracuse data; Connecticut state acronym in EMAP
d	distance
D	Delaware Bay acronym in EMAP
dbh	diameter at breast height (of a tree)
DBN	density of banana/plantain production in Puerto Rico
DCD	density of cultivated land in Puerto Rico
DCF	density of coffee production in Puerto Rico
DEM	digital elevation model
df	(the number of) degrees of freedom
DFAM	density of farm families in Puerto Rico
DFRM	density of farms in Puerto Rico
diag	diagonal matrix operation
DMLK	density of milk production in Puerto Rico
DO	dissolved oxygen in water
DSGR	density of sugarcane production in Puerto Rico
DTB	density of tobacco production in Puerto Rico
DTED	digital terrain elevation data
DTM	digital terrain model
E	an eigenvector matrix (in bold type)
EM	estimation-maximization algorithm for estimating missing data
E(✡)	expected value of a variable denoted by ✡ (in nonbold type)
EMAP	Environmental Monitoring an Assessment Program (U.S. EPA project)
EPA	Environmental Protection Agency
ESIC	Earth Science Information Center

ESSL	Engineering and Scientific Subroutine Library (for the IBM RISC/6000)
EXP	exponentiation (antilogarithm for base e)
GIS	geographic information systems
GLS	generalized least squares
GR	Geary Ratio
iid	independent and identically distributed
IMSL	International Mathematical and Statistical Library
JASA	Journal of the American Statistical Association
K-S	Kolmogorov-Smirnov normality test statistic
LANDSAT	a U.S. government satellite program collecting remotely sensed data for the earth's surface
LAT	latitude
LIGSN	EMAP acronym for near-coastal waters comprising Long Island, Great South Bay, and Narragansett Bay
LISA	local indicators of spatial association
LN	natural logarithm (base e)
LND	coastal lowlands/interior highlands regions of Puerto Rico
LONG	longitude
MA	Massachusetts state acronym in EMAP; moving average auto-regressive model
MANOVA	multivariate analysis of variance
MAUP	modifiable areal unit problem
MC	Moran Coefficient
ME	Maine state acronym in EMAP
ML	maximum likelihood
MSE	mean squared error
N	EMAP acronym for Nantucket Sound
NCEDR	National Center for Environmental Decision-making Research
NH	New Hampshire state acronym in EMAP
NJ	New Jersey state acronym in EMAP
NLS	nonlinear least squares
OLS	ordinary least squares
Pb	chemical symbol for lead
PCA	principal components analysis
PCB	any of a group of chlorinated isomers of biphenyl
pH	a quantitative index of water acidity
Pr	probability
r	semivariogram model range parameter
R-J	Ryan-Joiner normality test statistic
r*	effective semivariogram model range
RI	Rhode Island state acronym in EMAP
RSSE	relative sum of squared errors
S-W	Shapiro-Wilk normality test statistic

s.e.	standard error
SAR	simultaneous autoregressive (model specification)
SIDs	sudden infant deaths
SIN	trigonometric sine function
SMSA	standard metropolitan statistical area (defined by the U.S. Bureau of the Census)
SO_2	sulfur dioxide
SO_4	sulfate
SS	sums of squares
T	matrix transpose operator
TCDD	dioxin
TDBN	power-transformed density of banana/plantain production in Puerto Rico
TDCD	power-transformed density of cultivated land in Puerto Rico
TDCF	power-transformed density of coffee production in Puerto Rico
TDFAM	power-transformed density of farm families in Puerto Rico
TDFRM	power-transformed density of farms in Puerto Rico
TDMLK	power-transformed density of milk production in Puerto Rico
TDSGR	power-transformed density of sugarcane production in Puerto Rico
TDTB	power-transformed density of tobacco production in Puerto Rico
TIGER	topologically integrated geographic encoding and referencing line files
TM	LANDSAT "Thematic Mapper"
TR	matrix trace operator
TSP	total suspended particulate (air pollution)
U	the "east-west" georeferencing Cartesian coordinate
UR	urban-rural regions of Puerto Rico
USA	Understanding Spatial Autocorrelation (computer module)
USDA	U.S. Department of Agriculture
USDASC	U.S. Department of Agricultural sugar cane production figures
USGS	U.S. Geological Survey
UTM	Universal Transverse Mercator
V	the "north-south" georeferencing Cartesian coordinate
VAR(✿)	the variance of a variable denoted by ✿
VIF	variance inflation factor
VT	Vermont state acronym in EMAP
WLS	weighted least squares
γ(✿)	the semivariance of a variable at distance denoted by ✿

PART I

THEORETICAL BACKGROUND

I

INTRODUCTION

In the Introduction to *Practical Handbook of Spatial Statistics* (Arlinghaus 1996a, 1-15), Griffith stresses that spatial statistics is a crucial, rapidly growing, emerging area of statistics, rife with fascinating theoretical and applied research opportunities. This contention is attested to by attention given to the subdiscipline of spatial statistics in recent years by the U.S. National Science Foundation, the U.S. National Research Council, and myriad other national and international scientific and academic communities. (Spatial statistics even has a subsection devoted to it in *Rediscovering Geography* [National Research Council 1997, 65-9].) The methodological subject matter of spatial statistics is concerned with how to conduct a proper statistical analysis of georeferenced[1] data or of data for which observations may be ordered on a two-dimensional surface and tagged with Cartesian coordinates. Paralleling time series analysis, it differs from classical statistics in that the observations analyzed are not necessarily independent (violation of this single assumption is the crux of the problem). Moreover, observations are correlated strictly due to their relative locational positions (spatial autocorrelation), resulting in spillover of information from one location to another. This spillover in turn results in redundant information being present in georeferenced data values, and this redundancy increases as the degree of locational dependence increases—a notion analogous to that associated with

---

[1]Georeferencing coordinates are symbolic descriptors of place and can take on many forms. Surface partitionings—such as census tracts, blocks and block groups, or zip and postal codes—identify small portions of the earth's surface for which attribute values are tabulated. Often the partitioning cells (i.e., areal units) are georeferenced with the coordinates of their respective centroids. Other popular georeferencing systems include latitude/longitude and Universal Transverse Mercator (UTM) coordinates. Because advances in technology have produced increased georeferencing precision, coupled with the possibility of places migrating over time, georeferencing coordinates can be volatile, as may be seen when retrieving weather station coordinates in §5.4.

classical correlation between two variables. The duplication of information activates complications in the statistical analysis of georeferenced data that lie dormant in the statistical analysis of traditional data composed of independent observations and that are similar to those found in time series analysis; but they are exacerbated by the multidirectional, two-dimensional nature of spatial dependence (time series entail dependencies that are unidirectional along a single dimension). The net result is that classical statistics applied to georeferenced data frequently fail to capture locational information—the geography of the data—raising questions about certain statistics of estimator biasedness, efficiency, consistency, and sufficiency (e.g., see Griffith and Lagona 1997, 1998; Cordy and Griffith 1993): traditional mean response estimators frequently are unbiased, whereas traditional variance and covariance estimators often are biased; traditional standard error estimators increasingly lose efficiency as the degree of spatial autocorrelation increases; increasing domain asymptotics tends to render traditional mean response and variance estimators, but not spatial autocorrelation parameter estimators, consistent, whereas infill asymptotics do not (infill asymptotics tends to result in an increased degree of spatial autocorrelation); and sufficiency of traditional parameter estimators virtually always is at least partially lost (locational information is not captured). The correct estimators are those counterparts that more recently have been added to the statistics literature through spatial statistical developments.

In a search for tractability, mathematical statistical theory historically has posited the convenient assumption of independence (Cressie 1991, 3). A review of the standard derivation of the expected value of the conventional unbiased sample variance estimator for illustrative purposes demonstrates (see Freund 1992, 358-59),

$$(n-1)s^2 = \sum_{i=1}^{n} (x_i - \bar{x})^2$$

$$= \frac{1}{n^2} \sum_{i=1}^{n} (nx_i - \sum_{i=1}^{n} x_i)^2$$

$$= \frac{1}{n^2} \sum_{i=1}^{n} [(n-1)(x_i - \mu) + (n-1)\mu - \sum_{j=1, j\neq i}^{n} (x_j - \mu) - (n-1)\mu]^2$$

$$= \frac{1}{n^2} \sum_{i=1}^{n} \left[(n-1)^2 (x_i - \mu)^2 - 2(n-1)(x_i - \mu) \sum_{j=1, j\neq i}^{n} (x_j - \mu) + [\sum_{j=1, j\neq i}^{n} (x_j - \mu)]^2 \right]$$

$$= \frac{1}{n^2} \left[(n-1)^2 \sum_{i=1}^{n} (x_i - \mu)^2 - 2(n-1) \sum_{i=1}^{n} (x_i - \mu) \sum_{j=1, j\neq i}^{n} (x_j - \mu) + \sum_{i=1}^{n} [\sum_{j=1, j\neq i}^{n} (x_j - \mu)]^2 \right].$$

This assumption of independence means that the expected value of cross-product terms like $\sum_{i=1}^{n} (x_i - \mu) \sum_{j=1, j\neq i}^{n} (x_j - \mu)$ is 0. Moreover, the correlation between attribute values for any observation i and any other observation j ($i \neq j$) is 0, which rarely is the case when observations are areal units. For the expected value of s^2, then, this assumption results in the term $2(n-1)(x_i-\mu)\sum_{j=1,j\neq i}^{n}(x_j-\mu)$ being equated to 0, and hence the term $\sum_{i=1}^{n}[\sum_{j=1,j\neq i}^{n}(x_j-\mu)]^2$ equaling $n(n-1)\sigma^2$. These simplifications yield the following mathematical expectation results:

$$(n-1)E(s^2) = \frac{1}{n^2}\sum_{i=1}^{n}\left[(n-1)^2 E[(x_i-\mu)^2] - 0 + \sum_{j=1,j\neq i}^{n}E[(x_j-\mu)]^2\right]$$

$$= \frac{1}{n^2}\sum_{i=1}^{n}[(n-1)^2\sigma^2 + (n-1)\sigma^2]$$

$$= \frac{1}{n^2}n[(n-1)+1](n-1)\sigma^2$$

$$\therefore E(s^2) = \sigma^2 .$$

Durrett notes (1994, 4) that this arcane specification is an acceptable approximation to reality in many contexts but that "there is a growing list of examples of phenomena that must be treated by models that are spatially explicit." Hence, the independent and identically distributed (*iid*) assumption that has been so popular in theoretical statistics over the years and that operates as a statistical control in mathematical statistical analyses "should not be taken for granted, particularly when there are good physical reasons to abandon it" (Cressie 1989, 197). In other words, in the presence of spatial dependence, more than likely only some of the terms in the sum $\sum_{i=1}^{n}(x_i-\mu)\sum_{j=1,j\neq i}^{n}(x_j-\mu)$ are 0; although "everything is related to everything else [in geographic space], near things are more related than distant things" (Tobler 1970, 236).

To take geographic nearness into account, binary, 0-1 indicator variables can be created to differentiate between pairs of places in geographic space that are considered to be related: for areal units i and j ($i \neq j$), let indicator variable $c_{ij} = 1$ if these two areal units are related, and 0 otherwise. There are n^2 of these indicator variables, one for each areal unit pairing; the total set of them can be organized into a matrix, **C**, which is called the geographic connectivity or weights matrix.

Our primary objective in this book is to motivate and illustrative sound analyses of a variety of georeferenced data sets for which *iid* is not assumed. Methodology and data analysis interpretation are the emphasis; both spatial autoregressive and geostatistical methodologies are utilized and presented. Matrix **C**, with its depiction of the geographic configuration of areal units, plays a central role in these georeferenced data analyses.

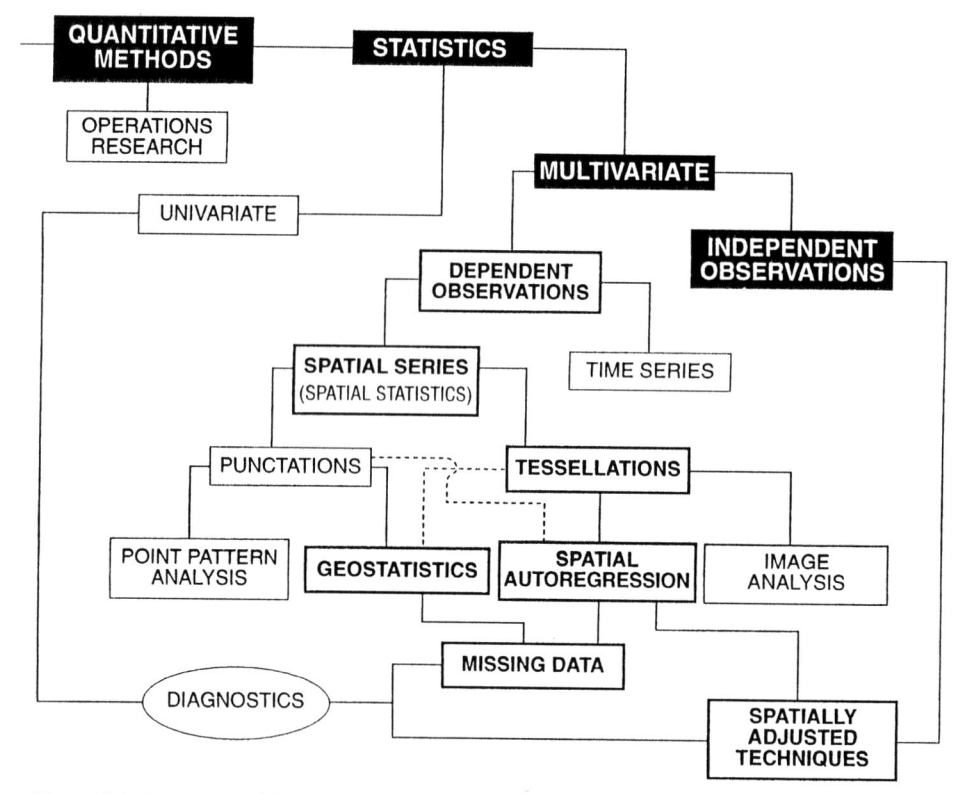

Figure 1.1. A conceptual framework placing the statistical analysis of georeferenced data within conventional data analytic methodology. See Map 5.1 for an example of a punctated geographic surface; see Map 1.1a for an example of a tessellated geographic surface.

1.1. PARALLELS BETWEEN SPATIAL AUTOREGRESSION AND GEOSTATISTICS

As occurs repeatedly in discussions of time series, the focus may be placed either on autocorrelation (serial correlation) or on partial autocorrelation. Discussions of spatial series are no different and tend to focus on either spatial autocorrelation (which is directly addressed in geostatistics and indirectly addressed in spatial autoregression) or partial spatial autocorrelation (which is directly addressed in spatial autoregression). The two classic treatises reviewing and extending spatial statistical theory are by Cliff and Ord (1981a), who have motivated research involving spatial autoregression, and Cressie (1991), who has crystallized research involving geostatistics. Although these two subfields have been evolving autonomously and in parallel, they are closely linked. Figure 1.1 presents part of a taxonomy that places spatial statistics within the context of conventional data analytic methodology. According to the articulation appearing in this figure, both spatial statistics and geostatistics reside in the realm of multivariate analysis—a map may be viewed as a multivariate sample of size 1. One difference highlighted in Figure 1.1 between geostatistics and spatial autoregression is that the former principally is concerned with more or less continuously occurring attributes, while

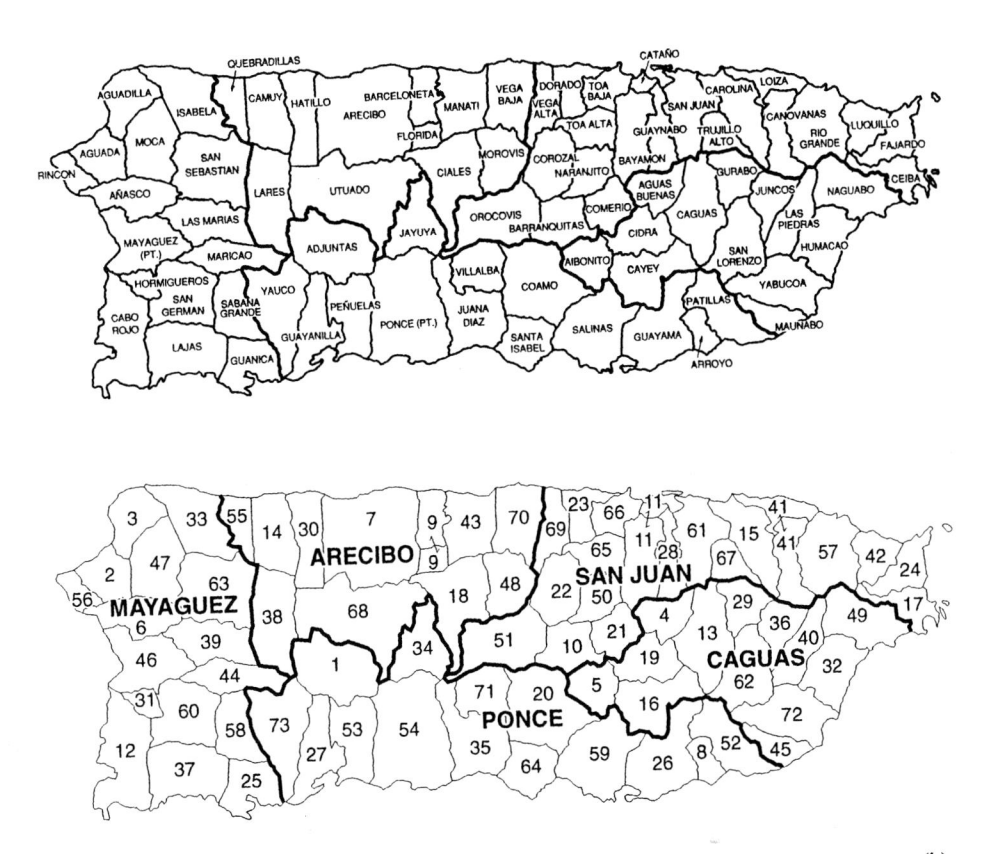

Map 1.1. Top: (a) Partitioning of Puerto Rico into 73 municipios, with their names. Bottom: (b) Assignment of ID numbers to the Puerto Rico municipios in Map 1.1a, and names to the 5 AARs (regional boundaries are in heavier lines).

the latter frequently is concerned with aggregations of phenomena into discrete regions (i.e., areal units). Another difference emphasized by Cressie (1989, 197) is that geostatistics usually is concerned with spatial prediction. In contrast, spatial autoregression often is concerned with enhancing statistical description and improving the inferential basis for statistical decision making (i.e., increasing precision). It is the prediction focus of geostatistics that provides the link between these two subdisciplines, a link being what may be labeled the missing data problem (Figure 1.1).

Generally speaking, the missing data problem (see Little and Rubin 1987) exploits redundant information contained in two variables, X and Y. In other words, values missing in variable Y are replaced by their conditional expectations through an iterative regression of Y on X, with the accompanying estimation being nonlinear. The popular *EM* algorithm has been formulated to solve this estimation problem. Letting the subscript o denote observed data values, the subscript m denote missing data values, and the superscript T denote the matrix transpose operation, such that $Y^T = <Y_o^T \vdots \hat{Y}_m^T>$, the missing data subvector solution may be written as

$$\hat{Y}_m = X_m\hat{\beta} \ . \tag{1.1}$$

When covariations among observations are present, equation (1.1) becomes

$$\hat{Y}_m = X_m\hat{\beta} - A_{mm}^{-1}A_{mo}(Y_o - X_o\hat{\beta}) \ , \tag{1.2}$$

where matrix A contains covariation-based terms depicting dependencies amongst observations, and may be written as the following partitioned matrix:

$$\begin{pmatrix} A_{oo} & \vdots & A_{om} \\ \cdots & & \cdots \\ A_{mo} & \vdots & A_{mm} \end{pmatrix}.$$

Matrix A_{oo} is a function of covariations among the observed values, matrices A_{om} and A_{mo} are functions of covariations between the observed and missing values, and matrix A_{mm} is a function of covariations among the missing values. [Note that in equation (1.1), given the *iid* assumption, $A_{mo} = A_{om}^T = 0$.] Accordingly, the missing spatial values problem exploits correlation between variables X and Y as well as correlation between some observation y_i and other observations y_j ($i \neq j$). In the case of spatial autoregression, matrix A is constructed from matrix ρC, where ρ is the spatial autocorrelation parameter that is a regression coefficient rather than a correlation coefficient per se. Matrix C is defined as before, an n-by-n binary matrix whose rows and columns are labeled by the same sequence of locations, with cell entry $c_{ij} = 1$ if locations i and j are neighbors, and $c_{ij} = 0$ otherwise. Consequently, for the conditional autoregressive (CAR) spatial statistical model, where using traditional multivariate notation the variance-covariance matrix may be written as $\Sigma = (I - \rho C)^{-1}\sigma^2$, equation (1.2) may be rewritten as

$$\hat{Y}_m = X_m\hat{\beta} + \rho(I - \rho C_{mm})^{-1}C_{mo}(Y_o - X_o\hat{\beta}) \tag{1.3}$$

(Haining et al. 1984, 1989). This result can be generalized to other popular spatial autoregressive specifications, such as the simultaneous autoregressive (SAR) spatial statistical model, by considering the inverse-covariance matrix $V\sigma^{-2}$, which is some function of matrix ρC and which allows the variance-covariance matrix to be written as $\Sigma = V^{-1}\sigma^2$, rendering

$$\hat{Y}_m = X_m\hat{\beta} - V_{mm}^{-1}V_{mo}(Y_o - X_o\hat{\beta}) \ . \tag{1.4}$$

Griffith (1988a, 1992a) outlines the specification of equation (1.4) for other popular spatial autoregressive models.

Meanwhile, in the case of geostatistics, matrix A (which now is equivalent to variance-covariance matrix Σ appearing in conventional multivariate analysis) is modeled from the semivariogram plot—a spatial covariation versus interpoint distances plot (see Map 1.2 for an example)—which depicts the nature, degree, and geographic extent of spatial autocorrelation effects (Cressie 1989). The accompanying spatial interpolation or kriging equation, which provides the functional form used to predict unknown attribute values in geostatistics, may be written as

$$\hat{Y}_m = X_m\hat{\beta} - \Sigma_{mo}\Sigma_{oo}^{-1}(Y_o - X_o\hat{\beta}) \ . \qquad (1.5)$$

The algebraic relationship between equations (1.4) and (1.5) can be established by inspecting standard partitioned matrix inverse results:

$$\Sigma_{oo} = (V_{oo} - V_{om}V_{mm}^{-1}V_{mo})^{-1}\sigma^2 \text{ and } \Sigma_{mo} = -V_{mm}^{-1}V_{mo}(V_{oo} - V_{om}V_{mm}^{-1}V_{mo})^{-1}\sigma^2$$

and

$$V_{mm} = (\Sigma_{mm} - \Sigma_{mo}\Sigma_{oo}^{-1}\Sigma_{om})^{-1}\sigma^2 \text{ and } V_{mo} = -(\Sigma_{mm} - \Sigma_{mo}\Sigma_{oo}^{-1}\Sigma_{om})^{-1}\Sigma_{mo}\Sigma_{oo}^{-1}\sigma^2 \ .$$

In other words, there is an exact algebraic correspondence between these two results (Griffith 1993a), highlighting that spatial autoregression directly deals with the inverse-variance-covariance matrix (i.e., $\Sigma^{-1} = V\sigma^{-2}$) while geostatistics directly deals with the variance-covariance matrix itself. This characteristic reveals how and why edge effects play different roles in these two approaches to spatial statistics, because edge effects are amplified through the operation of matrix inversion.

As an aside, solution equations (1.1) and (1.2) yield results that will suppress variation, because expected values are substituted for observed attribute values. If desired, derivation of these equations can be constrained to preserve s_o^2.

1.2. THE MANY FACES OF SPATIAL AUTOCORRELATION

As emphasized in the preceding section, the two unifying concepts linking geostatistics and spatial autoregression are the need to predict attribute values of locations for which data are not known and the notion of spatial autocorrelation. The latter parallels that of traditional statistical situations involving repeated measures and multicollinearity. Georeferenced data contain redundant information similar to the redundant information latent in repeated measurements. Given the conventional case of two repeated measures and the difference of means problem, for example, the redundant information in question is handled by incorporating the correlation between pairs of measures into the standard error. In the simplest repeated measures case of a common variance and a balanced design, $\frac{\sigma\sqrt{2}}{\sqrt{n}}$ (the standard error term assuming *iid* errors) becomes $\frac{\sigma\sqrt{2(1-\rho_{Y_1Y_2})}}{\sqrt{n}}$. As $\rho_{Y_1Y_2}$ approaches 1, the standard error decreases toward 0, and hence incorrectly using $\frac{\sigma\sqrt{2}}{\sqrt{n}}$ will cause the standard error to be bigger than it actually is; as $\rho_{Y_1Y_2}$ approaches -1, the standard error increases toward $\frac{\sigma \times 2}{\sqrt{n}}$, and hence incorrectly using $\frac{\sigma\sqrt{2}}{\sqrt{n}}$ will cause the standard error to be smaller than it actually is. This particular correlation is conceptually analogous to spatial autocorrelation and provides insight into what happens to the sampling distribution variance in the presence of positive or negative spatial autocorrelation.

Computationally, though, spatial autocorrelation functions more like the R^2 term in a variance inflation factor (the VIF of linear regression), which is of the form $\frac{1}{1-R^2}$ (see §2.1). For illustrative purposes, consider the trivariate regression problem

$$Y = \beta_0 + \beta_1 X_1 + \beta_2 X_2 + \epsilon_y ,$$

where, for simplicity but without loss of generality, Y, X_1, and ϵ_y are mutually orthogonal and have means of 0 and variances of 1 (e.g., they have been converted to z-scores of principal components scores). Let $X_2 = aX_1 + b\epsilon$, where ϵ is *iid* and independent of Y, X_1, and ϵ_y, and $a + b = 1$. Then VIF $= 1 + \frac{a^2}{b^2}$. The following tabulation exemplifies changes in VIF accompanying changes in a and b, where $R^2 = \frac{a^2}{a^2+b^2}$:

R^2	a	b	VIF
0	0	1	1 (i.e., X_1 and X_2 are orthogonal)
0.00010	0.01	0.99	1.00010
0.01235	0.10	0.90	1.01235
0.5	0.50	0.50	2
0.98780	0.9	0.1	82
0.99990	0.99	0.01	9802
1	1	0	∞ (i.e., X_1 and X_2 are identical)

In other words, as ρ increases, the variance estimate becomes increasingly inflated (due to the presence of multicollinearity). In either the repeated measures or the VIF case, one principal inferential impact is that of obscuring the correct probability of Type I errors.

Spatial autocorrelation can be viewed, or interpreted, in various ways (Griffith 1992b): as self-correlation, as map pattern, in terms of information content, as a spatial spillover effect, as an indicator of the appropriateness of areal unit demarcation, as a nuisance, as a missing variable surrogate, and as a diagnostic tool. Its literal definition is self-correlation arising from the geographic context of attribute values.[2] Consider the Pearson product moment correlation coefficient formula,

$$\frac{\sum_{i=1}^{n} \frac{(x_i - \bar{x})(y_i - \bar{y})}{n}}{\sqrt{\sum_{i=1}^{n} \frac{(x_i - \bar{x})^2}{n}} \sqrt{\sum_{i=1}^{n} \frac{(y_i - \bar{y})^2}{n}}}, \tag{1.6}$$

which indexes the average association between two variables. Now if this formula is applied to a single variable, variable Y is replaced by variable X. If the relationship under study is between pairs of individual observations—ignoring the perfect relationship between an observation and itself—the denominator of the covariation term must count the number of these pairs. For georeferenced data, again using binary indicator variable c_{ij} (the elements of matrix **C**) to denote relatively close locations, the converted Pearson product moment correlation coefficient formula becomes

$$\frac{\sum_{i=1}^{n}\sum_{j=1}^{n} c_{ij} \frac{(x_i - \bar{x})(x_j - \bar{x})}{\sum_{i=1}^{n}\sum_{j=1}^{n} c_{ij}}}{\sqrt{\sum_{i=1}^{n} \frac{(x_i - \bar{x})^2}{n}} \sqrt{\sum_{i=1}^{n} \frac{(x_i - \bar{x})^2}{n}}} = \frac{n \sum_{i=1}^{n}\sum_{j=1}^{n} c_{ij}(x_i - \bar{x})(x_j - \bar{x})}{\sum_{i=1}^{n}\sum_{j=1}^{n} c_{ij} \sum_{i=1}^{n} (x_i - \bar{x})^2}. \tag{1.7}$$

This modified Pearson product moment correlation coefficient—the expression on the right-hand side of equation (1.7)—actually is the Moran Coefficient (MC) of spatial statistics; it is a covariation measure of spatial autocorrelation. This interpretation of spatial autocorrelation should help keep researchers from confusing spatial correlation with statistical correlation between attributes.

The most common interpretation of spatial autocorrelation is in terms of trends, gradients, or patterns across a map. Analogous to the manner in which values for variables X and Y align to produce a Pearson product moment correlation coefficient value: (1) as the value of MC approaches 1, similar numerical values tend to cluster in geographic space (positive spatial autocorrelation); (2) as the value of MC approaches −1, dissimilar numerical

---

[2] Bamston (1993) provides a space-time data set that can be used to illustrate this interpretation in a context analogous to that for the USA (Hancock and Griffith 1996) pollution monitoring module.

values tend to cluster in geographic space (negative spatial autocorrelation); and, (3) as the value of MC approaches $-\frac{1}{n-1}$ (which asymptotically goes to 0), relatively speaking, numerical values are arranged haphazardly on a map. Unlike conventional correlation coefficients, however, MC is not restricted to the range [−1, 1]; rather, its range is dictated by what essentially are the extreme eigenvalues of C, a matrix that is constructed from the n^2 c_{ij} binary values (de Jong, Sprenger, and van Veen 1984). In fact, the accompanying eigenvectors represent a kaleidoscope of orthogonal map patterns of possible spatial autocorrelation (Griffith 1996b; see also §2.6.3). Tiefelsdorf and Boots (1995) allude to this very relationship between the eigenfunctions and MC values.

Spatial autocorrelation also can be viewed as an index of information content in georeferenced data (see Sullivan 1996 for an illuminating discussion of this notion in a standard statistical analysis context). Richardson and Hémon (1981) promote this viewpoint with respect to correlation coefficients computed for pairs of geographically distributed variables. Dutilleul (1993) has refined their work, showing that the effective number of degrees of freedom associated with a correlation coefficient computed for two georeferenced variables is given by

$$\frac{TR[(I-\frac{11^T}{n})V_x^{-1}] \times TR[(I-\frac{11^T}{n})V_y^{-1}]}{TR[(I-\frac{11^T}{n})V_x^{-1}(I-\frac{11^T}{n})V_y^{-1}]} - 1,$$

where TR denotes the matrix trace operator, and matrix V_j is defined as it was for equation (1.4). This expression reduces to the conventional (n − 2) degrees of freedom when $V_x^{-1} = V_y^{-1} = I$, which is the classical *iid* case. And it converges on 0 as positive spatial autocorrelation increases (e.g., $V_x^{-1} \rightarrow 11^T$, a correlation matrix of all ones). Meanwhile, Haining (1991) demonstrates an equivalency between this interpretation and the removal of spatial dependency effects with filters like those used in constructing time series impulse-response functions. The crucial statistical notion here is that of degrees of freedom, which can be interpreted as an index of information content in data. As spatial autocorrelation approaches 1, the effective degrees of freedom approaches 0. In other words, for a constant attribute value across a geographic landscape (i.e., $\rho = 1$), once this value is known in any areal unit i, it is not free to vary in any of the remaining (n − 1) areal units j ($i \neq j$). This same effect can be seen with the conventional repeated measures situation. If X is compared to itself, then $\rho = 1$, and hence the standard error, $\frac{\sigma\sqrt{2(1-\rho)}}{\sqrt{n}}$, becomes 0, regardless of n; there is no information left to conduct a difference of means test! As spatial autocorrelation approaches −1, the effective degrees of freedom increases beyond n. And, as spatial autocorrelation approaches 0, the effective degrees of freedom approaches n.

These three interpretations of spatial autocorrelation focus specific attention on spatial prediction, which involves estimating missing georeferenced data when undertaking spatial autoregressive analyses, and kriging (i.e., geographic interpolation) when undertaking geostatistical analyses. In other words, given a set of georeferenced information, there is a better than "equally likely" chance that neighboring numerical values can be used to predict each other. The spatial autocorrelation can be exploited for this purpose, holes (missing data values) in a map pattern can be interpolated (spatial autocorrelation is the backbone of geographic interpolation), and the degree of redundancy can indicate how much information is free to vary. These three interpretations also highlight the quantifying role that the MC can play in spatial autoregression and geostatistical analyses.

Spatial autocorrelation also can be interpreted in a more customary manner. As is well established in the statistics literature, the linear regression model serves as the workhorse of linear statistical analysis[3] (and the general linear model). Adhering closely to this perspective in particular, spatial econometrics (see Anselin 1988) views spatial autocorrelation within the contexts of missing regression variables and diagnostic tools, and more specifically as a diagnostic tool for spatial heterogeneity. Conventional linear regression theory reveals that when variables are missing from a mean response specification (i.e., matrix $\mathbf{X}$ of $\mathbf{X\beta}$ is incomplete) then, more than likely, the ordinary least squares (OLS) estimate $\mathbf{b}$ is biased, the accompanying error variance estimate (MSE) tends to be inflated, the accompanying estimated standard error of $\mathbf{b}$ almost always is affected, and the resulting residuals ($\mathbf{e}$) may well be highly correlated. Further, if the variance is nonconstant, the ordinary least squares estimate of the standard error of $\mathbf{b}$ no longer is guaranteed to be a minimum variance estimator. These conceptualizations link spatial autocorrelation to model specification issues and enable various effects of linear regression model misspecification to be shown with MC.

Suppose that p_M of the set of $(p + p_M)$ X_j variables are missing from the mean response specification, such that $\mathbf{X\beta} = \mathbf{X_I \beta_I} + \mathbf{X_M \beta_M}$, where subscript I denotes the set of p X_j variables included in the implemented specification and subscript M denotes the set of p_M X_j variables omitted from this specification. Using matrix notation, the expected value of MC becomes

$$-\frac{n}{\mathbf{1}^T \mathbf{C} \mathbf{1}} \times \frac{\beta_M{}^T \mathbf{X}_M{}^T [\mathbf{I} - \mathbf{X}_I(\mathbf{X}_I{}^T \mathbf{X}_I)^{-1}\mathbf{X}_I{}^T]\mathbf{C}[\mathbf{I} - \mathbf{X}_I(\mathbf{X}_I{}^T \mathbf{X}_I)^{-1}\mathbf{X}_I{}^T]\mathbf{X}_M \beta_M + \sigma^2 TR[\mathbf{X}_I(\mathbf{X}_I{}^T \mathbf{X}_I)^{-1}\mathbf{X}_I{}^T \mathbf{C}]}{\beta_M{}^T \mathbf{X}_M{}^T [\mathbf{I} - \mathbf{X}_I(\mathbf{X}_I{}^T \mathbf{X}_I)^{-1}\mathbf{X}_I{}^T]\mathbf{X}_M \beta_M + (n-p)\sigma^2},$$

where $\mathbf{1}$ is an n-by-1 vector of ones. When $\beta_M = 0$, this expression reduces to the well-known standard result of $-\dfrac{n}{\mathbf{1}^T \mathbf{C} \mathbf{1}} \times \dfrac{TR[\mathbf{X}_I(\mathbf{X}_I{}^T \mathbf{X}_I)^{-1}\mathbf{X}_I{}^T \mathbf{C}]}{(n-p)}$; it simplifies

---

[3] While spatial statistics also can employ linear regression, nonlinear regression is the workhorse of spatial autoregression and geostatistics.

somewhat when matrices $\mathbf{X}_I$ and $\mathbf{X}_M$ are mutually orthogonal and reduces to $-\frac{1}{n-1}$ when $\mathbf{X} = \mathbf{1}$ (the constant mean case). In other words, MC values are distorted by overlooking variables in a mean response specification, sometimes causing the wrong level of spatial autocorrelation to be detected in georeferenced data. Of course, the corresponding standard error also will be distorted. This standard expected value result reveals as well that as the number of variables entered into a regression equation increases, and hence the induced negative correlation among the n regression residuals e_i increases (see Montgomery and Peck 1982, 150-51, for a discussion of this negative correlation among residuals in the *iid* case), then the expected value of MC tends to become increasingly negative (for a set of orthogonal variables containing positive spatial autocorrelation, in extreme cases this expected value can be as low as roughly −0.64, regardless of whether n is very large). Therefore, in this context, indexing the nature and degree of spatial autocorrelation becomes an exercise in missing variables diagnostic analysis. Missing variables are overlooked for a variety of reasons, including ignorance, an inability to measure them, or a lack of resources for collecting data on them. Fortunately, inferential and descriptive impacts of this specification error for georeferenced data often can be minimized by unknowingly capturing the effects of missing variables through the introduction of spatial dependency parameters into a mean response specification (Cressie 1991, 436). But because "one [researcher's] deterministic mean structure may be another [researcher's] correlated error structure" (Cressie 1991, 114), significant spatial autocorrelation should not automatically be interpreted as a misspecified mean response.

Spatial autocorrelation also can arise from neglecting the presence of spatial heterogeneity, a situation that is more the rule than the exception for georeferenced data. Suppose the variance of georeferenced data is given by $\sigma^2 \mathbf{S}$, where $\mathbf{S}$ is a diagonal matrix with nonconstant s_{ii} values; again using matrix notation, the expected value of MC becomes

$$-\frac{n}{\mathbf{1}^T\mathbf{C}\mathbf{1}} \times \frac{TR[\mathbf{S}[2\mathbf{I} - \mathbf{X}(\mathbf{X}^T\mathbf{X})^{-1}\mathbf{X}^T]\mathbf{C}\mathbf{X}(\mathbf{X}^T\mathbf{X})^{-1}\mathbf{X}^T]}{TR(\mathbf{S}) - TR[\mathbf{S}\mathbf{X}(\mathbf{X}^T\mathbf{X})^{-1}\mathbf{X}^T]},$$

which reduces to the well-known standard result of

$$-\frac{n}{\mathbf{1}^T\mathbf{C}\mathbf{1}} \times \frac{TR[(\mathbf{X}^T\mathbf{X})^{-1}\mathbf{X}^T\mathbf{C}\mathbf{X}]}{(n-p)}$$

when $\mathbf{S} = \mathbf{I}$, and $-\frac{1}{n-1}$ when $\mathbf{X} = \mathbf{1}$. This result reveals that indexing the nature and degree of spatial autocorrelation becomes in part an exercise in spatial heterogeneity diagnostic analysis. In most georeferenced data analysis cases, though, it is unlikely that nonconstant spatial variance is the sole generator of detected spatial autocorrelation. Conventional weighted regression and variance stabilizing transformation analyses bear witness to this contention.

1.2.1. THE GEARY RATIO

The most powerful test statistic for spatial autocorrelation is MC, although corroboration of its implication by the Geary Ratio (GR) is desirable. But GR, which is a paired comparisons measure of spatial autocorrelation that relates more directly to the semivariogram plot of geostatistics than does MC, may be written as a function of MC:

$$\frac{n-1}{\sum_{i=1}^{n}\sum_{j=1}^{n}c_{ij}} \times \frac{\sum_{i=1}^{n}(x_i - \bar{x})^2 \sum_{j=1}^{n}c_{ij}}{\sum_{i=1}^{n}(x_i - \bar{x})^2} - \frac{n-1}{n}\text{MC}.$$

This expression highlights the negative relationship between these two indices and indicates one reason why MC is preferred over GR. The first term in this expression reveals that each squared deviation, $(x_i - \bar{x})^2$, is weighted by the number of neighbors it has, $\sum_{j=1}^{n} c_{ij}$. For irregular surface partitionings, differences in numbers of neighbors can be quite sizeable, resulting in an overemphasis of such deviations by the GR spatial autocorrelation index for areal units with relatively large numbers of neighbors. This situation would be further exacerbated if x_i were an outlier as well. For most regional data, those areal units along the edge of a region will tend to have relatively fewer neighbors, resulting in an underemphasizing of such deviations by the GR index. Because MC is embedded in the GR formula, GR does capture all of the locational information summarized by MC.

1.3. THE MORAN COEFFICIENT SCATTERPLOT TOOL

The link between spatial statistics and regression already has been mentioned in §1.2. This connection becomes more apparent when rewriting MC as a ratio of regression coefficients. The numerator regression coefficient is obtained with the following procedure:

> Step 1: center the variable under study, X, yielding $(\mathbf{X} - \mathbf{1}^T\mathbf{X}\mathbf{1}/n)$;
> Step 2: compute $\mathbf{C}(\mathbf{X} - \mathbf{1}^T\mathbf{X}\mathbf{1}/n)$;
> Step 3: using a no-intercept regression model, regress $\mathbf{C}(\mathbf{X} - \mathbf{1}^T\mathbf{X}\mathbf{1}/n)$ on $(\mathbf{X} - \mathbf{1}^T\mathbf{X}\mathbf{1}/n)$.

Denote the resulting regression coefficient with b_{XCX}. Essentially this is the slope of the regression line for a MC scatterplot, where the vertical axis is labeled with $\mathbf{C}(\mathbf{X} - \mathbf{1}^T\mathbf{X}\mathbf{1}/n)$ and the horizontal axis is labeled with $(\mathbf{X} - \mathbf{1}^T\mathbf{X}\mathbf{1}/n)$. Anselin (1995) argues that the vertical axis should be the row-standardized version of matrix **C**, namely matrix **W**; but, using the row-standardized counterpart of matrix **C** obscures this more natural interpretation of b_{XCX}. Meanwhile, the denominator regression coefficient is obtained by the following procedure:

Step 1: compute **C1**;
Step 2: using a no-intercept regression model, regress **C1** on **1**.

Denote this resulting regression coefficient with b_{1C1}. Consequently,

$$MC = \frac{b_{XCX}}{b_{1C1}},$$

with the numerator regression coefficient linking MC to a wealth of regression analysis possibilities and the denominator regression coefficient producing the standardizing coefficient $\frac{n}{\sum_{i=1}^{n}\sum_{j=1}^{n} c_{ij}}$. SAS code for executing these regressions, and plotting the MC scatterplot, appears in Appendix 1-A.

MC scatterplots for eight 1969 Puerto Rican agricultural productivity measures (see Griffith and Amrhein 1991, 22-23) are presented in Figure 1.2. Those displaying an approximately linear relationship are DFRM (density of farms), DCD (density of cultivated land), and DFAM (density of farm families). Those exhibiting apparent nonlinear relationships are DMLK (density of milk production), DSGR (density of sugarcane production), DCF (density of coffee production), and DBN (density of banana/plantain production). DTB (density of tobacco production) seems to display nonconstant variance, as is shown by the varying spread of points along the vertical axis for different selected points along the horizontal axis. Because both n and the surface partitioning (Map 1.1a-b) are the same for each of the eight agricultural productivity variables, $b_{1C1} = 4.7123$ in each case, and the regression results are as follows:

variable	b_{XCX}	MC	GR
DFRM	1.6419	0.34842	0.73242
DCD	0.5121	0.10867	0.76046
DMLK	0.9450	0.20054	0.68529
DSGR	2.4402	0.51783	0.69996
DCF	1.4636	0.31060	0.53739
DTB	2.0626	0.43771	0.78647
DBN	1.6668	0.35371	0.90296
DFAM	1.7213	0.36528	0.64916

Considering the asymptotic standard error of MC (Griffith and Amrhein 1991, 330), namely

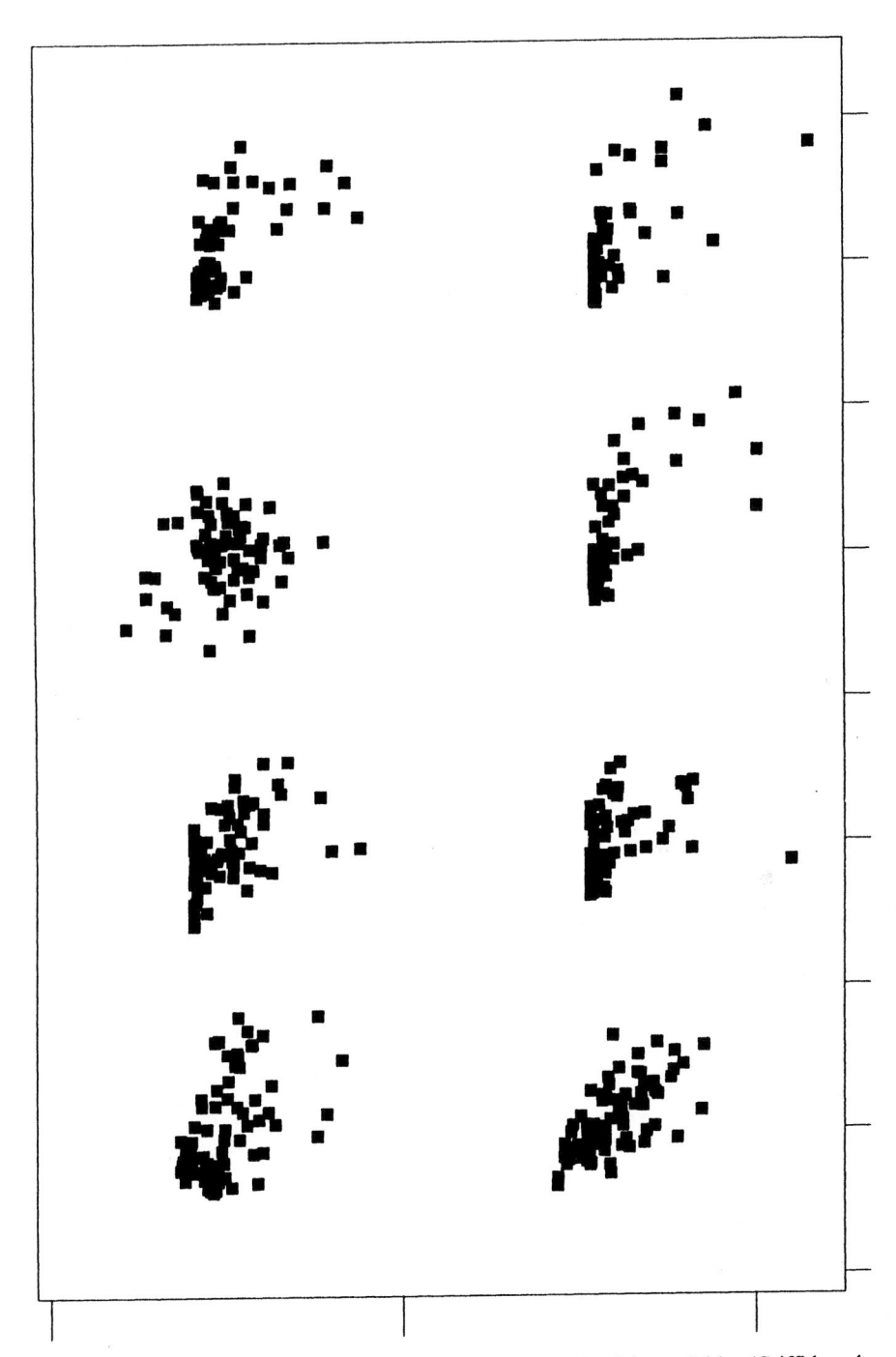

Figure 1.2. MC scatterplots of the eight 1969 agricultural productivity variables (Griffith and Amrhein's data). Variables are arranged in alphabetical order of acronyms—in a counter-clockwise direction, beginning with the upper left-hand scatterplot: C*DBN vs. DBN, C*DCD vs. DCD, C*DCF vs DCF, C*DFAM vs. DFAM, C*DFRM vs. DFRM, C*DMLK vs. DMLK, C*DSGR vs. DSGR, and C*DTB vs. DTB.

$$\sqrt{\frac{2}{\sum_{i=1}^{n}\sum_{j=1}^{n}c_{ij}}} = \sqrt{\frac{2}{344}} \approx 0.07625,$$

coupled with an expected value of −0.01389, only the MC statistic for DCD is not significant at a 5% level. But the nonlinearities and nonconstant variances insinuated by some of the MC scatterplots (Figure 1.2) suggest that these results may well be fraught with specification error. This latter notion is further corroborated by the failure of GR values to align in a consistent manner with their corresponding MC values (explored more in Chapter 2).

Regression diagnostic concerns span a wide range of issues besides linearity (Griffith and Amrhein 1997, Chapter 4). Because matrix inversion is done, impacts of multicollinearity latent in matrix **X** need to be evaluated. Addressing issues of attribute normality and variance homogeneity is deferred to Chapter 2; the more restrictive issues of induced tessellation heterogeneity and anisotropy will be introduced in this chapter. Also of diagnostic importance are observation-specific indices of spatial association, especially as they pertain to leverage, influential points, and spatial outlier detection (§1.5.1).

1.4. MULTIVARIATE SPATIAL ASSOCIATION

Two primary approaches have been taken to multivariate spatial autoregressive analysis. The first strategy reformulates univariate statistics of spatial autocorrelation, usually in terms of space-time data series (Griffith 1981a; Cliff and Ord 1981b). The second tactic formulates multivariate spatial autoregressive models (Streitberg 1979; Mardia 1988). The two multivariate techniques receiving the most attention are discriminant function (Mardia 1984) and principal components analysis (Switzer and Green 1984). Campbell (1981) illustrates empirically how spatial autocorrelation affects the accuracy of supervised classifications of land cover from remotely sensed data, while Griffith (1988a) illustrates empirically both attribute and observational features of the linear statistical structure latent in multivariate georeferenced data. Inadequate multivariate spatial statistical theory has been developed to date, and hence many issues remain unresolved. The purpose of this section, then, is to outline how the exploration of multivariate spatial association can be initiated within the context of spatial autoregression.

1.4.1. BIVARIATE GEOREFERENCED DATA ANALYSIS: A CLOSER INSPECTION

Consider a bivariate data set, the simplest case of multivariate analysis. Suppose the two variables under study are X and Y, which respectively can be rewritten as $X = X^* + \xi_X$ and $Y = Y^* + \xi_Y$, where X^* and Y^* denote the pure locational information (i.e., spatial autocorrelation) and ξ_X and ξ_Y denote the pure attribute information contained in these variables. Let ρ_{XY} denote the product moment correlation coefficient measuring the linear relationship between variates ξ_X and ξ_Y, r_{XY} denote the corresponding sample correlation coefficient, and ρ_X and ρ_Y respectively denote the spatial autocorrelation latent in variables X and Y. Imagine that two maps are being examined (Map 1.2), and that the values distributed over these maps constitute representative samples. Relatively speaking, as $r_{XY} \rightarrow 1$, x_i and y_i become increasingly similar (subject to translation and scale adjustments), whereas as $r_{XY} \rightarrow 1$, x_i and $(1) \times y_i$ become increasingly similar; as $r_{XY} \rightarrow 0$, x_i and y_i display a haphazard association. If $\rho_X = \rho_Y = 0$ (no spatial autocorrelation in either variable X or Y), the statistical decision based upon r_{XY} indeed is the one posited by classical statistics.

When $\rho_X \neq 0$ and/or $\rho_Y \neq 0$, spatial autocorrelation for the two maps in question can follow various scenarios. Consider the case of $\rho_X \rightarrow 1$ and $\rho_Y \rightarrow 1$. Two distinct map patterns can materialize in Map 1.2 for the geographic distribution of variables X and Y (see §1.5), namely, the set of observations {1, 3, 6, 8, 9, 11, 14, 16} can take on the eight largest values, and the set of observations {2, 4, 5, 7, 10, 12, 13, 15} can take on the eight smallest values, or vice versa. Negative spatial autocorrelation is optimized when these areal unit orderings also simultaneously represent the corresponding descending order of large values and ascending order of small values. When these set allocations are the same for X and Y, then $r_{XY} \rightarrow 1$; but when these set allocations are the opposite for X and Y, then $r_{XY} \rightarrow 1$. These outcomes say nothing about the value taken on by ρ_{XY} though. As $\rho_{XY} \rightarrow 1$, variables X and Y increasingly will tend to have the same set allocations; as $\rho_{XY} \rightarrow 1$, variables X and Y increasingly will tend to display the opposite set allocations. But, as $\rho_{XY} \rightarrow 0$, little can be said about the propensity of these set allocations to materialize.

Next consider the case of $\rho_X \rightarrow 1$ and $\rho_Y \rightarrow 1$. One possible map pattern that can materialize in Map 1.2 for the geographic distribution of variables X and Y

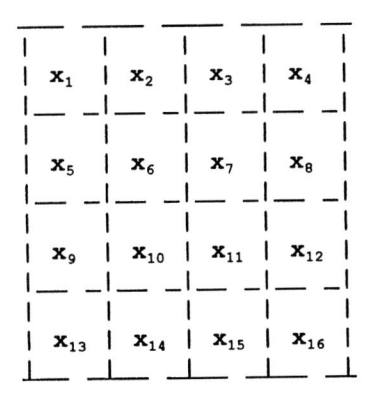

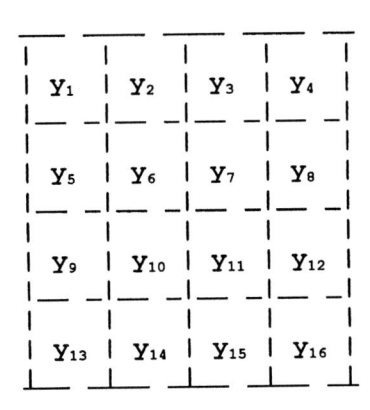

x_1	x_2	x_3	x_4
x_5	x_6	x_7	x_8
x_9	x_{10}	x_{11}	x_{12}
x_{13}	x_{14}	x_{15}	x_{16}

y_1	y_2	y_3	y_4
y_5	y_6	y_7	y_8
y_9	y_{10}	y_{11}	y_{12}
y_{13}	y_{14}	y_{15}	y_{16}

Map 1.2. Two hypothetical georeferenced variables. Left: geographic distribution of X. Right: geographic distribution of Y.

would be a gradient of increasing values moving from the southeast corner (observation #16) to the northwest corner (observation #1); such a pattern tends to maximize the positive spatial autocorrelation that is detected. Of course, this gradient could incline in the opposite direction (values increasing from the northwest to the southeast corner), in a 90° rotation of this pattern (values increasing from the southwest to the northeast corner), or in a 270° rotation of this pattern (values increasing from the northeast to the southwest corner). As $\rho_{XY} \to 1$, variables X and Y increasingly will tend to have the same map gradient. As $\rho_{XY} \to -1$, variables X and Y increasingly will tend to have opposite map gradients. Like before, as $\rho_{XY} \to 0$, little can be said about the propensity of these gradients to materialize. X and Y could incline in the same direction ($r_{XY} \to 1$), or in opposite directions ($r_{XY} \to -1$), or one could be a 90° rotation of the other ($r_{XY} \to 0$).

The third possibility is for one variable to be positively spatially autocorrelated while the other is negatively spatially autocorrelated, say $\rho_X \to -1$ and $\rho_Y \to 1$. The net result here would be for $r_{XY} \to 0$ regardless of the value taken on by ρ_{XY}.

Richardson (e.g., 1990) discusses a modified test of the null hypothesis H_0: $\rho_{XY} = 0$ (reviewed in §1.2). The critical issue here is securing a correct standard error. The conventional standard error (when $\rho_X = \rho_Y = 0$) is given by $\sqrt{[(1 - r_{XY}^2)/df]}$, where df (i.e., the number of degrees of freedom) is (n - 2) in classical statistics. The problem is that in the presence of nonzero spatial autocorrelation df = (n - 2) is incorrect, and hence the *effective* number of degrees of freedom must be determined (again, consulting Sullivan 1996 may be useful here). The preceding discussion indicates that $\rho_X \to 1$ and $\rho_Y \to 1$ causes $|r_{XY}| \approx 1$, tending to move VAR(r_{XY}) closer to a uniform distribution than would be the case for the unautocorrelated *iid* data counterpart, regardless of the value taken on by ρ_{XY}. Thus, the presence of spatial autocorrelation decreases the risk of rejecting the null hypothesis (H_0: $\rho_{XY} = 0$)—by increasing the width of the accompanying confidence interval—when such a conclusion is warranted at the stated level of significance. The *effective* dfs must increase from (n - 2) to compensate for this increase in the estimated standard error. The preceding discussion also indicates that $\rho_X \to 1$ and $\rho_Y \to 1$ increase the chances of $|r_{XY}|$ increasing, regardless of the value taken on by ρ_{XY}. As $|r_{XY}| \to 1$, $(1 - r_{XY}^2) \to 0$. Therefore, the presence of spatial autocorrelation increases the risk of rejecting the null hypothesis (H_0: $\rho_{XY} = 0$)—by decreasing the width of the accompanying confidence interval—when such a conclusion is not warranted at the stated level of significance. The *effective* dfs must decrease from (n - 2) to compensate for this decrease in the estimated standard error.

Determining the effective number of dfs is not a straightforward, uncomplicated task. Fortunately, Haining (1991) has found that the results are equivalent to those obtained by respectively partitioning X and Y into (X* + ξ_X) and (Y* + ξ_Y), and then calculating $r_{\xi_Y \xi_X}$, which actually is the estimate of ρ_{XY}. This process is known as prewhitening and requires the use of spatial autoregressive models like those treated in the ensuing chapters. A first step in achieving this end is to rewrite spatial autocorrelation indices in such a way that they capture and summarize both of these two data components. One such modification has been made to MC (Wartenberg 1985), which will be denoted here as MC_{XY}, the bivariate MC for variables X and Y, which becomes

$$MC_{XY} = \frac{\sum_{i=1}^{n}\sum_{j=1}^{n} c_{ij} \frac{(x_i - \bar{x})(y_j - \bar{y})}{\sum_{i=1}^{n}\sum_{j=1}^{n} c_{ij}}}{\sqrt{\sum_{i=1}^{n}\frac{(x_i - \bar{x})^2}{n}}\sqrt{\sum_{i=1}^{n}\frac{(y_i - \bar{y})^2}{n}}} . \qquad (1.8)$$

Equation (1.8) looks even more like the traditional product moment correlation coefficient than does equation (1.6).

The expected value of equation (1.8), when variables X and Y are normally distributed and 0 spatial autocorrelation prevails, is $\rho_{XY}/(n-1)$, which is consistent with the univariate MC expectation because $\rho_{XX} = \rho_{YY} = 1$ (i.e., a variable is perfectly positively correlated with itself). Clearly, this index is capturing both attribute and locational information. Its variance, based upon all possible permutations of values over the map, has been derived by Mantel (1967). Unfortunately, the form of its sampling distribution currently is unknown.

1.4.2. EXTENSIONS TO MORE THAN TWO GEOREFERENCED VARIABLES: MULTICOLLINEARITY AND MULTIVARIATE SPATIAL AUTOCORRELATION

Presently spatial autocorrelation complications in pairs of variables are both quite difficult to handle and poorly understood. Not surprisingly, then, impacts of spatial autocorrelation on the analysis of three or more variables are even more mystifying. But some progress can be made by generalizing the modified version of MC presented in equation (1.8).

Multivariate spatial autocorrelation associated with multicollinearity may be investigated with the cross-MC statistics computed with equation (1.8). As outlined in §1.4.1, of relevance here is the way spatial autocorrelation interacts with traditional correlations among attributes (Griffith 1991). Recall that Richardson (1990) notes the variance of the sampling distribution of the correlation coefficient is altered when at least one of a pair of attribute variables is spatially autocorrelated. For example, if two attributes geographically distributed across the island of Puerto Rico contain near maximum positive spatial autocorrelation, the accompanying map patterns could exhibit a trend or gradient along a northwest-southeast axis or a northeast-southwest axis. If both of the map patterns align with either one of these two axes, then the correlation between the accompanying attributes will tend to be pushed closer to ±1. If one map pattern aligns with one of the axes and the other map pattern aligns with the other of these axes, the correlation between the two attributes will tend to be pushed closer to 0. Thus, these different extreme possibilities (and there are an infinite number of intervening possibilities) introduce increased variability into the sampling distribution of the correlation coefficient calculated for a pair of attribute variables. Inflation of such correlation coefficients can increase detected multicollinearity in the regressor variables composing matrix **X**, further aggravating calculation error introduced by the matrix inversion $(X^TX)^{-1}$ used to compute regression coefficients (Griffith and Amrhein 1997, 97101). This notion is demonstrated empirically in Griffith (1988a, 214-18), where shifts in the principal components structure of a set of Buffalo crime data variables that are attributable to spatial autocorrelation are explored.

Returning to the 1969 Puerto Rican agricultural productivity measures, the attribute correlation matrix is

variable	DFRM	DCD	DMLK	DSGR	DCF	DTB	DBN
DCD	0.152						
DMLK	-0.182	0.050					
DSGR	0.451	0.236	-0.289				
DCF	-0.231	0.505	-0.131	-0.224			
DTB	0.404	0.009	0.020	-0.102	-0.275		
DBN	0.602	0.167	-0.253	0.801	-0.410	0.204	
DFAM	0.321	0.416	-0.026	0.658	0.024	-0.005	0.485

The accompanying varimax rotated principal components attribute structure matrix extracted from this correlation matrix is as follows (prominent loadings are double underlined):

variable	PCI	PCII	PCIII	PCIV
DFRM	0.511	0.007	<u>-0.660</u>	-0.178
DCD	0.306	<u>-0.859</u>	-0.111	0.135
DMLK	-0.129	0.026	0.017	<u>0.968</u>
DSGR	<u>0.938</u>	0.030	0.051	-0.191
DCF	-0.283	<u>-0.854</u>	0.245	-0.200
DTB	-0.082	0.065	<u>-0.931</u>	0.067
DBN	<u>0.844</u>	0.175	-0.299	-0.173
DFAM	<u>0.796</u>	-0.314	0.013	0.153
λ_j	2.6841	1.6022	1.4675	1.1225
percent of variance accounted	33.6	20.0	18.3	14.0

The attribute principal components structure matrix reveals that four eigenvalues are greater than 1, one common criterion used to select salient components (Griffith and Amrhein 1997, 169), and that the four accompanying principal components account for roughly 86% of the attribute variance.

In contrast, the cross-MC values for pairs of these variables are

variable	DFRM	DCD	DMLK	DSGR	DCF	DTB	DBN	DFAM
DFRM	0.348							
DCD	-0.012	0.109						
DMLK	-0.068	-0.156	0.201					
DSGR	0.182	0.141	-0.168	0.518				
DCF	-0.103	0.119	-0.107	0.000	0.311			
DTB	0.161	-0.044	0.020	-0.140	-0.264	0.438		
DBN	0.214	0.068	-0.122	0.329	-0.200	0.133	0.354	
DFAM	0.146	0.143	-0.177	0.394	0.113	-0.165	0.193	0.365

The accompanying varimax rotated principal components structure for this cross-MC matrix is[4] is as follows (prominent loadings are double underlined):

variable	PCI	PCII	PCIII
DFRM	0.276	0.387	-0.096
DCD	0.101	-0.032	-0.325
DMLK	-0.139	-0.064	0.425
DSGR	0.709	-0.025	-0.171
DCF	-0.111	-0.398	-0.399
DTB	-0.181	0.646	0.061
DBN	0.479	0.349	-0.023
DFAM	0.500	-0.101	-0.316
λ_j	1.0924	0.8226	0.5478
percent of variance accounted	38.2	29.1	19.8

Together these three principal components extracted from the cross-MC matrix are consistent with the original attribute correlation matrix criterion of 86% of the variance being accounted for.

The traditional principal components attribute structure reported above renders the following variable groupings:

group 1: DSGR, DBN and DFAM
group 2: DCD, DCF
group 3: DFRM, DTB
group 4: DMLK

Considering that the first three varimax rotated principal components composing the cross-MC structure accounted for an increased percentage of variance, essentially variance associated with the four original varimax rotated principal components have been reallocated to them. In addition, DFRM no longer loads

---

[4] A technical requirement for computing this solution was to add 0.04 to the diagonal entries of this matrix to ensure that the smallest eigenvalue is 0 (nonnegative); this amount then was subtracted from each of the resulting eigenvalues to obtain the correct ones.

the same component as DTB, and conceptually the second component disappears. Diagnosing this increase in multicollinearity (i.e., a larger percentage of variance accounted for by fewer components) is of primary concern with regard to multicollinearity impacts on regression results.

Such interplay between spatial autocorrelation latent in different attribute variables also is alluded to by the following algebraic relationship for orthogonal Xs

$$E(MC_Y) = -\frac{1 + \sum_{j=1}^{p-1} MC_{X_j}}{n - p},$$

where MC_{X_j} is the MC for attribute variable X_j, and 1 appears in order to account for the mean of Y. This re-expression is made possible by the matrix trace (*TR*) operator, which converts the matrix $[I - X(X^TX)^{-1}X^TC]$ into the sum of MCs, because each of the diagonal entries of $X(X^TX)^{-1}X^TC$ is of the form $(X - 1^TX1/n)^TC(X - 1^TX1/n)/[(X - 1^TX1/n)^T(X - 1^TX1/n)]$, while the cross-MC terms appear in the ignored off-diagonal cells. This expression reduces to $-\frac{1 + MC_X}{n - 2}$, the bivariate regression result reported in Cliff and Ord (1981, 202). If the X_j are random variables, too, then $E(MC_{X_j}) = -\frac{1}{n - 1}$, and $-\frac{1 + MC_X}{n - 2}$ reduces to $-\frac{1}{n - 1}$. If $p = 1$, which is the constant mean case, it again reduces to $-\frac{1}{n - 1}$. Meanwhile, Switzer and Green (1984) extend this general idea to a battery of attribute variables.

In terms of implementing the preceding cross-MC computations—and extending equation (1.8)—a matrix of z-scores, say **Z**, is constructed from the original raw data battery of georeferenced variables. An MC matrix (similar to a conventional correlation matrix) can be constructed from matrix **Z** by rewriting equation (1.8) as

$$\frac{nZ^TCZ}{\sum_{i=1}^{n}\sum_{j=1}^{n} c_{ij}(n - 1)}. \tag{1.9}$$

The diagonal entries of this matrix are the individual univariate MC values. The off-diagonal entries are the pairwise MC_{XY} values given by equation (1.8). The individual expectations under a null hypothesis of zero spatial autocorrelation yield $[-1/(n-1)]R$, where matrix **R** is the conventional correlation matrix computed for a battery of variables.

As mentioned in §1.4.2, the structure latent in the matrix defined by expression (1.9) is a convolution of spatial autocorrelation and attribute correlation, as is indicated by its expectation under a null hypothesis of zero spatial autocorrelation. Such a structure in a traditional correlation matrix may be summarized with principal components analysis, as has been done here with the Puerto Rican agricultural productivity data. A comparable eigenfunction analysis of expression (1.9) may be problematic due to the presence of negative eigenvalues (see footnote #2 of this chapter). Griffith (1988a) found that these two analyses may well give corroborating results, a finding that is compatible with those results reported in this section.

1.5. HETEROGENEITY AND LOCATIONAL INFORMATION

Locational information refers to the duplicate information content latent in attribute value x_i for areal unit i that is held in common with information contained in the x_j values ($i \neq j$) of nearby areal units j. As such, spatial autocorrelation measures index the nature and degree of this locational information. Further, it is captured by matrix **C**, which quantifies the surface partitioning that defines the geographic arrangement of areal units giving rise to this locational information. But a particular surface partitioning can both introduce and reveal an additional source of specification error, namely tessellation-induced heterogeneity and anisotropy.

Tessellation-induced heterogeneity injects two complications into georeferenced data analyses. The first, already mentioned in §1.2, concerns the limits of MC, which do not necessarily coincide with ±1. The extreme values of MC are determined by the eigenvalues of matrix

$$[\mathbf{I} - \mathbf{X}(\mathbf{X}^T\mathbf{X})^{-1}\mathbf{X}^T]\mathbf{C}[\mathbf{I} - \mathbf{X}(\mathbf{X}^T\mathbf{X})^{-1}\mathbf{X}^T], \qquad (1.10)$$

which when $\mathbf{X} = \mathbf{1}$ (the constant mean case) reduces to $(\mathbf{I} - \mathbf{11}^T/n)\mathbf{C}(\mathbf{I} - \mathbf{11}^T/n)$; as n increases, the eigenvalues of matrix expression (1.10) approach those of matrix **C**. For this constant mean case, the eigenvalues of matrix **C** roughly equal those of expression (1.10), once the principle eigenvalue extracted from matrix **C** is replaced with 0 (Table 1.1), with this approximation improving as n increases.[5] However, the principal eigenfunction of matrix **C** plays another important role elsewhere, namely that of providing an overall measure of topological connectivity for the surface partitioning in question (the eigenvalue), coupled with individual areal unit measures of relative topological centrality (the eigenvector elements). A contour plotting of this principal eigenvector of matrix **C** for the Puerto Rico municipio surface partitioning appears in Map 1.3a. As distance from the center of the island increases, the magnitude of the eigenvector values decreases (matrix theory ensures that all of these eigenvector values will be non-negative). The remaining eigenfunctions uncover a set of orthogonal and uncorrelated map patterns, with the eigenvalues being exactly proportional to the corresponding MCs

---

[5]Partial evidence supporting this contention is furnished by visually comparing Maps 1.3a and 1.3b, which respectively are contour mappings of the principal eigenvector of matrix **C** and the second eigenvector of the matrix given by expression (1.10). Roughly 31.2% of the variation across these two eigenvectors is held in common.

(Tiefelsdorf and Boots 1995) and the accompanying eigenvectors identifying corresponding map patterns that yield these various levels of spatial autocorrelation. Hence, these individual eigenvectors constitute ideal geographic map patterns affiliated with distinct levels of spatial autocorrelation. One pair of these orthogonal patterns, which is affiliated with two of the largest eigenvalues, comprises two roughly linear trend surfaces, one with an east-west-oriented gradient (eigenvector #1) and the other with a north-south-oriented gradient (eigenvector #4); in other geographic landscapes the orientations could be some rigid rotation of these orthogonal axes. Contour plottings of these two particular eigenvectors for the Puerto Rico example are presented in Maps 1.3a and 1.3b; $MC_{E_1} = \frac{73}{344} \times 5.20062 = 1.10362$ and $MC_{E_4} = \frac{73}{344} \times 4.03475 = 0.85621$. Each of the conspicuous linear trend surface gradients in these two geographic distributions accounts for approximately 55% of the variation displayed by the eigenvector.

Mathematical statistical analyses of the eigenvector map patterns reveal heterogeneity that arises from a particular surface partitioning—the tessellation heterogeneity being discussed here. This heterogeneity may be explored in a resampling fashion:

> Step 1: rank order the n values (i.e., elements) of a selected eigenvector;
> Step 2: sort a randomly selected set of attribute values in ascending/descending order;
> Step 3: because the sequence of eigenvector values geographically corresponds to the sequence of areal units used to label the rows of matrix **C**, allocate to each areal unit that attribute value whose rank order matches that of the respective eigenvector value's rank order;
> Step 4: repeat Steps 1-3 a number of times to approximate the sampling distribution for each areal unit attribute value.

This procedure may be illustrated by considering the sixteen municipios making up the Mayaguez region of Puerto Rico (see Map 1.1a). Elements of eigenvector E_2 of matrix **C** for this region, together with their rank orderings and ID numbers, are as follows:

```
ID #              2      3      6     12     25     31     33     37
E₂ element   -0.328 -0.276 -0.334  0.224  0.124  0.173 -0.271  0.237
rank order        3      5      2     14     10     12      6     15

ID#              39     44     46     47     56     58     60     63
E₂ element   -0.106  0.143  0.109 -0.413 -0.180  0.220  0.301 -0.306
rank order        8     11      9      1      7     13     16      4
```

Sorting a randomly selected set of 16 values from a uniform set of integers contained in the interval [10, 99] yielded the following set of simulated attribute values together with their allocation to the municipios in the Mayaguez region:

Table 1.1. Eigenvalues of matrices **C** and (1.10).

| positive eigenvalues | | negative eigenvalues | |
from C	from (1.10)	from C	from (1.10)
5.50193	*****	-0.11577	-0.11288
5.17230	5.20062	-0.12907	-0.12903
4.85617	4.93782	-0.19238	-0.19032
4.32879	4.32894	-0.29703	-0.28943
3.98436	4.03475	-0.47824	-0.45810
3.95364	3.96108	-0.49059	-0.47834
3.83436	3.87261	-0.53472	-0.53459
3.37842	3.37850	-0.60874	-0.60674
3.28026	3.28331	-0.67452	-0.66877
2.79956	2.80482	-0.81317	-0.80261
2.62981	2.68169	-0.91945	-0.91527
2.49041	2.53197	-0.94338	-0.94139
2.24493	2.37040	-1.03565	-1.03086
2.19156	2.22007	-1.10203	-1.10177
2.01379	2.01441	-1.18139	-1.17956
1.89030	1.89483	-1.22020	-1.22015
1.68543	1.68648	-1.29816	-1.29134
1.49305	1.52466	-1.38155	-1.38135
1.38026	1.39184	-1.43706	-1.43251
1.19199	1.21589	-1.50277	-1.50001
1.07740	1.10923	-1.53776	-1.52566
0.94561	0.95477	-1.58869	-1.57323
0.81514	0.82548	-1.62383	-1.62366
0.79106	0.79198	-1.73581	-1.73521
0.60500	0.62168	-1.85003	-1.84974
0.37613	0.37647	-1.87425	-1.86895
0.26905	0.27129	-1.92521	-1.92519
0.17413	0.19874	-1.96425	-1.96267
0.14188	0.15048	-1.97179	-1.97176
0.10095	0.10517	-2.02906	-2.02897
*****	0.00000	-2.07034	-2.06819
		-2.11464	-2.11179
		-2.17255	-2.17254
		-2.22608	-2.22590
		-2.26787	-2.25912
		-2.33547	-2.33518
		-2.40732	-2.40732
		-2.42148	-2.42044
		-2.47624	-2.47326
		-2.54057	-2.54057
		-2.66084	-2.66025
		-2.68042	-2.68033
		-2.76731	-2.76729

NOTE: Eigenvalue 5.50193 of matrix **C** is not present in the set for matrix (1.10), as denoted by *****; eigenvalue 0 of matrix

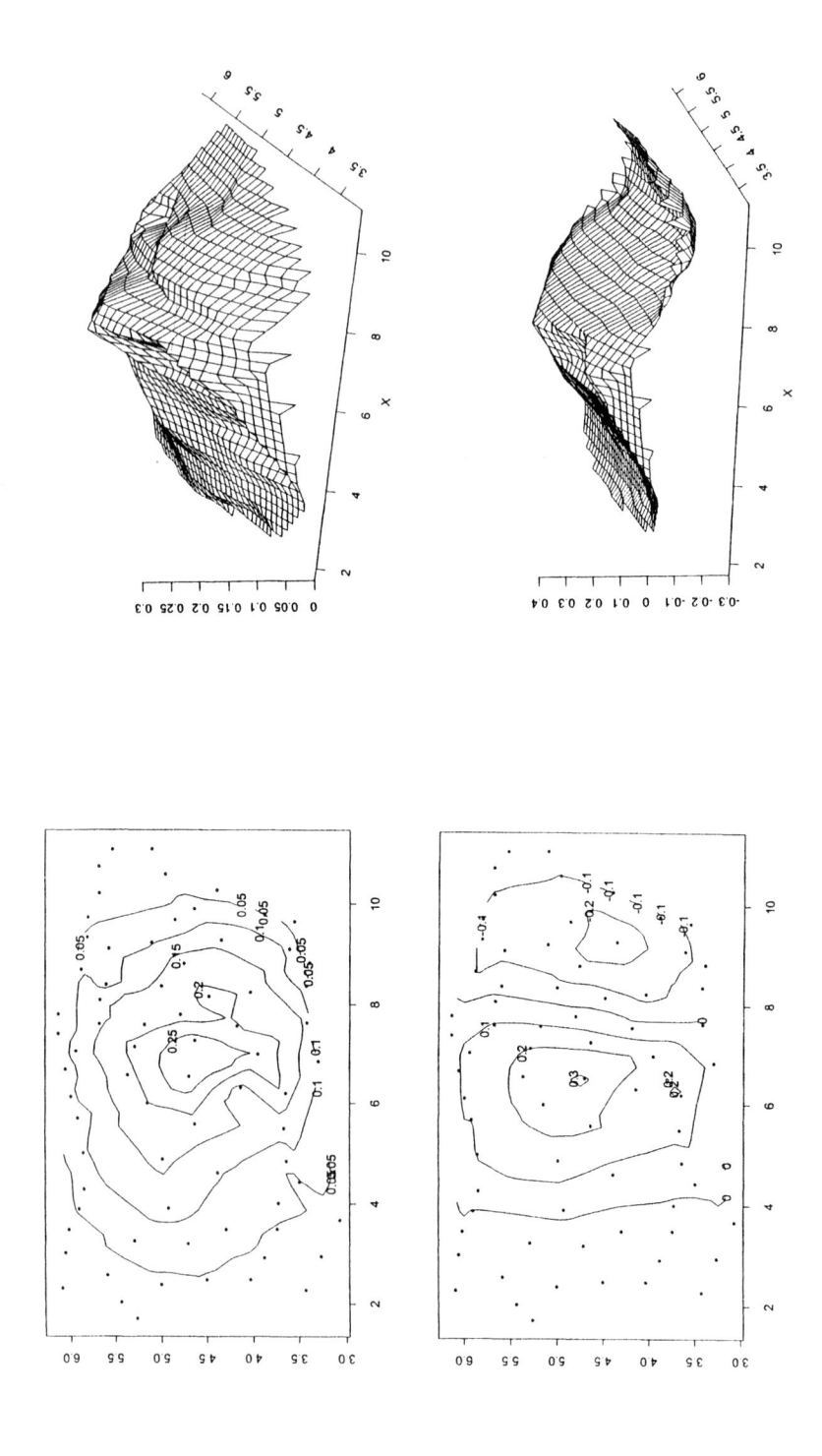

Map 1.3. Top: (a) Contour map of values from the principal eigenvector of matrix C; (c) three-dimensional plot of values from the principal eigenvector of matrix C. Bottom: (b) Contour map of values from the second eigenvector of the matrix given by expression (1.10); (d) three-dimensional plot of values from the second eigenvector of the matrix given by expression (1.10).

ID #	2	3	6	12	25	31	33	37	39	44	46	47	56	58	60	63
random number	15	46	13	92	66	70	48	98	61	69	64	10	51	71	99	18
rank	3	5	2	14	10	12	6	15	8	11	9	1	7	13	16	4

This procedure results in a map displaying approximately the level of spatial autocorrelation latent in the geographic distribution of the eigenvector values. For the eigenvector values of E_2 from matrix **C** for the Mayaguez region, MC = 0.91962 and GR = 0.15807; for the selected random numbers together with their accompanying geographic allocation reported immediately above, MC = 0.76402 and GR = 0.38406.

This conceptualization reveals that as spatial autocorrelation increases toward ±1, tessellation heterogeneity increases; a mathematical statistical analysis of the affiliated order statistics indicates that the extremes vary less than those near the mean (Table 1.2). As spatial autocorrelation decreases toward 0, tessellation heterogeneity decreases. Results in Table 1.2 also disclose that the average geographic distribution of tessellation heterogeneity is approximately proportional to the principal eigenvector of matrix **C** (Map 1.3a), relating this heterogeneity to the infamous edge effects of spatial analysis. Consequently, specification of the geographic connectivity matrix becomes important, because a carelessly specified one can introduce additional heterogeneity into a spatial analysis. Furthermore, if the presence of spatial autocorrelation is ascribed to a missing variable, these orthogonal and uncorrelated eigenvectors can be introduced into the mean response specification to compensate for missing variables, keeping spatial analysis in the realm of conventional statistical analysis.

The other feature of interest here is isotropy, or the lack of directional bias in spatial dependence. In some cases, spatial dependency in one direction is stronger than it is in another. Both the Wolfcamp aquifer (see Chapter 5) and the Mercer-Hall (see Chapter 6) data analyzed exhibit this characteristic. The presence of anisotropy can be handled by decomposing the spatial autocorrelation parameterization, in a spatial autoregressive or semi-variogram model specification, so that it captures directional dependencies. In terms of MC, this might be equivalent to computing, say, an east-west MC value and a north-south MC value

Table 1.2. Theoretical order statistic results for a 3-by-3 regular square tessellation.

areal unit	normal $\bar{X}$	s	exponential $\bar{X}$	s	sinusoidal $\bar{X}$	s	uniform $\bar{X}$	s
1	1.480	0.595	2.829	1.241	0.916	0.144	0.9	0.900
2	0.931	0.474	1.829	0.735	0.726	0.240	0.8	0.121
3	0.572	0.432	1.329	0.539	0.528	0.345	0.7	0.138
4	0.275	0.413	0.996	0.414	0.280	0.410	0.6	0.148
5	0	0.408	0.746	0.384	0	0.431	0.5	0.151
6	-0.275	0.413	0.540	0.274	-0.280	0.410	0.4	0.148
7	-0.572	0.432	0.377	0.229	-0.528	0.345	0.3	0.138
8	-0.931	0.474	0.236	0.174	-0.726	0.240	0.2	0.121
9	-1.480	0.595	0.111	0.111	-0.916	0.144	0.1	0.090

(e.g., the Mercer-Hall data), or a non-constant mean response (e.g., the Wolfcamp data). As before, this particular estimation problem may well involve some rigid rotation of the orthogonal axes initially utilized for georeferencing purposes (see Cressie 1989 for this treatment of the Wolfcamp data); the optimal orientation of these reference axes needs to be estimated, as well.

1.5.1. OBSERVATION-SPECIFIC INDICES OF SPATIAL ASSOCIATION

Both the Getis-Ord (1992) G and G_i statistics and the individual MC_i (Anselin 1995) values ($\sum_{i=1}^{n} MC_i = MC$)—what Anselin calls LISA (local indicators of spatial association)—have been proposed to index and better understand how variability in numerical contributions across a geographic landscape impact upon spatial autocorrelation detection. Serious spatial research should include follow-up work beyond the simpler diagnostic approach outlined here and involve a more thorough evaluation based upon the Getis-Ord and LISA statistics. Basically, the standard regression-based procedure presented here flags candidates for a more comprehensive evaluation of these statistics. In keeping with this perspective, and in further exploiting the MC scatterplot concept, the standard battery of regression diagnostics for detecting leverage, influential points, and outliers are inspected. These include z-scores of residuals, Cook's D, studentized residuals, the h_{ii} leverage measure, and DFfits (e.g., see Griffith and Amrhein 1997, 105). Because idiosyncrasies and nuances in georeferenced data can cause almost any areal unit attribute value to be flagged from time to time, of interest are cases in which each statistic in a given diagnostic set flags the same attribute value.

First consider a contour plot of the individual MC_i values, which helps to reveal any spatial patterns in the individual areal unit contributions to MC. Such a map for DSGR from the Puerto Rican example appears in Map 1.5a. It displays a pattern across the island that is somewhat reminiscent of the contour map of the first eigenvector (Map 1.4a) of matrix (1.10). In fact, this first eigenvector accounts for 59.3% of the variation across the island in these MC_i values for DSGR.

Next, consider selected diagnostics for detecting leverage, influential points, and outliers. Computation of these statistics for the eight variables included in the Puerto Rican example flag as aberrant areal units (areal unit numbers reference them to Map 1.1b)

variable	flagged by all four statistics	flagged by three of the four statistics
DFRM		Aguada (#2), Lares (#38), Las Piedras (#40), San Sebastian (#63)
DCD		San Juan/Rio Piedras (#61)
DMLK	Hatillo (#30)	Dorado (#23), Luquillo (#42)
DSGR	Jayuya (#34), Utuado (#68)	Lares (#38), Las Marias (#39), Maricao (#44)
DCF	Santa Isabel (#64)	Cabo Rojo (#12)

DTB	Barranquitas (#10), Morovis (#48
DBN	Aguas Buenas (#4), Cidra (#19), Comerio (#21) Jayuya (#34), Orocovis (#51)
DFAM	Jayuya (#34), Lares (#38), Penuelas (#53)

For the most part these identifications make intuitive sense. For instance, the density of cultivated land (DCD) should be lowest in the major metropolitan area of San Juan/Rio Piedras (#61); and with DCD displaying neither marked pattern nor rogue areal units in its MC scatterplot (Figure 1.2), there is no reason to expect that many areal units will be flagged. Similarly, with sugar cane's need for large tracts of relatively flat land, it is not surprising that flagged values for DSGR coincide with areal units located in the interior highlands. In each of the eight variable cases, the areal units that have been flagged appear to be spatial outliers (i.e., anomalous areal units with regard to their parent geographic distribution of agricultural productivity).

This regression-based approach can be extended to implement Anselin's LISA statistics by employing the following procedure:

Step 1: convert a given variable to its z-score counterpart, say vector **z**;
Step 2: calculate the matrix product **Cz**, denoting elements in the resulting vectors as cz_i;
Step 3: calculate $z_i \times cz_i$, denoting the resulting vector of numerical values as **z***;
Step 4: compute a no-intercept regression of **z*** on a vector **1**, also calculating studentized residuals.

The studentized residuals identify spatial outliers as described by Anselin (1995, 107). The regression coefficient obtained in Step 4 is proportional to MC; more specifically,

$$b = MC \times \frac{(n-1) \times \sum_{i=1}^{n} \sum_{j=1}^{n} c_{ij}}{n^2}$$

Meanwhile, Step 2 computes the n terms $\sum_{i=1}^{n} \sum_{j=1}^{n} c_{ij} z_j$, while Step 3 computes the n terms $z_i \times \sum_{i=1}^{n} \sum_{j=1}^{n} c_{ij} z_j$. Following are the results of this LISA regression for the eight variables included in the Puerto Rican example; these are the municipios flagged as spatially deviant areal units (areal unit numbers reference them to Map 1.1b):

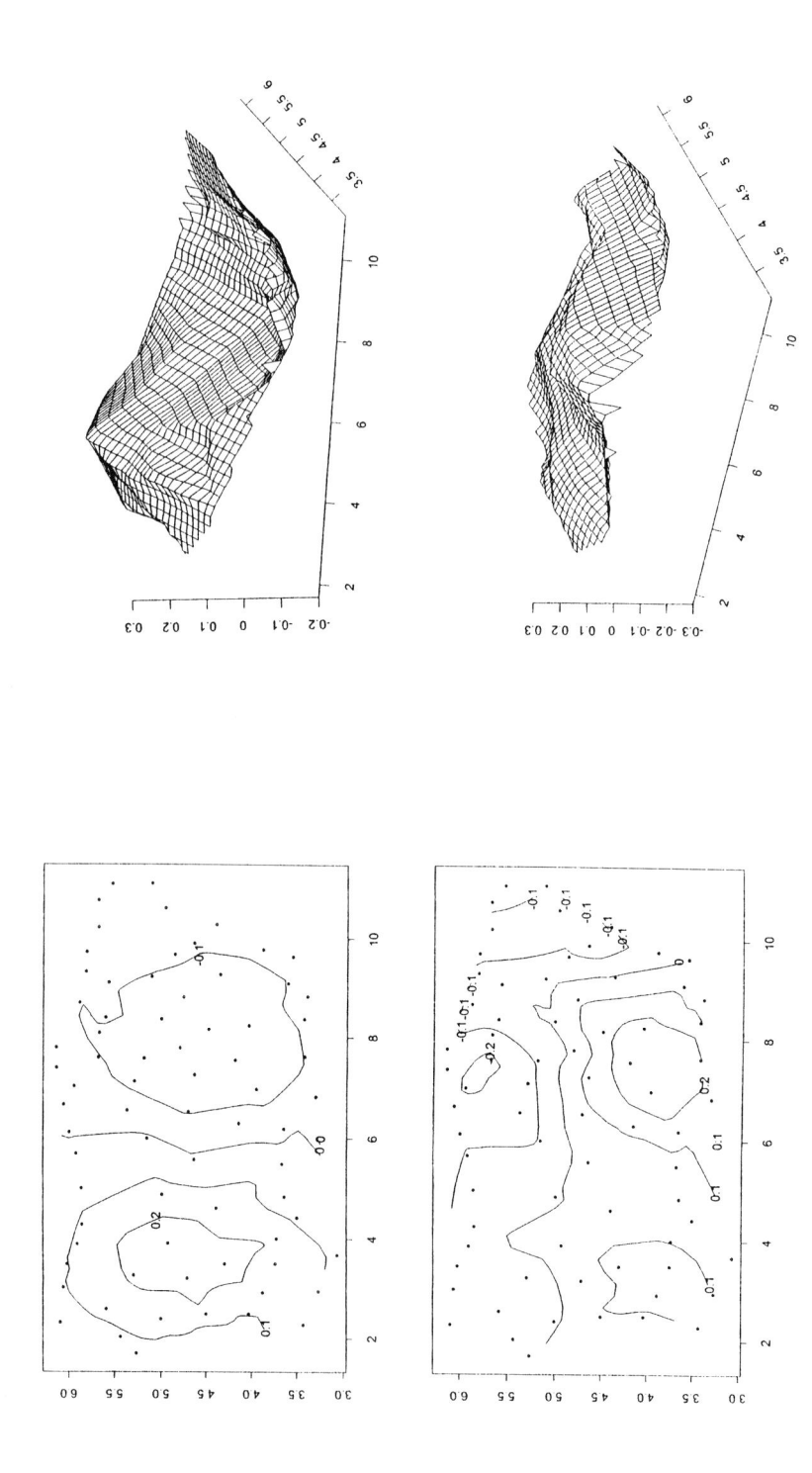

Map 1.4. Top: (a) Contour map of values from eigenvector #1 of the matrix given by expression (1.10); (c) three-dimensional plot of values from eigenvector #1 of the matrix given by expression (1.10). Bottom: (b) Contour map of values from eigenvector #4 of the matrix given by expression (1.10); (d) three-dimensional plot of values from eigenvector #4 of the matrix given by expression (1.10).

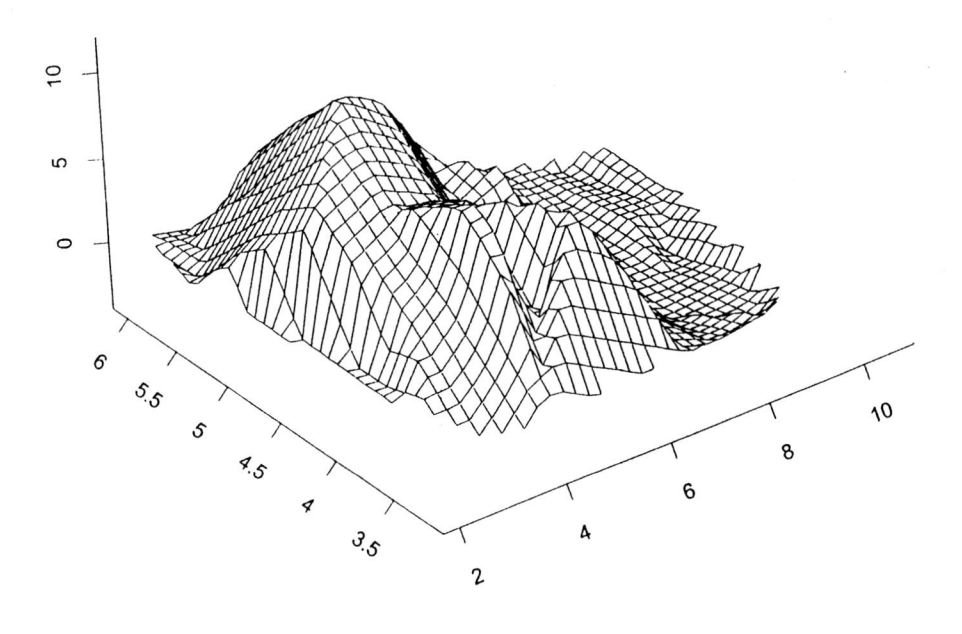

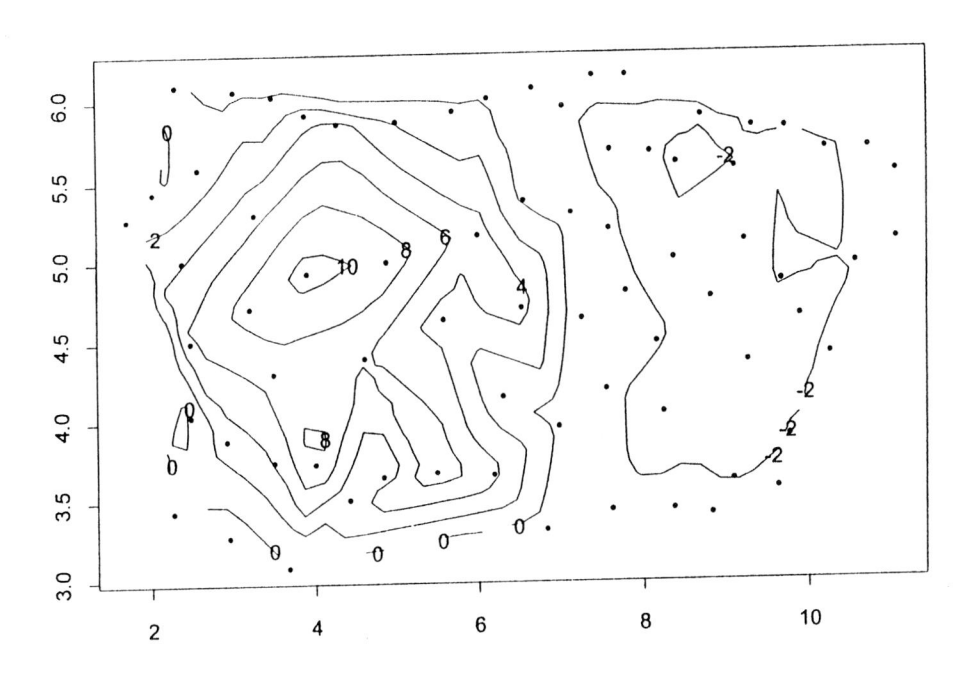

Map 1.5. Top: (a) Contour map of MC_i values for DSGR. Bottom: (b) Three-dimensional plot of MC_i values for DSGR.

variable	areal units with absolute standardized residuals ≥ 2
DFRM	Barranquitas (#10), Comerio (#21), <u>Lares</u> (#38), <u>San Sebastian</u> (#63)
DCD	Bayamon/Catano (#11), Guaynabo (#28), <u>San Juan/Rio Piedras</u> (#61)
DMLK	<u>Dorado</u> (#23), Gurabo (#29), <u>Hatillo</u> (#30), Juncos (#36), Toa Alta (#65), Toa Baja (#66)
DSGR	<u>Lares</u> (#38), <u>Las Marias</u> (#39), <u>Maricao</u> (#44), Utuado (#68)
DCF	<u>Cabo Rojo</u> (#12), Hormigueros (#31), Lajas (#37), San German (#60)
DTB	Barranquitas (#10), <u>Cidra</u> (#19), Comerio (#21)
DBN	Barranquitas (#10), <u>Jayuya</u> (#34), Lares (#38), Maricao (#44), <u>Orocovis</u> (#51)
DFAM	<u>Lares</u> (#38), Maricao (#44)

Although not complete, some agreement (designated by double underlined areal units) occurs between these LISA results and the results obtained with standard regression diagnostics. In virtually all (but not all) of the uncorroborated instances, areal units flagged by the LISA statistics have been flagged by one or two of the standard regression diagnostics. Certainly, those areal units flagged by all five statistics (i.e., z-scores of residuals, Cook's D, studentized residuals, the h_{ii} leverage measure, and DFfits) merit special consideration in a serious spatial analysis.

1.6. THE SEMIVARIOGRAM PLOT TOOL

Geostatistics, another approach to spatial data analysis, has developed primarily within the earth sciences. In keeping with this tradition, Kitanidis (1997) provides an excellent contemporary introduction to this subject for hydrogeology. Unlike spatial autoregression, geostatistics initially was designed to work with data that are distributed continuously across a surface rather than as attribute measures for discrete areal units on a surface. Also, the emphasis of geostatistics is that of description and prediction and not inferential testing per se, although prediction intervals—which are like confidence intervals—can be constructed for predicted values and model parameter estimates. An important tool of geostatistics is the semivariogram plot, which in many ways is similar to the MC scatterplot (§1.3). A semivariogram plot allows the investigator to examine the nature, degree, and extent of spatial autocorrelation latent in georeferenced data. The semivariogram plot, denoted by γ(h), is defined as half the average of the squared difference between paired values (making it similar to the GR index of spatial autocorrelation) and may be written as

$$\gamma(h) = \frac{1}{2N(h)} \sum_{(i,j)|h_{ij}=h} (y_i - y_j)^2 \; , \qquad (1.11)$$

where h_{ij} is the distance separating location coordinate pairs for selected points i and j, h is the distance class (or lagged distance), N(h) is the number of pairings within each distance class, and y_i and y_j are the attribute values respectively corresponding to the selected locations i and j. Division by 2 in equation (1.11) arises because, for a stationary and isotropic process,

$$\begin{aligned} E[(y_i - y_j)^2] &= E[(y_i - \mu) - (y_j - \mu)]^2 \\ &= E[(y_i - \mu)^2] - 2 \times E[(y_i - \mu)(y_j - \mu)] + E[(y_j - \mu)^2] \\ &= \sigma_i^2 - 2\sigma_{ij} + \sigma_j^2 = 2\sigma^2 - 2\sigma_{ij} \times \frac{\sigma^2}{\sigma^2} \\ &= 2\sigma^2 (1 - \frac{\sigma_{ij}}{\sigma^2}) = 2\sigma^2 (1 - \rho_{ij}) \ . \end{aligned}$$

Division by 2 simplifies algebra for mathematical statistics purposes associated with the semivariogram, $\gamma(h)$. The factor 2 also appears because $(y_i - y_j)^2 = (y_j - y_i)^2$; this identity allows the n^2 interdistance calculations to be reduced to $\frac{n \times (n-1)}{2}$ calculations:

$$\begin{aligned} \sum_{i=1}^{n} \sum_{j=1}^{n} (y_i - y_j)^2 &= \sum_{i=1}^{n} \sum_{j=1}^{i-1} (y_i - y_j)^2 + \sum_{i=1}^{n} \sum_{j=i}^{i} (y_i - y_j)^2 + \sum_{i=1}^{n} \sum_{j=i+1}^{n} (y_i - y_j)^2 \\ &= 2 \sum_{i=1}^{n} \sum_{j=1}^{i-1} (y_i - y_j)^2 + n \times 0 \ . \end{aligned}$$

The plot of a semivariogram is constructed using $\gamma(h)$ along the vertical axis and interpoint distance h_{ij} along the horizontal axis. Pairings of similar attribute values give smaller values of $\gamma(h)$ so that if positive spatial autocorrelation is present in georeferenced data, pairings of attribute values that lie in relatively close proximity of each other will have low $\gamma(h)$ values, with $\gamma(h)$ increasing as the distance between pairs increases. The actual scatterplot of $\gamma(h)$ versus distance can take on various functional forms, all of which give some description of spatial autocorrelation across a surface. A high degree of positive spatial autocorrelation will result in a semivariogram plot displaying a relatively shallow slope, while near-0 spatial autocorrelation will result in a steep slope. Inspection of a semivariogram plot allows visual identification of spatial outliers and detection of nonstationarity (e.g., a nonconstant mean) in georeferenced data. In the semivariogram plot, however, spatial outliers will be in terms of the $\gamma(h)$ values for a given distance class rather than a given data value. Also, the form of the semivariogram is inverted from that of a covariance function, where the covariance value tends to *decrease* as the distance between pairings of points increases.

In addition to assigning distance classes, direction tolerances may be specified to identify anisotropy on a surface. Anisotropic surfaces occur when the degree of spatial dependency is unequal in all directions. They usually are modeled as major and minor axes with a specified direction and tolerance determined for each axis (similar to a bivariate principal components analysis). Conversely, isotropic surfaces are those in which spatial dependency is approximately equal in all directions. When computing γ(h) values and constructing a semivariogram plot, isotropy is usually assumed first and anisotropy investigated later (Cressie 1991; Isaaks and Srivastava 1989). Direction tolerance for an isotropic surface can be defined with a direction of 0° and a tolerance of 90°. A more comprehensive treatment of anisotropy is presented in §3.2.4, with examples presented in Chapter 5.

Various valid semivariogram functions can be fitted to a semivariogram plot in order to summarize more concisely the spatial similarity that may be present in georeferenced data (Cressie 1991; Christensen 1991; Isaaks and Srivastava 1989). These functions then can be used to predict unknown attribute values for unsampled areas from the data collected for sampled locations; that is, kriging. Some semivariogram models have been shown to be valid and/or useful, namely, the linear, exponential, circular, spherical, Gaussian, Bessel, power and wave/hole; they routinely appear throughout the geostatistics literature. Selected models graphically portrayed in Figure 1.3 have been fitted to the Wolfcamp aquifer data. As an example, the equation for the exponential model may be written as

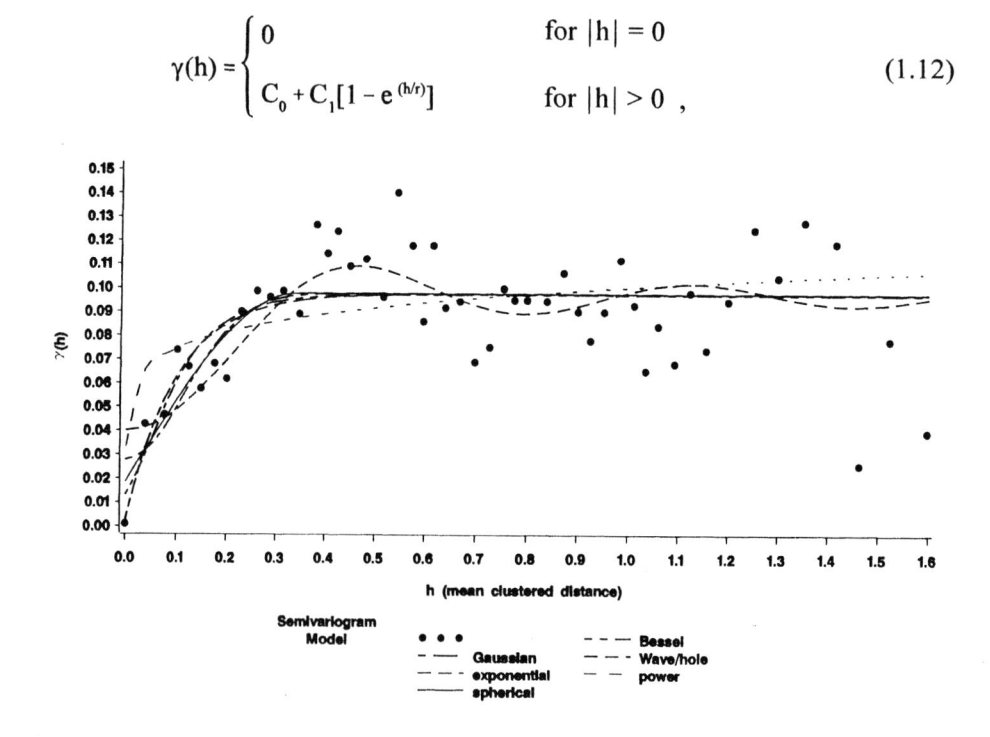

$$\gamma(h) = \begin{cases} 0 & \text{for } |h| = 0 \\ C_0 + C_1[1 - e^{(h/r)}] & \text{for } |h| > 0 \end{cases} \quad (1.12)$$

Figure 1.3. Semivariogram plot of piezometric-head pressure for Cressie's Wolfcamp aquifer data, with selected semivariogram model curves superimposed.

where C_0 is an intercept term, C_1 defines the slope of the semivariogram curve, and r is the range parameter[6] governing spatial dependency. Semivariogram models have only a few parameters that need to be estimated, and those can be interpreted from the plots. Theoretically, $\gamma(h)$ at lag 0 should be 0; however, a discontinuity may exist, called the nugget effect, which is represented as C_0 in equation (1.12). Since the semivariogram plot is similar to that of a covariance plot, the other two parameters that need to be estimated can be defined in terms of a covariance. The range is the distance at which the covariance becomes 0. The sill is the value of the covariance at distance 0 minus any nugget effect. Because a semivariogram plot is that of an inverted covariance plot, the range for the former is identified as the distance at which $\gamma(h)$ becomes approximately constant, and the sill is this value minus any nugget effect. Usually, several semivariogram models may be estimated with a given data set to assess which model has the best fit. The models are routinely fitted to data by means of weighted nonlinear least squares, and the resulting relative sum of squared errors (RSSE = $\frac{\text{error SS}}{\text{corrected total SS}}$) for each model then are compared to give a relative goodness of fit measure. However, selection of a semivariogram model should not be based solely on the smallest of a set of the RSSEs, because a number of competing models can produce very similar values; after all, the curve-fitting exercise is for $\sigma^2(1 - \rho_{ij})$, which limits the magnitude of the fit differences between models. Because the semivariogram deals with small covariance values, a visual inspection of a semivariogram plot along with the fitted function should be performed to verify selection of a particular model and is highly recommended by Cressie (1991). In addition, selection of final candidate competing models (i.e., models having essentially the same goodness-of-fit values) should be based upon cross-validation and perhaps even upon the Akaike information criterion (AIC). Also, although most of the models presented are useful for a surface containing positive spatial autocorrelation, the wave/hole model is the only useful function when dealing with surfaces containing negative spatial autocorrelation, while the power function is useful for capturing the presence of a nonconstant mean over a surface. Cressie (1991) and Haining (1990) should be consulted for detailed descriptions and individual ideal plots of each model.

One semivariogram model that shows considerable data analytic promise but that has not been extensively employed by practitioners is the modified Bessel function (motivated by Whittle 1954). Preliminary investigation by Griffith and Csillag (1993) and later by Griffith et al. (1996) found it to be an important geostatistical model that provides a link to the SAR model of spatial autoregression. Thus, it is presented in more detail here than the other semivariogram

---

[6]The concept of range denotes the distance beyond which spatial correlation essentially is 0. Parameter r is not necessarily equivalent to this distance value. While it is in the spherical semivariogram model (see §3.2.1), it merely determines the rate of convergence upon 0 spatial (auto)correlation in the exponential and Gaussian semivariogram models, which have infinite ranges. It does not even appear in the power function semivariogram model. A number of researchers confuse the notion of range with that of correlation length; generally speaking, these two values are not the same.

models previously discussed. This model is specified as

$$\gamma(h) = \begin{cases} 0 & \text{if } |h| = 0 \\ C_0 + C_1[(1 - \left(\frac{h}{r}\right) K_1\left(\frac{h}{r}\right)] & \text{if } |h| > 0 \end{cases} \qquad (1.13)$$

where K_1 is the modified Bessel function of the first-order, second-kind. The algorithm used in this book for computing K_1 has been adapted from that presented in Press (1986). Output produced from test runs of SAS was verified by using selected data values in Tables 32 and 34 of Spiegel (1968) and by comparing SAS output to that of the Bessel function in Maple V (MathSoft 1995). Although Heuvelink includes the modified Bessel function in his WLSFIT suite of semivariogram models, this function is not one of the semivariogram models found in most readily available geostatistical software packages.

When a model is fitted to a semivariogram plot, usually not all of the distance data points are used. The most important information content in a semivariogram plot is at the beginning of the plot, specifically that interval beginning at the intercept and up through the sill. As distances between locations increase toward the maximum distance possible (that is, the distance separating the two geographically most extreme points), fewer and fewer attribute value pairs are available to contribute to computing $\gamma(h)$. This decrease in N(h) usually leads to the spread of $\gamma(h)$ increasing as these lag distances increase. One complication of this spread in $\gamma(h)$ values leads to a poorer fit (larger RSSE) of any selected model. Such values in the semivariogram plot can be discarded because, again, they contribute only slightly (or not at all) to describing spatial dependency in the data set. Criteria by which to select a truncation point along the distance axis have been discussed by some authors, although there is no consensus about which is the more appropriate criterion. For instance, Issaks and Srivastava (1989) suggest using a rule of discarding $\gamma(h)$ values that are found at more than half of the maximum distance value. Graphically speaking, if the points on the semivariogram plot remain approximately constant and stay so for a substantial portion of the plot, then $\gamma(h)$ values at further lags can be safely discarded with little or no loss of information. Another useful criterion is to examine a plot of the number of pairs in a distance class versus distance. Where distances are relatively large, $\gamma(h)$ values that are computed when the number of pairs in a distance class falls below, say, 50 can be excluded. Cressie (1991) recommends a minimum of 30 distance pairs per distance class; the classical central limit theorem suggests as many as 100 pairs per class. Whatever method is used, however, care should be taken so that distance pairs are not excluded when the semivariogram plot is indicating a nonconstant mean situation. In this book we use a combination of examining the plot of the number of distance pairs versus lag distance and examining the shape of the distribution of $\gamma(h)$ values in the plot (the graphical method) to identify the relevant portion of a semivariogram for model-fitting purposes.

1.7. COMPUTER CODE FOR IMPLEMENTING SPATIAL STATISTICAL ANALYSES

Procedures for estimating spatial autoregressive or geostatistical models are not available in standard statistical software packages such as SAS, SPSS, and MINITAB, although recently MathSoft has included some of these functions in its S+ spatial statistics module, and SAS PROC GIS and PROC MIXED offer considerable potential. Users must find other specialty software packages when undertaking these types of analyses. Levine (1996) reviews some of the dedicated spatial autoregressive packages as well as a few other selected packages that relate to GIS interests. Legendre (1993) surveys 18 packages that are available for undertaking spatial and geostatistical analyses. With regard to semivariogram modeling, in addition to Heuvelink's WLSFIT package used to verify results reported in this book, there are GEO-EAS (particularly the UNIX version), GSLIB, SAGE95, and VARIOWIN. In fact, there is a proliferation of specialized, dedicated software packages for undertaking spatial autoregressive and geostatistical analyses. But, as is illustrated throughout this book, parameter estimation for spatial autoregressive and semivariogram models can be accomplished using the aforementioned standard statistical packages (cf., Griffith 1988b, 1989, 1992c for MINITAB examples). The principal advantage of utilizing implementations developed for standard statistical software packages is that a researcher does not have to leave the SAS, SPSS, or MINITAB environment to compute spatial statistics and hence is able to retain all of the other statistical technology contained in these reputable commercial statistical software packages. Spatial analysis conducted with S+ exemplifies this desirable situation. In terms of other popular software packages, Anselin and Hudak (1992) show how to implement spatial econometric model specifications in Gauss, Limdep, Rats, and Shazam.

Geostatistical models can be implemented in SAS or SPSS; SAS and SPSS code for these models appears in Appendices 3-E and 3-F at the end of Chapter 3. This SAS code uses PROC NLIN as the primary procedure for obtaining nonlinear weighted least squares estimates of parameters. An advantage of using PROC NLIN for estimating semivariogram parameters is that NLIN also will generate nonlinear regression diagnostics, whereas available dedicated packages for geostatistics usually give only the parameter estimates and sometimes their variances. As seen in §1.5.1, examination of regression diagnostics is helpful for identifying values that may cause problems in the analysis and interpretation of results. The reader needs to be aware, however, that preprocessing of georeferenced data by the user is necessary to efficiently and successfully use SAS and SPSS for this purpose.

1.7.1. SAS CODE FOR GEOSTATISTICAL ANALYSES

Before a model can be fitted to a semivariogram plot, the task of computing $\gamma(h)$ values used to construct the plot must be performed. Usually, a distance class within which georeferenced attribute value pairs are grouped is established by the user, and the semivariogram then is computed using equation (1.11). In fact, an appropriate semivariogram plot to model may be selected from a set constructed simply by varying the distance class. This same process also can be executed using SAS, but the method presented here is a two-step process where distance classes are not constructed until after computing γ values. Our implementation with SAS begins by computing a γ value, and corresponding distance, for each of the n^2 attribute value pairs. These values then are saved in a temporary SAS data set that contains the ID value for attribute value x_i, the ID value for x_j, distance (by default computed as a Euclidean distance separating the attribute value locations), frequency (1 in all cases), the γ value, and a sequence ID. Next, these data are clustered by distance using PROC CLUSTER where clustering defines the number of distance pairs that contribute to each distance class. The clustering method employed here uses Ward's algorithm, a minimum variance technique, and allows the dispersion within groups to be viewed as measurement error. Input required from the user is the number of clusters to use (default number is 50 clusters) and is specified in the macro variable NGROUPS.

Our method of computing and plotting semivariogram values consists of three SAS program files: a data processing file (Program #1), SEMI-VAR.SAS (Program #2) and SEMICLUS.SAS (Program #3). Program #1, an administrative program, contains code to read in the data, do the necessary preprocessing of data, and define macro variables used subsequently in SEMI-VAR.SAS and SEMICLUS.SAS. This file differs slightly for each data set analyzed but contains elements common to all input data sets. Program #1 is designed such that values for all macro variables are specified in the beginning of the program by the user, so that no further changes need to be made to the program, or to SEMI-VAR.SAS and SEMICLUS.SAS. Descriptions of each of the required inputs and options are described in the first comment box at the beginning of the program (see Program #1). In fact, all of the SAS programs in this book conform to this same layout. While at first glance the amount of information to be input by the user may seem intimidating, most of the values and options can remain unchanged for most data sets analyzed. The only changes that may need to be made in the code itself for this program occur if the analysis variable (DSVAR) is to be transformed in any manner and/or an indicator variable for outliers is to be created. Both of these tasks are done in the data step URAWDAT, although the user optionally can perform either or both of these tasks prior to using Program #1.

Several points about the preprocessing program (Program #1) merit explanation. The file extension of the input data set to be analyzed is required by this program to be "txt" (the file extension "dat" is reserved for an output file that is required for a GEO-EAS input data file). The first line of executable code provides a link to SAS macros used for converting data specified in terms of degrees to decimal degrees and decimal degrees to radians, and vice versa, and are listed in Appendix 1-C. These macros are required for two main reasons. Code in Program #1 is included that allows the user to specify explicitly direction and direction tolerance (in macro variables DIRDEG and TOLDEG) in order to investigate for anisotropy in georeferenced data, analogous to the methods outlined by Isaacks and Srivastava (1989, 93-106; Chapter 7). The default values—0 and 90 degrees for direction and tolerance, respectively—are set for an isotropic surface. The macros listed in Appendix 1-C are needed to convert angles specified in terms of degrees to radians since the trigonometric functions in SAS work only in terms of radians. Secondly, although a Euclidean interpoint distance is used for all data sets analyzed in this book, this set of programs also allows the option of using Great Circle distances,[7] again requiring the use of the SAS macros listed in Appendix 1-C to convert latitude-longitude measures to radians. Accordingly, optional code is included to convert decimal degrees to miles when computing distance where decimal degrees are used. However, for the geostatistical analyses of the data sets in this text, where decimal degrees are used as the coordinate reference system, because the geographic landscapes are relatively small, great circle distances are not utilized.

The next 17 lines of executable code contain the macro variables for specifying the path names of the program and data files, data file name, analysis variable name, and so on, as explained in the comment box preceding all of the code in this program. The data file is input using the first DATA URAWDAT data step and may need to be modified for each particular data set analyzed if the analysis (or other) variable is to be transformed or indicator variables for outliers created. Notice also in this data step that the analysis variable name (specified in the macro variable DSVAR) is changed by adding an "r" as the first letter and the original name is dropped, thus limiting the maximum number of characters in this variable to 7. The georeference variable names are required to be u and v for horizontal and vertical direction, respectively. Next, for each data set, the u-v coordinates are centered, because they always are measured with an interval and not a ratio scale. Centering of the u-v coordinates is accomplished using the first PROC STANDARD statement encountered in the file. Converting the analysis variable to z-scores (having $\bar{x} = 0$ and $s = 1$) is accomplished using the second PROC STANDARD statement in the file; instructions for doing so are provided as an option in the macro variable STDDSVAR. The section of code denoted with "*** begin map ***;" and "*** end map ***;" is used to create a map of the analysis variable data values after any transformations, standardizations and/or regressions have been performed, and can be modified to utilize the ID values of

---

[7] The SAS code for computing great circule distances is as follows:

```
p = abs(u{i} - u{j});
dsin = (sin(v{i}) * sin(v{j}));
dcos = (cos(v{i}) * cos(v{j}));
dlong = cos(p);
darc = dsin + (dcos * dlong);
if darc > 1 then darc=1;
if darc < -1 then darc = -1;
distdeg = arcos(darc);
dist = distdeg * 69; /* convert distance in decimal degrees to miles. */
```

the same data set. The last set of macro code at the end of the program, beginning with the line "`%if krigfile = yes %then %do;`", creates a data file with file extension "dat" and the same filename specified in macro variable DSVAR, which contains the non-centered u-v coordinates and values of the analysis variable (or, in most data sets analyzed in this text, the residuals from an OLS regression), which is used as a data input file for GEO-EAS, particularly for the kriging routines. One should note, however, that this data file must be modified to include necessary header information for input into GEO-EAS. Finally, the last two "%include" statements specify that Programs #2 and #3 (SEMI-VAR.SAS and SEMICLUS.SAS) should follow execution of the preprocessing program and be executed in tandem to it. An example file for Program #1 is presented below, which is the code used to pre-process the data used for computing the semi-variogram values and create the semivariogram plot for the Wolfcamp aquifer data set example (Chapter 5).

Program #1

```
/*---------------------------------------------------------------|
| WOLFCAMP.SAS                                                    |
| Preprocesses a data set for construction of a semivariogram. File|
| extension is required to be "txt'. This program centers the U-V |
| coordinates and provides options to standardize the analysis variable|
| and use the residuals from an OLS regression model. All user inputs|
| are provided in the first 17 macro variables:                   |
|    progpath - name of the subdirectory path where analysis programs|
|               are located.                                      |
|    dspath - name of the subdirectory path for the analysis data file.|
|    flname - filename of the analysis data file. File extension is|
|             required to be "txt".                               |
|    invars - variable names in the analysis data file in the correct|
|             order.                                              |
|    dsvar - name of the analysis data variable. This variable is |
|            limited to a maximum of 7 characters.                |
|    stddsvar - option to standardize the analysis variable first with a|
|               mean=0 and std=1 (yes or no).                     |
|    ngroups - number of distance groups that PROC CLUSTER will create.|
|    dirdeg - direction in degrees to specify for anisotropic analyses.|
|    toldeg - direction tolerance in degrees to specify for anisotropic|
|             analyses. For isotropic surfaces, dirdeg=0 and toldeg=90.|
|    hivalpo - placement of the uppermost value for each distance group|
|              in the high-low value plot.                        |
|    gcd - option to use Great Circle Distance (yes or no).       |
|    roundtol - rounding distance used by PROC CLUSTER in created |
|               distance groups (see text). Default value is 0.001.|
|    truncdis - truncation distance used in SEMIBVAR.SAS (see text).|
|    regress - option to use OLS regression residuals as the analysis|
|              data (yes or no).                                  |
|    miv - independent variables to use in the OLS regression.    |
|    plotmap - option to plot a map of the data analysis values, or|
|              point id's (yes or no).                            |
|    krigfile - option to output a text file used to input into GEO-EAS|
|               (yes or no). Output file will have the same filename as|
|               that specified in DSVAR and file extension "dat". |
|---------------------------------------------------------------*/

options mautosource sasautos="d:\lat_long" nocenter nodate pageno=1
    MPRINT SYMBOLGEN MLOGIC;
```

```
%global nrecs dsvar dspath flname ngroups dirdeg toldeg goutfq gouthl;

%let progpath = %str(c:\nsf_geog\tstprogs);
%let dspath = %str(c:\nsf_geog\tstprogs\wolfcamp);
%let flname = wolfcamp;
%let invars = %str(id u v piezo);
%let dsvar = piezo;
%let stddsvar = yes;
%let ngroups = 50;
%let dirdeg = 0;
%let toldeg = 90;
%let hivalpo = 0.02;
%let gcd = no;
%let roundtol = 0.001;
%let truncdis = 3.0;
%let regress = yes;
%let miv = %str(u v ioutlier);
%let plotmap = yes;
%let krigfile = yes;

goptions device=win norotate;           *<--- default graphics is to monitor;

TITLE "Semivariogram computations for variable &dsvar for file &flname";
%macro presem;
  data _null_;
   infile "&dspath.\&flname..txt" missover;
   input c1;
   call symput('nrecs',left(_n_));
  run;
  %put number of observations = &nrecs;

  data urawdat; infile "&dspath.\&flname..txt";
    input &invars;
    r&dsvar = &dsvar;
    if piezo = 3571 then ioutlier=1;
    else ioutlier=-1;
    drop &dsvar;
  run;
  %if &krigfile = yes %then %do;
    data krigdata; set urawdat;
      easting = u;            *<--- needed for GEO-EAS input data file;
      northing = v;           *<--- needed for GEO-EAS input data file;
      keep easting northing;
    run;
  %end;
  proc standard data=urawdat out=urawdat(replace=yes) mean=0;
    var u v;
  run;
  %if &stddsvar = yes %then %do;
    proc standard data=urawdat out=urawdat(replace=yes) mean=0 std=1;
      var r&dsvar;
    run;
  %end;
  data urawdat(replace=yes); set urawdat;
    maxu=abs(u); maxv=abs(v);
    maxgeo=max(maxu,maxv);
    mval=1; mergeval=1;
    drop maxu maxv;
  run;
  proc means max data=urawdat noprint;
    var maxgeo mval;
    output out=geomax max=maxuv mergeval;
  run;
```

```
data urawdat(replace=yes); merge urawdat geomax; by mergeval;
  u=u/maxuv; v=v/maxuv;
  drop mval mergeval maxuv maxgeo _type_ _freq_;
run;

%if &regress = yes %then %do;
  proc reg data=urawdat outest=reguv;
    model r&dsvar = &miv;
    output out=rawdat r=&dsvar;
  run;
%end;
%if &regress = no %then %do;
  data rawdat; set urawdat;
    &dsvar = r&dsvar;
    drop r&dsvar;
  run;
%end;
proc means var data=rawdat noprint;
  var &dsvar;
  output out=sigmavar var=standvar;
run;
data _null_; set sigmavar;
  call symput('standvar',standvar);
run;

%if &regress = yes %then %do;
  data adjuv; merge reguv(drop=_type_) geomax(drop=_type_);
    if u>. then u=u/maxuv;
    if v>. then v=v/maxuv;
  run;
  proc print data=adjuv;
    var intercep maxuv &miv;
  run;
%end;
%if &plotmap = yes %then %do;     /*** begin map ***/
  proc means min max data=rawdat noprint;
    var u v;
    output out=uvminmax min=umin vmin max=umax vmax;
  run;
  data minmax; set uvminmax;
    umin=umin-0.1;
    umax=umax+0.1;
    vmin=vmin-0.1;
    vmax=vmax+0.1;
    call symput('umin',umin);
    call symput('vmin',vmin);
    call symput('umax',umax);
    call symput('vmax',vmax);
  run;

  data anno;
    length function color style $ 8 text $ 5;
    retain xsys '2' ysys '2' hsys '1' when 'a';
    set rawdat;
    function='label'; color='black';
    size=3; text=left(put(&dsvar,5.2)); position='5';
    style='swissb'; x=u; y=v;
  run;

  axis1 label=(height=1 color=black font=swissb angle=90)
        minor=none
        order=(&vmin to &vmax by 0.1);
  axis2 label=(height=1 color=black font=swissb)
```

```
            minor=none
            order=(&umin to &umax by 0.1);
   symbol1 c=black v=none i=none;
   proc gplot data=rawdat;
      plot v*u / annotate=anno
                 vaxis=axis1
                 haxis=axis2;
      title1 "Plot of attribute values &dsvar by location for dataset"
                                                    " &flname ";
   run;
   %end;    /*** end map ***/
   %if &krigfile = yes %then %do;
      data _null_; merge krigdata rawdat(keep=&dsvar);
         file "&dspath.\&dsvar..dat";
         put easting northing &dsvar;
      run;
   %end;
   %include "&progpath.\semi-var.sas";
   %include "&progpath.\semiclus.sas";
%mend presem;
   %presem;
```

Following is the SAS code for SEMI-VAR.SAS, designed to compute the individual γ and distance values only, for all possible distance pairs. No modifications need to be made to this program by the user. Data from this program are then passed to SEMICLUS.SAS (Program #3).

Program #2

```
/*-----------------------------------------------------------------|
| SEMI-VAR.SAS                                                     |
| Computes semi-variogram values for all distance pairs.  Creates  |
| temporary SAS data set (rowvect) for processing by SEMICLUS.SAS. |
|-----------------------------------------------------------------*/

data standdat; set rawdat;
   attr = &dsvar;
run;

data u_n;    set standdat;      keep u;    run;
data v_n;    set standdat;      keep v;    run;
data attr_n; set standdat;      keep attr; run;
data id_n;   set standdat;      keep id;   run;

proc transpose data=u_n    out=tran_u   prefix=u;    var u;    run;
proc transpose data=v_n    out=tran_v   prefix=v;    var v;    run;
proc transpose data=attr_n out=tran_att prefix=attr; var attr; run;
proc transpose data=id_n   out=tran_id  prefix=id;   var id;   run;

proc datasets lib=work; delete u_n v_n attr_n id_n / mt=data; run;

data rowvect; merge tran_u(drop=_name_) tran_v(drop=_name_)
                    tran_att(drop=_name_) tran_id(drop=_name_);
   array u       u1-u&nrecs;
   array v       v1-v&nrecs;
   array attr    attr1-attr&nrecs;
   array id      id1-id&nrecs;
   meanattr = sum(of attr1-attr&nrecs)/&nrecs;
   gcd = "&gcd";
   seqid=0;
   %dectorad(&toldeg,tolrad);
```

```
%dectorad(&dirdeg,dirrad);
pi=arcos(-1);
min0tol=dirrad-tolrad; below0=min0tol;
max0tol=dirrad+tolrad;
if below0<0 then min0tol=(2*pi)+min0tol;
min1tol=(dirrad+pi)-tolrad;
max1tol=(dirrad+pi)+tolrad; above0=max1tol;
if above0>(2*pi) then max1tol=max1tol-(2*pi);
opdirrad=dirrad+pi;
do i=1 to &nrecs;
  do j=1 to &nrecs;
    idi=id{i}; idj=id{j};
    pointa=attr{i}; pointb=attr{j};
    originx=u{i};
    originy=v{i};
    px=u{j}-originx; py=v{j}-originy;
    r = sqrt(px**2 + py**2);
    if r=0 then do;
      angref=dirrad;
    end;
    else do;
      angref=arcos(px/r);
    end;
if (u{j} < originx & v{j} < originy) then angref=angref+2*(pi-angref);
if (u{j} > originx & v{j} < originy) then angref=angref+2*(pi-angref);
if (u{j} = originx & v{j} < originy) then angref=angref+(pi);
if below0<0 then do;
if ((0<=angref<max0tol) | (min0tol<=angref<(2*pi))) |
                                (min1tol<=angref<max1tol) then do;
      gamma = ((attr{i} - attr{j})**2)/2;
      dist = sqrt((u{i} - u{j})**2 + (v{i} - v{j})**2);
      if gcd = "yes" then do;
    p=abs(u{i} - u{j});
    dsin = (sin(v{i}) * sin(v{j}));
    dcos = (cos(v{i}) * cos(v{j}));
    dlong = cos(p);
    darc = dsin + (dcos * dlong);
    if darc > 1 then darc=1;
    if darc < -1 then darc = -1;
    distdeg = arcos(darc);
    dist = distdeg * 69;        /* convert distance in decimal degrees to
                                                                miles. */
      end;
      freq=1;
      seqid = seqid+1;
      mergeval=1;
      output;
    end;
  end;
  if above0>(2*pi) then do;
    if ((0<=angref<max1tol) | (min1tol<=angref<(2*pi))) |
                                (min0tol<=angref<max0tol) then do;
      gamma = ((attr{i} - attr{j})**2)/2;
      dist = sqrt((u{i} - u{j})**2 + (v{i} - v{j})**2);
      if gcd = "yes" then do;
    p=abs(u{i} - u{j});
    dsin = (sin(v{i}) * sin(v{j}));
    dcos = (cos(v{i}) * cos(v{j}));
    dlong = cos(p);
    darc = dsin + (dcos * dlong);
    if darc > 1 then darc=1;
    if darc < -1 then darc = -1;
    distdeg = arcos(darc);
```

```
            dist = distdeg * 69;      /* convert distance in decimal degrees to
                                                                    miles. */
        end;
        freq=1;
        seqid = seqid+1;
        mergeval=1;
        output;
      end;
    end;
    if (below0>=0 & above0<=(2*pi)) then do;
      if (min0tol<=angref<max0tol) | (min1tol<=angref<max1tol) then do;
          gamma = ((attr{i} - attr{j})**2)/2;
          dist = sqrt((u{i} - u{j})**2 + (v{i} - v{j})**2);
          if gcd = "yes" then do;
    p=abs(u{i} - u{j});
    dsin = (sin(v{i}) * sin(v{j}));
    dcos = (cos(v{i}) * cos(v{j}));
    dlong = cos(p);
    darc = dsin + (dcos * dlong);
    if darc > 1 then darc=1;
    if darc < -1 then darc = -1;
    distdeg = arcos(darc);
    dist = distdeg * 69;      /* convert distance in decimal degrees to
                                                                    miles. */
        end;
        freq=1;
        seqid = seqid+1;
        mergeval=1;
        output;
      end;
    end;
    keep idi idj dist freq gamma seqid mergeval pointa pointb;
   end;
 end;
run;

proc datasets lib=work; delete tran_u tran_v tran_att tran_id standdat /
                                                           mt=data; run;
```

Before presenting code for SEMICLUS.SAS, which, again, defines the distance intervals based on Ward's clustering algorithm via PROC CLUSTER in SAS, an explanation of how distance is preprocessed prior to clustering is given here. To drastically decrease the amount of computing time necessary to cluster a set of n^2 values, some of which may be very similar or identical (e.g., in the regular square tessellation case), a round-off index is constructed in which distances are standardized ($\bar{x} = 0$ and $s = 1$) and then rounded to the nearest 0.001 (this value can be increased or decreased as specified in the macro variable ROUNDTOL in Program #1, according to the size of a data set, but is the default value used throughout this book). These distance z-scores are kept in a separate index variable. Means for original distance and γ values are computed and their frequencies counted within each distance z-score group. Clustering then is performed on the distance means for these groups, weighting by frequency. After clustering, individual γ values are averaged by cluster class to give the weighted average γ(h) and distance values for each distance class, in effect yielding the same result that can be obtained from equation (1.11). One difference, which merits highlighting, between our clustering method and that of defining distance classes by the user is that defining distance classes by clustering will yield nonuniform distance intervals. Typically, default distance classes employed by the user are uniform distance intervals, for instance, in dedicated geostatistical packages such as GEO-EAS. Our method will tend to group more effectively the more similar distances. No modifications to this program are needed by the user.

Program #3

```
/*----------------------------------------------------------------|
| SEMICLUS.SAS                                                    |
| Clusters semi-variogram data file.                              |
|----------------------------------------------------------------*/
data tagain; set rowvect;              *** In this section standardize   ;
   stddist=dist;                       *** distance variable (dist)      ;
run;                                   *** to mean 0 and s.d. 1. This    ;
proc standard data=tagain out=srowvect mean=0 std=1;
                                       *** provides automated scaling    ;
var stddist;                           *** for any magnitude of distances. ;
run;                                   *** Standardized distances are    ;
proc datasets lib=work; delete tagain / mt=data; run;
data tone; set srowvect;               *** rounded and held in a temporary ;
   tdist=round(stddist,&roundtol);     *** distance variable. This       ;
run;                                   *** variable is used as the distance;
proc datasets lib=work; delete srowvect / mt=data; run;
proc sort data=tone; by tdist; run;    *** index. The mean of the raw    ;
proc means mean data=tone noprint; by tdist;
                                       *** distance by the rounded       ;
   var dist gamma;                     *** standardized distances is done. ;
   output out=tmean mean=mdist mgamma;
                                       *** The means of the raw distances ;
run;                                   *** are then used in the cluster  ;
data ttwo; set tmean;                  *** procedure so that time necessary;
   seqid=_n_;                          *** for clustering by Ward's method;
   dist=mdist; gamma=mgamma;           *** is substantially reduced.     ;
   drop mdist mgamma;
run;
data rowvect(replace=yes); set ttwo; run;
proc datasets lib=work; delete ttwo tmean / mt=data; run;
proc sort data=rowvect; by dist; run;

DATA STEP1; set rowvect;
    seqid=_n_;
    mergeval=1;
RUN;
proc datasets lib=work; delete rowvect / mt=data; run;
proc cluster data=step1 method=ward noprint outtree=tree;
    id seqid;
    var dist;
    COPY gamma;
    freq _freq_;
run;
proc tree data=tree noprint n=&ngroups out=temp1; id seqid; run;

proc sort data=temp1 out=temp1(replace=yes); by seqid; run;

data step1 (replace=yes);
    merge step1(drop=mergeval) temp1;
    by seqid;
run;
proc sort data=step1 out=step1(replace=yes);
```

```
         by cluster;
run;
pROC means DATA=sTEp1 NOPRINT; by cluster;
   VAR DIST gamma;
   freq _freq_;
   output out=results min=distmin gammamin max=distmax gammamax mean=distm
gammaM;
RUN;

symbol1 c=black v=dot i=none;
proc gplot data=results;
   axis1 c=black w=2
         label = (c=black font=swissb h=1 'h (mean clustered distance)')
         value = (c=black font=swissb h=1)
         major = (w=2 c=black)
         minor = none;
   axis2 c=black w=2
         label = (a=90 r=0 font=cgreek h=1 'g' font=swissb h=1 '(h)')
         value = (c=black font=swissb h=1)
         major = (w=2 c=black)
         minor = none;
   plot gammaM*distm /
                      haxis=axis1
                      vaxis=axis2;
   title1 h=1 "Semivariogram plot for variable &dsvar in file &flname";
   title2 h=1 "Number of clusters = &ngroups";
   title3 h=1 "Direction = &dirdeg with tolerance +/- &toldeg";
run;

proc sort data=results out=results(replace=yes); by distm; run;
data _null_;
     set results;
     if distm=0 then distm=0.00001;
     dir=&dirdeg;
     file "&dspath.\&dsvar..sem";
     if _n_ =1 then do;
        put '1' #2 '2' #3 '3' #4 '4' #5 '5' #6 '6' #7 '7' #8 '8';
     end;
     put distmin distmax _freq_ distm gammam;
     file "&dspath.\&dsvar..sam";
     put distm _freq_ gammam;
     file "&dspath.\&dsvar..ctr";
     put distm gammam dir;
run;

symbol1 c=black v=dot i=none;
proc gplot data=results;
   axis1 c=black w=2
         label = (c=black font=swissb h=1 'h (mean clustered distance)')
         value = (c=black font=swissb h=1)
         major = (w=2 c=black)
         minor = none;
   axis2 c=black w=2
         label = (a=90 r=0 font=swissb h=1 'Frequency')
         value = (c=black font=swissb h=1)
         major = (w=2 c=black)
         minor = none;
   plot _freq_*distm /
                      haxis=axis1
                      vaxis=axis2;
   title1 h=1
          "Frequency vs. distance plot for variable &dsvar in file &flname";
   title2 h=1 "Number of clusters = &ngroups";
```

```
   title3 h=1 "Direction = &dirdeg with tolerance +/- &toldeg";
run;

    proc sort data=step1 out=tstep1; by tdist; run;
    proc datasets lib=work; delete step1 / mt=data; run;
    data clusclas; merge tone tstep1(keep=cluster tdist); by tdist; run;
    proc sort data=clusclas; by cluster; run;
    proc sort data=results; by cluster; run;
    data distmean; merge clusclas results; by cluster; run;
    proc datasets lib=work; delete clusclas results / mt=data; run;
    data distmean(replace=yes); set distmean;
       maxattr=max(pointa,pointb);
       minattr=min(pointa,pointb);
    run;
    proc sort data=distmean; by distm; run;
    data hipoint(keep=distm maxgamma maxgattr mingattr gamma);
      set distmean; by distm;
      if first.distm then do;
        maxgamma=gamma;
        maxgattr=maxattr;
        mingattr=minattr;
      end;
      if gamma > maxgamma then do;
        maxgamma=gamma;
        maxgattr=maxattr;
        mingattr=minattr;
      end;
      if last.distm then output;
      retain maxgamma maxgattr mingattr;
    run;

data annoclus;
  length function color style $ 8 text $ 5;
  retain xsys '2' ysys '2' hsys '1' when 'a';
  set hipoint;
  function='label'; color='black';
  size=1.5; text=left(put(maxgattr,5.2)); position='2';
  style='swissb'; x=distm; y=maxgamma + &hivalpo;
  output;
  text=left(put(mingattr,5.2));
  y=maxgamma;
  output;
run;

symbol1 c=black v=none i=hilot;
proc gplot data=distmean;
  axis1 c=black w=2
        label = (c=black font=swissb h=1 'h (mean clustered distance)')
        value = (c=black font=swissb h=1)
        major = (w=2 c=black)
        minor = none;
  axis2 c=black w=2
        label = (a=90 r=0 font=cgreek h=1 'g' font=swissb h=1 '(h)')
        value = (c=black font=swissb h=1)
        major = (w=2 c=black)
        minor = none;
  plot gamma*distm /         annotate=annoclus
                             haxis=axis1
                             vaxis=axis2;
  title1 h=1 "Semi-variogram plot for the mean and high, low values.";
  title2 h=1 "For data set &flname and variable &dsvar";
  title3 h=1 "Number of clusters = &ngroups";
  title4 h=1 "Direction = &dirdeg with tolerance +/- &toldeg";
```

```
run;
    proc means mean var data=distmean noprint vardef=n; by distm;
      var gamma;
      output out=varmean mean=gammean var=gamvar;
    run;
    data ratiovm; set varmean;
      if gammean=0 then gammean=0.00001;
      gamvar = gamvar/&standvar;
      rvm = gamvar/gammean;
      noacorr = 2;
      file "&dspath.\&dsvar..rvm";
      put gammean gamvar rvm distm _freq_;
    run;
    proc means mean stderr data=ratiovm noprint;
      var rvm;
      output out=mmrvm mean=mrvm stderr=srvm;
    run;
    data _null_; set mmrvm;
      call symput('mrvm',mrvm);
      call symput('srvm',srvm);
    run;
symbol1 c=black v=none i=join w=2;
symbol2 c=black v=dot i=none;
proc gplot data=ratiovm;
  axis1 c=black w=2
        label = (c=black font=swissb h=1 'h (mean clustered distance)')
        value = (c=black font=swissb h=1)
        major = (w=2 c=black)
        minor = none;
  axis2 c=black w=2
        label = (a=90 r=0 font=swissb h=1 'Variance-to-mean ratio of '
                           font=cgreek h=1 'g' font=swissb h=1 '(h)')
        value = (c=black font=swissb h=1)
        major = (w=2 c=black)
        minor = none;
  plot noacorr*distm rvm*distm / overlay
                                 haxis=axis1
                                 vaxis=axis2;
  title1 h=1
          "Ratio of the variance to the mean for each distance group.";
  title2 h=1 "For data set &flname and variable &dsvar , "
                                 "Number of clusters = &ngroups";
  title3 h=1 "mean rvm = &mrvm , stderr rvm = &srvm , "
                                 "standvar = &standvar";
  title4 h=1 "Direction = &dirdeg with tolerance +/- &toldeg";
run;

quit;
```

The program SEMICLUS.SAS (Program #3) creates user-specified files that are used in subsequent analysis programs, particularly the programs that model the semivariogram, and are written to the same subdirectory pathname as that specified for the input data file. These output files have the same file name as that specified in the macro variable DSVAR in Program #1, but with file extensions "sam", "sem", "ctr", and "rvm". Files with the "sam" extension contain the weighted mean distance, frequency, and weighted mean semivariogram values for each distance class. This file is used in the SAS program SEMIGRAM.SAS employed to fit semivariogram models (see Appendix 3-E; Chapter 3). Files with the "sem" extension contain the minimum and maximum interpoint distances for each distance interval, together with the weighted mean distance, frequency, and weighted mean semivariogram values. This file can be read into WLSFIT to fit semivariogram models to the semivariogram plot. (Many times the "sem" file contains extra blank lines at the end of the file. These need to be removed before WLSFIT will correctly input the file. In fact, all of the output files produced should be checked and all blank lines removed to insure correct input files for further analyses.) For both the "sam" and "sem" files, if the distance value of the first distance class is equal to zero, the value is set to 0.00001 in order to circumvent computational problems with PROC NLIN and WLSFIT. The "ctr" file can be used in SAS to create a contour map of the semivariogram values to inspect visually for anisotropy (analogous to contour maps presented by Isaacks and Srivastava, 1989, pp 93-106). Finally, the "rvm" file contains data for computing the ratio of the variance of $\gamma(h)$ to the mean of $\gamma(h)$ for each distance interval and is used by the program SEMIPLOT.SAS (see Appendix 3-E); code in the program SEMICLUS.SAS computes and plots this information, too. For the "rvm" file, a mean gamma value of zero is converted to 0.00001 to prevent a division by zero error. The first GPLOT procedure produces a semivariogram plot without the model lines fitted, and the second GPLOT procedure produces a plot of the number of distance pairs versus lag distance. The third GPLOT procedure produces a plot of the high, low, and mean $\gamma(h)$ values and plots the maximum and minimum attribute values for each distance interval. The fourth GPLOT procedure produces a plot of the ratio of the variance of $\gamma(h)$ to the mean of $\gamma(h)$ for each distance interval, as well as a line representing the expected ratio value if no spatial autocorrelation is present in the data.

Finally, some data sets can become quite large when computing and storing the individual gamma values and distances for subsequent processing. The program SEMI-VAR.SAS is limited to a data set having a maximum of 1,024 observations. This is because we "trick" SAS into thinking it is working with data vectors in computing the gamma values and distances. In fact, this is accomplished in the data step by creating a single row vector (i.e., a single observation) with all the information necessary for computing gamma values and distances; the width of this single observation is limited by operating system constraints. For larger data sets whose number of observations exceeds this value, a modified version of the SEMI-VAR.SAS program file follows in Program #4, but does not include the aforementioned options for computing Great Circle distances and anisotropic specifications. An option to truncate distances greater than a specified amount (e.g., the default value of 3) is contained in the macro variable TRUNCDIS and the value set in Program #1. Truncation of the longer distances is useful for large data sets since most of the spatial autocorrelation in a semivariogram plot is contained in the shorter distance classes. Inclusion of longer distances generally does not add any useful information and can take up unnecessary amounts of disk space while the programs are executing.

Program #4

```
/*----------------------------------------------------------------|
|  SEMIBVAR.SAS                                                    |
|  Computes semi-variogram values and distances for all distance pairs. |
|-----------------------------------------------------------------*/

data standdat; set rawdat;
  attr = &dsvar;
  drop &dsvar;
run;

data rowvect; set standdat;
  u1=u; v1=v; attr1=attr; pointa=attr;
  if _n_ = 1 then seqid=0;
  do i=1 to last by 1;
    set standdat point=i nobs=last;
    if _error_ then abort;
    u2=u; v2=v; attr2=attr; pointb=attr2;
    gamma = ((attr1 - attr2)**2)/2;
    dist = sqrt((u1 - u2)**2 + (v1 - v2)**2);
    if dist > &truncdis then delete;
    freq=1;
    seqid = seqid+1;
    mergeval=1;
    output;
    keep dist freq gamma seqid mergeval pointa pointb;
  end;
  retain seqid;
  return;
  stop;
run;

proc datasets lib=work; delete standdat / mt=data; run;
```

This version of SEMI-VAR.SAS performs the same function of computing the individual gamma values and distances for each of the n^2 distance pairs. If Program #4 is to used, the file name SEMI-VAR.SAS in the first "%include" statement in Program #1 should be replaced with SEMIBVAR.SAS.

1.7.2. SAS CODE FOR SPATIAL AUTOREGRESSION ANALYSES

Analogous to the previous section, SAS code is presented in this section for preliminary examination of data using spatial autoregression. For spatial autoregression, the tedious task of constructing the geographic weights matrix (matrix **C** or **W**) must be completed, although for a regular square tessellation the SAS LAG function approach avoids this task (see Appendix 3-A for use of the LAG function). Several efficient approaches can be taken here; of course, brute force always is an option. First, if entry c_{ij} (or w_{ij}) is based upon distances separating areal unit centroids, a vector of coordinates (e.g., Cartesian, UTM, latitude-longitude) must be obtained; the calculation of interpoint distances d_{ij} is straightforward then. Distances less than some threshold value, say d*, can be coded unity, and others 0, to obtain matrix **C**. Or, following Upton (1990), $w_{ij} = (\frac{1}{d_{ij}^n})/(\sum_{j=1}^{n} \frac{1}{d_{ij}^n})$, with this computation approaching the more commonly used $w_{ij} = \frac{c_{ij}}{\sum_{j=1}^{n} c_{ij}}$ as η increases. Second, a GIS could be used to construct these matrices from polygonal boundary files. The GIS also can be used to generate a Thiessen polygon partitioning of a punctiform surface [e.g., the ArcInfo THIESSEN command (not available in the Windows version of PC ArcInfo); Appendix 1-B], whose boundary file then can be used to construct a corrected neighbors file, and matrix **C**, using Program #5. Third, a surface partitioning can be translated into a set of neighbors and then matrix **C** constructed from this condensed information. A combination of the last two approaches is used throughout this book.

All spatial autoregressive SAS programs presented in this book require a neighbors file, except for the program CONNCHK.SAS where it is an option. We prefer the use of a neighbors file rather than a connectivity matrix for several reasons. First, in terms of a text file, a connectivity matrix can become rather large and special considerations need to be made with respect to logical record length of the file for large matrices. Second, manually correcting mistakes in a neighbors file is easier than in a connectivity matrix file. Third, for many of the Thiessen polygon maps constructed from point data in this book, some connections between neighboring areal units were deleted explicitly; a neighbors file is easier to read and edit than is a connectivity matrix file.

The programs require the file extension of the neighbors file to be "nei" and stored in the same subdirectory as the data analysis file (again, whose file extension is "txt"). The neighbors file is a text file where each row (observation) contains all the connection information for a particular referent areal unit. Thus, the number of rows in the neighbors file is equal to the total number of areal units on the map. In addition, the values used in the neighbors file must be representative of the areal units numbered sequentially from 1 to the total number of areal units. The structure of the neighbors file, then, consists of the first value in a row representing a particular reference areal unit, and the set of values following this referent areal unit will be all of the areal units to which it is connected. All values in a row are separated by at least one space.

An important step in constructing either the neighbors or the connectivity matrix file is testing to make sure the file has been constructed properly. That is, the connectivity matrix itself (i.e., matrix **C**) is a square symmetric matrix containing all zeros on the diagonal, the sum of the eigenvalues computed from this matrix should equal zero, and the sum of the squared eigenvalues should equal the total number of 1s in the matrix. Program #5, CONNCHK.SAS, performs these checks, in addition to providing the option of reading in either a neighbors file or a **C** matrix file, converting between the two types of connectivity files, and providing the ability to correct the boundary file created from the Avenue program in Appendix 1-B[8]:

Program #5

```
/*---------------------------------------------------------------|
|  CONNCHK.SAS      [This program uses PROC IML]                 |
|  Checks neighbor file or connectivity matrix (binary weights matrix,|
|  or matrix C) for correctness. Options are provided for reading |
|  either a neighbor file (file extension must be "nei") or a C matrix|
|  file (file extension must be "con"). Options are provided to   |
|  convert a connectivity matrix from a neighbor file, or vice-verse. |
|  Both files are text files.                                    |
|  All user inputs are specified in the first 6 macro variables: |
|     dspath - name of the subdirectory path for the input and output |
|              files.                                            |
|     flname - filename of the neighbor or C matrix file.        |
|     filetype - file extension of the input file (con or nei).  |
|     conout - specify to create a connectivity matrix file from a |
|              neighbor file. Filename will be the same.         |
|     neiout - specify to create a neighbor file from a connectivity |
|              matrix file. Filename will be the same.           |
|     fixnei - specify to correct a neighborhood file produced from the|
|              Avenue program in Appendix 1-B. The name of the input |
|              file will be NEIGBOR.TXT and should be in the same path |
|              specified in the macro variable DSPATH. The options for |
|              FILETYPE should be set to "nei" and for NEIOUT to "no". |
|              The option for CONOUT can be "yes" or "no". Macro |
|              variable FIXNEI should be set to "yes". The output file |
|              name will be the name specified in the macro variable |
|              FLNAME with a "nei" file extension.               |
|----------------------------------------------------------------*/

%macro connchk;
%local dspath flname filetype conout neiout fixnei nrecs reclong;
%let dspath = %str(c:\nsf_geog\book\wolfcamp);
%let flname = %str(wolfcamp);
%let filetype = %str(con);
%let conout = %str(no);
%let neiout = %str(yes);
%let fixnei = %str(no);

options nocenter nodate pageno=1 MPRINT SYMBOLGEN MLOGIC;

%if &fixnei = yes %then %do;
  data fixnei; length neighrec $ 200;
    infile "&dspath.\neighbor.txt" missover;
    array a$ a1-a200;               * Construct corrected;
    array x x1-x200;                * neighbor file from ;
    input x1-x200;                  * that produced by   ;
    ai=1; neighrec = ' ';           * the ArcView Avenue ;
    do i=1 to 200;                  * program in Appendix;
      if x{i} > . then do;          * 1-B.               ;
        x{i} = x{i} - 1;
        a{ai}=x{i};
```

---

[8]Of note is that ArcInfo generated Thiessen polygons are numbered beginning with the ID 1 being assigned to the universal polygon that circumscribes a georeferenced data set. Hence, CONNCHK.SAS includes an option for correcting this peculiarity. The adjustment is not included in the ArcInfo scripts (Appendix 1-B) because in general individual areal units could be composed of separate polygons (e.g., coastal counties of the U.S. in, for instance, the Seattle, Maryland, and New Orleans areas), to which ArcInfo needs to assign distinct ID numbers.

```
      end;
      neighrec = trim(neighrec)||' '||left(trim(a{ai}));
      ai=ai+1;
    end;
    file "&dspath.\&flname..nei";
    put neighrec;
  run;
  proc datasets lib=work; delete fixnei / mt=data; run;
%end;

data _null_;                            *** read number of observations;
  infile "&dspath.\&flname..&filetype"; *** in data set and assign to;
  input c1;                             *** macro variable NRECS;
  call symput('nrecs',left(_n_));
run;
%let reclong = %eval(2 * &nrecs);

%if &filetype = nei %then %do;
  data grid; infile "&dspath.\&flname..nei" missover;
    array c{&nrecs} c1-c&nrecs;         * Construct matrix C from;
    array in_c{200} in_c1-in_c200;      * neighbor file;
    input in_c1-in_c200;
    do i = 1 to &nrecs;
      c{i} = 0;
    end;
    do i = 2 to 200;
      if in_c{i} > . then c{in_c{i}} = 1;
    end;
    drop i in_c1-in_c200;
  run;
%end;

%if &filetype = con %then %do;
  data c; length nei$ 200 con1$ 8;
    infile "&dspath.\&flname..con" lrecl = &reclong;
    input c1-c&nrecs;
    array c{&nrecs} c1-c&nrecs;
    nei = _n_;
    do i = 1 to &nrecs;
      if c{i} = 1 then do;
        con1 = i;
        nei = trim(left(nei))||' '||trim(left(con1));
      end;
    end;
  run;
  data grid; set c; keep c1-c&nrecs; run;
%end;

PROC IML; use grid;        /* Specify data set to bring into IML.    */
  read all into c;
  ncells = nrow(c);        /* Count number of rows in input matrix.  */
  print "Number of observations:" ncells;
  if trace(c) ^= 0 then do;
    print "Sum of diagonals is not zero.";
  end;
  start;                   /* Check diagonal for non-zeros. */
    do i=1 to &nrecs;
      if c[i,i] ^= 0 then do;
        print "Diagonal contains a non-zero value at observation:" i;
      end;
    end;
  finish; run;
  start;                   /* Check for asymmetry. */
```

```
    do i=1 to &nrecs;
      do j=1 to &nrecs;
        if c[i,j] ^= c[j,i] then do;
          print "Matrix not symmetric at observation:" i j;
        end;
      end;
    end;
  finish; run;
  start;                    /* Check eigen values.                    */
    symc=symsqr(c);         /* Note: If matrix C is not symmetric,    */
    eigvalc=eigval(c);      /*       this part will generate an       */
    eigsum=sum(eigvalc);    /*       error and not execute.           */
    print "Sum of eigen values should be approximately = to zero:" eigsum;
    print " ";
    clsum=sum(c);
    eig2 = eigvalc##2;
    eig2sum = sum(eig2);
    print "Sum of squared eigen values:" eig2sum;
    print "should be equal to sum of ones in matrix C:" clsum;
  finish; run;
%if &conout = yes %then %do;
  matCnam = "c1" : "c&nrecs";    /* Output matrix C to text file. */
  create matrixC from c [colname=matCnam];
  append from c;
  data _null_; set matrixC;
    file "&dspath.\&flname..con" lrecl = &reclong;
    put c1-c&nrecs;
  run;
%end;
%if &neiout = yes %then %do;
  data _null_; set c;
    file "&dspath.\&flname..nei";
    put nei;
  run;
%end;
%mend connchk;
  %connchk;
```

If the neighbors file is read into this program, a **C** matrix is created prior to checking the structure of the matrix. Just as the program will read in either a neighbors or a **C** matrix file, it also will write either type of file, effectively allowing the user to convert between the two types of connectivity files. Either type of file output has the same file name as the input file, specified in the macro variable FLNAME. In the case of a **C** matrix file, the file extension will be "con". In the case of a neighbors file, the file extension will be "nei".

Finally, for regular tessellations, such as square (useful for remotely sensed data) and hexagonal (useful with the U.S. EPA EMAP data set), latent repetitive structure can be exploited in such a way that matrix **C** can be constructed with little effort; for example,

```
regular rectangular lattice
%let p=4;           *number of rows;
%let q=4;           *number of columns;
%let dspath=%str(c:\nsf_geog\book\chp1);
%let flname=rook;
%let k = %eval(&p * &q);
```

```
%let reclong = %eval(2 * &k);
data regrect;   * regular rectangular lattice;
 array c{&k} c1-c&k;
 row=1; col=1;
 do cell=1 to &k;
   do i=1 to &k; c(i)=0; end;
   if col>&q then do; row=row+1; col=1; end;
     if row=1 then do;
       if col=1 then do;
         c(2) = 1;
         c(1 + &q) = 1;
       end;
       if col=&q then do;
         c(&q - 1) = 1;
         c(&q + &q) = 1;
       end;
       if (col>1 & col<&q) then do;
         c(col + 1) = 1;
         c(col + &q) = 1;
         c(col - 1) = 1;
       end;
     end;
     if (row>1 & row<&p) then do;
       if col=1 then do;
         c(cell + 1) = 1;
         c(cell - &q) = 1;
         c(cell + &q) = 1;
       end;
       if col=&q then do;
         c(cell - 1) = 1;
         c(cell - &q) = 1;
         c(cell + &q) = 1;
       end;
       if (col>1 & col<&q) then do;
         c(cell - &q) = 1;
         c(cell + 1) = 1;
         c(cell + &q) = 1;
         c(cell - 1) = 1;
       end;
     end;
     if row=&p then do;
       if col=1 then do;
         c(cell + 1) = 1;
         c(cell - &q) = 1;
       end;
       if col=&q then do;
         c(cell - 1) = 1;
         c(cell - &q) = 1;
       end;
       if (col>1 & col<&q) then do;
         c(cell - &q) = 1;
         c(cell + 1) = 1;
         c(cell - 1) = 1;
       end;
     end;
   end;
   output;
```

```
    file "&dspath.\&flname..con" lrecl=&reclong;
    put c1-c&k;
    col=col+1;
    retain row col;
  end;
run;
```

hexagon
```
%let p=4;              *number of rows;
%let q=6;              *number of columns in the first row;
%let dspath=%str(c:\nsf_geog\book\chp1);
%let flname=hex;
%let k = %eval(&p * &q);
%let reclong = %eval(2 * &k);
data hexrect;   * regular hexagonal lattice;
 array c{&k} c1-c&k;
 row=1; col=1;
 do cell=1 to &k;
    do i=1 to &k; c(i)=0; end;
    if col>&q then do; row=row+1; col=1; end;
      if col>1 then do;
        c(cell - 1) = 1;
      end;
      if col<&q then do;
        c(cell + 1) = 1;
      end;
      if row=1 then do;
        if col=1 then do;
           c(2) = 1;
           c(&q + 1) = 1;
        end;
        if col=&q then do;
           c(&q - 1) = 1;
           c(&q + (&q - 1)) = 1;
           c(&q + &q) = 1;
        end;
        if (col>1 & col<&q) then do;
           c(col - 1) = 1;
           c(col + 1) = 1;
           c(col + &q) = 1;
           c(col + (&q- 1)) = 1;
        end;
      end;
      if (row>1 & row<&p) then do;
        if col=1 then do;
           if mod(row,2)=0 then do;
              c(cell - &q) = 1;
              c(cell + &q) = 1;
              c(cell + (&q + 1)) = 1;
              c(cell - (&q - 1)) = 1;
           end;
           if mod(row,2)>0 then do;
              c(cell - &q) = 1;
              c(cell + &q) = 1;
           end;
        end;
```

```
          if (col>1 & col<&q) then do;
            c(cell - &q) = 1;
            c(cell + &q) = 1;
            if mod(row,2)=0 then do;
              c(cell + (&q + 1)) = 1;
              c(cell - (&q - 1)) = 1;
            end;
            if mod(row,2)>0 then do;
              c(cell + (&q - 1)) = 1;
              c(cell - (&q + 1)) = 1;
            end;
          end;
          if col=&q then do;
            if mod(row,2)=0 then do;
              c(cell - &q) = 1;
              c(cell + &q) = 1;
            end;
            if mod(row,2)>0 then do;
              c(cell - (&q + 1)) = 1;
              c(cell + (&q - 1)) = 1;
              c(cell - &q) = 1;
              c(cell + &q) = 1;
            end;
          end;
        end;
        if row=&p then do;
          if mod(row,2)=0 then do;
            c(cell - &q) = 1;
            if col<&q then do;
              c(cell - (&q - 1)) = 1;
            end;
          end;
          if mod(row,2)>0 then do;
            c(cell - &q) = 1;
            if col>1 then do;
              c(cell - (&q + 1)) = 1;
            end;
          end;
        end;
      output;
    file "&dspath.\&flname..con" lrecl=&reclong;
    put c1-c&k;
    col=col+1;
    retain row col;
  end;
run;
```

In the SAS code for rectangular tessellations, the user specifies the number of rows and columns in the lattice as p and q, respectively. The number of areal units is given as the product of the two. Matrix **C** is written to a user-specified file. For a regular hexagonal tessellation, the user specifies the number of rows (in macro variable p) and the number of columns (in macro variable q) in the first row. Each row of the hexagonal tessellation then will contain the same number of cells, with the first cell of each odd-numbered row shifted half a cell width to the left of the even-numbered rows.

Code for computing the Moran coefficient and creating an MC scatterplot is presented in Appendix 1-A. User input required is organized in the same way as that for the SAS programs presented in Programs #1 and #5, and is explained in the first text box in the program. Again, the type of connectivity file required for this program is the neighbors file, which must have the same filename as the data analysis file specified in the macro variable FLNAME, with file extension "nei".

APPENDIX I-A
SAS Code for Computing MC Using Standard Regression Techniques.

```
/*-------------------------------------------------------------|
| MORANC.SAS                                                   |
| Computes Moran coefficient using matrix C and creates a      |
| Moran scatterplot. Matrix C is input as a neighborhood       |
| file. Filenames of the analysis data file and neighborhood   |
| file are required to be the same. File extension of the      |
| data analysis file is required to be "txt", and for the      |
| neighborhood file "nei". Both files are required to be in    |
| the same subdirectory path. All user inputs are specified    |
| in the first 4 macro variables:                              |
|    dspath - name of the subdirectory path for the analysis   |
|             data and neighborhood files.                     |
|    flname - filename of the analysis data and neighborhood   |
|             files.                                           |
|    invars - variable names in the analysis data file in the  |
|             correct order.                                   |
|    dsvar  - name of analysis variable for which Moran        |
|             coefficient is computed.                         |
|-------------------------------------------------------------*/
%macro moranc;
 options nocenter nodate pageno=1 MPRINT SYMBOLGEN MLOGIC;
 %let dspath = %str(c:\nsf_geog\book\wolfcamp);
 %let flname = %str(wolfcamp);
 %let invars = %str(id u v piezo);
 %let dsvar  = %str(piezo);

  DATA indat; INFILE "&dspath.\&flname..txt";
    INPUT &invars;
    call symput('drecs',left(_n_));
  RUN;
  data _null_;
    infile "&dspath.\&flname..nei" missover;
    input c1;
    call symput('nrecs',left(_n_));
  run;
  %if %eval(&drecs ^= &nrecs) %then %do;
    %put Number of observations in data analysis file = &drecs;
    %put Number of observations in neighborhood file = &nrecs;
    data junk; length tw1$ 80;
      tw1 = "Number of observations in data file and"; output;
      tw1 = "   neighbor file are not equal."; output;
      tw1 = "Output is not correct. Check both input files.";
                                                          output;
      tw1 = "See SASLOG for number of records in each file.";
                                                          output;
    run;
    proc print data=junk; var tw1; run;
  %end;

  data c; infile "&dspath.\&flname..nei" missover;
    input in_c1-in_c200;            * Construct matrix C from;
    array c{&nrecs} c1-c&nrecs;     * neighborhood file;
```

```
   array in_c{200} in_c1-in_c200;
   do i = 1 to &nrecs; c{i} = 0; end;
   do i = 2 to 200;
     if in_c{i} > . then c{in_c{i}} = 1;
   end;
   crowsum=sum(of c1-c&nrecs);
   drop i in_c1-in_c200;
run;

PROC STANDARD MEAN=0 data=indat OUT=indat (REPLACE=YES);
                                            VAR &dsvar; RUN;
DATA indat (REPLACE=YES);
   merge indat c;
   ARRAY C{&nrecs} C1-C&nrecs;
   ARRAY CONY{&nrecs} CY1-CY&nrecs;
   DO i=1 TO &nrecs;
      CONY{i} = &dsvar * C{i};
   END;
RUN;
PROC MEANS DATA=indat NOPRINT;
   VAR CY1-CY&nrecs;
   OUTPUT OUT=CYOUT1 SUM=CY1-CY&nrecs;
RUN;
PROC TRANSPOSE DATA=CYOUT1 PREFIX=CY OUT=CYOUT2;
   VAR CY1-CY&nrecs;
RUN;
DATA indat (REPLACE=YES);
   SET indat;
   SET CYOUT2;
CY = CY1;
C1 = crowsum;
X0=1;
RUN;
PROC REG DATA=indat OUTEST=BYCY; MODEL CY = &dsvar / NOINT; RUN;
PROC REG DATA=indat OUTEST=B1C1; MODEL C1 = X0/NOINT; RUN;
DATA compmc;
   SET BYCY (KEEP=&dsvar);
   SET B1C1 (KEEP=X0);
MC = &dsvar / X0;
RUN;
PROC PRINT DATA=compmc;
   VAR MC;
   title "Moran coefficient for &dsvar from file "
                                  "&dspath.\&flname..txt";
RUN;

SYMBOL1 v=dot c=black l=1 i=rl0;
proc gplot data=indat;
   plot cy * &dsvar;
   title h=1 f=swissb
         "Moran scatterplot of CY on &dsvar for "
                                  "&dspath.\&flname..txt";
   run;
%mend moranc;
  %moranc;
```

APPENDIX 1-B
Directions for Using Arcinfo to Construct a Thiessen Polygon Surface Partitioning for a Set of Georeferenced Points.

Step	Description
1	create an ASCII file with the following structure:

    1,x,y
    2,x,y
    3,x,y
    ⋮
    END

2   in Arc/Info

    GENERATE <coverage>
    INPUT <ascii filename>
    POINTS
    QUIT

    BUILD <coverage> POINTS

    THIESSEN <in cover> <out cover>(workstation Arc/Info command)

    BUILD <out cover> LINES

3   in ArcView

```
open a View and add the theissen coverage as a theme
click project window
Save Project As...
double click Script button in project window
click on script pull-down menu
Load Text File
select NEIGHBOR.AVE
compile using the check mark icon
in project window single click on Tables
select Add
select .AAT table for the theissen coverage
select Project window
select Project pull-down menu
select Customize
select Table
select Buttons
select New
double click on 'click'
choose Script1
```

```
close customize window
click on att.dbf window to activate table
click on blank icon (to run neighbor.ave)
```

NOTE: This procedure writes NEIGHBOR.TXT to your ArcView work directory. In fact, the following program can be used to create a neighbors file from any Arc/Info polygon coverage (when opened in ArcView) or ArcView shape file containing a polygon theme.

7. Avenue script for NEIGHBOR.AVE (see Zhang and Griffith 1998)

```
theTable = av.GetActiveDoc
theVtab = theTable.GetVtab
Lpoly = theVtab.FindField("Lpoly_")
If ( lpoly = Nil ) then
  Msgbox.error("Can not find Left Polygon Field",
                                "Creating Polygon Neighbor")
  exit
end

Rpoly = theVtab.FindField("Rpoly_")
If ( Rpoly = Nil ) then
  Msgbox.error("Can not find Right Polygon Field",
                                "Creating Polygon Neighbor")
  exit
end

pnl = list.make
For each rec in theVtab
  lpv = theVtab.returnvaluenumber(Lpoly,rec)
  rpv = theVtab.returnvaluenumber(Rpoly,rec)
  if((lpv=1)or(rpv=1)or(lpv=rpv))then
    continue
  end
  m = lpv max rpv
  if( m > (pnl.count+1) ) then
    diff = m - pnl.count - 1
    For each i in 1..diff
       pnl.add(list.make)
    end
  end
  pnl.get(lpv-2).add(rpv)
  pnl.get(rpv-2).add(lpv)
end

'export the neighbor info to a text file
work_dir = av.getproject.GetWorkDir.asstring
msgbox.info(
" The neighborhood file will be saved to this directory: "
                                        + work_dir,"")
theFilename = Filename.merge(work_dir,"neighbor.txt")
theFile = LineFile.Make( theFilename, #FILE_PERM_WRITE)
```

```
For each i in 1..pnl.count
  pnl.get(i-1).removeduplicates
  pnl.get(i-1).sort(true)
  nblist=""
  size=pnl.get(i-1).count
  if (size > 1) then
    For each j in 1..(size-1)
      nblist=nblist+pnl.get(i-1).get(j-1).asstring+" "
    end
  end
  nblist=nblist+pnl.get(i-1).get(size-1).asstring
  theFile.WriteElt( (i+1).asstring++nblist )
end
```

APPENDIX I-C
SAS Macros Used for Converting among Degrees, Decimal Degrees and Radians.[9]

```
%MACRO DECTODMS (DEC, DMS) ;
  %LET DDD=%SUBSTR(&DEC            A,1,6) ;
    &DMS='1234567'; &DMS=' ' ;
    IF &DEC THEN DO;
      _D&DDD=INT(&DEC) ;
      _R&DDD=(&DEC-_D&DDD)*3600;
      _M&DDD=INT(_R&DDD/60);
      _S&DDD=ROUND(_R&DDD-_M&DDD*60,1);
      &DMS=PUT(_D&DDD*1,Z3.)||PUT(_M&DDD*1,Z2.)||PUT(_S&DDD*1,Z2.);
    END;
    DROP _D&DDD _R&DDD _M&DDD _S&DDD;
%MEND DECTODMS;


%MACRO DECTORAD (DEC, RAD) ;
    &RAD=&DEC*ARCOS(-1)/180;
%MEND DECTORAD;


%MACRO DMSTODEC (DMS, DEC) ;
    IF &DMS NE ' ' AND &DMS NE 'NA' AND LENGTH(&DMS)=7 THEN DO;
      IF SUBSTR(&DMS,1,1)='N' THEN SUBSTR(&DMS,1,1)='0';
      IF SUBSTR(&DMS,1,1)='W' THEN SUBSTR(&DMS,1,1)='0';
    IF (0 LE SUBSTR(&DMS,1,3) LE 360) AND (0 LE SUBSTR(&DMS,4,2) LE 60)
       AND (0 LE SUBSTR(&DMS,6,2) LE 60)
    THEN
       &DEC=SUBSTR(&DMS,1,3)+SUBSTR(&DMS,4,2)/60+SUBSTR(&DMS,6,2)/3600;
      ELSE &DEC=0;
    END;
%MEND DMSTODEC;


%Macro LLtoDEC ( vList );
  %* The variable list should be Latitude deg min sec followed by
      Longitude deg min sec, six words in that order. ;
  %Let vList = %upcase(&vList);
  %If &vList = %then %Let vList = LAD LAM LAS LOD LOM LOS ;
  %Let LAD = %scan(&vList,1);
  %Let LAM = %scan(&vList,2);
  %Let LAS = %scan(&vList,3);
  %Let LOD = %scan(&vList,4);
  %Let LOM = %scan(&vList,5);
  %Let LOS = %scan(&vList,6);
  LATDMS = substr('0'||left(&LAS),length(left(&LAS)),2);
  LATDMS = substr('0'||left(&LAM),length(left(&LAM)),2) || LATDMS;
  LATDMS = substr('00'||left(&LAD),length(left(&LAD)),3) || LATDMS;
  LONGDMS= substr('0'||left(&LOS),length(left(&LOS)),2);
```

---

[9]Each of these macros should be kept in a separate file with the same filename as the macro name and the file extension of ".sas".

```
    LONGDMS= substr('0'||left(&LOM),length(left(&LOM)),2) || LONGDMS;
    LONGDMS=substr('00'||left(&LOD),length(left(&LOD)),3) || LONGDMS;
    %DMSTODEC (LONGDMS,LONGDEC) ;
    %DMSTODEC (LATDMS,LATDEC)     ;
%Mend;


%Macro LLtoUTM ( vList );
    %* The variable list should be Latitude deg min sec followed by
       Longitude deg min sec, six words in that order. ;
    %Let vList = %upcase(&vList);
    %If &vList = %then %Let vList = LAD LAM LAS LOD LOM LOS ;
    %Let LAD = %scan(&vList,1);
    %Let LAM = %scan(&vList,2);
    %Let LAS = %scan(&vList,3);
    %Let LOD = %scan(&vList,4);
    %Let LOM = %scan(&vList,5);
    %Let LOS = %scan(&vList,6);
    LATDMS = substr('0'||left(&LAS),length(left(&LAS)),2);
    LATDMS = substr('0'||left(&LAM),length(left(&LAM)),2) || LATDMS;
    LATDMS =substr('00'||left(&LAD),length(left(&LAD)),3) || LATDMS;
    LONGDMS= substr('0'||left(&LOS),length(left(&LOS)),2);
    LONGDMS= substr('0'||left(&LOM),length(left(&LOM)),2) || LONGDMS;
    LONGDMS=substr('00'||left(&LOD),length(left(&LOD)),3) || LONGDMS;
    %DMSTODEC (LONGDMS,LONGDEC) ;
    %DMSTODEC (LATDMS,LATDEC)    ;
    %DECTORAD (LONGDEC,X) ;
    %DECTORAD (LATDEC,Y)  ;
    %RADTOUTM ;
%Mend LLtoUTM;


%MACRO RADTODEC(RAD,DEC);
    &DEC=&RAD*180/ARCOS(-1);
%MEND RADTODEC;
```

2

IMPORTANT MODELING ASSUMPTIONS

Popular conventional statistical assumptions invoked to guard against misleading or incorrect conclusions (from other than random error) include one positing independent and identically distributed (*iid*) values, one positing normality, and one positing linearity. Statistical inference can be either sampling design based or model based (de Gruijter and ter Braak 1990; Hahn and Meeker 1993). Sampling design-based inference exploits central limit theorems and considers the only marked source of variation to result from the sampling process employed. Given a large enough sample, then, a central limit theorem guarantees *iid* and normality; some researchers questionably argue that a sufficiently large sample will circumvent many nonlinearity complications, too. Model-based inference exploits the notion of random variables, treating a set of georeferenced values as just one realization of an underlying spatial process constituting a conceptual population. It requires the critical postulate that the process under study is statistically identical to that from which data have been collected. Attribute value locations need not be selected at random; rather, inference is based upon some specified model, with the soundness of inferences being dictated by the validity of model assumptions or the robustness of model results. Much of spatial statistics involves model-based inference, in part because the number of areal units available for analysis is fixed, in part because georeferenced data frequently are not collected utilizing classical sampling designs, and in part because most map realizations cannot be physically replicated. In other words, as such a map may be viewed as a multivariate sample of size 1.

The *iid* assumption enables strict statistical control and refers to statistical frequency distribution infrastructure supporting data analyses. A constant mean model specifies $\mu\mathbf{1}$ as the single term parameter describing a central tendency across n observations (vector **1** is n-by-1); a variable mean model replaces this specification with the standard linear regression one of **Xß**. A constant variance model specifies $\sigma^2\mathbf{I}$ as the single parameter term, describing variability across n observations (matrix **I** is n-by-n); a nonconstant variance model replaces this specification in one of two ways. Either a weighted least squares regression is specified that involves $\sigma^2\mathbf{H}$, where matrix **H** is a diagonal matrix having varying h_{ii}s (the weights h_{ii} must be known or be modeled with relatively few parameters), or a Box-Cox (1964) family of power transformations is utilized (this has been promoted for stabilizing the variance so that matrix **H** can be replaced with matrix **I**). This corrective action is necessary to retain the minimum variance property of least squares estimators and to provide statistical control in an analysis. Any of these three variance specifications postulates independence among values (i.e., the absence of non-0 spatial autocorrelation), with this independence being indicated by the diagonality of matrices **I** or **H**. The presence of non-0 spatial autocorrelation introduces off-diagonal values into these matrices, converting the variance specification to $\sigma^2\mathbf{V}^{1/2}\mathbf{H}\mathbf{V}^{1/2}$. The degree of redundancy present (§1.2) can be assessed by computing condition numbers for matrix **V** (Ababou et al. 1994), providing a measure of deviation from matrix **I** associated with the *iid* assumption.

Either before or after corrective actions are taken when adjusting for nonconstant variance, or even sometimes during the modeling stage of an analysis, extreme observations become conspicuous, especially when a boxplot is constructed. These anomalous, aberrant, rogue values may well come from a statistical population other than that from which the remaining observations come. An analyzer of data should be suspicious of such unusual values. But dismissing them, particularly by deleting them from a data set, is not necessarily good practice, because they could be phenomenonally significant. Of course, deletion makes sense if such atypical values are strictly due to coding error. There is no substitute for a subjective, intuitive understanding of the data under study to differentiate between these statuses. When extreme values are encountered in data sets analyzed for this book, they are retained in the analyses but distinguished from other observations with binary indicator variables.

Normality is by far the most common probability assumption made when statistical modeling is undertaken, whether or not data are georeferenced. In the case of spatial autocorrelation, a normal distribution conveniently results in the mean responses being linear functions. It also yields maximum likelihood estimators that are almost always computationally possible. And it involves a normalizing constant whose computation is tractable. But, as Geary (1947) stresses, "Normality is a myth. There never was and will never be, a normal distribution" for real-world data. And, just because assuming normality enables statistical computations to be made is no reason, in and of itself, to assume normality. Fortunately, and serendipitously, normality tends to be accompanied by constant variance, as well as by certain types of independence (e.g., lack of significant interaction terms in N-way ANOVA).

Linear models assume the presence of linearity; the specified model is linear in its parameters, meaning that $\frac{\partial Y}{\partial \beta_j}$ is not a function of β_j (Tietjen 1986, 48).

Suppose that the *iid* and normality assumptions at least approximately hold for variable Y, but that the relationship between X and Y is nonlinear. An appropriate Box-Tidwell (1962) power transformation should bring the relationship displayed between variables Y and X into better compliance with the linearity assumption as simply as possible. A basic difference between Box-Cox and Box-Tidwell transformations is that the former require a normalizing constant for proper computation, whereas the latter do not.

Some words of caution are warranted at this point. Because a given map is a single realization of some phenomenon, true tests for nonstationarity are not possible. Rather, diagnostic tools outlined in this section allow evidence of nonstationarity to be gleaned from georeferenced data. The inverse transformation can be used to directly convert findings from a power transformed measurement scale to its original measurement scale counterpart. But there is no guarantee that these inverse transformation results expressed in the original measurement units are the best possible ones. This sensitivity of back-transformed values to the power transformation applied must be weighed against the degree of specification error introduced into an analysis by heterogeneous or non-normal data.

2.1. THE RELATIVE IMPORTANCE OF THE PRINCIPAL ASSUMPTIONS

Violation of any combination of the principal assumptions introduces specification error into a statistical model and can result in seriously flawed conclusions. The magnitude of corruption introduced into an inferential basis attributable to this specification error leads to the following ranking of importance of these assumption: (1) constant variance, (2) non-0 spatial autocorrelation, and (3) normality. In fact, with regard to the adverseness of consequences generated by violation of these assumptions, there is a sizeable gap between those arising from non-normality and those arising from variance heterogeneity or spatial autocorrelation. This contention may be illustrated by inspecting the output from various RESAMPLE experiments (with 15,000 replications) conducted for the simple difference of two means test. Consider two distinct regions, each composed of 100 areal units forming a 10-by-10 square tessellation and each having an attribute with exactly the same mean (i.e., $\mu_1 = \mu_2$). Suppose these attributes conform to a normal distribution in their respective populations and that spatial autocorrelation is absent. Then

$\sigma_1 = 1$ & $\sigma_2 = 1$: $\overline{X}_{\bar{x}_1-\bar{x}_2} = -0.00166$ $S_{\bar{x}_1-\bar{x}_2} = 0.14134$

$\sigma_1 = 1$ & $\sigma_2 = 1.18071$: $\overline{X}_{\bar{x}_1-\bar{x}_2} = -0.00142$ $S_{\bar{x}_1-\bar{x}_2} = 0.15626$

$\sigma_1 = 1$ & $\sigma_2 = 10$: $\overline{X}_{\bar{x}_1-\bar{x}_2} = -0.00273$ $S_{\bar{x}_1-\bar{x}_2} = 1.01400$.

The second value for σ_2 (1.18071) was chosen because it relates directly to the critical value for $F_{99,99,0.95}$. These outcomes exemplify the well-known trend of increasing variability in the difference of means sampling distribution with increasing divergence of the two population variances. Furthermore, the corresponding analytical formula, $\sqrt{\dfrac{\sigma_1^2}{n_1} + \dfrac{\sigma_2^2}{n_2}}$, increasingly offers a poorer estimate of the standard error, respectively having deviations of 0.00008, 0.00153, and 0.00901 for these simulation results.

Returning to the original case involving constant variance ($\sigma_1 = \sigma_2 = 1$), the effects of spatial autocorrelation—indexed by the autoregressive parameter ρ—can be explored by evaluating the possible range of positive spatial dependencies:

$\rho = 0.00$: $\overline{X}_{\bar{x}_1-\bar{x}_2} = -0.00091$ $S_{\bar{x}_1-\bar{x}_2} = 0.1411$

$\rho = 0.10$: $\overline{X}_{\bar{x}_1-\bar{x}_2} = -0.00070$ $S_{\bar{x}_1-\bar{x}_2} = 0.1413$

$\rho = 0.60$: $\overline{X}_{\bar{x}_1-\bar{x}_2} = 0.00214$ $S_{\bar{x}_1-\bar{x}_2} = 0.1606$

$\rho = 0.99$: $\overline{X}_{\bar{x}_1-\bar{x}_2} = -0.01043$ $S_{\bar{x}_1-\bar{x}_2} = 0.4369$.

Two trends are apparent here. First, the standard error increasingly deviates from the analytical result of 0.14142 as ρ approaches its upper limit of 1. Second, the average difference of means seems to be increasing as ρ moves toward its upper limit. This deviation from the conventional analytical standard error is not surprising, because as Cliff and Ord (1975) show, the correct analytical formula is $\sqrt{\dfrac{\sigma_1^2}{n_1(1-\rho_1)} + \dfrac{\sigma_2^2}{n_2(1-\rho_2)}}$, where ρ_1 and ρ_2 respectively index the nature and degree of spatial autocorrelation present in the two regions. As revealed in this formulation, spatial autocorrelation operates like a variance inflation factor (the VIF statistic of regression analysis; see §1.2). Accordingly, because experience shows that the spatial autocorrelation parameter ρ often lies between 0.4 and 0.6, in general the resulting inflated variance will yield far more restricted regional variance differences than can materialize when σ_1 and σ_2 are free to vary.

Finally, returning again to the original case involving independence ($\rho_1 = \rho_2 = 0$), non-normality can be explored by assessing a wide range of underlying frequency distributions. In order of deviation from normality (as indexed by the Shapiro-Wilk [S-W], Anderson-Darling [A-D], and Kolmogorov-Smirnov [K-S] tests statistics; see § 2.3), consider the normal, uniform, sinusoidal (U-shaped), and exponential (marked positive skewness) distributions. RESAMPLE results for this example include, where the population variance for each frequency distribution was set to 1,

normal: $\overline{X}_{\bar{x}_1-\bar{x}_2} = -0.00166$ $S_{\bar{x}_1-\bar{x}_2} = 0.14134$

uniform: $\bar{X}_{\bar{x}_1-\bar{x}_2} = 0.00032$ $S_{\bar{x}_1-\bar{x}_2} = 0.14192$

sinusoidal: $\bar{X}_{\bar{x}_1-\bar{x}_2} = -0.00101$ $S_{\bar{x}_1-\bar{x}_2} = 0.14108$

exponential: $\bar{X}_{\bar{x}_1-\bar{x}_2} = -0.00019$ $S_{\bar{x}_1-\bar{x}_2} = 0.14177$.

Neither the simulated difference of means nor the simulated standard errors obtained for this case deviate very much from their respective theoretical counterparts.

The impact of spatial autocorrelation tends to be restricted by the limited range within which the value of ρ lies, and non-normality tends to generate relatively negligible impacts except in cases involving very small sample sizes. Therefore, because both nonconstant variance and spatial autocorrelation generate differential variation among georeferenced data, violations of these three assumptions render adverse consequences in accordance with their aforementioned rank ordering. Ignoring or glossing over these assumptions can result in a false sense of security, because the nonquantifiable uncertainty accompanying specification error introduces into an analysis is above and beyond sampling error, increasing variability upon which statistical inferences are based. As mentioned previously, though, their serendipitous behaviors result in a tendency for simultaneous improvement in all specification error consequences when remedial action is taken that brings data into better compliance with only one of the assumptions. Because weighted regression is needed to implement geostatistical models with ideal data, analyses reported in this book will employ Box-Cox power transformations when taking remedial actions to bring georeferenced data into closer compliance with model assumptions. Another approach would be to convert data values to their ordinal scale counterparts (i.e., rank the values) and then undertake nonparametric statistical analyses; for a number of reasons, this option is not exercised in this book, although it is favored by data analysts who argue that data values should not be altered in an attempt to force them to conform to the best inputs for statistical techniques.

2.2. VARIABLE TRANSFORMATIONS

The selection of a scale of measurement is only the first step in quantification; appropriateness of the units of measurement also is of concern to a spatial analyst. The goal of establishing an informative scale of measure is to simplify the structure latent in a data set and by doing so enhance interpretation, expedite detection of anomalous areal units, and facilitate prediction, interpolation, and extrapolation. But transforming data to such a scale, even when it proves to be beneficial, also can be accompanied by disadvantages. The new measurement scale may well be less familiar and, as such, may compromise intuitive understanding and an ability to interpret the new numerical values rendered for areal units. Hence, transformations should be employed only when benefits outweigh the inconveniences associated with implementing them. Choices here at least partly are a matter of judgment and frequently are justified with empirical trends.

The power transformation approach is favored by data analysts who argue that data values are always measured on arbitrarily chosen scales and that changing these scales merely changes the spacings between ordered attribute values. Such improvements may enhance data interpretability as well as promote desirable statistical characteristics of the data. For either the Box-Cox or the Box-Tidwell power transformation, the general equation is of the form

$$f_1(Y) = \alpha/J_1 + \beta(Y + \delta)^\gamma/J_1 \text{ for } \gamma \neq 0 \text{, or} \tag{2.1}$$

$$f_2(Y) = \alpha + \beta LN(Y + \delta) \times J_2 \text{ for } \gamma = 0 \text{,} \tag{2.2}$$

where $J_1 = \gamma \bar{y}^{\gamma-1}$ ($\bar{y}$ denotes the geometric mean of the data) and $J_2 = \bar{y}$ for the Box-Cox power transformation (Montgomery and Peck 1982, 94), $J_1 = J_2 = 1$ for the Box-Tidwell power transformation (Johnson and Wichern 1992, 165), and X would be substituted for Y if a regressor variable is being evaluated. In the Box-Cox context, the J_1 and J_2 terms directly relate to maximizing a normal probability density function (Johnson and Wichern 1992, 165-66). If the goal is not to move a frequency distribution as close as possible to a normal distribution, then even if a response variable (i.e., Y) is being treated, $J_1 = J_2 = 1$, the more commonly reported version of the Box-Cox transformation. Clearly, equations (2.1) and (2.2) are bivariate regression-type specifications. The term δ has been added here, although frequently not explicitly stated in the transformation literature, in keeping with the sentiment of matched transformations as well as to compensate for any georeferenced variable taking on the value of 0 (in keeping with Emerson and Stoto 1983). In either case, the parameters of interest that need to be estimated are δ and γ, with the selected estimates being those that minimize the sum of squared errors (see Hoaglin et al. 1983). Estimation of these parameters can be achieved with nonlinear regression techniques (see Griffith and Amrhein 1997).

Hinkley (1977) suggests a simple "quick and dirty" alternative to estimating δ and γ, noting that the mean and the median of the transformed frequency distribution should be as close as possible in value to minimize asymmetry. In other words, the estimates are a solution to the following optimization problem:

$$\text{MIN}: Z = \left| \frac{\sum_{i=1}^n (x_i + \delta)^\gamma}{n} - (x_m + \delta)^\gamma \right| \text{,}$$

where x_m is the sample median. Substituting $\frac{\partial Z}{\partial \delta} = 0$ into the differential equation $\frac{\partial Z}{\partial \gamma} = 0$ yields

$$\sum_{i=1}^{n} (x_i + \delta)^{\gamma} [LN(x_i + \delta) - \frac{x_m + \delta}{x_i + \delta} LN(x_m + \delta)] = 0 \ .$$

Computation of the estimates $\hat{\delta}$ and $\hat{\gamma}$ with this equation requires the use of nonlinear regression, which yielded for selected starting values

```
initial value:   δ₀=-0.31, γ₀=1      δ₀ = 0, γ₀=1       δ₀=x̄, γ₀=1
estimates:       δ̂=-0.29255          δ̂=-0.00058         δ̂=0.24485
                 γ̂= 0.34554          γ̂= 0.27991         γ̂=0.23827
S-W              0.99015              0.99171             0.98903 .
```

Because it is a single equation in two unknowns, it is fraught with local optima. These values are improvements over the raw data value S-W = 0.92353 [††] ($\delta = 0$ and $\gamma = 1$)[1] and are very close to Hinkley's crude estimates of $\hat{\delta} = 0$ and $\hat{\gamma} = ¼$.

When various criteria, such as normality, variance homogeneity, independence, and linearity, are being optimized, trade-offs can occur. There is no guarantee that employing a transformation to better satisfy one assumption will improve compliance with any of the other assumptions; in fact, there is no guarantee that a transformation exists leading to results that adequately imitate any single desired feature (i.e., symmetry, constant spread, or straightness of a scatterplot). Fortunately, as mentioned, experience suggests that a reexpressed measurement scale that improves on one of these criteria tends to improve on others, too. Even for purely descriptive statistical studies, the presence of symmetry, homoscedasticity, and linearity in attribute values enhances interpretability, whereas attributes that are non-normal in very different ways obscure interpretability of a data analysis.

2.3. TRANSFORMING IN SEARCH OF NORMALITY

Since Geary's 1947 statement (§2.1), many data analysts have initiated their work by testing their data for compliance with the normality assumption (Bera and Ng 1995). On the one hand, complications introduced into georeferenced data by the presence of latent non-0 spatial autocorrelation cause this common first step in data analysis to be far less straightforward (Dutilleul and Legendre 1992). On the other hand, the problem addressed here is the reverse of testing for normality and reflects the "starship" approach to data analysis (Owen 1988; Owen and Li 1988). The S-W statistic generally is recognized as the best diagnostic test statistic for evaluating normality. In the case of a perfect normal distribution, it takes on a value of 1. As its modified implementation in MINITAB reveals, S-W is essentially the

---

[1]Statistically significant values are denoted throughout by [†] for a 10%, [††] for a 5%, and [†††] for a 1% level.

correlation between the ranked attribute data values and their corresponding theoretical normal curve values for each rank's quantile. These approximate theoretical values are given by the normal curve z-score values for the percentiles

$$(r_i - 3/8)/(n + 1/4) \, , i = 1, 2, \ldots, n,$$

where r_i denotes the rank of the attribute value for areal unit i.[2] By regressing the set of z_is on the appropriate $f_1(Y)$ or $f_2(Y)$, using nonlinear regression, the values of δ and γ can be estimated. Drawing a systematic sample of size n=10,000 from the normal, uniform, sinusoidal, and exponential (markedly negatively skewed) distributions, this estimation exercise results in

Distribution	equation (2.1)				theoretical	
	δ	γ	RSS×10^{-2}	S-W	raw data S-W	
normal	0.00695	1.00000	0.50	0.98090	1.00000	
uniform	0.00000	0.80023	2.38	0.98094	0.97724	
sinusoidal	0.00000	0.64976	1.19	0.96667	0.94847	
exponential	0.00000	0.27162	0.92	0.96552	0.90335	

The accompanying test statistics for normality are the Anderson-Darling (A-D; the sum of the squared differences between empirical and normal frequency distribution z-score values, each inversely weighted by the product of the normal cumulative probability times its complement), Ryan-Joiner (R-J; a modified Shapiro-Wilk statistic that is the correlation accompanying a normal probability plot), and Kolmogorov-Smirnov (K-S; the maximum absolute difference between the cumulative empirical and normal frequency distributions) statistics. The computational results here are

Distribution	original variable			power-transformed variable		
	A-D	R-J	K-S	A-D	R-J	K-S
normal	0.024	1.000	0.000	***	***	***
uniform	111.187	0.978	0.057	103.391	0.979	0.062
sinusoidal	305.553	0.949	0.097	275.575	0.954	0.106
exponential	461.926	0.904	0.158	2.059	0.999	0.008 .

These results, when coupled with the preceding S-W results, indicate that a normal distribution does not need to be subjected to a power transformation and that a power transformation does not necessarily exist that will conform a given frequency distribution to the normal distribution. At least marginal improvement can be attained, though. Furthermore, as the exponential distribution reveals, such a power transformation should exist for most any unimodal distribution. This transformation essentially slides the modal hump along its horizontal axis while also squashing and/or flattening it. This process may be emulated by moving an air

---

[2]These (Blom) and other versions (Tukey, VW) of normal scores are an option in the PROC RANK procedure of SAS.

bubble around in a piece of folded plastic.

Consider estimation of the power transformation parameter values for equation (2.1) that maximize the symmetry of each attribute's frequency distribution, employing selected 1969 agricultural productivity density measures for Puerto Rico reported by Griffith and Amrhein (1997; see Map 1.1a for the surface partitioning involved).[3] Using PROC NLIN in SAS yields results that may be organized into the following tabulation:

variable:	S-W		equation (2.1)			
		δ	γ	RSS×10⁻²	S-W
DFRM:	0.95195 (0.0192)	-0.64384	0.63454	0.50	0.98090
DCD:	0.97800	-3.52053	1.18064	2.38	0.98094
DMLK:	0.67895 (0.0001)	0.00000	0.26998	1.19	0.96667
DSGR:	0.56356 (0.0001)	0.00005	0.14472	0.92	0.96552
DCF:	0.82225 (0.0001)	0.00000	0.39709	2.74	0.93389
					(0.0010)
DTB:	0.54864 (0.0001)	0.00000	0.20591	0.64	0.91128
					(0.0001)
DBN:	0.66762 (0.0001)	0.03652	0.00000	0.65	0.97686
DFAM:	0.89523 (0.0001)	0.05592	0.36299	1.32	0.97105

Symmetry impacts of these transformations can be portrayed with boxplots (Figure 2.1). For the most part, these transformations have removed many outliers, expanded interquartile spreads, and moved medians and means closer together.

Seven of the eight original, raw data variables display marked non-normality. Even in the two power-transformed data cases that continue to display significant non-normality, modest to dramatic improvement in symmetry is achieved by using a judiciously selected power transformation. These two rogue cases also experience movement toward a more normally distributed frequency distribution.

Supplementing the preceding S-W results, a more comprehensive evaluation of these data yielded

variable	untransformed			transformed		
	A-D	R-J	K-S	A-D	R-J	K-S
DFRM:	0.659	0.984	0.084	0.149	0.997	0.053
	(0.08)	(0.07)				
DCD:	0.709	0.986	0.104	0.550	0.988	0.094
	(0.06)	(0.10)	(0.05)			
DMLK:	7.818	0.818	0.242	0.384	0.994	0.061
	(0.00)	(0.01)	(0.01)			
DSGR:	11.563	0.751	0.282	0.358	0.995	0.076
	(0.00)	(0.01)	(0.01)			
DCF:	3.425	0.919	0.149	1.125	0.986	0.096
	(0.00)	(0.01)	(0.01)	(0.01)		(0.10)
DTB:	12.253	0.760	0.293	1.474	0.996	0.061

---

[3]The results for the Hinkley (1977) example data are $\hat{\delta}$ = -0.05637 and $\hat{\gamma}$ = 0.28025 (S-W = 0.99229), which in fact are superior to those estimates reported in §2.2.

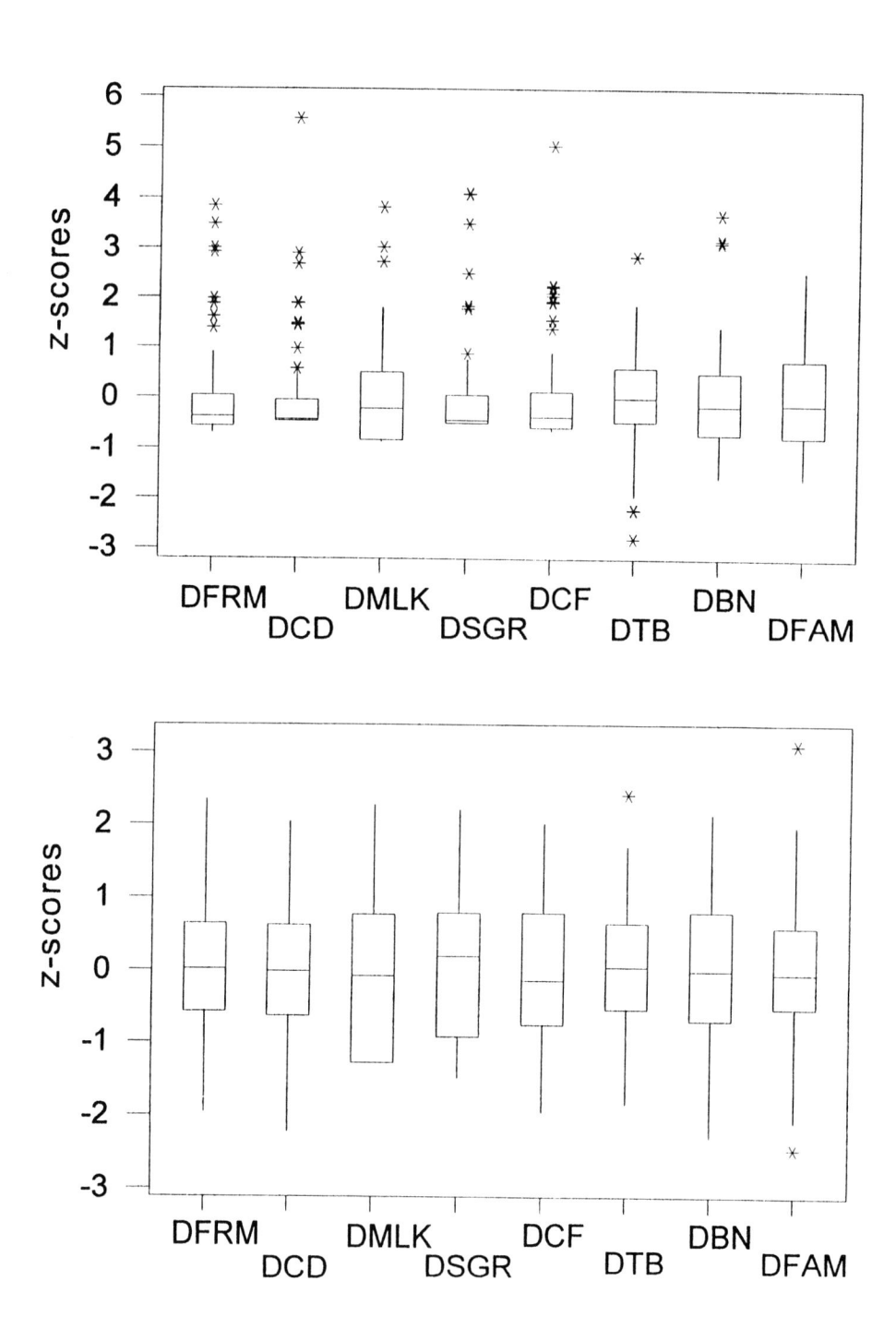

Figure 2.1. Top: (a) Boxplots of the raw agricultural productivity variables (Griffith and Amrhein's data). Bottom: (b) Boxplots of the optimally transformed agricultural productivity variables.

		(0.00)	(0.01)	(0.01)	(0.00)		
DBN:		8.511	0.815	0.250	0.134	0.997	0.039
		(0.00)	(0.01)	(0.01)			
DFAM:		1.583	0.950	0.106	0.295	0.993	0.057
		(0.00)	(0.01)	(0.04)			

Since different statistics hone in on different idiosyncrasies and nuances in data (data are noisy, messy, and dirty), this battery of statistics furnishes stronger evidence in support of the overall success of the Box-Cox power transformations in moving the Puerto Rican agricultural productivity data to measurement scales whose corresponding frequency distributions conform far more closely to a normal distribution. Even the power transformation applied to DCF appears to be successful, given this second investigation of the data. But the power transformation derived for DTB still fails to completely achieve its goal (a single outlier persists), although all three of its statistics exhibit substantial improvement.

2.4. TRANSFORMING IN SEARCH OF CONSTANT VARIANCE

The assumption of homogeneous (i.e., constant) variance is a basic requirement of most traditional statistical techniques. Although it does not compromise the unbiasedness of estimators, it does inflate the variance of their sampling distributions, causing confidence intervals to be less precise. One common reason leading to the violation of this assumption is that the mean response and spread of an attribute are correlated. For georeferenced data this type of correlation can occur in two different ways. First, regional means and variances can be correlated, a possibility that will be assessed throughout this book by dividing georeferenced data into either prespecified or contrived regions. When prespecified regions are not provided with a data set, a map will be divided into the four common quadrants of a plane, and each quadrant will serve as a (contrived) region for homogeneity of variance evaluation purposes. The second possibility is that the variance fluctuates as the level of a response variable changes; this situation relates to the conventional notion of constancy of variance and can be evaluated by inspecting a plot of residuals versus predicted values, without regard to regional location. For georeferenced data, the more important of these two circumstances is constant variance across regions. In either case, however, relating variance to mean response is a useful tool for guiding the selection of an appropriate power transformation, if one exists for the data under study. An accompanying goal is to reduce or eliminate the dependency between a mean and its associated variance, thus acquiring numerical values that are better suited for comparative purposes.

As a general rule of thumb, the use of a transformation is warranted (i.e., has a noticeable effect) when a dependency displayed between the mean and the variance of attribute Y across a geographic landscape is accompanied by a ratio of the extreme data values, say y_{max}/y_{min}, that exceeds 3.[4] In such a situation, this dependency may well be reduced or eliminated by employing a power transformation of Y. The threshold of 3 is selected because a transformation applied over a relatively narrow range of values tends to have little effect on the results of an analysis, whereas a transformation applied over a wide range of values can have a dramatic effect. In either case, the magnitude of this impact is dictated by how far the exponent of the power transformation deviates from 1. The slope of the regression line (b_1) of subgroup variances regressed on subgroup means indicates the value of the power transformation exponent [which in equation (2.1) is denoted by γ] as follows:

$$\text{power transformation exponent} = 1 - [b_1 \text{ from } LN(s) \text{ regressed on } LN(\bar{y})], \quad (2.3)$$

and is based upon the model

$$\sigma = \alpha\mu^\beta \times \epsilon,$$

where α, β, and ε respectively are the two bivariate regression parameters and the error term. These equations suggest that given a set of regions that partitions a geographic landscape into mutually exclusive and collectively exhaustive regions, preferably compact ones, the bivariate regression slope from regressing $LN(s)$ on $LN(\bar{y})$ determines, by substitution into equation (2.3), the estimate of the power transformation exponent needed to help stabilize geographic variance across these regions.

Because constant variance with regard to georeferenced data pertains to homogeneity of variation across a geographic landscape—variation from one region to the next is assumed to be constant—in the case of Puerto Rico, three regionalization schemes become germane: (1) five agricultural administrative regions (AAR); (2) ten groupings from a cross-classification of AAR with coastal lowlands/interior highlands (LND); and, (3) ten groupings from a cross-classification of AAR with an urban-rural dichotomy (UR). The LN(standard deviation)-LN(mean) regressions for each agricultural productivity attribute, using both the AAR-by-LND and AAR-by-UR cross-classifications, render the following results:

---

[4]Other criteria can be used for making this decision about employing a power transformation. Hair et al. (1995, 71) suggest the guideline of $\frac{\bar{x}}{s} < 4$. Sen and Srivastava (1990, 198) suggest the following procedure: (1) divide a data set into three equal groups, one comprising high values, one comprising medium values, and one comprising small values; (2) calculate the median for each of the three groups; (3) find the slope of the line joining the largest two of these medians, and find the slope of the line joining the smallest two of these medians; and, (4) if these slopes are unequal, a transformation is applied to the data until these slopes become roughly equal.

variable	Y_{max}/Y_{min} ≥ 3	$LN(s) \| LN(\bar{x})$ slope AAR×LND	AAR×UR	implied exponent
DFRM	38.5	0.495 (R²=0.494)	0.702 (R²=0.734)	0.402
DCD	21.3	-1.028 (R²=0.166)	-1.172 (R²=0.096)	1.000
DMLK	60460.0	1.002 (R²=0.952)	0.996 (R²=0.942)	0.002
DSGR	28184.4	0.916 (R²=0.964)	0.940 (R²=0.985)	0.072
DCF	14104.5	0.764 (R²=0.945)	0.646 (R²=0.849)	0.295
DTB	6306.8	0.850 (R²=0.933)	0.855 (R²=0.894)	0.148
DBN	637.2	1.048 (R²=0.899)	1.178 (R²=0.867)	-0.113
DFAM	49.1	0.748 (R²=0.782)	0.665 (R²=0.717)	0.293

Each implied exponent is calculated from the average of the two cross-classification results. These findings suggest that the variable DCD is not in need of a variance-stabilizing transformation (neither of the slope coefficients is significant), corroborating the preceding finding based upon optimizing normality. In fact, basic agreement between the two sets of results is obtained for DFRM, DSGR, DCF, DBN, and DFAM. These results suggest that DMLK should be subjected to a logarithmic rather than a non-0 power transformation and that DBN should be subjected to an inverse rather than a logarithmic transformation.

These findings may be corroborated by conducting formal tests of equality of variance across geographic regions. Two popular ones are Bartlett's test, which is not robust to the frequency distribution of georeferenced data departing from a normal distribution, and Levene's test, which requires an interval/ratio measurement scale but is robust for underlying frequency distributions that are not markedly skewed, but which does become somewhat unreliable when any group size, n_j, is less than 10. The following results document the presence of non-constant variance across the island of Puerto Rico for each agricultural productivity attribute and determine whether the power transformation that optimizes normality helps to stabilize these variances (probability levels for unlikely statistics are reported in parentheses beneath them):

variable	AAR Bartlett	AAR Levene	AAR×LND Bartlett	AAR×LND Levene	AAR×RU Bartlett	AAR×RU Levene
DFRM	3.569	0.452	11.089	0.677	11.680	0.902
transform	4.364	0.774	6.677	0.501	8.344	0.687
DCD	8.553	1.699	16.994	1.559	15.900	1.322
	(0.07)		(0.05)		(0.07)	
transform	7.338	1.350	16.064	1.288	15.004	1.122
			(0.07)		(0.09)	
DMLK	75.442	3.510	78.289	1.738	75.069	2.993
	(0.00)	(0.01)	(0.00)	(0.01)	(0.00)	(0.01)
transform	2.703	0.499	3.155	0.368	6.682	0.738
DSGR	77.715	2.561	141.113	4.133	124.210	1.988
	(0.00)	(0.05)	(0.00)	(0.00)	(0.00)	(0.06)
transform	7.843	1.639	9.251	1.138	11.953	1.128
	(0.10)					
DCF	24.504	3.007	78.893	2.407	32.129	1.507
	(0.00)	(0.02)	(0.00)	(0.02)	(0.00)	
transform	4.202	0.840	19.906	1.561	8.465	1.099
			(0.02)			

DTB	85.339	2.884	147.020	4.429	139.604	5.078	
	(0.00)	(0.03)	(0.00)	(0.00)	(0.00)	(0.00)	
transform	7.858	1.828	13.817	1.541	15.748	1.038	
	(0.10)				(0.07)		
DBN	15.263	0.465	84.534	3.494	70.077	1.396	
	(0.00)		(0.00)	(0.00)	(0.00)		
transform	8.286	1.208	12.008	1.273	20.792	1.215	
	(0.08)				(0.01)		
DFAM	20.390	2.033	20.178	0.620	20.938	1.366	
	(0.00)	(0.01)	(0.020)		(0.01)		
transform	4.308	1.234	4.170	0.991	5.287	1.177	

In all but one case (DFRM for AAR), both the Bartlett and Levene statistics decrease, sometimes marginally and sometimes dramatically. In general, then, spread stabilization is promoted by symmetry-enhancing power transformations. But as DFRM in AAR illustrates, there is no guarantee that improvement in one statistical feature (e.g., symmetry) will be accompanied by improvement in others (e.g., constant spread). Meanwhile, as Bartlett's statistic—which has been accused of being too sensitive to trivial differences in regional variances—repeatedly reveals, there is no guarantee that a transformation will completely remove any detected differences in variances across regions. Indisputable candidates for power transforming here are DMLK, DSGR, DTB, and DBN. Especially for both cross-classification regional groupings, the power transformations employed here remove the outlier nature of a number of areal units, shrink large regional spreads, swell small spreads, and even induce better symmetry in the within-regions frequency distributions.

The following S-W statistics evidence supports this last contention (probability levels for unlikely statistics are reported in parentheses beneath them):

variable	Shapiro-Wilk statistics for AAR frequency distributions				
	San Juan	Arecibo	Mayaguez	Ponce	Caguas
DFRM	0.85189	0.91057	0.97678	0.93321	0.95115
	(0.006)				
transform	0.89906	0.94381	0.98862	0.97105	0.96308
	(0.047)				
DCD	0.96150	0.96033	0.88958	0.94906	0.93683
			(0.055)		
transform	0.95312	0.95758	0.87381	0.94548	0.94980
			(0.031)		
DMLK	0.77548	0.66159	0.80125	0.85498	0.87543
	(0.000)	(0.000)	(0.002)	(0.025)	(0.060)
transform	0.94914	0.95799	0.88801	0.91757	0.94778
			(0.052)		
DSGR	0.56346	0.68903	0.67559	0.69795	0.66690
	(0.000)	(0.001)	(0.000)	(0.000)	(0.000)
transform	0.95987	0.92174	0.97110	0.93084	0.89865
DCF	0.60515	0.83216	0.97525	0.80832	0.88956
	(0.000)	(0.026)		(0.006)	(0.094)
transform	0.88453	0.85069	0.84670	0.95941	0.97531
	(0.025)	(0.044)	(0.012)		
DTB	0.57293	0.61336	0.35337	0.55218	0.71346

	(0.000)	(0.000)	(0.000)	(0.000)	(0.001)
transform	0.85575	0.95401	0.86575	0.88020	0.93027
	(0.007)		(0.023)	(0.058)	
DBN	0.71975	0.63230	0.60813	0.65781	0.87483
	(0.000)	(0.000)	(0.000)	(0.000)	(0.059)
transform	0.96047	0.93383	0.97461	0.93498	0.97274
DFAM	0.89055	0.93070	0.84925	0.86020	0.95106
	(0.032)		(0.013)	(0.030)	
transform	0.89415	0.96559	0.93600	0.95159	0.92751
	(0.038)				

DSGR and DTB display the most dramatic improvements in the S-W statistic that are attributable to power transforming. Although in general the power transformation applied to each variable tends to cause the regional frequency distributions to better conform to a normal distribution, once again some trade-offs can be identified (e.g., DCD). If a Bonferroni adjustment is employed here, all but one of the regional statistics implies normality for $\alpha = 0.05$ (the Bonferroni value here is $\frac{\alpha}{5}$ = 0.01).

2.5. LINEARITY: EXPLOITATION OF LINEAR RELATIONSHIPS BY LINEAR STATISTICAL MODELS

When relationships between attributes are of interest, traditional linear statistical models assume that these relationships can be described by or captured with straight lines; pairwise scatterplots often are constructed for visual validation of straight line trends and correlation coefficients calculated in order to index the nature and degree of these linear relationships. When the relationship is nearly linear, the correlation coefficient provides a useful and enlightening quantitative index of it; as the trend becomes increasingly curvilinear, the correlation coefficient increasingly fails to quantify adequately the portrayed relationship. Other advantages arising with a linear relation include a better ability to evaluate the fit of a regression equation and the availability of a sophisticated body of conventional statistical theory for data analysis purposes. When linear relations are not apparent in scatterplots, transforming the Y variable, the X variable, or both (in fact, it usually is best to explore simultaneously transformations of X and Y; see van Gastel and Paelinck 1995) may result in the scatterplot trend being straighter than some curved trend displayed by the original, raw data values. The goal here is to promote straight line relationships between two attributes. In an ideal situation, the trend line should be a straight line forming a 45° angle with the horizontal axis. A possible trade-off encountered when employing linearizing power transformations is the corruption of constant variance or normality characteristics of the data.

Guidance has been developed from algebraic series expansions and differential calculus for inducing linearity with power transformations. The classic paper on this topic was penned by Box and Tidwell (1962), who outline a procedure that depends on the relationship between X and Y being quite conspicuous. Their procedure encounters convergence problems in cases where the error variance is large (a common occurrence in georeferenced data) or where the range of Y is very small relative to its mean. This situation is further complicated by a frequent need in multiple linear regression analysis to have linear relationships between a pair of attributes only after the effects of a third variable have been regressed out of one or both of these variables (i.e., added variable plots).

Returning to the illustrative 1969 Puerto Rican data set, the correlation coefficients matrix for the eight variables, where the lower left triangular section of the tabulation refers to correlations among the raw data attribute values (see §1.4.1) and the upper right triangular section of the tabulation refers to correlations among the transformed attribute values, is as follows (correlation coefficients that have increased in absolute value due to the use of a power transformation are double underlined):

	DFRM	DCD	DMLK	DSGR	DCF	DTB	DBN	DFAM
DFRM		0.122	-0.248	0.600	-0.275	0.431	0.687	0.362
DCD	0.152		-0.018	0.239	0.441	-0.119	-0.020	0.455
DMLK	-0.182	0.050		-0.538	-0.020	0.176	-0.349	-0.060
DSGR	0.451	0.236	-0.289		-0.217	0.135	0.724	0.556
DCF	-0.231	0.505	-0.131	-0.224		-0.366	-0.614	0.103
DTB	0.404	0.009	0.020	-0.102	-0.275		0.338	0.070
DBN	0.602	0.167	-0.253	0.801	-0.410	0.204		0.233
DFAM	0.321	0.416	-0.026	0.658	0.024	-0.005	0.485	.

In this example, when compared with the correlation coefficients for raw data attribute counterparts, those coefficients for pairs of transformed variables have increased in absolute magnitude 19 times, decreased in absolute magnitude 9 times, and had a sign change 5 times. Of special note is the change in coefficients affiliated with DCD, because its power transformation exponent essentially is 1: 5 coefficients have increased in absolute magnitude, 1 has decreased, and 3 have changed signs. Corresponding scatterplot matrices appear in Figure 2.2. The power-transformed attributes yield pairwise scatterplots that exhibit more nearly linear scatterings of points than do their respective raw data scatterplots.

The accompanying varimax rotated principal components structure matrix—a linear statistical outcome—extracted from the correlation matrix calculated with the power-transformed variables is (prominent loadings are double underlined)

variable	PCI	PCII	PCIII	PCIV	PCV	PCVI
TDFRM	-0.186	-0.157	0.908	-0.204	0.053	0.234
TDCD	0.187	-0.010	0.062	-0.240	0.944	-0.052
TDMLK	0.005	0.970	-0.113	0.005	0.022	0.108
TDSGR	-0.360	-0.571	0.365	-0.528	0.182	0.044
TDCF	0.905	-0.056	-0.079	-0.052	0.292	-0.174

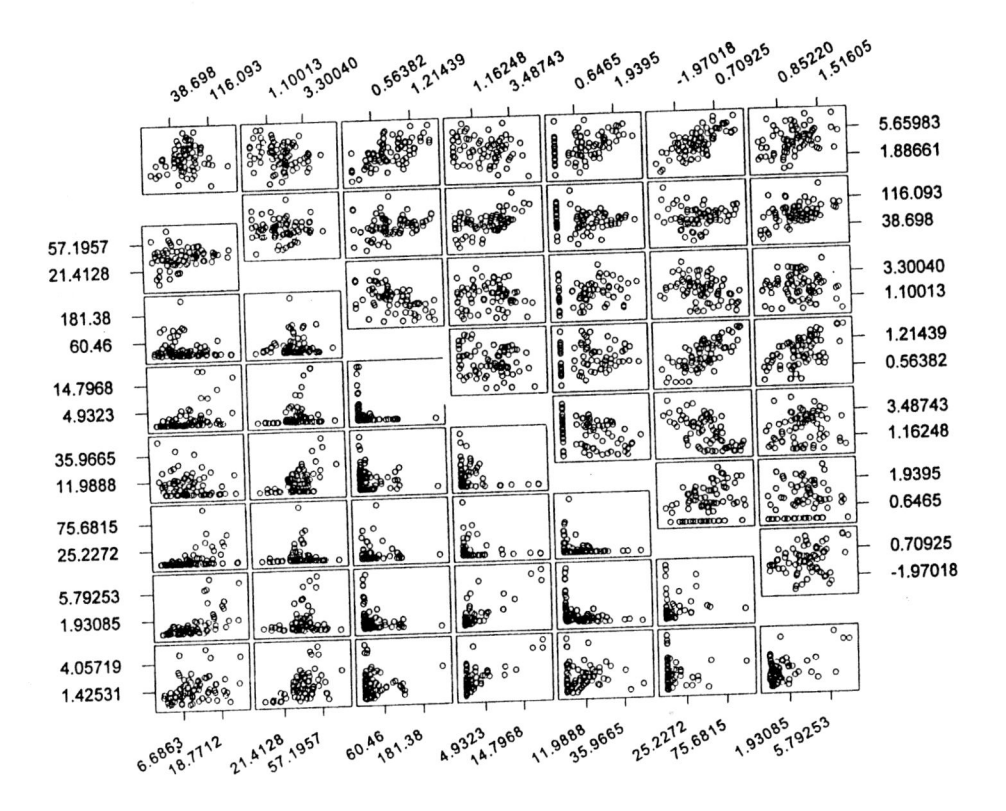

Figure 2.2. Lower triangular region: (a) Pairwise scatterplots of the raw agricultural productivity variables (Griffith and Amrhein's data). Upper triangular region: (b) Pairwise scatterplots of the power-transformed agricultural productivity variables.

TDTB	-0.185	0.100	0.206	-0.032	-0.055	<u>0.953</u>
TDBN	<u>-0.715</u>	-0.329	0.514	-0.152	0.072	0.134
TDFAM	0.027	-0.021	0.157	<u>-0.951</u>	0.216	0.029
λ_j	1.5653	1.4132	1.3118	1.3097	1.0682	1.0292
percent of variance accounted	0.196	0.177	0.164	0.164	0.134	0.129

This transformed attributes principal components structure matrix reveals that now six eigenvalues (rather than the four found with the raw data) are greater than 1 and that the six accompanying principal components account for roughly 96% of the collective attribute variance. Most agricultural productivity measures are distinct in this power-transformed situation. Transformed density of coffee (TDCF) and banana/plantain (TDBN) production group together, with opposite relations, to PCI; this result is intuitively sensible, because these two types of agricultural production are nearly disjoint. Meanwhile, transformed density of milk (TDMLK) and sugarcane (TDSGR) production group together, with opposite relations, to PCII; this result, too, is intuitively sensible, because milk production has been increasingly displacing sugarcane production as urbanization occurs across the island.

Consequently, the serendipitous nature of power transformations materializes again. Pairwise relationships exposed in this section are more robust than their raw data companions because now scatterplots better display linear patterns. The most conspicuous finding here is that the cost of bolstering this linearity assumption is a decreased ability to summarize more parsimoniously the statistical information content of the data.

2.6. AN ABSENCE OF INDEPENDENCE: THE PRESENCE OF SPATIAL AUTOCORRELATION IN GEOREFERENCED DATA

Chapter 1 addresses the notion of spatial autocorrelation in considerable detail. This feature of georeferenced data distinguishes it from all other types of data, although the presence of serial correlation in time series data moves that category of data somewhat closer to georeferenced data (see Figure 1.1). Arguments spelled out in §2.1 rank spatial autocorrelation second only to nonconstant variance as a source of complications introduced into a statistical analysis when it is overlooked. But spatial autocorrelation also interacts with the assumptions of constant variance and normality, just like it does with linearity and multicollinearity (see §1.4.1). Accordingly, transformations that cause data to conform better to these other two statistical assumptions also can alter initial evaluations of latent spatial autocorrelation.

2.6.1. IMPLICATIONS FROM INSPECTIONS OF MC SCATTERPLOTS

Returning to the Puerto Rican example, MC scatterplots for the transformed versions of the eight 1969 agricultural productivity measures are present in Figure 2.3. Those raw data cases displaying an approximately linear relationship (i.e., DFRM, DCD, and DFAM) continue to do so with their transformed scales. Those raw data cases exhibiting apparent nonlinear relationships (i.e., DMLK, DSGR, DCF, and DBN) now display far more linearized scatters of points. And DTB, whose raw data measures seem to display nonconstant variance, appears to undergo variance stabilization with its transformation. The accompanying revised spatial autocorrelation regression results, for which again $b_{1CI} = 4.7123$ in each case, are as follows:

MODELING ASSUMPTIONS   87

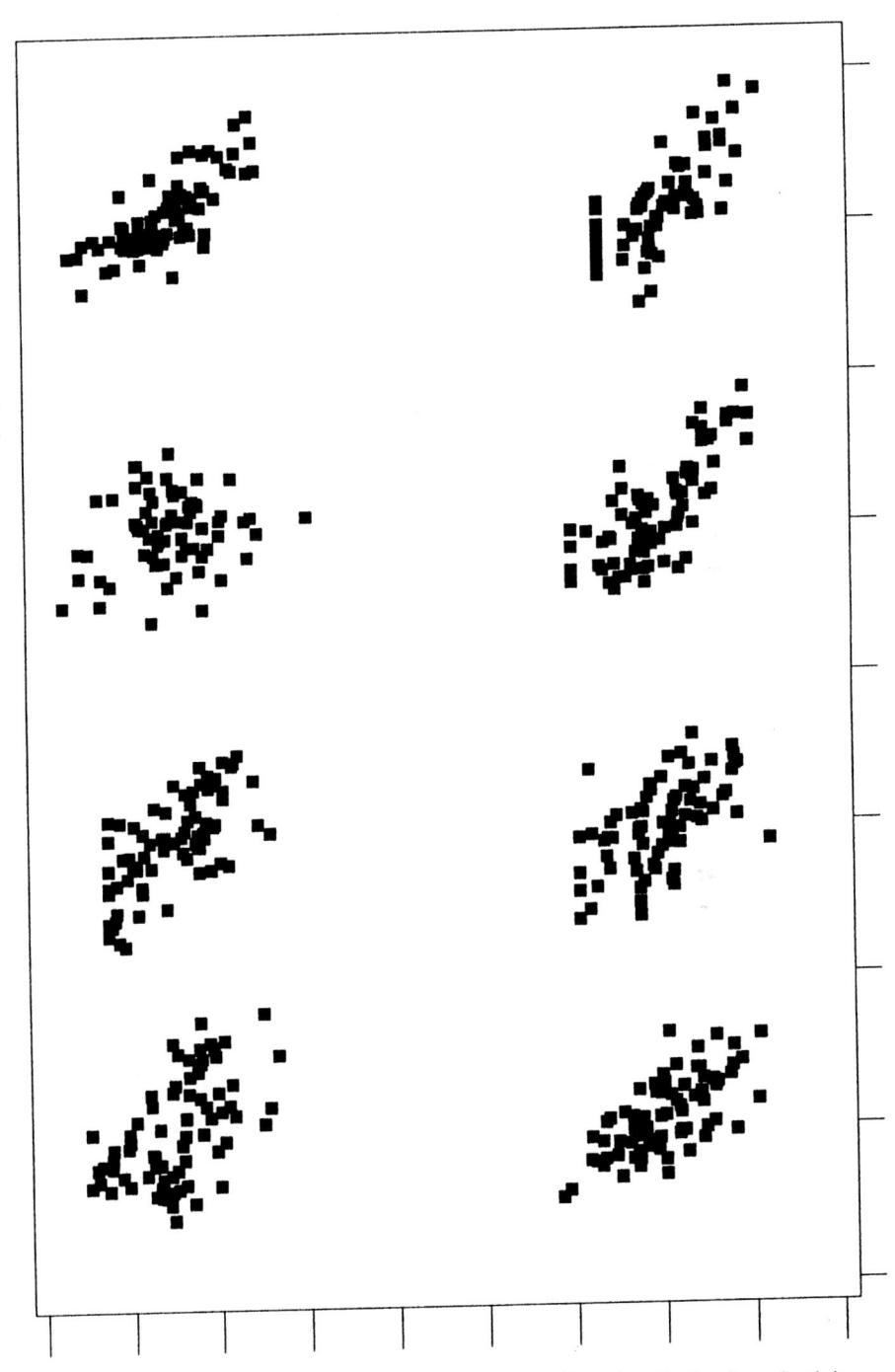

Figure 2.3. MC scatterplots of the eight 1969 power-transformed agricultural productivity variables (Griffith and Amrhein's data). Variables are arranged in alphabetical order of acronyms—in a counter-clockwise direction, beginning with the upper left-hand scatterplot: C*DBN vs. DBN, C*DCD vs. DCD, C*DCF vs DCF, C*DFAM vs. DFAM, C*DFRM vs. DFRM, C*DMLK vs. DMLK, C*DSGR vs. DSGR, and C*DTB vs. DTB.

transformed variable	b_{xcx}	MC	GR	original MC
TDFRM	1.7216	0.36534	0.77310	0.34842
TDCD	0.4261	0.09042	0.76657	0.10867
TDMLK	1.6174	0.34323	0.60002	0.20054
TDSGR	2.5743	0.54629	0.45841	0.51783
TDCF	2.0645	0.43810	0.58066	0.31060
TDTB	2.3056	0.48928	0.55554	0.43771
TDBN	1.8606	0.39483	0.62682	0.35371
TDFAM	1.8061	0.38327	0.56047	0.36528

For seven of these eight variables MC has increased; for two variables (i.e., DMLK and DCF) it has increased by quite a substantial amount. In addition, the MC and GR pairs of values align much better after transformations have been applied to the eight variables (the percentage of variance in common between MC and GR increases from 0.2% to 62.1%).

Recalculating the selected regression diagnostics (a strategy promoted in §1.4.2), used to detect leverage, influential points, and outliers, for the eight transformed variables included in the Puerto Rican example, flag as aberrant areal units (areal unit identification numbers reported in parentheses reference them to Map 1.1b)

variable	flagged by all four statistics	flagged by three of the four statistics
DFRM		Aguada (#2), Lares (#38), San Sebastian (#63)
DCD		San Juan/Rio Piedras (#61)
DMLK	Hatillo (#30)	Anasco (#6), San Jan/Rio Piedras (#61)
DSGR	Santa Isabel (#64)	Ceiba (#17), Lares (#38)
DCF	Santa Isabel (#64)	Arroyo (#8), Barranquitas (#10), Corozal (#22), Ponce (#54)
DTB		Barranquitas (#10), Cidra (#19), Comerio (#21), Trujillo Alto (#67)
DBN	Orocovis (#51)	Barranquitas (#10), Guanica (#25), Hatillo (#30), Lares (#38), Salinas (#59), Santa Isabel (#64)
DFAM		Humacao (#32), Jayuya (#34), Lares (#38).

When compared with the original raw data results, these identifications reveal that some variables have fewer anomalous areal unit values, while some have more. A number of the areal units have been reclassified, because their either are identified as outliers in the raw data but not in the transformed data, fail to be identified as outliers in the raw data but are identified as such in the transformed data, or switch between the two groupings of "flagged by all four statistics" and "flagged by three of the four statistics." The intuitive interpretation for the density of cultivated land (DCD) is retained; this is not surprising, because the power transformation applied to DCD employs an exponent very close to 1. As for the case of sugarcane production, results based on the transformed variables seem less intuitively obvious.

With regard to the LISA (local indicators of spatial association) regression evaluations for the eight transformed variables included in the Puerto Rican example, those municipios flagged (with double underlining) as spatially deviant areal units are (again, areal unit numbers reference them to Map 1.1b):

variable	areal units with absolute standardized residuals ≥ 2
DFRM	Barranquitas (#10), Comerio (#21), <u>Lares</u> (#38), Las Marias (#59), Santa Isabel (#64)
DCD	Bayamon/Catano (#11), Guaynabo (#28), <u>San Juan/Rio Piedras</u> (#61)
DMLK	<u>Anasco</u> (#6), Gurabo (#29), Las Marias (#39), Maricao (#44), <u>San Juan/Rio Piedras</u> (#61)
DSGR	Adjuntas (#1), Jayuya (#34), <u>Lares</u> (#38), Las Marias (#39), Maricao (#44), Utuado (#68)
DCF	*Barranquitas (#10), Bayamon/Catano (#11), <u>Corozal</u> (#22), Hatillo (#50), Hormigueros (#51)
DTB	<u>Barranquitas</u> (#10), <u>Cidra</u> (#19), <u>Comerio</u> (#21)
DBN	<u>Barranquitas</u> (#10), <u>Lares</u> (#38), <u>Orocovis</u> (#51)
DFAM	<u>Lares</u> (#38), Maricao (#44).

The articulation of agreement (designated with an asterisk) between these sets of diagnostics also has been altered. Certainly those areal units flagged in both raw data evaluations and these transformed data evaluations merit special consideration in a serious spatial analysis.

2.6.2. IMPLICATIONS FROM INSPECTIONS OF SEMIVARIOGRAM PLOTS

Similar to spatial autoregressive modeling, the preceding discussion of *iid* and frequency distribution assumptions is found in semivariogram scatterplots; however, an important distinction between the Moran scatterplots and semivariogram plots refers to the ability to identify individual data points in Moran scatterplots. In a semivariogram plot, data points are replaced with distance classes, or lags, and the scatter of points on such a plot represents groupings of distances. Thus, identification of aberrant points on a semivariogram plot is more obtuse because of the lack of a one-to-one correspondence with individual data points. Nevertheless, *iid* and statistical frequency distribution assumptions for geostatistical data analyses are important for estimating the semivariogram, fitting a model to the semivariogram scatterplot, and using the semivariogram model for estimation of nonsampled locations (kriging), as well as constructing their accompanying prediction intervals. These assumptions can be examined with univariate diagnostic statistics together with assessment of spatial stationarity. Discussion of impacts of violating these assumptions on the shape of a semivariogram is presented in this section.

Several assumptions about spatial data usually are posited in geostatistics because they make parameter estimation and the drawing of inferences possible within the context of classical statistics. Spatial data can be viewed as a multivariate[5] sample of size 1. Likewise, each attribute value (Y_i) is assumed to be the result of a random process, allowing the assumption $E(Y_i) = \mu$ to be made. Additionally, the frequency distribution for attribute values at each location usually is assumed to be a normal distribution; repeated sampling of a geographic surface using the same locations for each sampling is assumed to generate a normal distribution of attribute values at each location. Assuming that random processes produce normality is important, in part, because a normal process is characterized by its mean and covariance function. When normality is present, estimation, prediction, and statistical distribution theory are greatly simplified. Also, by the central limit theorem, the sum of many small-order effects is approximately normal. Finally, errors that are normally distributed imply the presence of second-order stationarity. (Notice that the assumption of a normal random process at each location should be kept distinct from the univariate distribution of a set of attribute values for a map.) Likewise, an additional assumption can be made to estimate optimal linear predictors (for kriging), namely, $COV(Y_i, Y_j) = C(s_i - s_j)$, where $C(s_i - s_j)$ is the covariogram. Without these assumptions, drawing inference becomes next to impossible.

Before examining the attribute values themselves, consideration should be given to the placement of sampling locations on a surface. The geographic distribution of sampling locations across a region is important not only for capturing spatial dependency via the semivariogram but also for controlling the size of prediction intervals associated with kriged values calculated for unsampled locations. The primary goal is to arrange a set of sampling locations in such a way that the study region is covered in a representative fashion and that estimations of $\gamma(h)$ at distances near 0 can be made. Typically, sample locations are selected using systematic sampling either by itself or in combination with simple random sampling. Locations selected with a systematic sampling scheme usually yield a more representative depiction of a surface and have a lower mean squared deviation per class than do locations chosen with simple random sampling. Unfortunately, uniform spacing imposes a limit on the minimum distance between points, the distance interval in which capturing spatial dependency is most important. A random placement of locations on a surface provides better estimates of $\gamma(h)$ values near zero distances but often a less representative depiction of a surface (some subregions are oversampled while others are undersampled; a Poisson distribution results). If a choice must be made between the two types of sampling, a systematic scheme is preferred as a more accurate representation of a surface, which is more important when prediction errors are computed from

---

[5] If this multivariate distribution is the normal one, the associated simple kriging estimator is equivalent to the conditional mean, while the associated kriging variance is the same as the variance of the conditional distribution. Univariate normality is not sufficient to guarantee this property.

kriging. While Russo (1984) and Bellhouse (1977) give methods for optimal placement of sampling locations, a spatial scientist at the design stage of a study also should consult Stehman and Overton (1996). For the data sets used in this book, however, estimating optimal sampling locations is not particularly helpful because sampling locations already have been determined, and data already have been collected.

Whereas traditional statistical graphics, such as quantile plots and histogram shapes, are well understood, what should the expected shape be for an ideal semivariogram plot (§1.7)? In a well-behaved plot, where positive spatial autocorrelation is present, the scatter of $\gamma(h)$ values should increase at a decreasing rate toward the range and then remain approximately constant across the remaining distance classes, randomly fluctuating around an invisible horizontal line whose vertical position is governed by the sill. The rate of convergence upon this invisible line is determined by the degree of spatial autocorrelation present in the georeferenced data under study. Recall that in practice, though, $\gamma(h)$ values may fluctuate wildly toward the maximum distances due, in part, to the smaller number of pairs of points available when computing $\gamma(h)$ for these large distances. Accordingly, emphasis is to be placed on the semivariogram plot pattern between the smallest distances and up to roughly half of the maximum distance. To illustrate a semivariogram scatterplot with spatially independent data, 100 values from a normal distribution with mean 0 and unit variance were randomly generated and allocated to the 100 grid locations on a 10-by-10 regular lattice (Map 2.1). Examination of the selected data reveals a normal frequency distribution with no aspatial outliers, a constant geographic variance (as described in §2.4) across the surface, and no spatial trend on the surface. A semivariogram plot (using the methods described in §1.7) was constructed, along with a plot containing the mean and high and low values for each distance class and a plot of the ratio of the variance to the mean for each distance class (Figures 2.4a-c). As expected for spatially independent data, the distribution of $\gamma(h)$ values in the first few distance classes of the semivariogram plot depicts a steep slope toward the sill and a relatively constant distribution of points across the rest of the distance classes, with a widening of the scatter of points toward the higher distance classes. Note that the pattern of $\gamma(h)$ values in the higher distance classes can be somewhat unpredictable because, as mentioned, fewer distance pairs contribute to the estimation of $\gamma(h)$. Inspection of the semivariogram plot including the minimum and maximum γ values for each distance class reveals that no single value (either high or low) is dominating the set of extreme values for the distance classes. When each attribute value on a map is assumed to come from a normal distribution, and because squaring a normal variable converts it to a χ^2 distribution whose mean is given by the number of degrees of freedom and whose variance is given by twice the number of degrees of freedom, $\frac{\sigma^2}{\mu} = 2$. Consequently, in the plot of the ratio of the variance to the mean, points should, and indeed do in this numerical example, fluctuate around the expected ratio value of 2.

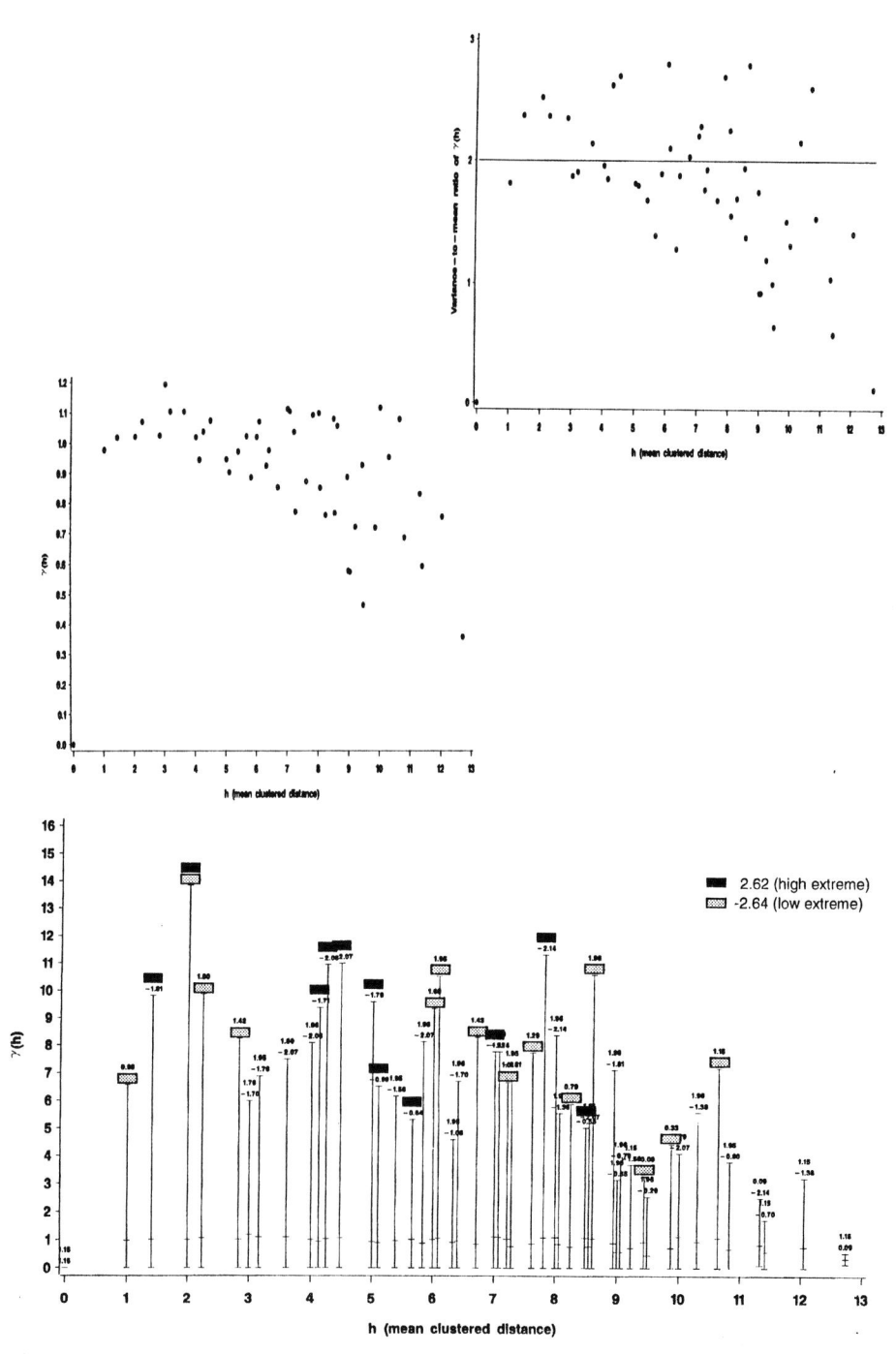

92   A CASEBOOK FOR SPATIAL STATISTICAL DATA ANALYSIS

Figure 2.4. Top left: (a) Semivariogram plot of simulated spatially independent normal data (n = 100). Bottom: (b) Semivariogram plot displaying the mean and extreme values for Figure 2.4a. Top right: (c) Semivariogram-type plot of the distance class variance-to-mean ratio for Figure 2.4a.

Geographic trend contamination (i.e., second-order stationarity cannot be assumed) is perhaps the most easily identifiable trait causing deviation away from the ideal shape of a semivariogram. When a trend is present across a surface, the shape of the resulting semivariogram plot mimics an exponential growth function plot. The semivariogram plot in Figure 2.5 depicts this characteristic shape, where the trend, in this case, was created by adding respective values of the horizontal axis to each of the corresponding attribute values (i.e., a linear trend in the east-west direction). Presence of a trend also may be verified by examining the OLS regression of an attribute variable on the accompanying georeferenced coordinate variables. Significance of an estimated slope for either georeferenced variable indicates a trend across the study surface. For the example values derived from Map 2.1, a significant slope is found in the horizontal direction but not in the vertical direction, as expected. Trend removal is necessary because the presence of a trend in data can mask the actual underlying spatial dependency in these data (§1.2). In addition, it could suggest the wrong semivariogram model for describing them.

One of the more important and desirable characteristics of geostatistical data is stationarity. Christensen (1991), Cressie (1991), and Isaaks and Srivastava (1989) emphasize the importance of second-order (or weak) stationarity and strict (or strong) stationarity. Second-order stationarity, in its simplest form, occurs when the mean and variance are constant across a geographic surface. Strict stationarity results when all moments are constant across the surface and, when present, implies second-order stationarity. Because spatial data can be viewed as an n-variable multivariate sample of size 1, inference is impossible without positing these assumptions. The most important relationship preserved by second-order stationarity is that $\gamma(h) = C(0) - C(h)$, where $C(0)$ is an n-by-n matrix containing the variance in its diagonal, and 0 elsewhere, and $C(h)$ is an n-by-n matrix containing the covariances. That is, the semivariogram is a function of the variance and covariance of a stationary spatial process. While stationarity is not easily testable directly, characterization of the data, both spatially and aspatially, can indicate absence of stationarity. For instance, absence of stationarity is seen in Figure 2.5 and is explained by the presence of a trend. Deviations from *iid* and posited statistical frequency distributions also can result in a failure of the stationarity assumption to be valid.

As shown for spatial autoregressive processes, extreme values in a data set can be classified as outliers within the context of a univariate frequency distribution (aspatial), the set of neighboring values (spatial), or both. Aspatial outliers are seen as extreme deviations of values in terms of a referent normal distribution (i.e., deviations greater than, say, ±3 standard deviations). Aspatial outliers can result in a dramatic influence on estimates of $\gamma(h)$ because $\gamma(h)$ is not a resilient estimator; lack of resiliency results from the squared term in the summand of equation (1.11). The main impact of an aspatial outlier is inflation of $\gamma(h)$ estimates across all distance classes. Again, presence of an aspatial outlier can be most easily detected by examining the frequency distribution of a data set.

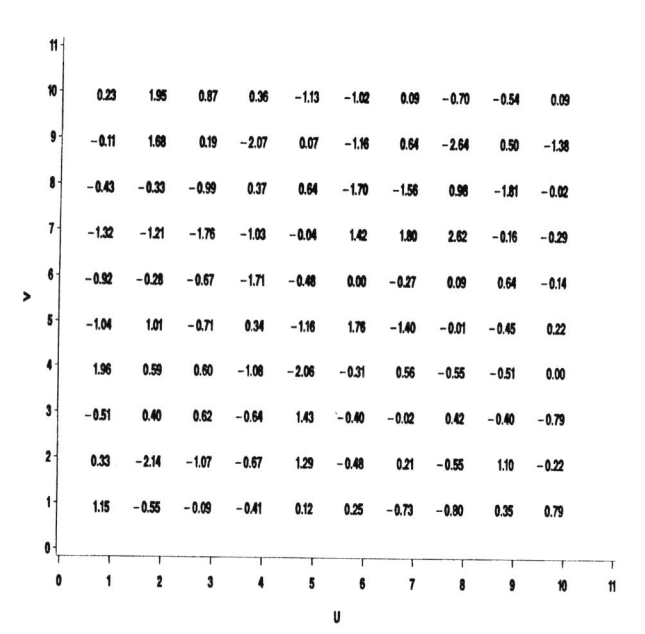

94   A CASEBOOK FOR SPATIAL STATISTICAL DATA ANALYSIS

v \ u	1	2	3	4	5	6	7	8	9	10
10	0.23	1.95	0.87	0.36	-1.13	-1.02	0.09	-0.70	-0.54	0.09
9	-0.11	1.68	0.19	-2.07	0.07	-1.16	0.64	-2.64	0.50	-1.38
8	-0.43	-0.33	-0.99	0.37	0.64	-1.70	-1.56	0.98	-1.81	-0.02
7	-1.32	-1.21	-1.76	-1.03	-0.04	1.42	1.80	2.62	-0.16	-0.29
6	-0.92	-0.28	-0.67	-1.71	-0.48	0.00	-0.27	0.09	0.64	-0.14
5	-1.04	1.01	-0.71	0.34	-1.16	1.76	-1.40	-0.01	-0.45	0.22
4	1.96	0.59	0.60	-1.06	-2.06	-0.31	0.56	-0.55	-0.51	0.00
3	-0.51	0.40	0.62	-0.64	1.43	-0.40	-0.02	0.42	-0.40	-0.79
2	0.33	-2.14	-1.07	-0.67	1.29	-0.48	0.21	-0.55	1.10	-0.22
1	1.15	-0.55	-0.09	-0.41	0.12	0.25	-0.73	-0.80	0.35	0.79

Map 2.1. 100 randomly generated values from a normal distribution with mean 0 and unit variance randomly allocated to the 100 grid locations forming a 10-by-10 regular square lattice.

A value identified as an aspatial outlier more than likely will be identified as a spatial outlier as well. A spatial outlier occurs when a relatively high attribute value is surrounded by lower values on a map, and vice versa for a low-value outlier. For a regular lattice of locations, Cressie (1991) suggests constructing a bivariate plot of attribute values separately by row (vertical position) and by column (horizontal position) to identify spatial outliers. The plot is constructed by pairing a referent attribute value with another attribute value one lag distance away, within a row or within a column. (For locations on an irregular lattice, rows and columns may be formed by assigning locations to row and column classes [Cressie 1991]). Thus, the axes of the plot are formed from the range of attribute values on the horizontal axis and the range of attribute values plus one lag distance on the vertical axis. Any spatial outliers are readily identified on such a bivariate plot as points that lie away from a main cluster of points. Similarities exist between this procedure and the Moran scatterplot graphic.

In addition to similar effects on estimation of $\gamma(h)$ exhibited by an aspatial outlier, a spatial outlier can have visible effects on the shape of a semivariogram plot, depending on the location of the value with respect to the edge of a map and regions of values that are similar to it. To illustrate this, an extreme value of 10 standard deviations was inserted into the geographic distribution of attribute values in Map 2.1, first being positioned in the lower right-hand corner of the map and then sequentially moved $\sqrt{2}$ lags diagonally toward the center of the map. A semivariogram plot was constructed for each position of the extreme value (Figures 2.6a-e). Plots containing the minimum and maximum $\gamma(h)$ values for each distance class and plots containing the ratio of the variance to the mean appear in Figures

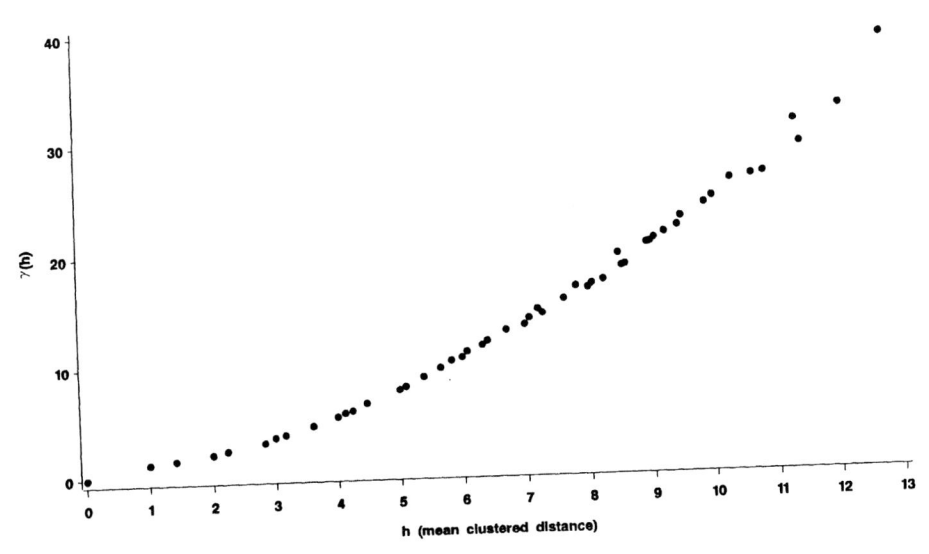

Figure 2.5. Semivariogram plot of simulated data for Figure 2.4a with a linear geographic trend added.

2.7a-b and 2.8a-b, the contrasting cases of the position of the extreme value being in the corner or in the center of the map. When an outlier is located on the edge of a map, the semivariogram plot resembles one constructed for data containing a nonconstant mean (compare Figure 2.6a to Figure 2.5); the main difference between the two is that the slope of the semivariogram plot in Figure 2.5 is steeper for lower values of distance classes. This particular shape of the plot is realized for two reasons: (1) the extreme value is found in all distance pairs, including the largest distance pair; and (2) the extreme value occurs on the edge of the map and, hence, it is unknown whether the extreme value is truly a spatial outlier or the beginning of a new set of values from a different population that would occur if the map edges were extended beyond their present position. Moving the location of the extreme value from the edge of a map toward its center results in an increase in $\gamma(h)$ only in the smaller distance classes (Figures 2.6a-e), because the extreme value is used in fewer and fewer larger distance class computations. In fact, one characteristic of a semivariogram plot containing a spatial outlier is a noticeable decrease in $\gamma(h)$ values at the higher distance classes, where the extent of the decreasing $\gamma(h)$ values is related to the location of the outlier relative to the center of a map. Likewise, comparing the extreme values for each distance class when an outlier is on the edge versus in the center of a map (Figures 2.7a-b) reveals that peripheral locations are accompanied by domination of the maximum value per distance class by the extreme value (Figure 2.7a). When an outlier lies toward the center of a map, it dominates the maximum value per distance class only at the lower distances, whereas other data values determine the maximum values at larger heteroscedasticity is not readily detectable merely by inspection of a semivariogram

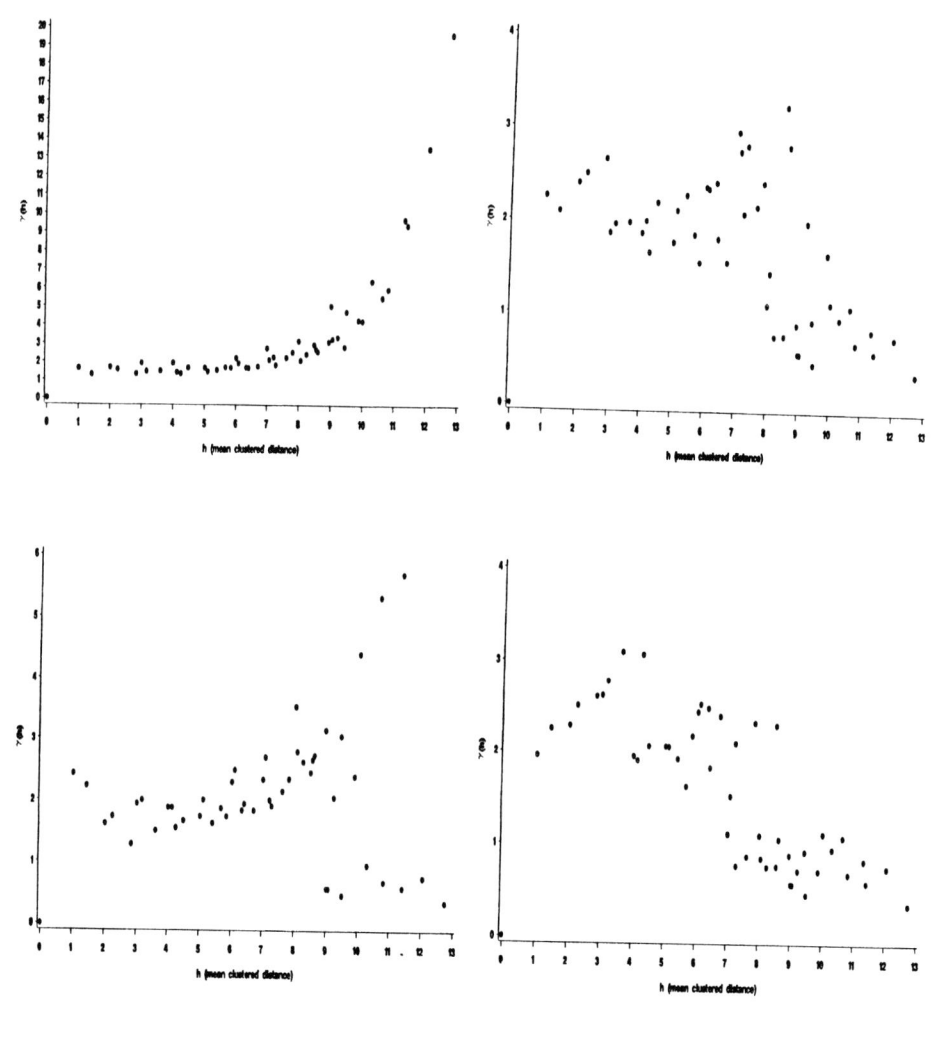

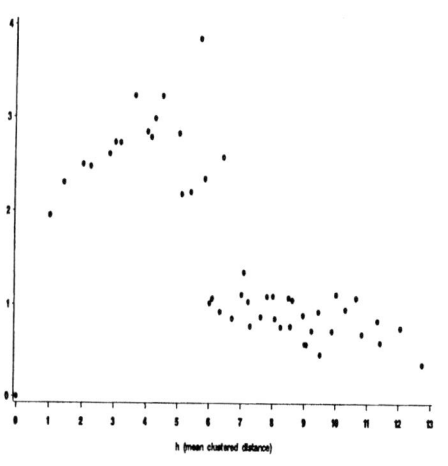

Figure 2.6. Top right: (a) Semivariogram plot of simulated data for Figure 2.4a with an outlier inserted at lattice position (10,1) of Map 2.1. Middle right: (b) Semivariogram plot of simulated data for Figure 2.4a with an outlier inserted at lattice position (9,2) of Map 2.1. Top left: (c) Semivariogram plot of simulated data for Figure 2.4a with an outlier inserted at lattice position (8,3) of Map 2.1. Middle left: (d) Semivariogram plot of simulated data for Figure 2.4a with an outlier inserted at lattice position (7,4) of Map 2.1. Bottom left: (e) Semivariogram plot of simulated data for Figure 2.4a with an outlier inserted at lattice position (6,5) of Map 2.1.

MODELING ASSUMPTIONS 97

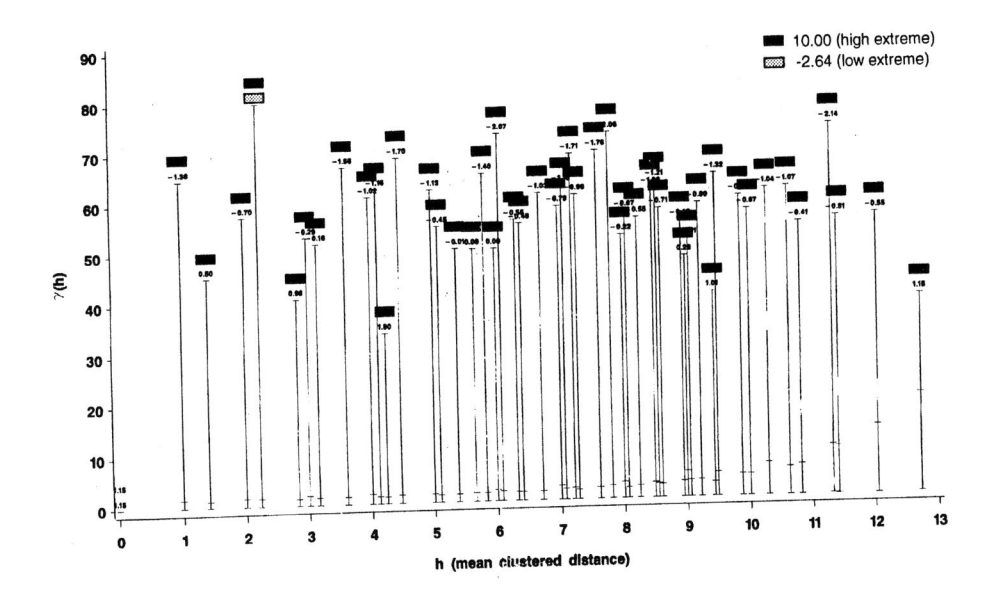

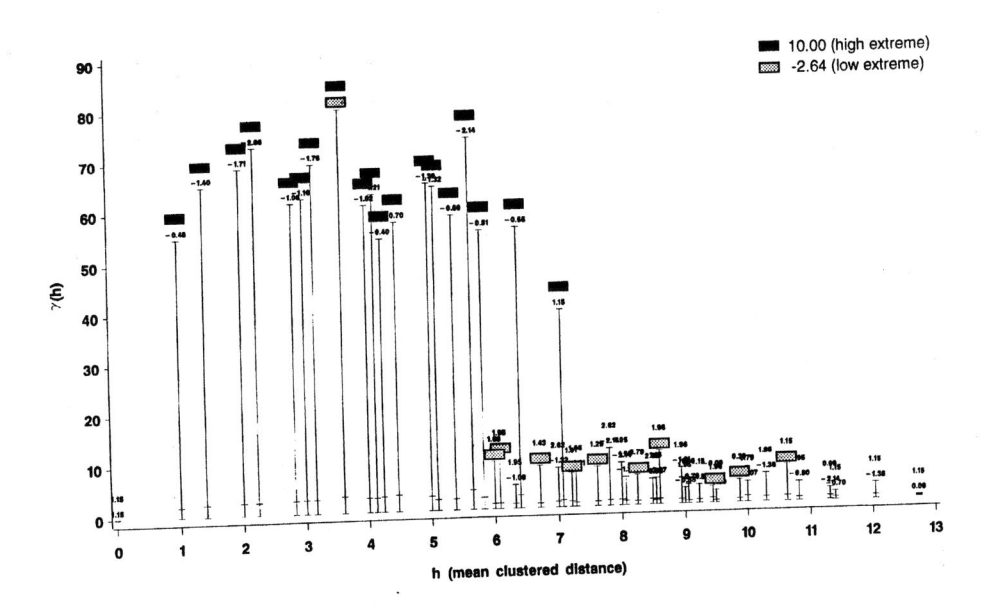

Figure 2.7. Top: (a) Semivariogram plot displaying the mean and extreme values for Figure 2.6a. Bottom: (b) Semivariogram plot displaying the mean and extreme values for Figure 2.6e.

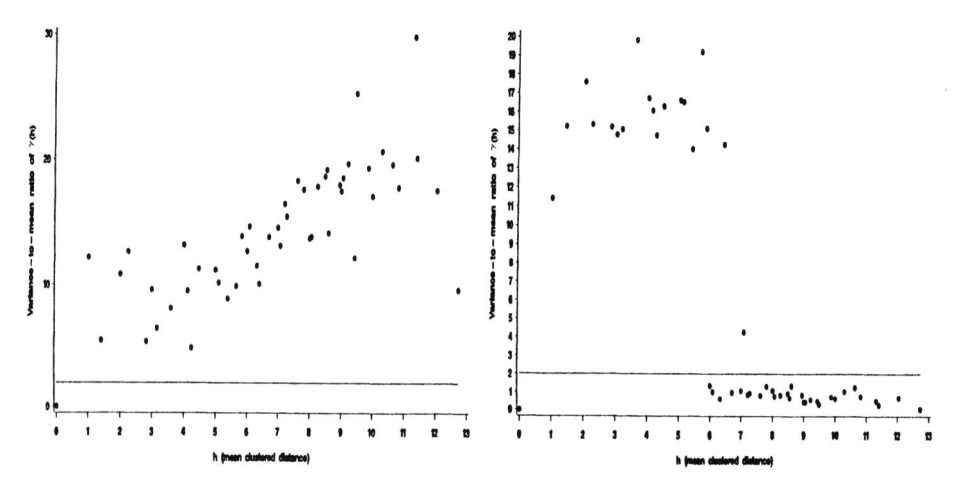

Figure 2.8. Left: (a) Semivariogram-type plot of the distance class variance-to-mean ratio for Figure 2.6a. Right: (b) Semivariogram-type plot of the distance class variance-to-mean ratio for Figure 2.6e.

distance classes (Figure 2.7b). Patterns observed for the plots of the variance-to-mean ratio (Figures 2.8a-b) also indicate that an important role is played by the relative position of the outlier: this ratio decreases quite noticeably at higher distance classes as the location of the outlier is moved toward the center of a map.

This crude series of numerical examples shows how the shape of a semivariogram plot can be used to explore for the presence of a spatial outlier. Some spatial outlier detection rules of thumb based on this shape interpretation may be summarized: (1) a convex or concave scattering of $\gamma(h)$ values at the higher distance classes; (2) a single attribute value (either high or low) dominating the extreme value for most distance classes below some threshold distance, with the threshold value depending on the position of the outlier with respect to the edge of a map; and (3) the variance-to-mean ratio for at least the smaller distance classes is noticeably different from 2. All three of these traits are indicative of non-stationarity over a surface.

The effects of geographic heteroscedasticity (§2.4) on a semivariogram plot are similar to those exhibited by the presence of an aspatial outlier: $\gamma(h)$ values for a heteroscedastic surface are inflated relative to $\gamma(h)$ estimates when no such heteroscedasticity exists, and variability increases within distance classes. To illustrate this condition, the values in Map 2.1 were divided into four equal-sized groups on the surface (the quadrants of the plane; §2.4). Values in quadrant I were multiplied by 2, then all values on the map were standardized again to 0 mean and unit variance. All tests for homoscedasticity are found to be significant. As seen in Figure 2.9a, $\gamma(h)$ estimates increase above those found in Figure 2.4a. Increased variability (represented by minimum and maximum attribute values) within distance classes is clearly evident when variability detectable in Figure 2.9b is compared with that found in Figure 2.4b. Although presence of geographic

MODELING ASSUMPTIONS 99

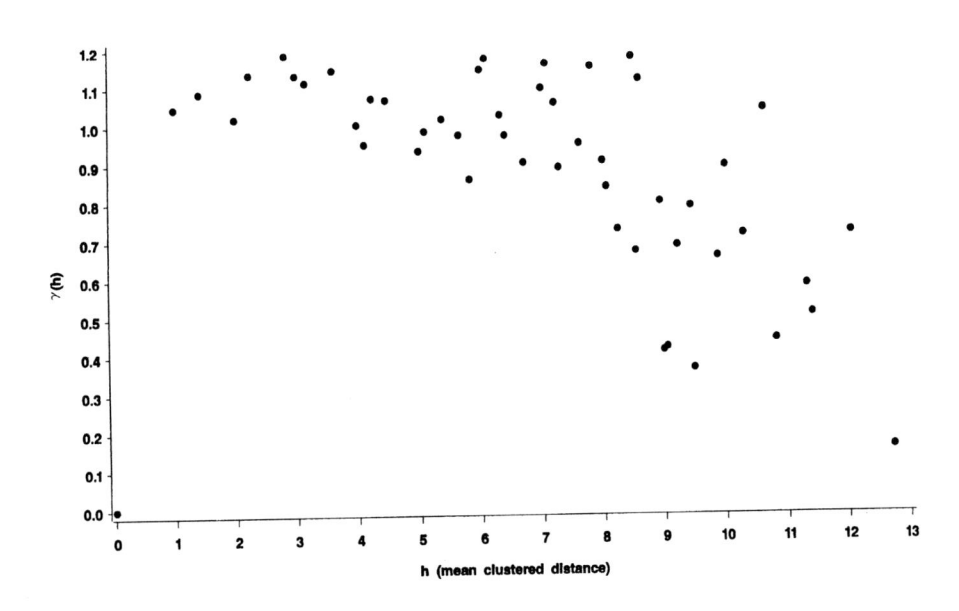

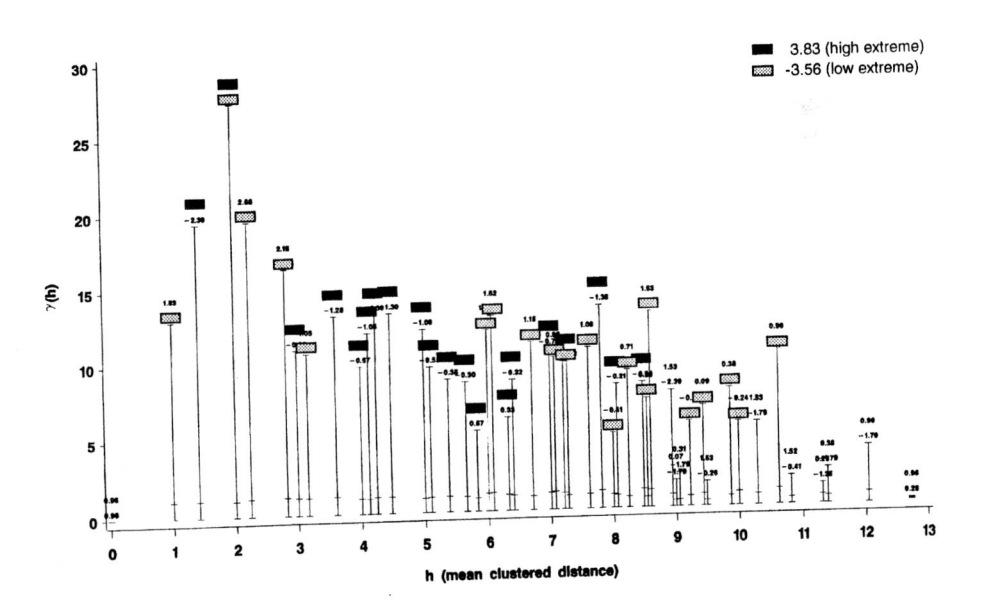

Figure 2.9. Top: (a) Semivariogram plot of simulated data for Figure 2.4a with nonconstant geographic variation induced by doubling the variance in the first quadrant of the plane for Map 2.1. Bottom: (b) Semivariogram plot displaying the mean and extreme values for Figure 2.9a.

Table 2.1. Eigenvectors affiliated with significant positive spatial autocorrelation.

eigen-vector #	MC	eigen-vector #	MC	eigen-vector #	MC
1	1.10362	9	0.59521	16	0.35789
2	1.04785	10	0.56908	17	0.32355
3	0.91864	11	0.53731	18	0.29536
4	0.85621	12	0.50302	19	0.25802
5	0.84058	13	0.47112	20	0.23539
6	0.82180	14	0.42748	21	0.20261
7	0.71695	15	0.40210	22	0.17518
8	0.69675			23	0.16807

plot, expectations of the shape of a semivariogram plot can be made by imagining the effects not of a single outlier but of a group of (perhaps contiguous) outliers. Decreasing $\gamma(h)$ values at the larger distance classes, as can be seen in Figure 2.9, result because the extreme values (both positive and negative) produced by simulating a heteroscedastic surface are located approximately midway between the edge and center of the map.

For cases involving spatially dependent data, semivariograms plots were constructed using the Puerto Rican data. Point data for the Puerto Rico map can be obtained by using the location coordinates of the municipio centroids (Griffith and Amrhein 1991, 116; these can be computed with, for instance, Arc/Info). The same attribute values for each of the eight variables used in §2.6.1 can be attached to these location coordinates to allow geostatistical analysis of these data. Semivariogram values and plots were obtained using the methods described in §1.7, with the plots for each of the nontransformed variables presented in Figure 2.10a. The desired semivariogram plot pattern—with regard to convexity/concavity, dominance of a single attribute value, and the variance-to-mean ratio—may be seen by inspecting the cases for DFRM, DMLK, DSGR, and DCF.

Based upon geographic distributions coupled with concerns for having constant spread and symmetric frequency distributions for data, the Puerto Rican data explored here also were subjected to a power transformation using the same δ and γ values reported in §2.3. Their corresponding semivariogram plots appear in Figure 2.10b. Generally speaking, the semivariogram patterns do not change drastically from their respective counterparts exhibited in Figure 2.10a, although scale obviously does. The most conspicuous change in the semivariogram scatterplot patterns for the transformed data is a tighter grouping of the semivariogram values (i.e., the vertical distance among the points in the plots decreases—indicative of variance stabilization), yielding a clearer picture of the spatial autocorrelation component latent in these data. Now, TDFRM, TDMLK, TDSGR, and TDCF exhibit patterns suggesting stationarity across the map (Figure 2.10b). TDTB as well as TDBN suggest the presence of a nonconstant mean whose

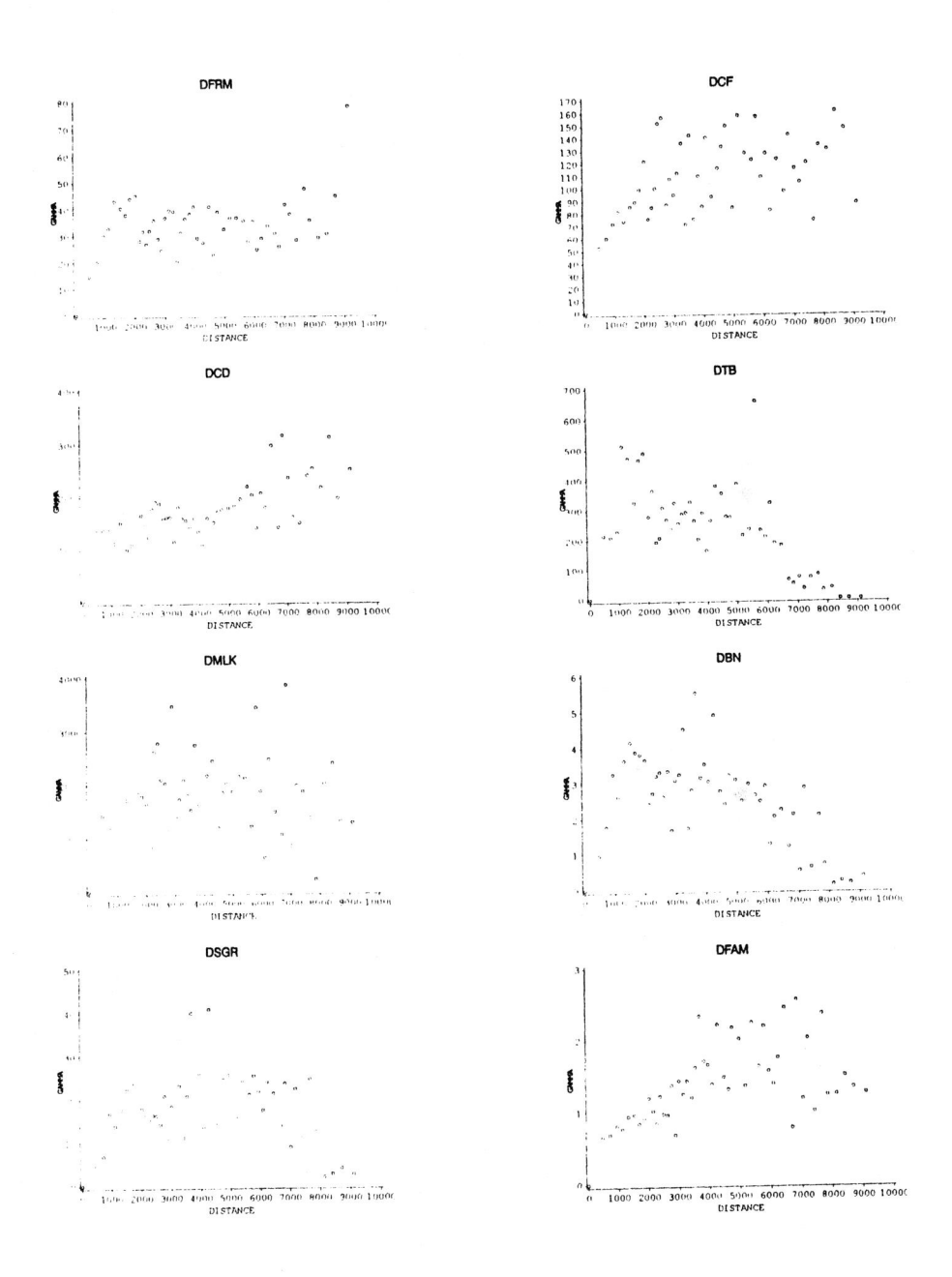

Figure 2.10a. Semivariogram plots for the eight 1969 raw agricultural productivity variables (Griffith and Amrhein's data).

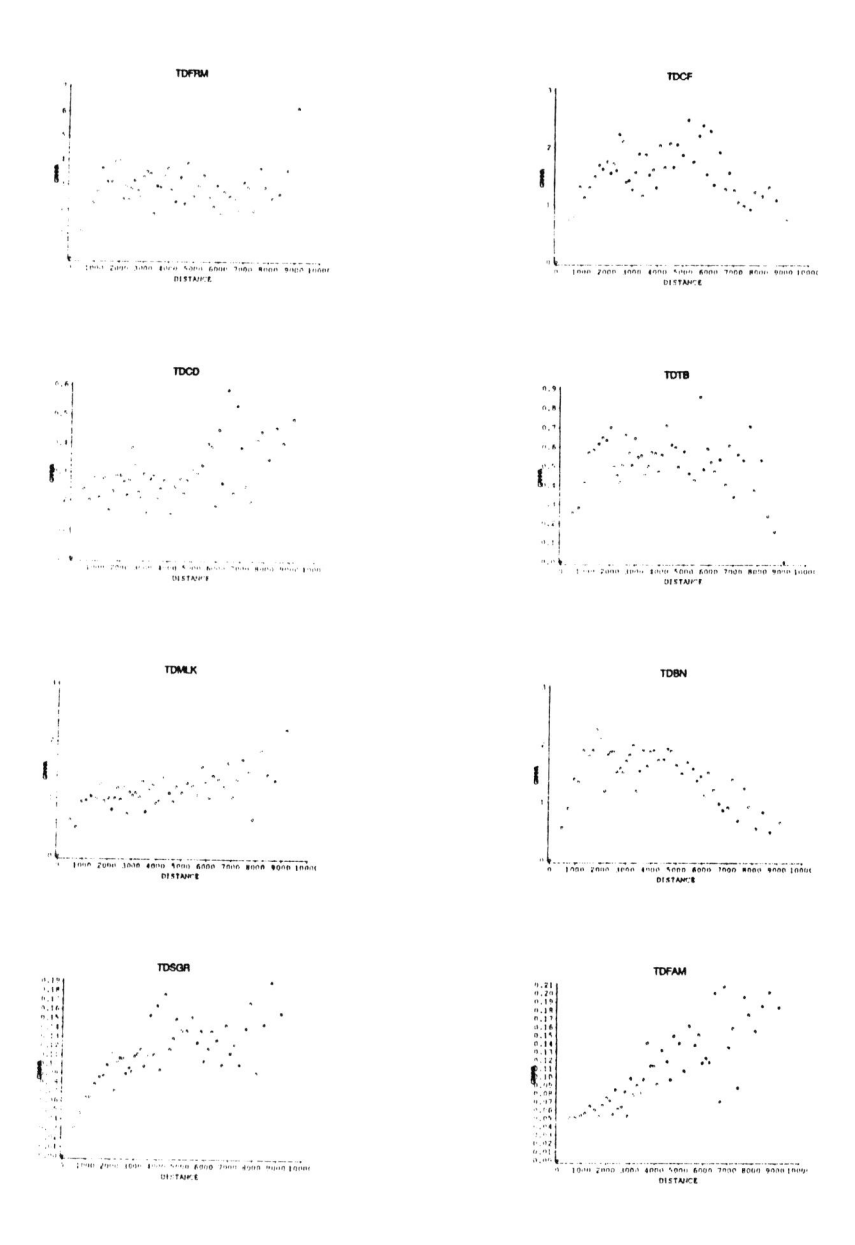

102 A CASEBOOK FOR SPATIAL STATISTICAL DATA ANALYSIS

Figure 2.10b. Semivariogram plots for the eight 1969 power-transformed agricultural productivity variables (Griffith and Amrhein's data).

geographic pattern either resembles a mountain or a sink. Meanwhile, since γ(h) values for TDFAM continuously increase as distance increases, a nonstationary mean displaying a linear geographic gradient across the island is suggested. A similar situation is suggested for TDCD, particularly after comparing this plot with that for DCD (Figure 2.10a).

Semivariogram computations for these Puerto Rican data sets were performed assuming an isotropic surface. Anisotropic surfaces are not uncommon, however, and any detailed geostatistical analysis should include a search for evidence of anisotropy in the georeferenced data (§3.2.4, §3.3.3). Most commonly, this is accomplished by examining semivariogram plots constructed for mutually exclusive subsets of a georeferenced data set where a direction tolerance is defined and then used to determine these subsets. This procedure commonly takes the form of examining plots in specified directions separately, such as examining data where pairing directions lie in a northwest-southeast direction versus the southwest-northeast direction. In other words, directions for pairs of points are included in the respective semivariogram calculation only if the point with which they are paired lies in the same direction. Anisotropy is further explored in later chapters.

2.6.3. EXPLORING AN EXAMPLE OF MAXIMUM POSITIVE SPATIAL AUTOCORRELATION

The eigenvalues associated with the range of possible orthogonal map patterns for Puerto Rico, extracted from equation (1.10), are presented in Table 1.1 (§1.5). The largest of these eigenvalues, 5.20062, has an eigenvector with $\bar{x} = 0$ and $s = \dfrac{1}{\sqrt{n-1}}$ ≈ 0.11785. The Moran Coefficient for this eigenvector is 1.10362 (Table 2.1), with a corresponding GR = 0.12662. MC exceeds 1, as explained in §1.2; this value essentially is determined by the largest eigenvalue of matrix **C**. This principal eigenvector also displays a conspicuous linear trend across the island of Puerto Rico (§1.2), as is apparent by inspecting Map 1.4a. This linear trend may be described by the regression equation, where (U, V) denotes the georeferencing Cartesian coordinates,

$$\hat{E}_1 = 0.19503 - 0.03283 \times (U/1000) + 0.00357 \times (V/1000) \, , \, R^2 = 0.557 \, . \, (2.4)$$

This map pattern may be described as a linear-type gradient that increases from a sink in the mideastern part of the island to a peak in the midwestern part of the island, a general pattern captured by the linear trend surface regression model. But this gradient also slopes down around the entire edge of the island, much like a tablecloth draping over an incline plane. This edge effect is what deteriorates the linear statistical description. These two tendencies can be identified in a semivariogram plot for these data (Figure 2.11). Meanwhile, the residuals from this regression yield

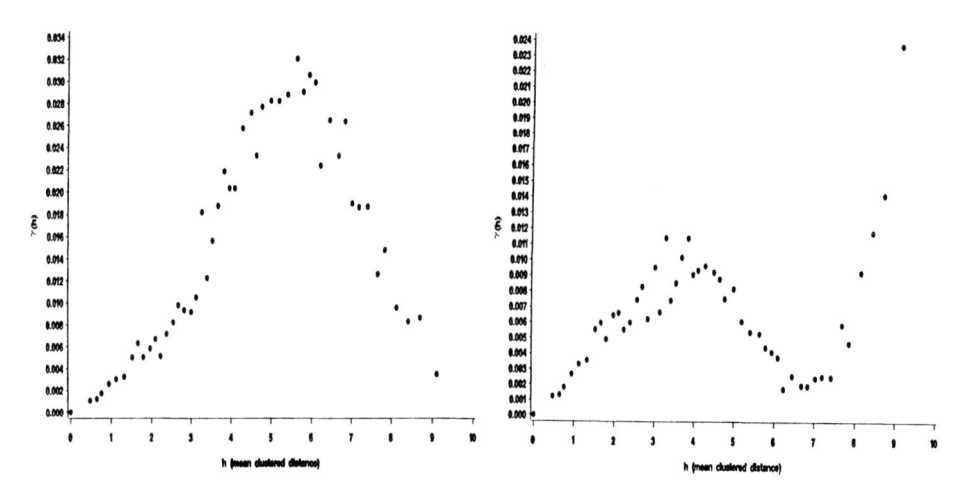

Figure 2.11. Left: (a) Semivariogram plot for eigenvector #1 of matrix expression (1.10) constructed for Puerto Rico (Map 1.1a). Right: (b) Semivariogram plot for residuals from fitting a linear trend surface model to eigenvector #1 portrayed in Figure 2.11a.

$$MC = 0.83305, \ GR = 0.28824$$
$$A\text{-}D = 0.347, \ R\text{-}J = 0.9939, \ K\text{-}S = 0.084.$$

The decrease of 0.27057 in MC between the original eigenvector and the residual values is attributable to the presence of a nonconstant mean across the geographic landscape. In other words, this particular linear trend component of the map pattern accounts for roughly 25% of the original MC value. Normality diagnostics suggest that the residuals from this regression adequately conform to a Gaussian frequency distribution.

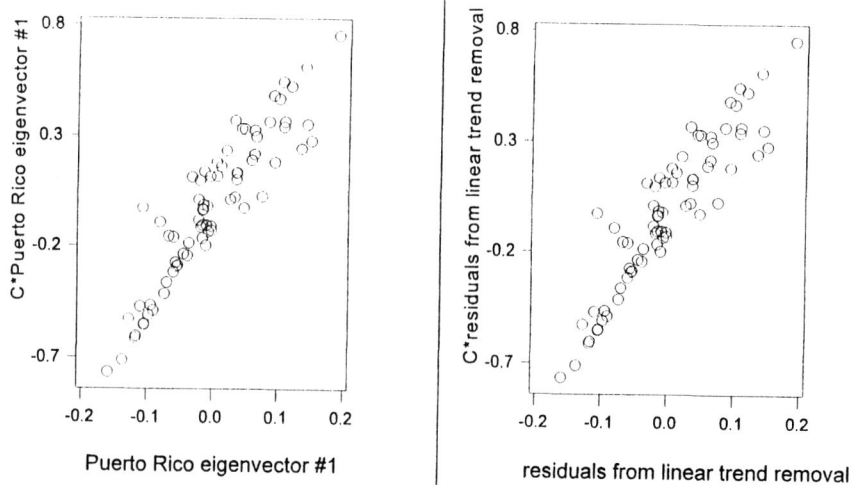

Figure 2.12. Left: (a) MC scatterplot for eigenvector #1 of matrix expression (1.10) constructed for Puerto Rico (Map 1.1a). Right: (b) MC scatterplot for residuals from fitting a linear trend surface model to eigenvector #1 portrayed in Figure 2.11a.

The accompanying MC scatterplots (Figure 2.12) provide evidence of marked positive spatial autocorrelation in this particular geographic distribution. The nature and degree of this specific relationship is not a function of the nonconstant mean across the geographic landscape.

The corresponding semivariograms in Figure 2.11 also suggest the presence of strong positive spatial autocorrelation in this eigenvector's geographic distribution. This interpretation is based upon the visual observation of a noticeable, marked positive relationship between $\gamma(h)$ and distance appearing in the first half of each plot. The exponential-like increase displayed by the first half of the graph in Figure 2.11a, which is absent in Figure 2.11b, is attributable to a nonconstant mean across the island; this variable mean is principally the trend surface gradient detected with regression equation (2.4). This nonlinearity is removed with OLS regression analysis, leaving residuals that display positive spatial autocorrelation. There is a decline following an initial incline in both plots due to the edge effects present on the map surface (Map 1.4b). A subsequent increase of semivariogram values is seen after the decline in Figure 2.11b, which may be attributed to two factors. As mentioned, fewer pairs of points are available at these greater distances. In addition, trend surface models provide very poor fits for attribute values located along the edges of geographic regions (because $\bar{u}$ and $\bar{v}$ are located in the middle of the map, the best regression fits also will tend to occur in the center of the map). Hence, the absence of this second increase of semivariogram values in Figure 2.11a presumably is due to a dampening effect from the linear trend that is present across the island; its appearance in Figure 2.11b may well be due to a magnification of edge effects resulting from the fitting of a trend surface to the geographic distribution.

2.7. Exploring Residuals in Spatial Analysis

Thomas's (1968) seminal piece outlines how mapping residuals from regression can be used to foster more informed and comprehensive spatial analysis. Extending this argument in the context of Thomas's missing variables discussion, regression residuals analysis can be used to identify ideal map patterns that capture latent spatial autocorrelation and to compute a model-based estimate of the standard error of spatial autocorrelation. For this second use of residuals, the objective is to acquire a better understanding of the sampling distribution of statistics, such as MC. This goal may be achieved through Monte Carlo simulation, the bootstrap technique (Efron 1982), or other resampling procedures.

For benchmark purposes, consider the analytical standard error of MC in the presence of 0 spatial autocorrelation and assume an underlying normal frequency distribution, which is $\sigma_{MC} = 0.07542$ (Cliff and Ord 1981, 44); its asymptotic counterpart approximation is given by $\sqrt{\dfrac{2}{\sum_{i=1}^{n}\sum_{j=1}^{n} c_{ij}}} = 0.07625$, which is almost exactly the same value. In addition, consider the analytical standard error of MC based on randomization (all n! possible permutations of a given set of distinct x_is over a partitioned surface). The formula for this standard error, given the Puerto Rico surface partitioning, is

$$\sqrt{\frac{244299456 - \frac{m_4}{m_2^2} \times 3273776}{42345353216}} - \frac{1}{72^2} \quad \text{(Cliff}$$

and Ord 1981, 46), where m_j is the j-th sample moment. The standard errors rendered by this formula for both the raw and the power-transformed attributes are as follows:

variable	untransformed		transformed	
	$\frac{m_4}{m_2^2}$	σ_{MC}	$\frac{m_4}{m_2^2}$	σ_{MC}
TDFRM	2.65589	0.07329	2.50670	0.07337
TDCD	3.80940	0.07268	3.75060	0.07271
TDMLK	10.92766	0.06879	2.49942	0.07337
TDSGR	10.92803	0.06879	2.37307	0.07344
TDCF	5.65428	0.07169	1.94972	0.07366
TDTB	15.83389	0.06597	2.07268	0.07359
TDBN	7.97000	0.07043	2.59516	0.07332
TDFAM	4.66840	0.07226	2.69661	0.07327

If a variable is perfectly normally distributed, $\frac{m_4}{m_2^2} = 3$ and $\sigma_{MC} = 0.07311$. In other words, regardless of whether or not the data are transformed or the inferential basis is built upon randomization, the resulting standard errors are very close to the asymptotic normal approximation one of 0.07625. But this standard error of MC is for the case of 0 spatial autocorrelation. What happens in the more common case of non-0 spatial autocorrelation? An answer to this question is sought in the next section.

2.7.1. A Monte Carlo Estimation of the Standard Error of MC

The name Monte Carlo simulation, derived from the European gambling resort, appropriately conveys an element of randomness and an element of mock-up. This approach uses a computer to create many samples of artificial data (i.e., pseudo-random numbers[1]) drawn from a selected hypothetical population/frequency distribution invented by the spatial analyst (and hence known to be true). A sampling distribution is emulated by calculating a statistic for each of the fabricated samples. Virtually any sampling distribution can be simulated if its properties are well understood. Arithmetic means of empirical sampling distribu-

---

[1] The generators of these artificial numbers are equations that can produce large amounts of random data from any theoretical distribution.

tions constructed by Monte Carlo simulation provide a way to assess statistical unbiasedness, whereas their standard deviations provide a way to assess statistical efficiency. One classical use of Monte Carlo methods is to simulate the sampling distribution for MC given 0 spatial autocorrelation (the randomization perspective described in the preceding section). The residuals of concern here equal the difference between each pseudorandom number and its sample mean—the constant mean model specification. Theoretical properties of MC become less clear, particularly when n is small, if the pseudorandom numbers fail to be randomly allocated to those areal units on a map. Monte Carlo results reported in this section are based upon $\Re$ (denoting the number of replicates) = 100 samples of size n = 73 drawn from a normal distribution, allocating these values according to their rankings in the rank order of the observed attribute values to the areal units on the map of Puerto Rico, and then calculating MC for each of these allocations (following the steps listed in §1.5). Essentially, this Monte Carlo simulation combines the properties of the observed level of spatial autocorrelation and a normal frequency distribution (which is mimicked by applying the appropriate power transformation to the original attribute). These results apply to either the original raw data or the power transformed data, because a power-transformation does not alter the rank order of attribute values (i.e., it is a monotonic function).

For the Puerto Rican agricultural productivity data, Monte Carlo results are

attribute	normal distribution s_{MC}	$\bar{x}_{MC}$	raw data MC	transformed data MC
TDFRM	0.00898	0.35613	0.34842	0.36534
TDCD	0.01867	0.05172	0.10867	0.09042
TDMLK	0.01559	0.34864	0.20054	0.34323
TDSGR	0.02832	0.52241	0.51783	0.54629
TDCF	0.02848	0.36465	0.31060	0.43810
TDTB	0.02232	0.47660	0.43771	0.48928
TDBN	0.01220	0.37950	0.35371	0.39483
TDFAM	0.02035	0.37743	0.36528	0.38327

These simulation results render average MC values that are closer to those for the transformed data, as should be expected, because sampling is from a normal distribution. They also suggest that the standard error may be related to the underlying level of spatial autocorrelation, that far more replications than 100 are needed to obtain sound Monte Carlo results, and that a better understanding of georeferenced data properties is needed if a simple Monte Carlo estimate of the standard error of MC is to be attained. The discrepancies between these standard errors and the analytical ones are too large.

A Monte Carlo simulation experiment can be designed in a way that reflects the modeling format underlying §1.2. Suppose n *iid* values ϵ_i are drawn from a normal distribution having mean μ and variance σ^2. A covariance matrix containing a prespecified level of spatial autocorrelation, ρ, can be formulated and used to construct a georeferenced attribute vector $\mathbf{X} = (\mathbf{I} - \rho\mathbf{W})^{-1}\epsilon$ (Haining,

Griffith and Bennett 1983).[2] Matrix $(I - \rho W)^{-1}$ does not change from replication to replication, when a new sample of ϵ_i values is drawn, in this case from a normal distribution, for each replication. For the Puerto Rico agricultural productivity data, Monte Carlo results for this experiment, again conducting 100 replications, are

ρ	$\bar{x}_{MC}$	s_{MC}	ρ	$\bar{x}_{MC}$	s_{MC}
0.0	-0.02512	0.07834	0.6	0.32579	0.12637
0.1	0.02542	0.07427	0.7	0.40808	0.11677
0.2	0.06531	0.08786	0.8	0.52403	0.11579
0.3	0.10998	0.09408	0.9	0.64761	0.10518
0.4	0.18659	0.08456	0.99	0.83403	0.11055
0.5	0.24219	0.10587			

This second set of results corroborates the first set in part by also suggesting that the standard error may be related to the underlying level of spatial autocorrelation and that far more replications than 100 are needed to obtain sound Monte Carlo results (for $\rho = 0$, $\bar{x}_{MC}$ should be closer to -0.01389 and s_{MC} should be closer to 0.07542).

One important finding from these Monte Carlo simulation experiments is that the use of analytical standard errors for MC when non-0 spatial autocorrelation is present in georeferenced data might introduce some specification error into a spatial analysis.

2.7.2. A BOOTSTRAP ESTIMATE OF THE STANDARD ERROR OF MC

Mathematical reasoning based upon a set of posited assumptions yields theoretical sampling distributions for many classical statistics, ones whose properties are heavily dependent upon these assumptions (the model-based approach to statistical inference). As mentioned in Chapter 1, assumptions such as 0 spatial autocorrelation and a normal frequency distribution are made *only* for mathematical convenience. Without these assumptions, mathematical reasoning may become infeasible. The resampling technique called bootstrapping, whose name is reminiscent of the phrase "pulling oneself up by one's bootstraps," replaces such knowledge with understanding based upon brute force computing, just as Monte Carlo simulation does. Experience shows that it offers a superior alternative to analytical results when the usual underlying statistical assumptions seem dubious; it is a powerful statistical tool that liberates spatial analysts from the restrictions of unrealistic assumptions. Once again the sampling distribution is simulated but now the real data at hand are used to draw random samples of size n (with replacement) from a set of *iid* values that constitute the sample, without presuming that a true data model is known. No specific form of the parent frequency distribution need be assumed, although the observed sample is expected to be representative of the

---

[2] This specification of the variance-covariance matrix is based upon the simultaneous autoregressive (SAR) model, which is presented in Chapter 3.

population. Constant mean and variance and 0 spatial autocorrelation are assumed for the sample that is being reused to select the bootstrap samples. These properties can be induced in an attribute by power transformations coupled with a regression employing a composite variable that yields residuals with 0 spatial autocorrelation; these surrogate variables can be the eigenvectors of matrix **C** [or more precisely as transformed expression (1.10)] discussed in §1.4.2. More specifically,

$$Y = E\beta + \epsilon \rightarrow b = (E^TE)^{-1}E^TY = E^TY ,$$

indicating—as a parallel to the handling of redundant information by principal components analysis—that geographically orthogonal components of information are being extracted (E^TY) and then used to predict spatially autocorrelated attribute values (**Y**).[3] The bootstrap actually is applied to the residuals from this regression, whose resampled values then are added to the predicted regression values to obtain replicate areal unit attribute values; failing to remove spatial autocorrelation effects first can render misleading bootstrap results. Variability across these bootstrap samples serves as a substitute for sample-to-sample differences that would be obtained if the parent population were repeatedly sampled. Practice indicates that, for the bootstrap, $\Re$ should be somewhere between 50 and 200; for standard error estimates, $\Re=100$ usually is adequate.

Returning once again to the Puerto Rico agricultural productivity example, bootstrapping of the individual attribute variables yields the following results:

attribute	s_{MC}	$\bar{x}_{MC}$	E(MC)	residual MC	transformed data MC
TDFRM	0.04585	0.34484	-0.05949	0.04946	0.36534
TDCD	0.07679	0.05021	-0.02578	0.02578	0.09042
TDMLK	0.07013	0.37792	-0.07789	-0.10211	0.34323
TDSGR	0.03740	0.56420	-0.11171	-0.07255	0.54629
TDCF	0.06864	0.39720	-0.08870	0.07394	0.43810
TDTB	0.03334	0.50380	-0.10188	-0.07972	0.48928
TDBN	0.03760	0.38547	-0.10564	0.04158	0.39483
TDFAM	0.09443	0.41951	-0.05815	-0.06654	0.38327

These results disclose that the standard error of MC may well vary even for a fixed geographic surface partitioning, that those standard errors of MC suggested by Monte Carlo simulation seem too large or too small, and that spatial autocorrelation often can be effectively, although artificially, filtered out of regression residuals by including judiciously selected eigenvectors in the mean response specification (all of the residual MC values are considerably closer to their respective expected values).

The regression equations supporting these bootstrap results utilize the

---

[3] Getis (1990, 1995) proposes an alternative spatial filtering procedure that also could be used here.

following sets of eigenvectors as regressors:

attribute	eigenvectors (in order of importance)	R^2
TDFRM	E_{12}, E_{11}, E_5, E_9, E_{10}	0.6326
TDCD	E_4	0.0804
TDMLK	E_4, E_{14}, E_1, E_{12}, E_{16}, E_{10}, E_{17}	0.5901
TDSGR	E_{12}, E_1, E_4, E_2, E_{11}, E_{18}, E_{19}, E_{14}, E_{23}, E_5	0.7609
TDCF	E_6, E_1, E_2, E_{10}, E_{14}, E_{16}, E_{21}, E_{20}	0.5390
TDTB	E_{12}, E_1, E_9, E_{10}, E_8, E_6, E_{16}, E_{13}, E_{15}	0.7607
TDBN	E_{12}, E_{14}, E_{13}, E_{16}, E_6, E_2, E_{11}, E_{21}, E_7, E_{10}	0.7136
TDFAM	E_1, E_{12}, E_2, E_{16}	0.4527

These results reveal that the orthogonal and uncorrelated ideal spatial autocorrelation patterns latent in geographic connectivity matrix **C** [they actually have been extracted from matrix expression (1.10)] can account for large amounts of spatial autocorrelation through a mean response model specification. If spatial autocorrelation is relatively small (e.g., TDCD), these synthetic variates account for little variation in the response variable; if spatial autocorrelation is relatively large (e.g., TDSGR), these synthetic variates tend to account for a large amount of the variation in the response variable. Furthermore, using the expected value generalization of MC for the response variable Y, presented in §1.4.2, namely

$$E(MC_Y) = -\frac{1 + \sum_{j=1}^{p-1} MC_{X_j}}{n - p},$$

coupled with the eigenvalue results reported by Tiefelsdorf and Boots (1995), the expected values for these variables can be easily calculated directly from the standardized eigenvalues, and evaluated using the asymptotic analytical standard error approximation to calculate the margin of error. These standardized eigenvalue results are reported in Table 2.1 and are the converted eigenvalues reported in Table 1.1.

Only 23 eigenvalues are reported in Table 2.1. Because all of the Puerto Rican agricultural productivity variables display positive spatial autocorrelation, attention has been restricted to those 28 eigenvectors associated with positive spatial autocorrelation. This set was further reduced by focusing attention only to those eigenvectors exhibiting significant positive spatial autocorrelation according to a 95% z-score criterion (z = 1.96). Accordingly, the 23 eigenvectors were selected because they satisfy the inequality

$$MC \geq 1.96 \times (0.07542) - \frac{1}{73 - 1} \quad \text{or} \quad MC \geq 0.13393.$$

Corroborating examples of these sorts of bootstrap results can be found in Griffith and Amrhein (1997, 52-53).

2.8. OTHER STATISTICAL FREQUENCY DISTRIBUTION ASSUMPTIONS

As emphasized in §2.3, discussion in this book mostly focuses on either the univariate or the multivariate normal distribution. Ripley (1990, 10) notes that this emphasis at least in part arises from the assumption of a normal distribution rendering computable estimators: "although maximum likelihood estimation [for the auto-normal model] is almost always possible computationally, we must not assume that it is necessarily statistically desirable." Recognizing this contention, elementary statistics courses often stress that other frequency distributions can prove very useful as well when undertaking data analyses. The Poisson distribution, for example, supports analysis of events that are both discrete and rare, and often counts, such as points on a surface in point pattern analysis. Specifically, the nearest neighbor statistic (Griffith and Amrhein 1991, Chapter 5) allows the geographic distribution of sampling locations to be quantified for geographic coverage assessment purposes. One drawback of the spatial autoregressive Poisson model is that its normalizing constant is intractable (Cressie 1991, 462), which is not the case for the auto-normal model (§3.1.1, §3.1.2). Another is that its specification forces spatial dependency parameters to be negative (Besag 1974), allowing the presence only of negative spatial autocorrelation (e.g., competitive spatial processes) and not positive spatial autocorrelation (e.g., cooperative spatial processes). The mean response specification and approach outlined and promoted in §1.4.2 and §2.7.2 is one attempt to circumvent these two drawbacks.

The binomial distribution, a second popular alternative to the normal distribution, supports analysis of binary and percentage data and often can be directly related to the normal distribution through its logit specification, $LN\left[\dfrac{\dfrac{count_i}{total_i}}{1 - \dfrac{count_i}{total_i}}\right]$ for areal unit i, which, for example, can be used to analyze the gray levels on an image (Haining 1990, 100). One encumbrance of the auto-binomial model is that it has nonconstant variance; for areal unit i, the variance is given by $n_i \times (\dfrac{count_i}{total_i}) \times (1 - \dfrac{count_i}{total_i})$. As with the auto-Poisson model, another drawback of the spatial autoregressive binomial (i.e., auto-logistic) model is that its normalizing constant is intractable (Cressie 1991, 432), making exact maximum likelihood estimation cumbersome. Again the mean response specification and approach outlined and promoted in §1.4.2 and §2.7.2 attempts to circumvent this second drawback.

112   A CASEBOOK FOR SPATIAL STATISTICAL DATA ANALYSIS

2.8.1. REPLACING THE NORMAL FREQUENCY DISTRIBUTION ASSUMPTION WITH A POISSON ASSUMPTION

The Poisson frequency distribution often characterizes situations involving rare events, such as geographic distributions of points over a surface, flows of physical items or information among areal units, or disease incidences within areal units. One example of a georeferenced data analysis of this type is supplied by Haining (1981), who explores the frequency distribution of settlement data (houses per quadrat) collected by Dacey (1968). A similar data set is furnished by Matui (1968) and analyzed by Larsen and Marx (1986, 412-14); Matui even includes a density contour map of the region. These Japanese data are further explored here.

The study area is a 3 km-by-4 km rectangular region situated in a large alluvial plain west of Tokyo. A 30-by-40 rectangular grid was superimposed upon this region, partitioning its 12 square kilometers into 1,200 square quadrats. The locations of 911 houses then were georeferenced to this grid. The frequency distribution for this distribution of houses by quadrat is:

# of houses	0	1	2	3	4	5	6	7	9
# of quadrats	584	398	168	35	9	4	0	1	1

Larsen and Marx show that a chi-square test of goodness-of-fit to a Poisson distribution calculated for these data, with $\bar{x} = 0.75917$, results in a failure to reject the null hypothesis that they come from a population frequency distribution that conforms to a Poisson one. Notably, $s^2 = 0.89024$, which would yield as another test statistic, $\frac{\bar{x}}{s^2} = 0.85276$ ($H_0: \frac{\mu}{\sigma^2} = 1$). A spatial autocorrelation analysis for these data yields MC = 0.06014 (GR = 0.94660). As Griffith and Amrhein (1991, 109-11) underscore for the set of specialized classical correlation coefficients, the calculated MC here is valid, whereas its standard error is not the traditionally stated one. This lack of pronounced spatial autocorrelation is confirmed by inspection of a Moran scatterplot (Figure 2.13). The estimated standard error based upon randomization is $\sigma_{MC} = 0.02071$ (§2.7), which is very similar to the asymptotic approximation of $\sqrt{\dfrac{2}{\sum_{i=1}^{n}\sum_{j=1}^{n} c_{ij}}} = \sqrt{\dfrac{2}{4660}} = 0.02072$. This result implies that positive spatial autocorrelation is present in the population for which the map of houses by quadrat is a realization. But, clearly, the degree of this autocorrelation is rather trivial, being very close to 0. In other words, there seems to be no reason why these georeferenced data cannot be analyzed with conventional statistical techniques.

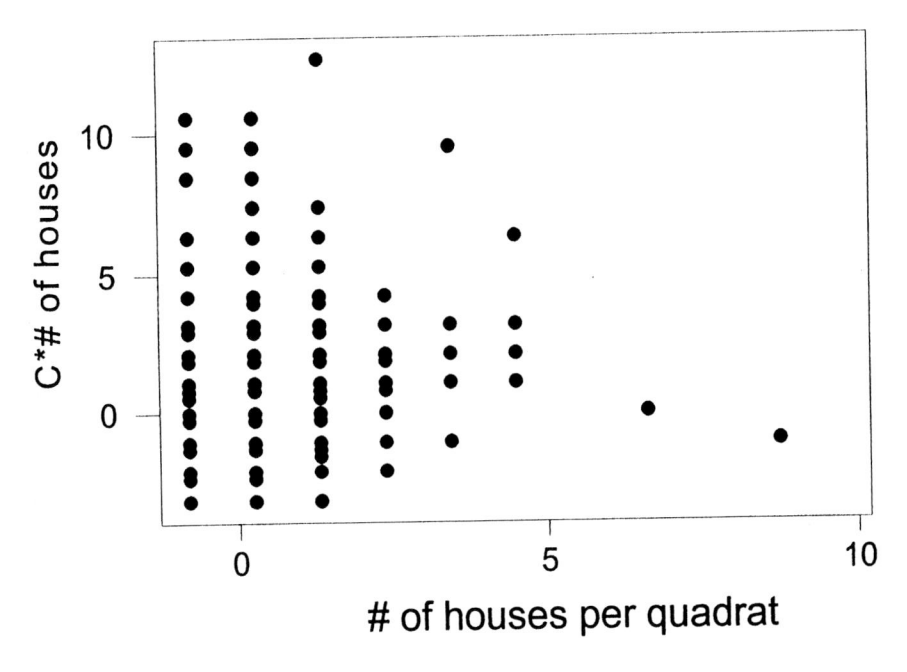

Figure 2.13. MC scatterplot for Matui's data.

2.8.2. REPLACING THE NORMAL FREQUENCY DISTRIBUTION ASSUMPTION WITH A BINOMIAL ASSUMPTION

Binary or percentage response variables often relate to a binomial distribution; even a multinomial distribution can be related to a binomial. Georeferenced data of this type include land use classifications, presence/absence of a disease case, and percent of population in a given ethnic group or age cohort. The example explored here is land use coverage from a 384-by-384 pixels (i.e., picture elements) remotely sensed image, which is included in the IDRISI tutorial, for an area west of Worcester, Massachusetts. The measurement of each coverage type is binary (a pixel classified as a given land use type is denoted by 1, and a pixel classified as any of the remaining 13 types is denoted by 0). As with the preceding Poisson illustration, the MC calculated here is valid, whereas its standard error is not the traditionally stated one. To evaluate the statistical significance of spatial autocorrelation in the geographic distribution of binary land use measures, then, z-scores of the join count statistics (z_{1-0} and z_{1-1}; Griffith 1987, 31-33) are computed.

The segment of the IDRISI image selected for analysis is a subregion 128-by-99 pixels in size (n = 128×99 = 12672), situated in the southeast section of the remotely sensed image. These pixels were classified into 14 different land use categories, with the following multinomial distribution:

114   A CASEBOOK FOR SPATIAL STATISTICAL DATA ANALYSIS

land use class	code	# pixels	% pixels	Z_{1-1}	Z_{1-0}	MC
Water	1	181	1.43	111.58	- 8.18	0.77215
Deciduous #1	2	1493	11.78	25.12	- 9.34	0.26768
Deciduous #2	3	1986	15.67	29.37	-14.94	0.34891
Deciduous #3	4	2975	23.48	35.10	-31.68	0.50769
Conifer #1	5	476	3.76	24.96	- 3.35	0.19534
Conifer #2	6	70	0.55	16.86	- 0.65	0.11031
Grass/Suburb	7	2945	23.24	21.23	-19.04	0.30673
Agriculture	8	187	1.48	28.24	- 2.07	0.19589
Urban Residential	9	497	3.92	29.88	- 4.27	0.23625
Urban Commercial	10	102	0.80	37.74	- 1.94	0.25115
Pavement #1	11	1384	10.92	28.06	- 9.77	0.29183
Pavement #2	12	216	1.70	33.37	- 2.67	0.23451
Gravel	13	101	0.80	51.02	- 2.61	0.33949
Barren	14	59	0.47	13.79	- 0.50	0.08978

While the joint frequency distribution of these land use categories is multinomial, each individual class, when contrasted with the remaining 13 categories, is a marginal binomial distribution. The 2 land use classes chosen for analysis here are Deciduous #3 and Grass/Suburb, because each covers nearly 25% of the total area under study. The geographic distributions of these land use classifications of pixels appear in Map 2.2a-b.

Maximum likelihood estimation based on logistic regression yields for Deciduous #3,

parameter	df	parameter estimate	standard error	Wald chi-square	Pr > chi-square
b_0	1	1.3323	0.0237	3154.9888	0.0001
b_u	1	0.0121	0.0006	389.1547	0.0001
b_v	1	0.0223	0.0008	741.7068	0.0001

where Pr denotes probability. While a geographic trend is detected across this landscape, the accompanying pseudo-R^2 value[4] equals 0.098, suggesting that it is not too pronounced. Removal of this geographic trend reduces MC from 0.50769 (GR = 0.49272) to 0.45482 (GR =0.54612). In other words, only a small part of the detected spatial autocorrelation can be attributed to map pattern.

Meanwhile, maximum likelihood estimation based on logistic regression yields for Grass/Suburb,

parameter	df	parameter estimate	standard error	Wald chi-square	Pr > chi-square
b_0	1	1.2642	0.0223	3213.3313	0.0001
b_u	1	-0.0086	0.0006	211.9574	0.0001
b_v	1	-0.0143	0.0008	340.9221	0.0001

---

[4] In keeping with Buse (1973). Upton and Fingleton (1985, 298) note that this equals the squared correlation between Y and $\hat{Y}$.

Logistic Regression

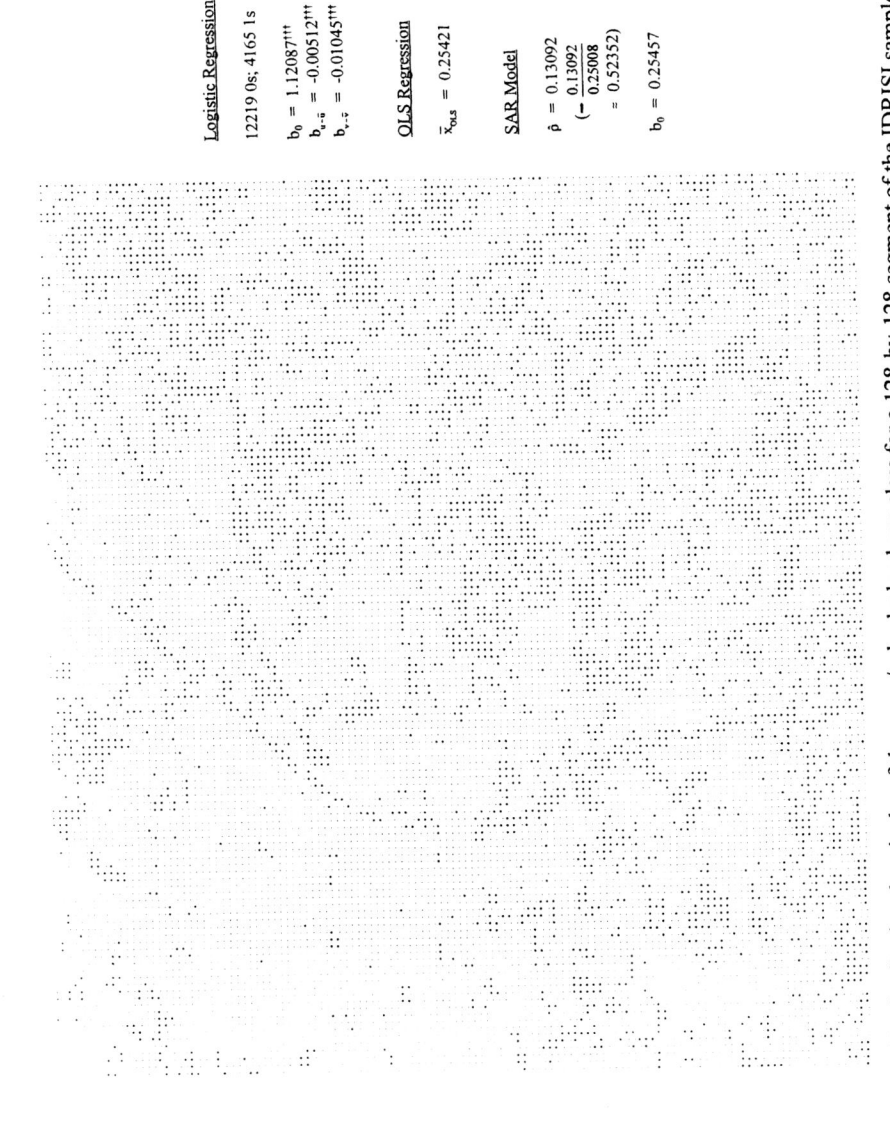

12219 0s; 4165 1s

$b_0 = 1.12087^{\dagger\dagger\dagger}$
$b_{u-\bar{u}} = -0.00512^{\dagger\dagger\dagger}$
$b_{v-\bar{v}} = -0.01045^{\dagger\dagger\dagger}$

OLS Regression

$\bar{x}_{OLS} = 0.25421$

SAR Model

$\hat{\rho} = 0.13092$
$\left(\dfrac{0.13092}{0.25008} = 0.52352\right)$

$b_0 = 0.25457$

Map 2.2a. Geographic distribution, by pixels, of the grass/suburbs land use class for a 128-by-128 segment of the IDRISI sample remotely sensed data for the Boston, Massachusetts, region.

Logistic Regression

13108 0s; 3276 1s

$b_0 = 1.60994$†††
$b_{u-\bar{u}} = 0.01259$†††
$b_{v-\bar{v}} = 0.02071$†††

OLS Regression

$\bar{x}_{OLS} = 0.19995$

SAR Model

$\hat{\rho} = 0.17854$
$\quad\quad 0.17854$
$(-\,\overline{\quad 0.25008\quad}$
$\quad\quad = 0.71394)$

$b_0 = 0.19865$

Map 2.2b. Geographic distribution, by pixels, of the deciduous #3 land use class for a 128-by-128 segment of the IDRISI sample remotely sensed data for the Boston, Massachusetts, region.

Again, while a geographic trend is detected across this landscape, the accompanying pseudo-R^2 equals 0.042, suggesting that it is not too pronounced. Removal of this geographic trend reduces MC from 0.30673 (GR = 0.69194) to 0.27664 (GR = 0.72197). Once more, only a small part of the detected spatial autocorrelation can be attributed to map pattern.

Although MC can be computed for binary variables, as mentioned, statistical interpretation of the resulting numerical value is more difficult. Just as the phi coefficient is the interpretable correlation coefficient for two binary variables in classical statistics (Amrhein and Griffith 1991, 267-73), the join count statistics are the appropriate spatial autocorrelation indices for binary georeferenced data. The foregoing summary tabulations for the 14 and use classes includes both the z-scores for the "1-1" (traditionally labeled WW) and "0-1" (traditionally labeled BW) joint count statistics, as well as the accompanying MC values. These values may be related with the following regression equation:

$$\hat{MC} = 0.00848 + 0.00614 \times z_{1-1} - 0.00936 \times z_{1-0}; \quad R^2 = 0.996.$$

(Note: none of the regression model assumptions are satisfied here, and n = 14 is a very small sample size.) This approximate rule of correspondence suggests that the foregoing MC values for Deciduous #3 and Grass/Suburb are, indeed, significant.

The preceding pair of logistic regression result is very disappointing. Not only does the mean vary across the IDRISI scene, but so does the variance. Aggregating these binary data into 12 regions of size 32-by-33 pixels ($n_i = 1,056$) yields

513	329	159
399	417	105
277	222	99
286	56	113

frequency of Deciduous #3 pixels

0.250	0.214	0.128
0.235	0.239	0.090
0.194	0.166	0.090
0.197	0.050	0.096

associated binomial variance estimates

These results reinforce the viewpoint that the classified remotely sensed data are messy (e.g., nonlinear) as well as noisy (e.g., containing various types of error).

Only the first eigenvector of expression (1.10) exhibits a marked relationship with these classification results. It does not, however, account for much of the detected level of spatial autocorrelation in either of these geographic distributions, negligibly reducing the MCs, respectively, to 0.45389 and 0.27282 and negligibly increasing the pseudo-R^2s, respectively, to 0.099 and 0.047. Although the performance of these eigenfunction spatial autocorrelation filters is disappointing in this case, it is consistent with the notion of a nonhomogeneous geographic process operating over the IDRISI scene, which should not be unexpected.

These findings suggest that a serious analysis of remotely sensed data should (1) work directly with the reflectance values prior to classification (see Chapter 5), (2) zoom in on smaller regions that geographically are more homogeneous, or (3) aggregate pixels into larger groups (i.e., change the scale of analysis). Consider, for example, the 32-by-33 upper left-hand quadrat subregion of the IDRISI scene, which has a probability of 513/1056 of having a pixel classified as Deciduous #3. This is a more homogeneous segment of the scene—its four quadrants have variances of, in a clockwise ordering, 0.152, 0.156, 0.244, and 0.250. Spatial autocorrelation indices calculated for these raw data include MC = 0.49553 (GR = 0.50408), z_{1-1} = -34.88, and z_{1-0} = -151.12. The Cartesian coordinates coupled with 28 of the geographic connectivity matrix eigenvectors yield logistic regression results that include a pseudo-R^2 = 0.392, and for the misclassified pixels for this model, MC = 0.18707 (GR = 0.81249). In other words, these regressors reduce the spatial autocorrelation in the residuals by a substantial amount, as is expected. There is a reasonable chance that this residual spatial autocorrelation is related to pixel misclassification and/or the underlying heterogeneity of land use types that is overlooked when a pixel is classified as being of a single type—again, a reason for reverting to original reflectance measures.

The original 128-by-99 IDRISI scene investigated here also can be aggregated in a 16-by-11 coarser scale surface partitioning, where each aggregation comprises an 8-by-9 compact set of original pixels. Percentages of each set of 72 pixels that are classified as Grass/Suburb, for example, then can be analyzed. The Cartesian coordinates coupled with 24 of the geographic connectivity matrix eigenvectors yield logistic regression results that include a pseudo-R^2 = 0.541 and a reduction from 0.50385 to 0.16053 in the MC (accompanied by an increase from 0.49545 to 0.82182 in the GR). In other words, these regressors reduce the spatial autocorrelation in the residuals by a substantial amount, as is expected. The aggregated pixels yield a MC value that is substantially larger than the MC value for the detailed binary georeferenced data set. As before, one may well anticipate that the residual spatial autocorrelation arises from an error process.

This last result illustrates an important feature of remotely sensed data. A researcher needs to temper this aggregation approach with an understanding that many statistical inference problems are associated with aggregating georeferenced data (e.g., the modifiable areal unit problem [MAUP]; see Amrhein 1993; Amrhein and Reynolds 1996, 1997; Reynolds and Amrhein 1998). Nevertheless, as is the case with producing histograms and dot plots constructed for conventional univariate statistical analyses, remotely sensed images may be too detailed, thus masking pattern (similar to many dotplots). By aggregating pixels (similar to establishing intervals for constructing histograms), obscuring detail can be averaged out, allowing pattern to be uncovered.

3

POPULAR SPATIAL AUTOREGRESSIVE AND GEOSTATISTICAL MODELS

Parallels between spatial autoregression and geostatistics are outlined in §1.2. To recapitulate, discussions of spatial series tend to focus on either spatial autocorrelation (geostatistics) or partial spatial autocorrelation (spatial autoregression). Although these two subfields have been evolving autonomously and in parallel, they are closely linked. Recalling Figure 1.1, both spatial statistics and geostatistics fall under the heading of multivariate analysis. The former frequently is concerned with aggregations of phenomena into discrete regions, while the latter principally is concerned with more or less continuously occurring attributes. By exploiting latent spatial autocorrelation in georeferenced data, spatial autoregression usually seeks to enhance statistical description and increase statistical precision, whereas geostatistics often seeks to generate spatial predictions. The prediction focus of the latter provides a link between these two subdisciplines through the missing data problem. The exact algebraic correspondence established with this spatial prediction/missing data linkage emphasizes that spatial autoregression directly deals with the inverse-variance-covariance matrix while geostatistics directly deals with the variance-covariance matrix itself (§1.1).

Historically, the initial focus of spatial autocorrelation analysis paralleled that of classical descriptive statistics. Can its nature and degree be quantified meaningfully for a given georeferenced data set? Once basically valid, accurate, and precise quantification occurred, repeatedly, moderate levels of positive spatial autocorrelation were uncovered. This general revelation naturally refocused attention on inference and diagnostics, and then on modeling. Today, spatial statistical modeling efforts focus on specification, estimation, and diagnostics of models for the inverse-variance-covariance matrix when in the realm of spatial autoregression, and for the variance-covariance matrix when in the realm of geostatistics. These efforts relate to incorporating spatial autocorrelation into a parameterization of these variance-covariance matrices, as well as into the accompanying mean response specifications.

3.1. SPATIAL AUTOREGRESSIVE MODELS

Just as the regression technique is the workhorse of data analysis in general, nonlinear regression is the workhorse of spatial autoregressive data analysis. Statistical methodology for spatial autoregressive modeling of georeferenced data, in which complexities due to the presence of spatial autocorrelation lie dormant, has only recently begun to reach the level of sophistication achieved early in this century in traditional multivariate statistical analysis and just several decades ago in time series analysis. Much of this methodological development has been accomplished by rewriting conventional statistical techniques as regression models, including a spatial autocorrelation parameter, and then estimating parameters with generalized least squares procedures. As is done in factor analytic modeling, the specification and estimation of some n-by-n matrix **L** is sought to orthogonalize the redundant statistical information contained in georeferenced data. Matrix **L** then is used in the rewriting of conventional linear statistical models.

Suppose areal units can be grouped together according to the amount of redundant information they contain, this duplicate information being embedded in their respective attribute values by some spatial autocorrelation process (e.g., a diffusion process, competition for space, or other forms of spatial interaction). Clearly these groups will be overlapping. Then all areal units within a particular group are highly correlated among themselves but have relatively small correlations with areal units in disjoint groups. The role of matrix **L** is to determine particular, uncorrelated linear combinations of the original n areal unit attribute values in such a way that any latent redundant statistical information is accounted for only once. Matrix **L** is concerned with statistically explaining the variance-covariance matrix for n areal units and has a parameterization that relates directly to the Moran Coefficient (MC). Synthetic attribute values rendered by matrix **L** are expected to display *iid* (independent and identically distributed) variability across a geographic landscape, whereas the commonality component of attribute values that is filtered out of georeferenced data by matrix **L** is attributed to some underlying spatial autocorrelation process.

Suppose ρ is the spatial autocorrelation parameter; the n-by-n matrix $L_j(\rho)$ is constructed from a geographic weights matrix that represents the geographic configuration of areal units channeling spatial autocorrelation through a surface tessellation. Accordingly, a general spatial autoregressive model can be written as

$$\frac{L_1(\rho)Y}{EXP[J^{\frac{1}{2}}(\rho)]} = \frac{L_2(\rho)X\beta}{EXP[J^{\frac{1}{2}}(\rho)]} + \frac{L_3(\rho)\epsilon}{EXP[J^{\frac{1}{2}}(\rho)]} \qquad (3.1)$$

(see Griffith 1988b), where $EXP[J^{1/2}(\rho)]$ is a normalizing constant (i.e., the Jacobian term) that ensures that the total of the probabilities associated with the normal frequency distribution probability density function for this equation equals 1. The parameters of equation (3.1) must be estimated using nonlinear procedures because the spatial autocorrelation parameter ρ appears in both the numerator, as a multiplier of vector ß, and the denominator, as an argument of the normalizing constant, of this equation.

Instead of attempting to specify the operator matrix **L** directly in equation (3.1), we specify it indirectly through spatial interaction schemes. Three popular model specifications of the triplet of matrices $L_j(\rho)$ dominate the spatial autoregressive literature, namely the conditional autoregressive (CAR; first-order), the simultaneous autoregressive (SAR; second-order), and the autoregressive response (AR; second-order) models. Descriptions of these models are presented in Anselin (1988), Cressie (1991), Griffith (1988a), Haining (1990), Ripley (1988), and Upton and Fingleton (1985), among others. The most conspicuous algebraic difference between these three model specifications arises from how they respectively characterize the inverse-variance-covariance matrix. If the error structure serves as the vehicle by which spatial autocorrelation appears in georeferenced data and the geographic dependency effect affiliated with this spatial autocorrelation constitutes a relatively weak local (first-order in magnitude) spatial field, then

$$L_1(\rho) = I, \quad L_2(\rho) = I, \quad \text{and} \quad L_3(\rho) = E\Lambda^{-1/2} \text{ where } E\Lambda E^T = (I - \rho C),$$

where the n-by-n matrix **C** is the geographic contiguity matrix used to calculate the MC (§1.2). This specification renders the CAR model, which focuses on spatial autocorrelation latent in the error term of equation (3.1),

$$\frac{\Lambda^{1/2} E^T Y}{EXP\ [J^{\frac{1}{2}}(\rho)]} = \frac{\Lambda^{1/2} E^T X \beta}{EXP\ [J^{\frac{1}{2}}(\rho)]} + \frac{\epsilon}{EXP\ [J^{\frac{1}{2}}(\rho)]}. \quad (3.2)$$

Phenomena that appear to conform to this CAR specification include errors in remotely sensed images (e.g., the scattering of reflectance). The argument for its normalizing constant is given by

$$J^{\frac{1}{2}}(\rho) = \frac{1}{n} \sum_{i=1}^{n} LN(1 - \rho \lambda_i)^{1/2}, \quad (3.3)$$

where λ_i is the ith of the n eigenvalues of matrix **C**. These eigenvalues relate to orthogonalization of the georeferenced data (§1.5, §2.7.2). This CAR model has the appealing properties of

(1) being linked to a simpler, more natural Markov-type dependency

structure;
(2) conceptually being linked to the exponential semivariogram model (§3.2);
(3) yielding minimum variance interpolations of georeferenced attribute variable Y;
(4) being directly extendable to space-time model specifications; and
(5) achieving the maximum entropy among the set of all stationary models with a given finite set of variances and covariances.

If the error structure continues to serve as the vehicle by which spatial autocorrelation appears in georeferenced data and the geographic dependency effect affiliated with this spatial autocorrelation constitutes a relatively strong local (second-order in magnitude) spatial field, then

$$L_1(\rho) = I, \quad L_2(\rho) = I, \quad \text{and} \quad L_3(\rho) = (I - \rho W)^{-1},$$

where the n-by-n matrix W is the row-standardized version of the geographic contiguity matrix C (i.e., $w_{ij} = \dfrac{c_{ij}}{\sum_{j=1}^{n} c_{ij}}$). This specification renders the SAR model, which focuses on spatial autocorrelation latent in the error term,

$$\frac{Y}{EXP[J^{\frac{1}{2}}(\rho)]} = \frac{\rho W Y + (I - \rho W) X \beta}{EXP[J^{\frac{1}{2}}(\rho)]} + \frac{\epsilon}{EXP[J^{\frac{1}{2}}(\rho)]}. \quad (3.4)$$

Phenomena that appear to conform to its specification include multiple linear regression models misspecified because of missing variables and situations where measurement error exhibiting a geographic pattern is present (e.g., digital elevation models). The argument for its normalizing constant is similar to that given by expression (3.3) but with the term $(1 - \rho\lambda_i)$ raised to the power 1 instead of ½, where now λ_i is the ith of the n eigenvalues of matrix W. This SAR model has the appealing properties of

(1) always being able to be rewritten as a CAR specification (but a given CAR specification cannot necessarily be rewritten as an SAR specification);
(2) conceptually being linked to the Bessel function semivariogram model;
(3) being analogous to the way autoregressive models are formulated for time series analysis;
(4) yielding an unbiased ordinary least squares (OLS) estimator **b** and a relatively simple asymptotic covariance matrix; and
(5) having a more natural interpretation of the linear combinations of

georeferenced attribute values, because each areal unit value x_i is cast as a function of a weighted average of its juxtaposed areal unit values (for the CAR specification, each areal unit value x_i is cast as a conditional function of its juxtaposed areal unit values).

If the response variable itself serves as the vehicle by which spatial autocorrelation appears in georeferenced data, another specification becomes

$$L_1(\rho) = I, \quad L_2(\rho) = (I - \rho W)^{-1}, \text{ and } L_3(\rho) = (I - \rho W)^{-1},$$

rendering the AR model specification, which focuses on spatial autocorrelation latent in variable Y's values,

$$\frac{Y}{EXP\,[J^{\frac{1}{2}}(\rho)]} = \frac{\rho W Y + X\beta}{EXP\,[J^{\frac{1}{2}}(\rho)]} + \frac{\epsilon}{EXP\,[J^{\frac{1}{2}}(\rho)]}. \tag{3.5}$$

Phenomena that appear to conform to its specification include housing prices, median family income, and population density, as well as other socioeconomic attributes of areal units. The argument for its normalizing constant is similar to that given by equation (3.3); again the term $(1 - \rho\lambda_i)$ is raised to the power 1 rather than ½ now. This AR model has the appealing properties of

(1) a specification that is relatively simple to estimate; and
(2) conceptually being linked to the power semivariogram model ($\lambda \geq 1$).

One should keep in mind that the CAR model, equation (3.2), is more frequently utilized in remote sensing and image analysis, that the SAR model, equation (3.4), is more often used by spatial statisticians, and that the AR model, equation (3.5), is more frequently used by spatial econometricians. Rarely will statistical diagnostics suggest a preference for any of these specifications. In practice, selection should be based upon substantive, conceptual, or theoretical considerations.

3.1.1. IMPLEMENTING SPATIAL AUTOREGRESSIVE MODELS: PARAMETER ESTIMATION

Gaussianity has been central to spatial autoregressive analyses, in part because the accompanying probability density function yields tractable maximum likelihood estimation results and in part because of the versatility of regression. In general, the nonconstant mean and constant variance spatial autoregressive log-likelihood function based upon any inverse-covariance matrix $L^T L \sigma^{-2}$ may be written as

$$\text{constant} - \frac{n}{2} LN(\sigma^2) + \frac{1}{2}LN[\det(\mathbf{L}^T\mathbf{L})] - (\mathbf{Y} - \mathbf{X}\boldsymbol{\beta})^T\mathbf{L}^T\mathbf{L}(\mathbf{Y} - \mathbf{X}\boldsymbol{\beta})/(2\sigma^2) \quad (3.6)$$

where $\mathbf{L}^T\mathbf{L} = \mathbf{V}$ of §1.1. The normalizing constant (the Jacobian of the transformation from a spatially autocorrelated to a spatially unautocorrelated mathematical space), $\frac{1}{2}LN[\det(\mathbf{L}^T\mathbf{L})]$ in this case, complicates spatial autoregressive parameter estimation. Fortunately for the Gaussian distribution, it can be easily rewritten in terms of the eigenvalues of matrix $\mathbf{L}^T\mathbf{L}$ (Ord 1975), reinforcing the preceding linkage between spatial autocorrelation and map patterns displayed by the eigenvectors of matrix $\mathbf{C}$ (§1.5, §2.7.2). Griffith and Sone (1995) also propose a very good and considerably simplified approximation for this normalizing constant, one that facilitates the analysis of extremely large georeferenced data sets.

For the CAR spatial autoregressive model, matrix $\mathbf{L}^T\mathbf{L}$ in expression (3.6) becomes $(\mathbf{I} - \rho\mathbf{C})$, and the n eigenvalues for the normalizing constant are $(1 - \rho\lambda_i)$, i=1, 2, ..., n. Estimation of the parameters $\boldsymbol{\beta}$, σ^2, and ρ is complicated because there is no closed form estimator for the spatial autoregressive parameter ρ. The simplest way to estimate these three parameters is to use a nonlinear algorithm to solve simultaneously the set of first derivative equations, $\frac{\partial LN(L)}{\partial \beta_j} = 0$ (j=0, 1, .., p), $\frac{\partial LN(L)}{\partial \sigma^2} = 0$, and $\frac{\partial LN(L)}{\partial \rho} = 0$, where $LN(L)$ is the log-likelihood function:

$$LN(L) = \text{constant} - \frac{n}{2}LN(\sigma^2) + \frac{1}{2}LN[\det(\mathbf{I} - \rho\mathbf{C})] - \frac{(\mathbf{Y}-\mathbf{X}\boldsymbol{\beta})^T(\mathbf{I}-\rho\mathbf{C})(\mathbf{Y}-\mathbf{X}\boldsymbol{\beta})}{2\sigma^2}$$

$$\frac{\partial LN(L)}{\partial \boldsymbol{\beta}} = -2\mathbf{X}^T(\mathbf{I}-\rho\mathbf{C})\mathbf{Y} + 2\mathbf{X}^T(\mathbf{I}-\rho\mathbf{C})\mathbf{X}\boldsymbol{\beta} = 0$$

$$\rightarrow \mathbf{X}^T\mathbf{Y} = \mathbf{X}^T\mathbf{X}\boldsymbol{\beta} + \rho\mathbf{X}^T\mathbf{C}(\mathbf{Y}-\mathbf{X}\boldsymbol{\beta})$$

$$\frac{\partial LN(L)}{\partial \sigma^2} = -\frac{n}{2}\frac{1}{\sigma^2} - (-1)\frac{(\mathbf{Y}-\mathbf{X}\boldsymbol{\beta})^T(\mathbf{I}-\rho\mathbf{C})(\mathbf{Y}-\mathbf{X}\boldsymbol{\beta})}{2\sigma^4} = 0$$

$$\rightarrow \mathbf{Y}^T\mathbf{Y} = n\sigma^2 + 2\mathbf{Y}^T(\mathbf{I}-\rho\mathbf{C})\mathbf{X}\boldsymbol{\beta} - \boldsymbol{\beta}^T\mathbf{X}^T(\mathbf{I}-\rho\mathbf{C})\mathbf{X}\boldsymbol{\beta} + \rho\mathbf{Y}^T\mathbf{C}\mathbf{Y}$$

$$\frac{\partial LN(L)}{\partial \rho} = \frac{1}{2}\sum_{i=1}^{n}\frac{-\lambda_i}{1-\rho\lambda_i} - \frac{(\mathbf{Y}-\mathbf{X}\boldsymbol{\beta})^T(-\mathbf{C})(\mathbf{Y}-\mathbf{X}\boldsymbol{\beta})}{2\sigma^2} = 0$$

$$\rightarrow \mathbf{Y}^T\mathbf{C}\mathbf{Y} = 2\mathbf{Y}^T\mathbf{C}\mathbf{X}\boldsymbol{\beta} - \boldsymbol{\beta}^T\mathbf{X}^T\mathbf{C}\mathbf{X}\boldsymbol{\beta} + \sigma^2\sum_{i=1}^{n}\frac{\lambda_i}{1-\rho\lambda_i}.$$

In a pure spatial autoregressive situation, which would be equivalent to a constant

126 A CASEBOOK FOR SPATIAL STATISTICAL DATA ANALYSIS

mean model specification, for which $X\beta = \mu\mathbf{1}$, this triplet of equations reduces to

$$\sum_{i=1}^{n} y_i = n\mu + \rho(\sum_{i=1}^{n}\sum_{j=1}^{n} c_{ij}y_j - \mu\sum_{i=1}^{n}\sum_{j=1}^{n} c_{ij})$$

$$\sum_{i=1}^{n} y_i^2 = n\sigma^2 + 2\mu\sum_{i=1}^{n} y_i - n\mu^2 + \rho(\sum_{i=1}^{n} y_i\sum_{j=1}^{n} c_{ij}y_j - 2\mu\sum_{i=1}^{n}\sum_{j=1}^{n} c_{ij}y_j + \mu^2\sum_{i=1}^{n}\sum_{j=1}^{n} c_{ij})$$

$$\sum_{i=1}^{n} y_i\sum_{j=1}^{n} c_{ij}y_j = \sigma^2\sum_{i=1}^{n} \frac{\lambda_i}{1-\rho\lambda_i} + 2\mu\sum_{i=1}^{n}\sum_{j=1}^{n} c_{ij}y_j - \mu^2\sum_{i=1}^{n}\sum_{j=1}^{n} c_{ij} \quad ,$$

which by centering Y (i.e., replacing y_i with $y_i - \bar{y}$) further simplifies to

$$0 = n\mu + \rho(\sum_{i=1}^{n}\sum_{j=1}^{n} c_{ij}y_j - \mu\sum_{i=1}^{n}\sum_{j=1}^{n} c_{ij}) \tag{3.7a}$$

$$\sum_{i=1}^{n} y_i^2 = n\sigma^2 - n\mu^2 + \rho(\sum_{i=1}^{n} y_i\sum_{j=1}^{n} c_{ij}y_j - 2\mu\sum_{i=1}^{n}\sum_{j=1}^{n} c_{ij}y_j + \mu^2\sum_{i=1}^{n}\sum_{j=1}^{n} c_{ij}) \tag{3.7b}$$

$$\sum_{i=1}^{n} y_i\sum_{j=1}^{n} c_{ij}y_j = \sigma^2\sum_{i=1}^{n} \frac{\lambda_i}{1-\rho\lambda_i} + 2\mu\sum_{i=1}^{n}\sum_{j=1}^{n} c_{ij}y_j - \mu^2\sum_{i=1}^{n}\sum_{j=1}^{n} c_{ij} \quad . \tag{3.7c}$$

Efficient initial nonlinear iteration values here are $\mu_{\tau=0} = 0$ (because the variables have been centered), $\sigma_{\tau=0} = s$, and $\rho_{\tau=0} = b_{y|Cy}$. In general, then, the parameter estimates can be obtained by (1) shifting the X_j and Y data to means of 0 and then calculating an OLS estimate **b**, (2) using the accompanying MSE (mean squared error) as an estimate for σ^2, and (3) including vector **CY** in the regression to obtain an initial estimate for ρ. The nonlinear estimate for β then can be applied to each of the respective X_j variable means and added to $\bar{y}$ to compute the b_0 estimate for the original data. Both SAS (Appendix 3-A) and SPSS (Appendix 3-B) code for estimating equations (3.7a)-(3.7c) is listed at the end of this chapter.

For either the AR or SAR spatial autoregressive model, matrix L^TL in expression (3.6) becomes $(I - \rho C)^T(I - \rho C)$, or $(I - \rho W)^T(I - \rho W)$ if the row-standardized (stochastic) version of matrix **C** is preferred. But for the AR model, the likelihood expression (3.6) also is modified to

$$\text{constant} - \frac{n}{2}LN(\sigma^2) + \frac{1}{2}LN[\det(L^TL)] - (LY - X\beta)^T(LY - X\beta)/(2\sigma^2) \quad .$$

The simplest way to estimate the parameters now is to solve the pair of differential equations $\frac{\partial LN(L)}{\partial \beta_j} = 0$ (j = 0, 1, ..., p) and $\frac{\partial LN(L)}{\partial \sigma^2} = 0$, substitute the resulting estimators into the likelihood function, and then employ a nonlinear algorithm to optimize the reduced likelihood function. For the SAR model, employing matrix

W,

$$LN(L) = \text{constant} - \frac{n}{2}LN(\sigma^2) + \frac{1}{2}LN[\det(I - \rho W)^T(I - \rho W)]$$

$$- \frac{(Y - X\beta)^T(I - \rho W)^T(I - \rho W)(Y - X\beta)}{2\sigma^2}$$

$$\frac{\partial LN(L)}{\partial \sigma^2} = -\frac{n}{2}\frac{1}{\sigma^2} - (-1)\frac{(Y - X\beta)^T(I - \rho W)^T(I - \rho W)(Y - X\beta)}{2\sigma^4} = 0$$

$$\rightarrow \hat{\sigma}^2 = \frac{(Y - X\beta)^T(I - \rho W)^T(I - \rho W)(Y - X\beta)}{n},$$

Substituting this maximum likelihood estimate into the likelihood equation yields

$$L = \text{constant} \times EXP\left[-\frac{(Y - X\beta)^T(I - \rho W)^T(I - \rho W)(Y - X\beta)}{2\frac{(Y - X\beta)^T(I - \rho W)^T(I - \rho W)(Y - X\beta)}{n}}\right] \times$$

$$DET[(I - \rho W)^T(I - \rho W)]^{\frac{1}{2}} \times [\frac{(Y - X\beta)^T(I - \rho W)^T(I - \rho W)(Y - X\beta)}{n}]^{-\frac{n}{2}}$$

$$= (\text{constant}) \times EXP(-\frac{n}{2}) \times (\frac{1}{n})^{-\frac{n}{2}} \times [\prod_{i=1}^{n}(1 - \rho\lambda_i)^2]^{\frac{1}{2}}[(Y - X\beta)^T(I - \rho W)^T(I - \rho W)(Y - X\beta)]^{-\frac{n}{2}}$$

$$= (\text{constant}) \times \left[\prod_{i=1}^{n}(1 - \rho\lambda_i)\right]^{-\frac{2}{n}}[(Y - X\beta)^T(I - \rho W)^T(I - \rho W)(Y - X\beta)]\right]^{-\frac{n}{2}}.$$

Because the product $[\prod_{i=1}^{n}(1 - \rho\lambda_i)]^{-\frac{2}{n}}[(Y - X\beta)^T(I - \rho W)^T(I - \rho W)(Y - X\beta)]$ is raised to a negative power, maximization of the likelihood function L is achieved by minimizing this product. But this function, to be minimized, is equivalent to

$$\left[\frac{1}{e^{[\frac{1}{n}\sum_{i=1}^{n}LN(1 - \rho\lambda_i)]}}(I - \rho W)(Y - X\beta)\right]^T \left[\frac{1}{e^{[\frac{1}{n}\sum_{i=1}^{n}LN(1 - \rho\lambda_i)]}}(I - \rho W)(Y - X\beta)\right],$$

(where e is synonymous notation for *EXP*) which is the same as the regression equation given by equation (3.4). Accordingly, this substitution results in the following nonlinear regression model for the SAR autoregression specification:

$$Y/EXP[-\frac{1}{n}\sum_{i=1}^{n}LN(1-\rho\lambda_i)] = [\rho W + X\beta - \rho WX\beta]/EXP[-\frac{1}{n}\sum_{i=1}^{n}LN(1-\rho\lambda_i)] +$$

$$\epsilon/EXP[-\frac{1}{n}\sum_{i=1}^{n}LN(1-\rho\lambda_i)], \quad (3.8)$$

where the eigenvalues are from matrix **W**. Of course, ρ**C** and the eigenvalues of matrix **C** could be used instead. The AR nonlinear regression counterpart is

$$\mathbf{Y}/EXP[-\frac{1}{n}\sum_{i=1}^{n} LN(1-\rho\lambda_i)] = [\rho\mathbf{W} + \mathbf{X\beta}]/EXP[-\frac{1}{n}\sum_{i=1}^{n} LN(1-\rho\lambda_i)] +$$

$$\epsilon/EXP[-\frac{1}{n}\sum_{i=1}^{n} LN(1-\rho\lambda_i)] . \quad (3.9)$$

In both cases the denominator term, $EXP[-\frac{1}{n}\sum_{i=1}^{n} LN(1-\rho\lambda_i)]$, is the normalizing constant. Both SAS (Appendix 3-C) and SPSS (Appendix 3-D) code for estimating equations (3.8) and (3.9) is provided at the end of this chapter.

The nonlinear regression formulation of these two spatial autoregressive models also uncovers the interesting feature that a systematic set of ρ values spanning the feasible parameter space for ρ, namely $(1/\lambda_{min}, 1/\lambda_{max})$, can be sequentially substituted into either equation, each time followed by the execution of a conventional linear regression routine. The estimated value of ρ is that value for which the accompanying linear regression in this series has the smallest MSE. A nonlinear algorithm merely automates this sequence of linear regressions, being able to converge more efficiently on a more accurate estimate.

3.1.2. COMPUTING THE GAUSSIAN NORMALIZING CONSTANT

The first step in undertaking a spatial autoregressive analysis is to deal with the normalizing constant. The importance of this constant is illustrated in Griffith and Amrhein (1997, 278-79). The normalizing constant acts as a constraint on $\hat{\rho}$, safeguarding against physically meaningless spatial autocorrelation coefficient values (recall that the value of a correlation coefficient cannot exceed 1). This normalizing constant, operating much like a weight in weighted least squares (WLS), also ensures that the correct MSE curve is being evaluated. The smooth, systematic nature of the normalizing constant between the feasible parameter space limits of $\hat{\rho}$ is a property that can be exploited in spatial autoregressive analysis.

Figure 3.1 displays plots obtained with simulated data; 81 random normal values were selected from a normal distribution with $\mu = 0$ and $\sigma = 1$ (A-D = 0.583, R-J = 0.991, K-S = 0.090), and then geographically distributed across a 9-by-9 regular square tessellation. Spatial autocorrelation levels spanning the range of -1 to 1 were convoluted with these simulated values using an SAR model. As mentioned, the normalizing constant (J) has a well-behaved, smooth curve. The correct (SAR) mean squared errors has a markedly concaved upward curve, one that reaches a minimum very close to $\rho = 0$. But the incorrect (OLS) mean squared errors has a shallow curve, one that reaches a minimum near $\rho = -0.2$.

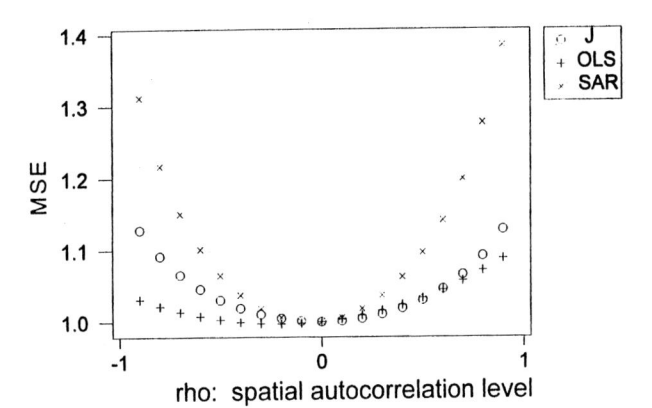

Figure 3.1. MSE curve plots and Jacobian curve plot for simulated data.

One instinctive and straightforward approach to the task of calculating equations (3.8) and (3.9) is to compute the n eigenvalues of matrix **C**. Several tricks can be employed to dupe a standard commercial statistical software package into thinking that it is dealing with a correlation matrix. One matrix that can be analyzed is $(\mathbf{I} + \mathbf{C})$, which then can be treated as a correlation matrix input by a factor analysis routine; the identity matrix **I** is added to the binary geographic weights matrix because each of the diagonal entries of a correlation matrix must equal 1 (i.e., any variable must perfectly and positively correlate with itself). This modification simply adds 1 to each of the resulting eigenvalues, which then can be subtracted from each of the calculated eigenvalues. If the geographic weights matrix **C** is converted to its row-standardized version, the asymmetric matrix **W**, the corresponding eigenvalue problem may be rewritten as follows:

$$\begin{aligned}DET\,[(\mathbf{I}+\mathbf{W}-\lambda\mathbf{I}] &= DET\,(\mathbf{I})DET\,[(\mathbf{I}+\mathbf{CD})-\lambda\mathbf{I}]\\ &= DET\,(\mathbf{D}^{\frac{1}{2}} \times (\mathbf{D}^{-\frac{1}{2}}DET\,[(\mathbf{I}+\mathbf{CD})-\lambda\mathbf{I}]\\ &= DET\,(\mathbf{D}^{\frac{1}{2}})DET\,[(\mathbf{I}+\mathbf{CD}-\lambda\mathbf{I}]DET\,(\mathbf{D}^{-\frac{1}{2}})\\ &= DET\,[(\mathbf{I}+\mathbf{D}^{\frac{1}{2}}\mathbf{CD}^{\frac{1}{2}})-\lambda\mathbf{I}]\\ &= 0\ .\end{aligned}$$

This version yields the correct eigenvalues of matrix **W**, again plus 1, but extracts them from an equivalent symmetric matrix.

The computation of eigenvalues for equations (3.8) and (3.9) is numerically intensive, and when n gets sufficiently large, this computational chore becomes infeasible, even with a supercomputer. To circumvent this restriction, and in doing so exploit the smooth, systematic curvature displayed by the normalizing constant (see Figure 3.1) across the range of possible ρ values, the approximation equation (3.10) has been formulated as a substitute for the normalizing constant i.e., the Jacobian term):

$$J^{\frac{1}{2}}(\rho) \approx -\frac{\rho(\alpha_{2,n}\lambda_{min}^2 - \alpha_{1,n}\lambda_{max}^2)}{2n}$$

$$\frac{\alpha_{2,n}|\lambda_{min}|LN\left(\frac{\gamma_{2,n}}{|\lambda_{min}|}\right) + \alpha_{1,n}\lambda_{max}LN\left(\frac{\gamma_{1,n}}{\lambda_{max}}\right)}{n}$$

$$-\frac{\alpha_{2,n}|\lambda_{min}|LN\left(\frac{\gamma_{2,n}}{|\lambda_{min}|} + \rho\right) - \alpha_{1,n}\lambda_{max}LN\left(\frac{\gamma_{1,n}}{\lambda_{max}} - \rho\right)}{n}$$

(3.10)

This equation can be calibrated directly with the eigenvalues of matrix **C** or **W** to obtain estimates of its four unknown coefficients $\alpha_{1,n}$, $\alpha_{2,n}$, $\gamma_{1,n}$, and $\gamma_{2,n}$; λ_{max} and λ_{min} are the extreme eigenvalues obtained from the preceding eigenfunction problem. Principal benefits and costs of employing equation (3.10) are outlined in Griffith and Sone (1995). A comprehensive treatment of eigenvalues for matrices **C** and **W** appears in Martin and Griffith (1995).

3.1.3.   A MEAN RESPONSE APPROACH

Cressie (1991, 428, 431) notes that determining the normalizing constant is an intractable problem for auto-binomial (or auto-logistic) and auto-Poisson models. One approach that attempts to circumvent this difficulty is alluded to in §2.7.2, namely, using the eigenvectors of the projected version[1] of matrix **C** as spatial autocorrelation surrogate variables in the mean response specification of a spatial statistical model. Because these eigenvectors can be linked directly to mutually orthogonal spatial patterns associated with different levels of spatial autocorrelation, they have a more natural interpretation.

Consider an auto-logistic model for the 1990 percentage of urban population in the municipios of Puerto Rico (see Griffith and Amrhein 1997, 307-8). The raw percentages (for which 0 was replaced by 1 in urban population counts data) yield MC = 0.32017 and GR = 0.67712, indicating a marked level of positive spatial autocorrelation (Figure 3.2). Converting these raw percentages to weighted log-ratios, namely,

---

[1] Recall that these eigenvectors asymptotically converge on those of matrix **C** once the principal eigenvector is replace with a vector of constants

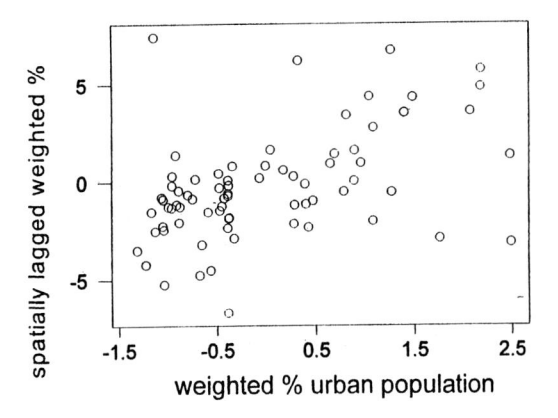

Figure 3.2. MC scatterplot for weighted percentage of urban population by municipio in Puerto Rico, 1990 (U.S. Census of Population data).

$$y_i = \sqrt{\text{total population}_i \times \frac{\text{urban population}_i}{\text{total population}_i} \times \left(1 - \frac{\text{urban population}_i}{\text{total population}_i}\right)} \times LN\left[\frac{\frac{\text{urban population}_i}{\text{total population}_i}}{1 - \frac{\text{urban population}_i}{\text{total population}_i}}\right],$$

to adjust for the heterogeneity latent in binomial variables as well as the restricted scale of 0%-100%, results in MC = 0.27177 and GR = 0.77024. The differences between these respective spatial autocorrelation indices would be attributable to specification error (i.e., nonconstant variance). Regressing this equation on the set of eigenvectors using a stepwise procedure resulted in the following sets of selected eigenvectors (which are uncorrelated by construction) as desirable regressors:

eigenvectors (in order of importance)	R^2	MC	GR
$E_{12}, E_4, E_{13}, E_1, E_6$	0.2979	0.07554	0.97236
$E_{12}, E_4, E_{13}, E_1, E_6, E_2$	0.3440	0.00721	1.03092

The expected value of MC for the six-regressor model is $-\dfrac{1+4.80363}{73-7} = -0.08793$

(see §1.4.2). Therefore, the six eigenvectors used in this regression effectively remove spatial autocorrelation from the data values, transforming their errors to unautocorrelated ones. Furthermore, normality diagnostics (A-D = 0.839, p = 0.029; R-J = 0.979, p = 0.030; K-S = 0.092, p = 0.125) imply that the frequency distribution of these residuals closely conforms to a normal distribution. Because

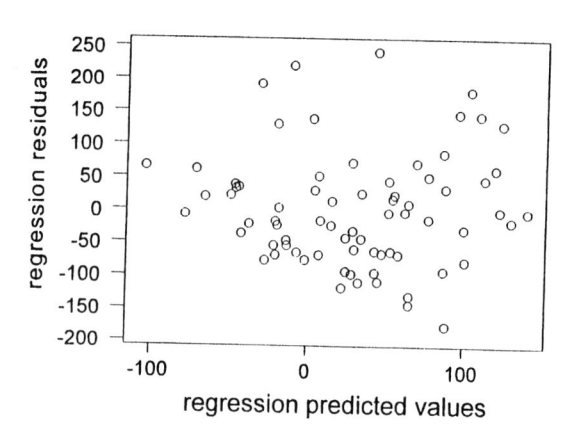

Figure 3.3. Traditional homogeneity of variance plot for the auto-binomial regression model description of the percentage of urban population in Puerto Rico, 1990.

the residuals from this regression already have been weighted to control for heterogeneity, in accordance with theoretical considerations, a plot of the residuals versus their corresponding predicted values not surprisingly produces a scatter of points that fails to display any conspicuous variance heterogeneity (Figure 3.3). Eigenvector E_2 accounts for less than 5% of the variation in the constructed Y variable and hence is not selected initially for entry into the regression equation. But, the five-regressor model yields residuals having an MC value that verge upon being statistically significant (given the asymptotic standard error of 0.07542), suggesting a need to include an additional surrogate variable. Apparently, spatial autocorrelation accounts for roughly one-third of the variation exhibited by variable Y.

This same strategy can be followed for an auto-Poisson problem. Consider the geographic distribution of Kansas oil well locations (i.e., a point pattern) presented in Jones and Wrigley (1995; §5.1). Standardizing the study region such that its area is unity and such that both northing (vertical axis) and easting (horizontal axis) coordinates fall into the interval [0, 1] yields the map appearing in Map 3.1. Superimposing a quadrat mesh on this region for which the quadrat size is optimal (Griffith and Amrhein 1991, 131) renders an 8-by-8 regular square tessellation of quadrats. Descriptive statistics for this situation include $\bar{x} = 1.93750$, $s^2 = 6.56746$, MC = 0.24359 (asymptotic $s_{MC} = 0.09449$), and GR = 0.82999. Because the mean and variance of a simple Poisson random variable are equal, the ratio 6.56746/1.93750 = 3.38966 suggests that these quadrat counts are not compatible with those generated by an independent Poisson process. This conclusion is corroborated by submitting these data to a maximum likelihood Poisson regression analysis (using the STATA software package), which yields the estimate $EXP(0.66140) = 1.93750$ and the accompanying goodness-of-fit probability $Pr(\chi^2 = 166.546) = 0$. Both MC and GR imply that potentially the problem is a lack of independence; in other words, the culprit is the presence of latent spatial autocorrelation in these count data. Eigenvectors extracted from

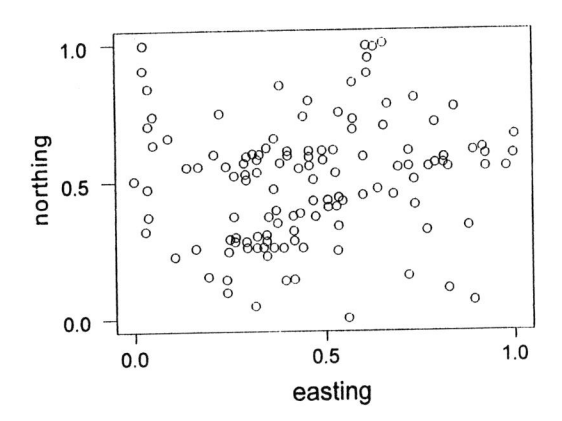

Map 3.1. Geographic distribution of Kansas oil well locations after standardizing georeferenced coordinates presented in Jones and Wrigley.

matrix C characterizing an 8-by-8 regular square tessellation produced

prominent eigenvectors	pseudo-R^2	MC	GR
E_6, E_{17}, E_{18}, E_4, E_{21}, E_{20}, E_{15}, E_5	0.3531	-0.19516	1.26834

The Poisson regression analysis (using the STATA software package) in this nonconstant mean case has the accompanying goodness-of-fit probability $P(\chi^2 = 64.848) = 0.1708$, implying that the adjusted quadrat counts are in compliance with a Poisson process. Spatial autocorrelation accounts for about a one-third of the variance in these counts. As before, the expected value of MC decreases from -0.01587 (if this were a standard linear regression problem, the new expected value of MC would be approximately -0.09426). A word of caution is warranted: the MC and GR results reported here are exploratory and experimental in nature; a much better understanding of their behavior is needed when these indices are applied to non-normal and discrete data.

3.2. GEOSTATISTICAL MODELS

Moving from a discrete surface to a continuous surface requires employment of geostatistics; however, the approach to fitting models to semivariogram plots is similar in concept to that of spatial autoregressive model fitting. When second-order stationarity can be assumed, the most important relationship in semivariogram modeling occurs between $\gamma(h)$ and the variance-covariance matrix Σ, namely, $\gamma(h) = C(0) - C(h)$, where $C(0)$ is the variance component in matrix Σ (diagonal elements) and $C(h)$ is the covariance component (off-diagonal elements). Because both $\gamma(h)$ and $C(0)$ can be estimated from georeferenced data, $C(h)$ can be estimated not only for locations where measurements are made but also for location points where measurements are not made (i.e., kriging). This interpolation is the link between geostatistics and spatial autoregression (§1.1).

Although one role played by the semivariogram plot is to allow selection of a model that describes the spatial dependency across a two-dimensional surface, in practice model fitting and parameter estimation of a semivariogram model employed for kriging purposes are the more routinely performed computations involving a semivariogram plot. Kriging requires that Σ be estimated using a semivariogram model, but Σ must also be positive definite (i.e., all of its eigenvalues are positive) to ensure invertibility and to yield a set of equations usable as the kriging functions. Thus, positive definiteness of Σ is given as a condition for fitting a model to a semivariogram, though it is not a necessary condition. Although we could specify any function to model any particular semivariogram plot, several parametric functions have been developed that are known to be positive definite for matrix Σ. These models (presented in the next section) reflect a set of basic models for simple processes, namely, second-order stationarity, for an isotropic surface.

Semivariogram models can be divided into two basic groups: (1) transition models, where the sill reaches a "plateau" [values of $\gamma(h)$ become approximately equal for the mid- to larger-distance classes]; and (2) nontransition models, whose sill does not reach a plateau [$\gamma(h)$ increases as the magnitude of distance increases; see Figure 2.5]. Models of the second type are somewhat problematic because variance goes to infinity; we prefer to use transition models. Nontransition models can be useful when a geographic trend (§2.6.2) is latent in the data but not explicitly modeled, although geographic trend in this book is treated as a different component of spatial autocorrelation and removed by OLS regression, while spatial autocorrelation is modeled.

3.2.1. MODELS DESCRIBING SEMIVARIOGRAM PLOTS

Popular models—which have been found to be the most useful ones for georeferenced data analysis purposes—surveyed and employed most frequently throughout this book are the spherical, exponential, Gaussian, Bessel, power, and wave/hole. Others not considered here include the linear, circular, rational quadratic, and De Wijsian (Cressie 1991; Haining 1990). The functions for the former models are defined as follows, and except for the power semivariogram model (which already has been fit by Cressie 1989), each of these models is fitted to the Wolfcamp aquifer data, with its graph appearing in Figure 1.3:

SPHERICAL

$$\gamma(h) = \begin{cases} 0 & \text{if } |h| = 0 \\ C_0 + C_1[1.5\left(\dfrac{h}{r}\right) - 0.5\left(\dfrac{h}{r}\right)^3] & \text{if } 0 < |h| < r \\ C_0 + C_1 & \text{if } |h| \geq r \end{cases} \quad (3.11)$$

EXPONENTIAL

$$\gamma(h) = \begin{cases} 0 & \text{if } |h| = 0 \\ C_0 + C_1[1 - \text{EXP}\left(-\dfrac{h}{r}\right)] & \text{if } |h| > 0 \end{cases} \quad (3.12)$$

GAUSSIAN

$$\gamma(h) = \begin{cases} 0 & \text{if } |h| = 0 \\ C_0 + C_1[1 - \text{EXP}\left(-\dfrac{h^2}{r^2}\right)] & \text{if } |h| > 0 \end{cases} \quad (3.13)$$

BESSEL

$$\gamma(h) = \begin{cases} 0 & \text{if } |h| = 0 \\ C_0 + C_1[1 - \left(\dfrac{h}{r}\right) K_1\left(\dfrac{h}{r}\right)] & \text{if } |h| > 0 \end{cases} \quad (3.14)$$

POWER[2]

$$\gamma(h) = \begin{cases} 0 & \text{if } |h| = 0 \\ C_0 + C_1[h^\lambda] & \text{if } |h| > 0 \end{cases} \quad (3.15)$$

---

[2]This specification is equivalent to the linear semivariogram model when $\lambda = 1$.

WAVE/HOLE

$$\gamma(h) = \begin{cases} 0 & \text{if } |h| = 0 \\ C_0 + C_1[1 - \left(\dfrac{\text{SIN}(rh)}{r}h\right)] & \text{if } |h| > 0 \end{cases} \quad (3.16)$$

where C_0 is the nugget effect, C_1 is the sill, r is the range parameter, equal to the range only for the spherical model, h is the distance class, or lag, and λ (for the power model) is the exponent of the function. With the exception of the power model, all of the above are transition models. Also noteworthy is that no range parameter (r) is included in the power model, again because this model is for semivariogram plots where $\gamma(h)$ values continuously increase as distance classes increase. The wave/hole function is of special interest because $\gamma(h)$ values decreasingly oscillate around some constant $\gamma(h)$ value with increasing distance classes, and it is the only useful semivariogram model when negative spatial autocorrelation is present in georeferenced data. Except for the Gaussian and Bessel functions, each of the other transition functions increases roughly linearly from the origin until a plateau is reached. In the case of the Gaussian and Bessel functions an inflection, or cusp, is present at the beginning of this curve.

3.2.2. ESTIMATING SEMIVARIOGRAM MODEL PARAMETERS

Parameters for semivariogram models almost always are estimated using one of two different methods, namely, maximum likelihood (ML) or weighted least squares. In addition to frequent reliance on the normality assumption, the main problem with the ML method is that estimates are biased, particularly with smaller sample sizes, and restrictively so for very small sample sizes. A remedy for this drawback is provided by the generalized least squares (GLS) approach, which does not make assumptions about the entire distribution of the data but rather uses only the second-order structure of the variogram estimator (Cressie 1991). Also, as shown by Carroll and Ruppert (1982), when the distribution of the attribute variable is misspecified, GLS is more robust than ML estimation. A problem with GLS, however, is that estimation of the variance-covariance matrix for the semivariogram is required, which is not always easy (Cressie 1991). An alternative to using GLS would be to use an OLS procedure, but OLS estimation does not take into account the distributional variation and covariation of a semivariogram estimator (Cressie 1991). A compromise between these two least squares approaches is to use WLS. Zimmerman and Zimmerman (1991) compare several methods of semivariogram-parameter estimation, and Cressie (1989, 198) concisely summarizes their results from a previously unpublished version of the paper by stating that "the weighted least squares approach usually performs well and never does poorly."

The form of the WLS procedure can be stated as the minimization of

$$\sum_{i=1}^{K} \frac{N(h_i)}{\{\gamma_m(h_i);\theta\}^2} \{\gamma(h_i) - \gamma_m(h_i);\theta\}^2 \quad , \qquad (3.17)$$

where N(h) is the number of distance pairs in distance class h, $\gamma_m(h)$ is a model function, $\gamma(h)$ are the estimated semivariogram values for h, and θ is the set of model parameters being estimated. The usual set of parameters contained in θ consists of the nugget (C_0), the range (r), and the sill (C_1), although in the case of the power model the range parameter is not included in this set, and in the case of an anisotropic model specification additional parameters are included (§3.2.4). We implement equation (3.17) with PROC NLIN in SAS. Both SAS and SPSS code for the previously specified semivariogram models is listed at the end of this chapter (Appendices 3-E and 3-F).

A serious drawback that merits emphasis here is that, presently, sampling distribution properties of the semivariogram model parameter estimates are not well understood, regardless of the estimation methods used (Cressie 1991, 99). Consequently, significance testing is rarely done on these estimates.

3.2.3. ASSESSING THE GOODNESS-OF-FIT OF SEMIVARIOGRAM MODELS

As seen in Appendix 3-A, the NLIN procedure requires that initial starting values be supplied for parameters before estimation can begin. These can be obtained by inspection of the semivariogram plot. The initial value for the nugget (C_0) always can be set to 0, the theoretical value. Starting values for r and C_1 can be obtained from the horizontal and vertical axes, respectively, where an inflection of an imagined "best-fitting" curved line through the scatterplot occurs on the graph (cf., Figure 1.3). Obviously, for any given semivariogram plot, the starting values for any transition model will remain the same, regardless of the model chosen to be fit to the semivariogram plot. We employ a goodness-of-fit criterion based on comparing the RSSE (relative sum of squared errors) among models fitted to a semivariogram plot. The RSSE is obtained by dividing the sum of squared errors by the corrected total sum of squared errors. The model selected for a particular semivariogram plot is the one with the lowest RSSE. A cautionary note should be heeded here because the lowest RSSE for a set of fitted models does not necessarily reflect a "good" model fit, since the RSSE is bounded between 0 and 1. That is, although an RSSE value of, say, 0.7 may be the lowest RSSE for a set of models fitted to a semivariogram plot, such a value does not indicate a very good fit to the plot. Another cautionary note concerns sets of model fits that achieve very low, but not compellingly different, RSSE values. To circumvent this dilemma, other model selection criteria, such as cross-validation or the Akaike information criterion (AIC), often are resorted to.

isotropic semivariance

$$\rho = \frac{0.225}{2}$$

anisotropic semivariance

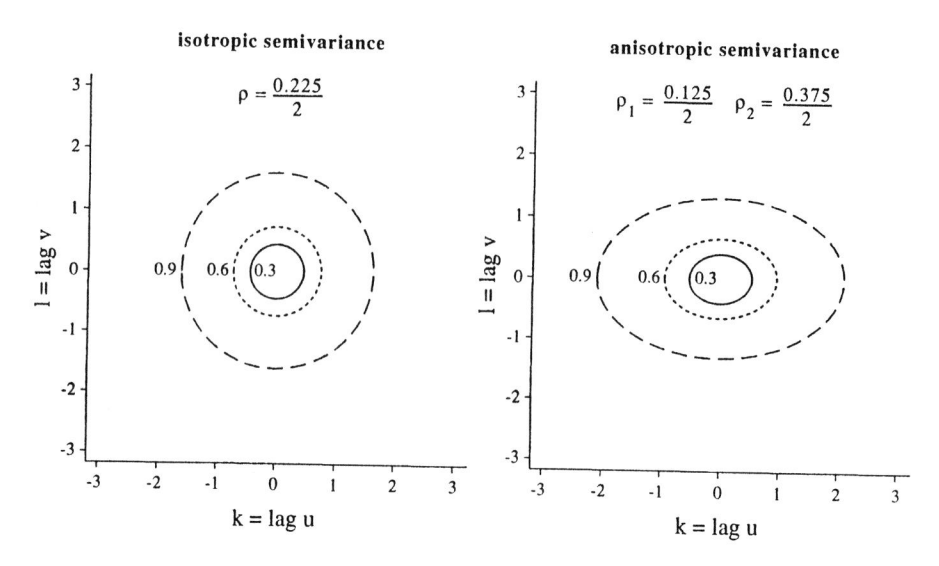

$$\rho_1 = \frac{0.125}{2} \quad \rho_2 = \frac{0.375}{2}$$

Figure 3.4. Left: (a) Isosemivariance contours indicative of an isotropic surface. Right: (b) Isosemivariance contours indicative of an anisotropic surface.

3.2.4. ANISOTROPIC MODEL SPECIFICATION

Semivariogram plots and models presented thus far assume isotropy over a surface. Anisotropic surfaces are not, however, uncommon in practice (see Chapter 6), because spatial dependency does not necessarily vary equally in all lateral directions. Overlooking this feature of georeferenced data can result in semivariogram plots that appear to be ill-behaved. An anisotropic surface occurs when spatial dependency differs directionally; it usually is modeled with two orthogonal direction components, such as north-south versus east-west. Various authors present different methods for detecting anisotropy, which generally consist of uncovering anisotropic axes analytically (Cressie 1989) or by trial and error, where directional semivariogram plots are computed using a specified directional slice through a landscape and a direction tolerance (Isaaks and Srivastava 1989). In the trial-and-error approach, several directional semivariogram plots may be constructed for (typically) north-south, east-west, southwest-northeast, and northwest-southeast directions, generally using a direction tolerance of ±22.5°. Isaaks and Srivastava (1989) can be consulted regarding selection of optimal directional tolerances. The shapes of these directional semivariogram plots then can be compared visually, as can the slopes of their curves, and a decision made as to whether the directional semivariograms are similar or different enough to posit the presence of anisotropy on a surface. Although both methods actually are experimental in nature—optimization of a nonlinear function containing trigonometric function often involves trial and error—uncovering anisotropic axes analytically is our preferred method. Another detection method is to plot a contour map of the surface, where isolines generalize the spatial variability of semivariogram values across lags (Figure 3.4). Isosemivariance contours that are approximately circular are indicative of an isotropic surface, whereas any deviation from circularity (typically an elliptical form) is more indicative of anisotropy across a surface. One drawback to this last approach is that the set of available distance lags, particularly the shorter lags, from sampled geographic locations might not be sufficiently representative of the underlying semivariogram landscape to support meaningful contour construction.

Modeling an anisotropic surface using a semivariogram typically is performed by separately fitting a given model to each directional semivariogram, then specifying a single semivariogram model as the sum of the two separate directional models, where model types must be the same (Isaaks and Srivastava 1989). Such specifications involve distances computed with a modified version of the Euclidean distance formula, namely,

$$d_{uv} = \sqrt{\left(\frac{u}{r_u}\right)^2 + \left(\frac{v}{r_v}\right)^2},$$

where r_u is the range parameter for the u coordinate axis and r_v is the range parameter for the v coordinate axis. When $r_u = r_v = r$, this range parameter factors out of the right-hand side square root expression, reducing this case to the isotropic one. Because similar distances are aggregated into groups represented by an average distance h (see §1.6), the previously mentioned orthogonal semivariogram plots commonly are used to implement this decomposition. The pair of orthogonal distance lag axes used are such that the axis $d_{uv} = \sqrt{\left(\frac{u}{r_u}\right)^2 + \left(\frac{v}{r_v}\right)^2}$, aligning with the greater magnitude of spatial dependency defines the major axis of the ellipse contour plots, with the minor axis of this ellipse aligning with the smaller magnitude of spatial dependency. These anisotropic axes may coincide with those directly computable from the data coordinate system (defined as north-south and east=west directions) or result from a rotation of these original coordinate axes, where the angle of rotation, θ, is defined clockwise from the north-south direction (Isaaks and Srivastava 1989). Whether or not rotation is performed, the resulting model also may be a linear combination of different semivariogram model types, such as

$$\gamma(h) = C_0 + C_{1,m_1}\gamma_{m_1}(h_1) + C_{1,m_2}\gamma_{m_2}(h_2),$$

where the C_j parameters are defined as before [e.g., equation (1.12)], the subscript m_k denotes semivariogram model type k (§1.6, §3.2.1), and h_k denotes the average distance for distance group h having two different directional ranges in model type k [as presented in the above hybrid equation for γ(h)]. That is, such a model may consist of a spherical, exponential, and perhaps even Bessel model semivariance component for both axes (Isaaks and Srivastava 1989; Deutsch and Journel 1992). When the nugget [i.e., C_0 in the above hybrid equation for $\gamma(h)$] and sill remain the same [i.e., $C_{1,m_1} = C_{1,m_2}$ in the above hybrid equation for $\gamma(h)$] but the range changes [i.e., $r_u \neq r_v$ in the above hybrid equation for $\gamma(h)$] as direction changes, geometric anisotropy is said to be present (Isaaks and Srivastava 1989). In the case of geometric anisotropy, only one nugget effect is estimated. Zonal anisotropy occurs when the range remains the same [i.e., $r_u = r_v$ in the above hybrid equation for $\gamma(h)$] but the sill changes [i.e., $C_{1,m_1} \neq C_{1,m_2}$ in the above hybrid equation for $\gamma(h)$; Isaaks and Srivastava 1989]. In the latter case, the nugget effect from the model describing the greater spatial dependency is used in thee resulting anisotropic model.

A word of caution: in some empirical analyses, best-fitting directional semivariogram models for an anisotropic surface may consist of distinctly different types for each axis, perhaps an exponential semivariogram model along one axis and a Gaussian model along the other orthogonal axis. The sum of these differing model types does not necessarily result in a valid semivariogram because only one dimension for each model is actually being described (Myers and Journel 1990) and as such does not satisfy the property of conditional negative definiteness in two dimensions, which possibly may lead to noninvertible kriging matrices (Myers and Journel 1990; Rouhani and Myers 1990). Therefore, the fitting of different model types for each anisotropy axis is not recommended. For our purposes, whenever a model is fit to a semivariogram plot for an anisotropic surface, only one model type is used (§6.1).

If anisotropy is detected, the angle determining the major and minor axes of the contour ellipse may be included as a parameter in the semivariogram model specification and hence estimated (e.g., Cressie 1989). Modeling anisotropy, then, consists of estimating the range parameters as well as the optimal direction of the anisotropy, described previously as the angle away from the north-south direction. This effectively results in the georeferencing Cartesian coordinates being transformed to polar coordinates and is analogous to the rotation of axes in a bivariate principal components analysis, where the rotation results in orthogonal axes. In other words, given a set of n Cartesian coordinate pairs, say (u_i, v_i), which constitute an arbitrary initial georeferencing coordinate system, the analytical problem is to determine an angle, θ, for axis rotation such that the new coordinate pairs, say (u_i^*, v_i^*), maximize the semivariance accounted for. These two coordinate systems are alternative representations of the same georeferencing scheme. The most convenient way of describing these rotated coordinates in terms of the original coordinates, retaining $\gamma(0)$ of the isotropic semivariance spatial distribution as the origin for each of these two coordinate systems, is to use the cosines of functions of the angle of rotation, θ, which is actually the angle between the two sets of orthogonal axes. Accordingly,

$$\begin{bmatrix} u* \\ v* \end{bmatrix} = \begin{bmatrix} \cos(\theta) & \cos(90° - \theta) \\ \cos(90° + \theta) & \cos(\theta) \end{bmatrix} \times \begin{bmatrix} u \\ v \end{bmatrix} .$$

Moreover the rotated axes are linear combinations of the arbitrary original orthogonal axes. Because the appropriate rotation seeks to maximize semivariance, its geometric interpretation is as follows: the axis coinciding with the maximum semivariance is also the major axis of the isotropic semivariance ellipse and corresponds to the direction of maximum spatial dependence, whereas the axis orthogonal to this semivariance-maximizing one coincides with the minor axis of the isotropic semivariance ellipse. This second rotated axis corresponds to the direction of weaker spatial dependence, and accounts for the conditionally maximum semivariance (the maximum semivariance of any axis that is uncorrelated with the first rotated axis). In essence, then, this is a principal axis rotation.

Analytical estimation of the angle of rotation θ requires a more detailed set of data, including horizontal and vertical distance directions (both positive and negative) and angles between pairs of points (to convert the geocodes to polar coordinates). Therefore, required computations from the original set of georeferenced coordinates (u_i, v_i) include

$$\theta_u = \cos^{-1}\left(\frac{\Delta_u}{d_{uv}}\right)$$

$$\theta_v = \cos^{-1}\left(\frac{\Delta_v}{d_{uv}}\right)$$

$$d_{uv} = \sqrt{\Delta_u^2 + \Delta_v^2}$$

$$\phi = \sqrt{\cos(\theta_u + \theta)^2 + \cos\left(\theta_v + \theta + \frac{\pi}{2}\right)^2} ,$$

where Δ_u and Δ_v, respectively, are the differences between the original u and v axis locations for a pair of points. When incorporated into the exponential semivariogram model, for example, these expressions yield

$$\gamma(d_{uv}) = C_0 + C_1 \left\{1 - EXP\left[-\phi \times \sqrt{\left(\frac{\Delta_u}{r_u}\right)^2 + \left(\frac{\Delta_v}{r_v}\right)^2}\right]\right\} , \quad -\frac{\pi}{4} < \theta < \frac{\pi}{4} .$$

As before, one complication arises from the common practice of using aggregate versus individual distances. This problem should not be as severe in the clustering approach we promote in this book, because maximally similar distances are grouped together to construct a semivariogram plot.

Determining the rotation angle θ in geostatistical analyses is far easier than it is in spatial autoregressive analyses, where θ impacts upon the eigenvalues appearing in the Jacobian term. This drawback suggests that perhaps θ should be determined with geostatistical techniques first, with the result used as an input into spatial autoregression analyses. This sequential approach argues strongly for articulating relationships between spatial autoregressive and geostatistical models.

3.3. ARTICULATING RELATIONSHIPS BETWEEN SPATIAL AUTOREGRESSIVE AND GEOSTATISTICAL MODELS

One motivation for pursuing the theme of this section is the quest for a unified body of knowledge. Another is the establishment of a common umbrella for a growing body of seemingly disparate work (e.g., Rey et al. 1994; Griffith 1992a; Little and Rubin 1987). The extension outlined here builds on results reported by Griffith (1993a) and Griffith and Csillag (1993). Explicit numerical connections between autoregressive and geostatistical models—beyond the missing values problem emphasized in §1.1—can be established and explored by constructing theoretical surfaces for both the rook's and the queen's adjacency cases (using chess move analogies) defined on a regular square tessellation, and employing the following spectral density function for a discrete surface:

$$\frac{\int_0^\pi \int_0^\pi \frac{\cos(k \times x)\cos(l \times y)}{\left[1 - 2[\rho_p \cos(x) + \rho_q \cos(y) + \sqrt{2}\rho_{pq}\cos(x)\cos(y)]\right]^t} \, dx \, dy}{\int_0^\pi \int_0^\pi \frac{1}{\left[1 - 2[\rho_p \cos(x) + \rho_q \cos(y) + \sqrt{2}\rho_{pq}\cos(x)\cos(y)]\right]^t} \, dx \, dy} \qquad (3.18)$$

(after Bartlett 1975). An isotropic surface is specified when $\rho_p = \rho_q$; otherwise it is anisotropic. k represents the easting lag distances and l the northing lag distances. The CAR model is specified when the exponent $t = 1$ (rendering it a first-order model) and the SAR specified when $t = 2$ (rendering it a second-order model), correcting a misleading claim by King and Smith (1988). Higher-order models could be specified by increasing t. The term $\sqrt{2}\rho_{pq}\cos(x)\cos(y) = 0$ for the rook's case and is specified for the queen's case so that the juxtaposed diagonals are weighted by $1/\sqrt{2}$. The values yielded by this spectral density function are the entries in the aforementioned variance-covariance matrix, Σ, whose inverted form is affiliated with spatial autoregression.

To illustrate this situation, consider a 10-by-10 lag correlation grid from which matrix Σ is constructed. Its construction is described here using a set of arbitrary data values arranged on a regular square lattice (Table 3.1) and its corresponding lag correlation grid, which can be constructed from those values.

A schematic representation of a spatial lag correlation grid for the 10-by-10 square lattice of data values is presented in Figure 3.5. Each intersection of this grid represents a lag-specific correlation coefficient for pairs of values based on each referent value and its corresponding pairing value, determined by the distance and direction of the paired value from the referent value. For instance, the (0,0)

Table 3.1. Attribute values given for a 10-by-10 square lattice.

3.	8.	5.	6.	2.	3.	9.	8.	1.	4.
2.	3.	7.	2.	8.	3.	1.	0.	5.	6.
8.	9.	1.	3.	1.	9.	9.	9.	2.	5.
5.	2.	6.	9.	3.	4.	6.	5.	4.	6.
3.	6.	4.	5.	7.	6.	5.	2.	3.	3.
7.	8.	2.	2.	3.	9.	1.	0.	2.	1.
8.	6.	4.	4.	2.	3.	1.	2.	3.	3.
6.	6.	6.	4.	2.	3.	9.	3.	2.	4.
7.	1.	0.	6.	3.	2.	2.	3.	9.	3.

location on the grid denotes the case in which each value in Table 3.1 is paired with itself then entered in turn as a term in the correlation coefficient formula, with the computed result being placed at the (0,0) location. By definition this value is 1. Moving one lag to the right in this grid, each referent value in Table 3.1 is paired with the value to its immediate right, with the affiliated covariance term being entered in turn into the correlation coefficient formula, and the computed result is placed at the (1,0) location. By definition this value is less than 1, and for an isotropic situation it also appears at the (0,1), (-1,0), and (0,-1) locations. This process continues until all of the intersections in Figure 3.5 have correlation coefficients attached to them. Equation (3.18) is used to compute the theoretical correlation coefficients that would occupy each of the intersections in Figure 3.5.

The denominator term in equation (3.18), namely $2[\rho_p \cos(x) + \rho_q \cos(y) + \sqrt{2}\rho_{pq} \cos(x)\cos(y)]$, comprises the asymptotic eigenvalues of connectivity matrix **C**. Given a P-by-Q rectangular region, the finite counterparts of these eigenvalues are, for the rook's adjacency case,

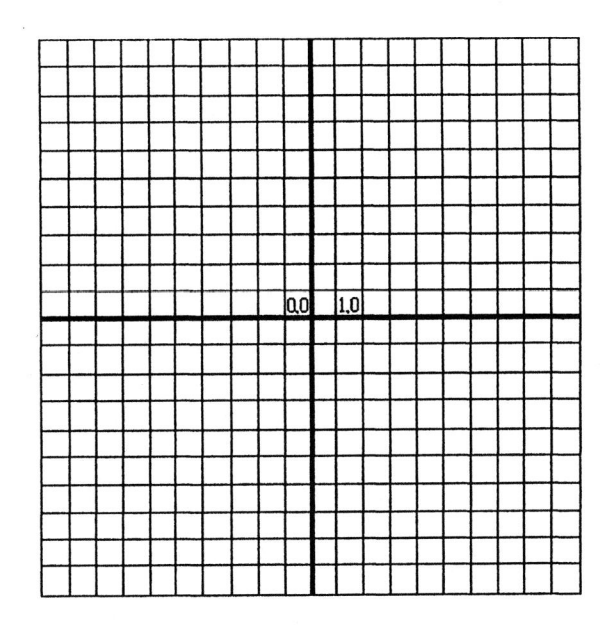

Figure 3.5. A spatial lag correlation grid for a 10-by-10 square lattice of data values.

$$2[\rho_p\cos(\frac{k\pi}{P+1}) + \rho_q\cos(\frac{l\pi}{Q+1})], \quad k=1,2,...,P \text{ and } l=1,2,...,Q,$$

and for the queen's adjacency case,

$$2[\rho_p\cos(\frac{k\pi}{P+1}) + \rho_q\cos(\frac{l\pi}{Q+1}) + \sqrt{2}\rho_{pq}\cos(\frac{k\pi}{P+1})\cos(\frac{l\pi}{Q+1})],$$
$$k=1,2,...,P \text{ and } l=1,2,...,Q.$$

These eigenvalues further reinforce previously noted connections (§1.1) between spatial autocorrelation and the eigenfunctions of matrix **C**.

Bartlett (1975) numerically evaluates equation (3.18) for the CAR model and a rook's connection. Extending his work to the SAR model yields

Rook's connection

	CAR			SAR		
ρ	$\varrho_{1,0}$	$\varrho_{1,1}$	$\varrho_{2,0}$	$\varrho_{1,0}$	$\varrho_{1,1}$	$\varrho_{2,0}$
-0.5	-0.136	0.036	0.019	-0.277	0.107	0.060
0.0	0.0	0.0	0.0	0.0	0.0	0.0
0.1	0.025	0.001	0.001	0.050	0.004	0.002
0.2	0.051	0.005	0.003	0.102	0.015	0.008
0.3	0.077	0.012	0.006	0.155	0.035	0.018
0.4	0.105	0.022	0.011	0.213	0.065	0.035
0.5	0.136	0.036	0.019	0.277	0.107	0.060
0.6	0.171	0.056	0.031	0.350	0.166	0.098
0.7	0.213	0.084	0.048	0.437	0.248	0.156
0.8	0.266	0.126	0.078	0.547	0.366	0.252
0.9	0.346	0.201	0.136	0.700	0.556	0.432
0.99	0.537	0.415	0.341	0.944	0.908	0.859

where $\rho_p = \rho_q = \rho/\lambda_{max} = \rho/4$. Extending his work to the queen's connection case yields

Queen's connection

	CAR			SAR		
ρ	$\varrho_{1,0}$	$\varrho_{1,1}$	$\varrho_{2,0}$	$\varrho_{1,0}$	$\varrho_{1,1}$	$\varrho_{2,0}$
-0.5	-0.096	0.070	0.017	-0.214	0.160	0.056
0.0	0.0	0.0	0.0	0.0	0.0	0.0
0.1	0.015	0.011	0.000	0.031	0.022	0.001
0.2	0.032	0.023	0.002	0.067	0.048	0.006
0.3	0.051	0.036	0.005	0.108	0.078	0.016
0.4	0.072	0.052	0.009	0.156	0.114	0.031
0.5	0.096	0.070	0.017	0.214	0.160	0.056
0.6	0.125	0.092	0.027	0.285	0.218	0.095
0.7	0.160	0.120	0.036	0.374	0.296	0.156
0.8	0.207	0.160	0.072	0.494	0.408	0.258
0.9	0.281	0.227	0.127	0.667	0.588	0.449
0.99	0.467	0.414	0.316	0.943	0.919	0.876

where $\rho_p = \rho_q = \rho/\lambda_{max} = \rho/(4 + 2\sqrt{2})$ and $\rho_{pq} = \rho_p \times \rho_q$. Negative spatial autocorrelation renders periodic negative interlocation correlations across a geographic landscape, with the same absolute value of correlation as is produced by its positive spatial autocorrelation counterpart. Further, as either ρ or the order of the model increases, interlocation correlations across a geographic landscape increase. In the limit, all interlocation correlations equal 1, which is the degenerate case of ρ = 1.

One feature of interest here concerns boundary or edge effects. Griffith (1983) addresses the general problem of edge effects in quantitative geographical analyses, underscoring their seriousness and acknowledging that one solution found in the cartographic literature involves working with surface derivatives. Rathbun (1995) provides an excellent example of the type of distortion that overlooked edge effects can create, showing how properly handling them allows better interpolation of salinity in Chesapeake Bay. Results reported in Griffith and Csillag (1993) especially exhibit an affinity between the magnitude of prevailing spatial dependencies and the dominance of regional edge effects. This factor emerges because semivariogram models directly deal with the covariance matrix for a set of observations, whereas spatial autoregressive models deal with the inverse of this matrix. The procedure of matrix inversion is what amplifies and propagates edge effects. To date, these effects have defied the formulation of general corrective techniques. Insight into this edge effects problem is gained here by holding the degree of spatial dependence constant and changing the size of the region within which estimation is conducted. Returning to the case of a queen's set of neighbors spatial linkage structure, an isotropic SAR specification, letting ρ = 0.2491 and using Heuvelink's weighted least squares software yields

	Bessel	function	parameter	estimates
maximum lag	C_0	C_1	range	RSSE
5	0.0093	0.906	9.35	0.000
10	0.0103	0.965	9.86	0.000
15	0.0109	0.978	10.00	0.000
25	0.0116	0.987	10.10	0.000
35	0.0119	0.987	10.10	0.000
5000	0.0200	0.987	10.50	0.000 .

Theoretically, one would expect $C_0 = 0$ and $C_1 = 1$. As the region used for parameter estimation purposes increases in size, the fit of the modified Bessel function of the first-order and second-kind remains near perfect, while the parameter estimates themselves change. The range of the spatial autocorrelation field increases to some upper limit, and C_1 approaches 1, as expected. Somewhat surprisingly C_0 (the nugget effect) does not move toward 0, which may indicate a liability associated with using a nonlinear model specification and/or weighted least squares estimation.

Identifying the correct spatial autoregressive or semivariogram model is essential to optimal interpolation of georeferenced data that yield a geographic surface representation. Such a surface representation promotes scientific visualization. Unfortunately, surface visualization is plagued by edge effects, which can render biased model parameter estimates as well as produce misleading surface portrayals. Results reported here reinforce this limitation.

3.3.1. NUMERICAL LINKAGES FROM SPATIAL AUTOREGRESSIVE TO GEOSTATISTICAL MODELS

Previous work by Griffith and Csillag (1993) indicates a correspondence between selected semivariogram and autoregressive models: the linear semivariogram pairs with the moving average (MA) autoregressive model, the exponential semivariogram pairs with the conditional autoregressive (CAR) model, and the modified Bessel (first-order, second-kind) semivariogram pairs with the simultaneous autoregressive (SAR) model. Each of these connections is established by analyzing spatial autoregressive structures with semivariogram models. The autoregressive structures were constructed for both the rook's and the queen's adjacency cases using the aforementioned spectral density function [equation (3.18)] for a discrete isotropic surface presented in §3.3.

Along these same lines, a lag correlation matrix for a 10-by-10 square lattice of data was generated using the appropriate spectral density functions for both the rook's and the queen's cases CAR and SAR models, specifying $\rho = 0.7$, where $\rho_p/\lambda_{max} = \rho_q/\lambda_{max} = 0.7/\lambda_{max}$. This matrix then was inverted and standardized with the center value from the inverted matrix. The pattern of diagonal values from this inverted matrix may be examined in Maps 3.2a-b and 3.3a-b, where expected values are 1. Likewise, the pattern using the center row of values (representing the spatial dependency field) from the inverted matrix may be examined. Expected values here are given as ρ/λ_{max}, where $\lambda_{max} = 4$ for the CAR and SAR rook's case, and $\lambda_{max} = 4 + 2\sqrt{2}$ for the CAR and SAR queen's case.

Correspondences among the semivariogram and autoregressive model pairings reported in Table 3.2 are good but not perfect, as indicated by the RSSE (§3.23) being slightly greater than 0, and the parameter estimates for C_0 and C_1 being different from 0 and 1, respectively. Even though expected dependency fields are recovered for both the rook's and the queen's case CAR and SAR models, for the CAR model, all values of the diagonal are equal to the expected value of 1 except for a single ring of boundary cells around the edge (Maps 3.2a-b). These lower values represent edge effects amplified by the matrix inversion and are limited to a single ring because the CAR is a first-order model. Likewise for the case of the SAR model, two rings of boundary cells around the edge are less than unity (Maps 3.3a-b) for the same reasons as the single ring in the CAR model, because the SAR model is a second-order model.

When considering the appropriateness of equation (3.18) as a descriptor of geographic covariation, the following numerical evidence (computed with Heuvelink's weighted least squares software) based on an isotropic dependency structure [using the spatial autoregressive parameter ρ value of $0.9964/n_i$, where n_i is the number of articulated neighbors given an infinite surface; for the rook's

```
.9286560  .9657578  .9657601  .9657602  .9657602  .9657602  .9657602  .9657602  .9657601  .9657578  .9286560
.9657578 1.0000000 1.0000000 1.0000000 1.0000000 1.0000000 1.0000000 1.0000000 1.0000000 1.0000000  .9657578
.9657601 1.0000000 1.0000000 1.0000000 1.0000000 1.0000000 1.0000000 1.0000000 1.0000000 1.0000000  .9657601
.9657602 1.0000000 1.0000000 1.0000000 1.0000000 1.0000000 1.0000000 1.0000000 1.0000000 1.0000000  .9657602
.9657602 1.0000000 1.0000000 1.0000000 1.0000000 1.0000000 1.0000000 1.0000000 1.0000000 1.0000000  .9657602
.9657602 1.0000000 1.0000000 1.0000000 1.0000000 1.0000000 1.0000000 1.0000000 1.0000000 1.0000000  .9657602
.9657602 1.0000000 1.0000000 1.0000000 1.0000000 1.0000000 1.0000000 1.0000000 1.0000000 1.0000000  .9657602
.9657602 1.0000000 1.0000000 1.0000000 1.0000000 1.0000000 1.0000000 1.0000000 1.0000000 1.0000000  .9657602
.9657601 1.0000000 1.0000000 1.0000000 1.0000000 1.0000000 1.0000000 1.0000000 1.0000000 1.0000000  .9657601
.9657578 1.0000000 1.0000000 1.0000000 1.0000000 1.0000000 1.0000000 1.0000000 1.0000000 1.0000000  .9657578
.9286560  .9657578  .9657601  .9657602  .9657602  .9657602  .9657602  .9657602  .9657601  .9657578  .9286560


.9485358  .9731814  .9732596  .9732615  .9732616  .9732616  .9732616  .9732615  .9732596  .9731814  .9485358
.9731814 1.0000000 1.0000000 1.0000000 1.0000000 1.0000000 1.0000000 1.0000000 1.0000000 1.0000000  .9731814
.9732596 1.0000000 1.0000000 1.0000000 1.0000000 1.0000000 1.0000000 1.0000000 1.0000000 1.0000000  .9732596
.9732615 1.0000000 1.0000000 1.0000000 1.0000000 1.0000000 1.0000000 1.0000000 1.0000000 1.0000000  .9732615
.9732616 1.0000000 1.0000000 1.0000000 1.0000000 1.0000000 1.0000000 1.0000000 1.0000000 1.0000000  .9732616
.9732616 1.0000000 1.0000000 1.0000000 1.0000000 1.0000000 1.0000000 1.0000000 1.0000000 1.0000000  .9732616
.9732616 1.0000000 1.0000000 1.0000000 1.0000000 1.0000000 1.0000000 1.0000000 1.0000000 1.0000000  .9732616
.9732615 1.0000000 1.0000000 1.0000000 1.0000000 1.0000000 1.0000000 1.0000000 1.0000000 1.0000000  .9732615
.9732596 1.0000000 1.0000000 1.0000000 1.0000000 1.0000000 1.0000000 1.0000000 1.0000000 1.0000000  .9732596
.9731814 1.0000000 1.0000000 1.0000000 1.0000000 1.0000000 1.0000000 1.0000000 1.0000000 1.0000000  .9731814
.9485358  .9731814  .9732596  .9732615  .9732616  .9732616  .9732616  .9732615  .9732596  .9731814  .9485358
```

Map 3.2. Top: (a) Geographic distribution of the diagonal values from the inverted n-by-n matrix R, where r_{ij} is defined by the CAR spectral density function: rook's adjacency case, $\rho = 0.7/\lambda_{max}$. Bottom: (b) Geographic distribution of the diagonal values from the inverted n-by-n matrix R, where r_{ij} is defined by the CAR spectral density function: queen's adjacency case, $\rho = 0.7/\lambda_{max}$.

move case this value is 0.2491 ($n_i = 4$), the absolute value of the preceding negative autocorrelation spatial autoregressive parameter value, whereas for the queen's case this value is 0.12455 ($n_i = 8$)] indicates that the two critical features leading to a Bessel function semivariogram model specification appear to be intensity and degree of articulation of a geographic structure:

geographic	autoregressive model							
articulation	conditional				simultaneous			
	C_0	C_1	range	RSSE	C_0	C_1	range	RSSE
square	0.350	0.617	2.56	0.007	0.005	0.994	8.27	0.000
hexagon[3]	0.443	0.489	2.19	0.041	-----	-----	-----	-----
queen	0.442	0.520	2.88	0.004	0.011	0.988	10.1	0.000

---

[3] Spatial correlations were estimated here by inverting a 75^2-by-75^2 matrix having $\rho = 0.9964/6$ and then employing those values computed for the central unit. This matrix inversion required 20 hours of dedicated RISC/6000 computing time and contains noticeable numerical error. Assessment suggests that the correlations are correct to 3 decimal places.

.7658135	.8886013	.8897404	.8897404	.8897404	.8897404	.8897404	.8897404	.8897404	.8886013	.7658135
.8886013	.9977397	.9988719	.9988719	.9988719	.9988719	.9988719	.9988719	.9988719	.9977397	.8886013
.8897404	.9988719	1.0000000	1.0000000	1.0000000	1.0000000	1.0000000	1.0000000	1.0000000	.9988719	.8897404
.8897404	.9988719	1.0000000	1.0000000	1.0000000	1.0000000	1.0000000	1.0000000	1.0000000	.9988719	.8897404
.8897404	.9988719	1.0000000	1.0000000	1.0000000	1.0000000	1.0000000	1.0000000	1.0000000	.9988719	.8897404
.8897404	.9988719	1.0000000	1.0000000	1.0000000	1.0000000	1.0000000	1.0000000	1.0000000	.9988719	.8897404
.8897404	.9988719	1.0000000	1.0000000	1.0000000	1.0000000	1.0000000	1.0000000	1.0000000	.9988719	.8897404
.8897404	.9988719	1.0000000	1.0000000	1.0000000	1.0000000	1.0000000	1.0000000	1.0000000	.9988719	.8897404
.8897404	.9988719	1.0000000	1.0000000	1.0000000	1.0000000	1.0000000	1.0000000	1.0000000	.9988719	.8897404
.8886013	.9977397	.9988719	.9988719	.9988719	.9988719	.9988719	.9988719	.9988719	.9977397	.8886013
.7658135	.8886013	.8897404	.8897404	.8897404	.8897404	.8897404	.8897404	.8897404	.8886013	.7658135

.8310190	.9188367	.9193642	.9193667	.9193668	.9193668	.9193668	.9193667	.9193642	.9188367	.8310190
.9188368	.9967468	.9984483	.9984509	.9984510	.9984510	.9984510	.9984509	.9984483	.9967468	.9188368
.9193642	.9984483	1.0000000	1.0000000	1.0000000	1.0000000	1.0000000	1.0000000	1.0000000	.9984483	.9193642
.9193667	.9984509	1.0000000	1.0000000	1.0000000	1.0000000	1.0000000	1.0000000	1.0000000	.9984509	.9193667
.9193668	.9984510	1.0000000	1.0000000	1.0000000	1.0000000	1.0000000	1.0000000	1.0000000	.9984510	.9193668
.9193668	.9984510	1.0000000	1.0000000	1.0000000	1.0000000	1.0000000	1.0000000	1.0000000	.9984510	.9193668
.9193668	.9984510	1.0000000	1.0000000	1.0000000	1.0000000	1.0000000	1.0000000	1.0000000	.9984510	.9193668
.9193667	.9984509	1.0000000	1.0000000	1.0000000	1.0000000	1.0000000	1.0000000	1.0000000	.9984509	.9193667
.9193642	.9984483	1.0000000	1.0000000	1.0000000	1.0000000	1.0000000	1.0000000	1.0000000	.9984483	.9193642
.9188368	.9967468	.9984483	.9984509	.9984510	.9984510	.9984510	.9984509	.9984483	.9967468	.9188368
.8310190	.9188367	.9193642	.9193667	.9193668	.9193668	.9193668	.9193667	.9193642	.9188367	.8310190

Map 3.3. Top: (a) Geographic distribution of the diagonal values from the inverted n-by-n matrix R, where r_{ij} is defined by the SAR spectral density function: rook's adjacency case, $\rho = 0.7/\lambda_{max}$. Bottom: (b) Geographic distribution of the diagonal values from the inverted n-by-n matrix R, where r_{ij} is defined by the SAR spectral density function: queen's adjacency case, $\rho = 0.7/\lambda_{max}$.

the exponential semivariogram model[4] furnishes a better description of the CAR cases than does the particular modified Bessel function estimated here. An SAR structure, which involves second-order dependencies, is better characterized by the popular Bessel function (first-order, second-kind) than is the CAR structure, which involves only first-order dependencies.

3.3.2. NUMERICAL LINKAGES FROM GEOSTATISTICAL TO SPATIAL AUTOREGRESSIVE MODELS

The backward transformation problem, which is the invertibility problem accompanying the linkage problem presented in the preceding section, begins with a variance-covariance matrix generated by a semivariogram model, inverts this covariance matrix, standardizes the diagonal values of this inverted matrix by its

---

[4]This actually is a Bessel function of the 1/2 order.

geographically central value, and then evaluates whether or not the corresponding autoregressive models can be identified. Following this procedure, lag correlation matrices were generated using semivariogram functions, then inverted, and finally standardized. For the exponential semivariogram function case, patterns observed in the table of diagonal values and in values in the spatial dependency field do not correspond exactly to their CAR counterparts generated with the spectral density function (Maps 3.4a-d). In the rook's case (Map 3.4a), three rings of edge effects are observed where only one ring would be expected, suggesting the possible need to specify a third-order model. These edge effects lessen when the rook's case spatial dependency field is replaced with the queen's case (Map 3.4c). Likewise, values in the spatial dependency field (Maps 3.4b and 3.4d) for the juxtaposed cells do not add to the specified level of $\rho = 0.7$, and values for all other cells are not equal to the expected value of 0. Also, values appear in the diagonals of the map field, which is expected for the queen's case but not for the rook's case. These same patterns are observed for the inverted, standardized matrix using the modified Bessel function (first-order, second-kind) for the SAR model. In these cases, more edge effects are present, with a diminishing of these effects when the queen's case is substituted for the rook's case. In general, all observed patterns suggest autoregressive models of a higher order than the corresponding first- and second-order ones implied by the semivariogram modeling fittings.

Expected values for the matrices are obtained from $(I - \rho C)$ and $(I - \rho C + \rho^2 C^2)$ for the CAR and SAR models, respectively, where, as before, I is the identity matrix containing the diagonal values from the inverted, standardized correlation matrix and C is the connectivity matrix indicating the adjacency of surrounding cells to any referent cell. The particular coefficients of interest here, highlighted with double underlining, for the rook's case are as follows:

central areal unit values for
an 11-by-11 regular square tessellation

map of coefficients from C^1	map of coefficients from C^2
0 0 0 0 0 0 0 0 0 0 0	0 0 0 0 0 0 0 0 0 0 0
0 0 0 0 0 0 0 0 0 0 0	0 0 0 0 0 0 0 0 0 0 0
0 0 0 0 0 0 0 0 0 0 0	0 0 0 0 0 0 0 0 0 0 0
0 0 0 0 0 0 0 0 0 0 0	0 0 0 0 0 1 0 0 0 0 0
0 0 0 0 0 1 0 0 0 0 0	0 0 0 0 2 0 $\underline{\underline{2}}$ 0 0 0 0
0 0 0 0 1 0 $\underline{\underline{1}}$ 0 0 0 0	0 0 0 1 0 $\underline{\underline{4}}$ 0 1 0 0 0
0 0 0 0 0 1 0 0 0 0 0	0 0 0 0 2 0 2 0 0 0 0
0 0 0 0 0 0 0 0 0 0 0	0 0 0 0 0 1 0 0 0 0 0
0 0 0 0 0 0 0 0 0 0 0	0 0 0 0 0 0 0 0 0 0 0
0 0 0 0 0 0 0 0 0 0 0	0 0 0 0 0 0 0 0 0 0 0
0 0 0 0 0 0 0 0 0 0 0	0 0 0 0 0 0 0 0 0 0 0

Accordingly, for the CAR model, denoting the (i, j) cell entry in the standardized inverse-covariance matrix V by v_{ij}, let $v_{00} = 1$, $v_{10} = -\hat{\rho}$, and all other $v_{ij} = 0$. These results are gleaned from matrix C^1. For the SAR model, $v_{00} = 1 + 4\hat{\rho}^2$, $v_{10} = -2\hat{\rho}$, $v_{11} = 2\hat{\rho}^2$, $v_{20} = \hat{\rho}^2$, and all other $v_{ij} = 0$, result from matrices C^1 and C^2. In each case, these sets of equations must be solved simultaneously to calculated the

150   A CASEBOOK FOR SPATIAL STATISTICAL DATA ANALYSIS

```
.9357491  .9692228  .9692463  .9692463  .9692463  .9692463  .9692463  .9692463  .9692228  .9357491
.9692228  .9998677  .9999343  .9999344  .9999344  .9999344  .9999344  .9999343  .9998677  .9692228
.9692463  .9999343  .9999998  .9999999  .9999999  .9999999  .9999999  .9999998  .9999343  .9692463
.9692463  .9999344  .9999999 1.0000000 1.0000000 1.0000000 1.0000000  .9999999  .9999344  .9692463
.9692463  .9999344  .9999999 1.0000000 1.0000000 1.0000000 1.0000000  .9999999  .9999344  .9692463
.9692463  .9999344  .9999999 1.0000000 1.0000000 1.0000000 1.0000000  .9999999  .9999344  .9692463
.9692463  .9999344  .9999999 1.0000000 1.0000000 1.0000000 1.0000000  .9999999  .9999344  .9692463
.9692463  .9999344  .9999999 1.0000000 1.0000000 1.0000000 1.0000000  .9999999  .9999344  .9692463
.9692463  .9999343  .9999998  .9999999  .9999999  .9999999  .9999999  .9999998  .9999343  .9692463
.9692228  .9998677  .9999343  .9999344  .9999344  .9999344  .9999344  .9999343  .9998677  .9692228
.9357491  .9692228  .9692463  .9692463  .9692463  .9692463  .9692463  .9692463  .9692228  .9357491
```

```
-.00002 -.00002 -.00002 -.00002 -.00002 -.00002 -.00002 -.00002 -.00002 -.00002
-.00002 -.00001 -.00001 -.00001 -.00001 -.00001 -.00001 -.00001 -.00001 -.00002
-.00002 -.00001 -.00002  .00001  .00013  .00020  .00013  .00001 -.00002 -.00002
-.00002 -.00001  .00001  .00088  .00277  .00532  .00277  .00088  .00001 -.00002
-.00002 -.00001  .00013  .00277 -.03410 -.15068 -.03410  .00277  .00013 -.00002
-.00002 -.00001  .00020  .00532 -.15068 1.00000 -.15068  .00532  .00020 -.00002
-.00002 -.00001  .00013  .00277 -.03410 -.15068 -.03410  .00277  .00013 -.00002
-.00002 -.00001  .00001  .00088  .00277  .00532  .00277  .00088  .00001 -.00002
-.00002 -.00001 -.00002  .00001  .00013  .00020  .00013  .00001 -.00002 -.00002
-.00002 -.00001 -.00001 -.00001 -.00001 -.00001 -.00001 -.00001 -.00001 -.00002
-.00002 -.00002 -.00002 -.00002 -.00002 -.00002 -.00002 -.00002 -.00002 -.00002
```

```
.9408617  .9715732  .9715886  .9715885  .9715885  .9715885  .9715885  .9715886  .9715732  .9408617
.9715732  .9999025  .9999519  .9999518  .9999518  .9999518  .9999518  .9999519  .9999025  .9715732
.9715886  .9999519 1.0000002 1.0000001 1.0000001 1.0000001 1.0000001 1.0000002  .9999519  .9715886
.9715885  .9999518 1.0000001 1.0000000 1.0000000 1.0000000 1.0000000 1.0000001  .9999518  .9715885
.9715885  .9999518 1.0000001 1.0000000 1.0000000 1.0000000 1.0000000 1.0000001  .9999518  .9715885
.9715885  .9999518 1.0000001 1.0000000 1.0000000 1.0000000 1.0000000 1.0000001  .9999518  .9715885
.9715885  .9999518 1.0000001 1.0000000 1.0000000 1.0000000 1.0000000 1.0000001  .9999518  .9715885
.9715886  .9999519 1.0000002 1.0000001 1.0000001 1.0000001 1.0000001 1.0000002  .9999519  .9715886
.9715732  .9999025  .9999519  .9999518  .9999518  .9999518  .9999518  .9999519  .9999025  .9715732
.9408617  .9715732  .9715886  .9715885  .9715885  .9715885  .9715885  .9715886  .9715732  .9408617
```

```
-.00019 -.00014 -.00014 -.00014 -.00014 -.00014 -.00014 -.00014 -.00014 -.00019
-.00014 -.00010 -.00010 -.00010 -.00010 -.00010 -.00010 -.00010 -.00010 -.00014
-.00014 -.00010 -.00011 -.00008  .00003  .00009  .00003 -.00008 -.00011 -.00014
-.00014 -.00010 -.00008  .00070  .00241  .00475  .00241  .00070 -.00008 -.00014
-.00014 -.00010  .00003  .00241 -.03305 -.14545 -.03305  .00241  .00003 -.00014
-.00014 -.00010  .00009  .00475 -.14545 1.00000 -.14545  .00475  .00009 -.00014
-.00014 -.00010  .00003  .00241 -.03305 -.14545 -.03305  .00241  .00003 -.00014
-.00014 -.00010 -.00008  .00070  .00241  .00475  .00241  .00070 -.00008 -.00014
-.00014 -.00010 -.00011 -.00008  .00003  .00009  .00003 -.00008 -.00011 -.00014
-.00014 -.00010 -.00010 -.00010 -.00010 -.00010 -.00010 -.00010 -.00010 -.00014
-.00019 -.00014 -.00014 -.00014 -.00014 -.00014 -.00014 -.00014 -.00014 -.00019
```

Map 3.4. Top: (a) Geographic distribution of the diagonal values from the inverted n-by-n matrix R, where r_{ij} is defined by the exponential semivariogram model: rook's adjacency case, $\rho = 0.7/\lambda_{max}$, r = 0.62363. Second from top: (b) Geographic distribution of the values for the spatial dependency field associated with the central cell of a 10-by-10 lattice for Map 3.4a. Second from bottom: (c) Geographic distribution of the diagonal values from the inverted n-by-n matrix R, where r_{ij} is defined by the CAR spectral density function: queen's adjacency case, $\rho = 0.7/\lambda_{max}$, r = 0.62363. Bottom: (d) Geographic distribution of the values for the spatial dependency field associated with the central cell of a 10-by-10 lattice for Map 3.4c.

estimate $\hat{\rho}$. Expected values for higher-order models are reported in Griffith and Layne (1997).

As these results show, there are no perfect correspondences among pairings of semivariogram and autoregressive models, but good approximations seem possible, as seen by the RSSE. Ancillary numerical experiments, however, suggest that correspondence deteriorates somewhat as $\rho/\lambda_{max} \to 1$. Some discrepancies appear, in part, because autoregressive models are specified for discrete surfaces, while semivariogram models have been developed for continuous surfaces. Also, examination of the spatial dependency field (Map 3.4d) for the queen's case suggests that a weight specification other than $1/\sqrt{2}$ shows some promise. Finally, higher-order (> 2nd-order) autoregressive models with near-negligible values of ρ_j (j > 1) are suggested here. Employing such higher-order models would, however, be counter to the operational notion of scientific parsimony.

3.3.3. REFLECTIONS ON §3.3.2 BY PROFESSOR J. K. ORD[5]

Time series analysts have an easy life! They deal in but one dimension, with a clear direction of causality and usually with regularly spaced data. Thus, for series of reasonable length, end effects are negligible, and it is possible to establish relatively straightforward connections among the spectral, covariance, and autoregressive-moving average representations. In addition, when the original series is nonstationary, differencing will often serve to recover that desirable state of affairs.

By contrast, the spatial modeler has a tough task. Edge effects in two dimensions impact the autocorrelations to $O(n^{-\frac{1}{2}})$ rather than $O(n^{-1})$ so that adjustments for edge effects are much more critical. Further, spatial data, whether aggregated or punctated, are often irregularly spaced. If we restrict attention to very large data sets located on a regular grid, such as satellite images, it becomes possible to link the CAR and SAR schemes to variogram models using the exponential and Bessel forms respectively, as noted by Griffith and Layne. The natural question that arises is how close the linkages remain when the data set is of modest size. The authors have provided an interesting numerical exploration of this question and show that the correspondence does indeed deteriorate at the edges in a fairly consistent manner, with the side effect that one may be led to fit a more elaborate model.

Stepping back from these results a little, we must ask what are their implications? When the study area is a "sample" embedded in a larger region, as with satellite images and in agricultural trials, it will often be possible to use "guard rows" of observations and to condition upon these values, thereby reducing edge effects (cf. Cressie 1991, 433-47). When such an approach is not possible, we

---

[5]This commentary was drafted as a review of Griffith and Layne (1997).

need to consider models for which boundary cells or observations have higher conditional variances to allow for their lower degree of connectivity; the unconditional variances specified in the variogram approach are then simpler to deal with, although the model assumptions may still be inappropriate. What does it mean to specify stationarity or a regionalized variable structure for a study area with well-defined boundaries, where the assumptions break down when the boundaries are crossed?

When we turn to irregular patterns of observations, new problems arise. Economic or political boundaries may provide an innate form of nonstationarity, and we would do well to recognize that no model ever represents reality but that spatial models are often based on even more heroic assumptions than most others. In a sense, this admission affords a degree of relief, because we can approach data analysis in a more descriptive vein and seek a model as a convenient summary rather than a basis for rigorous inferences. The authors' results may then be interpreted to show that we can adopt whichever representation is most appropriate, without doing too much violence to the overall descriptive picture that emerges.

If spatial models are to be useful in practice, it is essential that we understand their interrelationships and their limitations. Griffith and Layne have helped us take a major step down that road.

3.3.4. TWO-PARAMETER AUTOREGRESSIVE FUNCTIONS AND DIRECTIONAL SEMIVARIOGRAM MODELS

Anisotropy (§3.2.4) alludes to nonstationary, multiparameter spatial autoregressive models. Martin (1990) notes that the use of even biparametric spatial autoregressive inverse-covariance matrix specifications is extremely rare in applications, although well established for very special cases. This avoidance of the two-parameter model in part is due to numerical difficulties affiliated with its estimation. The critical simplicity feature of a specification is that the pair of geographic weights matrices either is commutative or can exploit the matrix algebra notion of similarity. In this first instance, $C_1 C_2 = C_2 C_1$ for the pair of n-by-n geographic weights matrices C_1 and C_2. Although matrix commutativity does not hold in general—and it can be overlooked here for the trivial case of diagonal matrices, because they would imply the absence of areal unity configuration articulation—it does hold for the regular square tessellation. In this second instance, one would like to find a rotation matrix T, where $T^T T = I$, such that $C_1 = E_1 \Omega_1 E_1^T$, $C_2 = E_2 \Omega_2 E_2^T$, and hence $E_1(\alpha I + \beta \Omega_1^\delta)^{1/2} \approx E_2 \Omega_2^{1/2} T$, where E_j is the matrix of eigenvectors, Ω_j is the diagonal matrix of eigenvalues for matrix C_j, and δ is an exponent. This second approach holds considerable potential generalization utility but to date has been limited to the case of $C_2 = C_1^\delta$. In contrast, specification and estimation of directional semivariogram functions is far more common.

Theoretical correlograms can be constructed for the biparametric spatial autoregressive model matched with a regular square tessellation for the rook's set of neighbors spatial linkages; closed form spectral density function counterparts to equation (3.18) are known. The theoretical two-dimensional correlograms generated by these equations do not need to be rotated, as their major and minor axes coincide with the corresponding horizontal and vertical axes (i.e., the angle of rotation would be 0).

Results of biparametric semivariogram model fits are reported in Griffith and Layne (1997). For the isotropic case, the exponential semivariogram model relates almost perfectly to the spatial statistical CAR model case, the Bessel function semivariogram model relates almost perfectly to the spatial statistical SAR model case, and in all cases explored here the RSSE essentially equal 0. In other words, these linkages persist in the presence of anisotropy. Differences in the computed directional ranges are consistent with the differentials between the autoregressive parameters ρ_p and ρ_q. Furthermore, as the size of the region under study increases, an inspection of the change in semivariogram parameter estimates reveals the presence of edge effects, as it does in the isotropic situation. Given the isotropic tendency noted earlier, these edge effects become more pronounced as the two levels of spatial autocorrelation increase.

3.4. COMPUTER CODE FOR SPATIAL AUTOREGRESSIVE AND SEMIVARIOGRAM MODELING

Paralleling the content of §1.7, an explanation for implementing SAS code used for spatial autoregressive and semivariogram modeling is presented in this section. Again, the principal advantage of utilizing implementations developed for standard statistical software packages is that a researcher does not have to leave the package's environment to compute spatial statistics, and hence all of the other statistical technology contained in the statistical software package remains available to the researcher.

Spatial autoregressive procedures, as well as other spatial statistics, can be implemented in SAS, in MINITAB (Griffith 1988b, 1989, 1992c), and in SPSS (Appendix 3-B and 3-D). Griffith (1988b) discusses implementation of this maximum likelihood estimation in IMSL, which, with little effort, can be converted to ESSL. To implement spatial autoregressions, a given standard statistical technique is rewritten in terms of regression. Hence a regression routine is the key to properly analyzing georeferenced data in this fashion. Because a normalizing constant (the Jacobian term, J; §3.1.1, §3.1.2) is present and is a function of the eigenvalues of the geographic weights matrix (usually **C** or **W**), an eigenvalue routine also is needed, preferably for a real asymmetric matrix because **W** usually is of this type; fortunately a symmetric counterpart for eigenfunction extraction purposes exists when a symmetric matrix is row-standardized. Spatial lag variables, such as **CX** and **WY**, appear in the regression equation; thus, matrix multiplication is required to conduct a spatial autoregressive analysis. Finally, because the commonly used multivariate normal probability density function is not linked to a linear model in the presence of spatial autocorrelation, a nonlinear minimization routine is required. Thus, the essential procedures in SAS include PROC FACTOR, PROC TRANSPOSE, PROC NLIN, and the LAG function for regular square tessellations.

The SAS code presented in this section and in Appendices 3-A, 3-C and 3-E uses PROC NLIN as the primary procedure for obtaining nonlinear weighted least squares estimates of parameters, while also allowing generation of nonlinear regression diagnostics, helpful for identifying values that may cause problems in the analysis and interpretation of results. Again, the reader needs to be aware that preprocessing of georeferenced data by the user is necessary to efficiently and successfully use SAS, or other commercial statistical software, for this purpose.

Of note is that when the "Marquardt" method of nonlinear estimation is specified for PROC NLIN, SAS version 6.12, and beyond, will automatically compute the analytical derivative statements (specifically the first derivatives for each of the parameters estimated, as specified in the MODEL statement) if they are not explicitly supplied by the user. We utilize this functionality of PROC NLIN in cases where the analytical derivative statements computed by SAS are correct. This is the case for the code in Program #6 below, for the CAR model in Appendix 3-A, and for the semivariogram models appearing in Appendix 3-E. In the case of the constant mean CAR model (i.e., no regressor variables are included in the model specification), which is the version implemented here, the derivative statements created by SAS are correct; the same parameters are estimated each time in the CAR model. In the case of the SAS code presented in Appendix 3-C, for the AR and SAR models, SAS does not compute all of the analytical derivatives correctly[6]; hence, they need to be explicitly supplied by the user. For these two models, both the model and derivative statements are explicitly supplied, but are done through the use of macro coding within the program. Differing numbers of parameters need to be estimated, depending on whether or not regressor variables are included in the model. The macro code takes these differences into account and creates the correct model and derivative statements. Thus, it is not incumbent on the user to explicitly supply these statements.

3.4.1. SAS CODE FOR SPATIAL AUTOREGRESSIVE MODELS

The first step in spatial autoregressive modeling is to obtain the eigenvalues of matrices **C** and **W**. Within SAS, they must be declared to be correlation matrices, reinforcing the notion that spatial autocorrelation refers to redundant information. Because the diagonal entries of a correlation matrix always are unity, both c_{ii} and w_{ii} must be set equal to 1; hence, the resulting eigenvalues are $(\lambda_i + 1)$ and must have 1 subtracted from them. Once these preliminary adjustments are made, PROC FACTOR can be employed (PROC PRINCOMP will not work in general because negative eigenvalues often will be present). For matrix **W**, its symmetric equivalent

---

[6]Specifically the partial derivative with respect to ρ is incorrect.

matrix must be used. Meanwhile, deficiencies of SAS/BASE include (1) the need to work in terms of data vectors (rows), which makes necessary data manipulations for spatial statistics quite awkward, (2) the implementation of the CAR model being in terms of the log-likelihood function differential equations, and (3) the LAG function being restricted to at most 99 lags. Some of these problems can be circumvented with SAS IML, a topic not treated in this book (except for the program CONNCHK.SAS, §1.7.2), while others can be resolved by trading off an increase in computing time with a decrease in efficiency (see §5.7).

Program #6 (EIGEN_J.SAS), presented below, is used to compute the eigenvalues from both matrix **C** and **W**, and output the minimum and maximum eigenvalues (and their respective inverses) for either **C** or **W**. (This program does not need to be utilized if the user is analyzing data spatially arranged on a regular square lattice and implementing the CAR model). The format for user input of this program follows the same organization as that used for Programs #1 and #5 (§1.7.1 and §1.7.2), and the code presented in Appendix 1-A, and is explained in the first text box at the beginning of the program. Again, for this program, and the SAS programs presented in Appendices 3-A and 3-C, the geographic weights matrix (or connectivity matrix) is input in the form of a neighbors file.

Program #6

```
/*-------------------------------------------------------------|
|   EIGEN_J.SAS                                                 |
|   This program computes the eigenvalues, inverses of the minimum and |
|   maximum eigenvalues and estimates of coefficients A1,A2,D1,D2 used |
|   in the Jacobian approximation for either the C or W matrix. |
|   Inputs required by the user are contained in the macro variables: |
|       dspath - path location of the neighbors file.          |
|       flname - filename of the neighbors file. File extension must |
|                be "nei".                                      |
|       contype - type of connectivity matrix to work with (C or W). |
|       parmstr - contains the starting values if the Jacobian |
|                 approximation will be used in subsequent CAR or SAR/AR |
|                 models. The current values are appropriate if the |
|                 number of areal units is about 1000 or less. These |
|                 parameter estimates can beignored if the exact |
|                 solution of the Jacobian term will be used.   |
|       adbound - specified as: 0.001<A1<=n,0.001<A2<=n,1.00001<D1<2, |
|                 1.00001<D2<2 although the lower bounds of D1 and D2 |
|                 might need to be decreased if the estimate equals the |
|                 lower bound.                                  |
|-------------------------------------------------------------*/
options nocenter nodate pageno=1;

%let DSPATH = %str(c:\nsf_geog\book\uranium);
%let FLNAME = %str(uranium);
%let CONTYPE = C;

data _null_;                        * Count number of observations;
  infile "&dspath.\&flname..nei";   * in neighborhood file.        ;
  input c1;
  call symput('nrecs',left(_n_));
run;
%put number of observations = &nrecs;
```

```
/*------------------------------------------------------------------|
 | Coefficients A1, A2, D1 and D2 may need to be adjusted in order  |
 | to get a global optimum. Otherwise use starting values:          |
 | A1 = A2 = 5, D1 = 1.1 and D2 = 1.5.                              |
 |------------------------------------------------------------------*/
%let PARMSTR = %str(PARMS A1=5   A2=5    D1=1.1    D2=1.5);
%let ADBOUND = %str(BOUNDS 0.001<A1<=&nrecs , 0.001<A2<=&nrecs ,
                           1.00001<D1<2 , 1.00001<D2<2);

 data c; infile "&dspath.\&flname..nei" missover;
   input in_c1-in_c200;            * Construct matrix C from;
   array c{&nrecs} c1-c&nrecs;     * neighborhood file.      ;
   array in_c{200} in_c1-in_c200;
   do i = 1 to &nrecs; c{i} = 0; end;
   do i = 2 to 200;
     if in_c{i} > . then c{in_c{i}} = 1;
   end;
   crowsum=sum(of c1-c&nrecs);
   do i=1 to &nrecs;
     if _n_=i then c{i} = 1;
   end;
   drop i in_c1-in_c200;
 run;
 data w; set c;                    * Construct matrix W;
   array c{&nrecs} c1-c&nrecs;     * from matrix C.     ;
   array w{&nrecs} w1-w&nrecs;
   do i=1 to &nrecs;
     w{i} = c{i}/sqrt(crowsum);
   end;
 run;
 proc transpose data=w prefix=trsum out=temp1;
   var crowsum;
 run;
 data iw; set w;
   if _n_ = 1 then set temp1;
   array rmtx{&nrecs} trsum1-trsum&nrecs;
   array w{&nrecs} w1-w&nrecs;
   do i=1 to &nrecs;
     w{i} = w{i}/sqrt(rmtx{i});
   end;
   do i=1 to &nrecs;
     if _n_=i then w{i} = 1;
   end;
 drop _name_ crowsum trsum1-trsum&nrecs;
 run;
 data correl(type=corr); set c;
   _type_ = 'corr';
 run;                              * Compute ieigenvalues.;
 proc factor data=correl method=prin n=0 outstat=outeigc;
   var c1-c&nrecs;
   TITLE1 "IGNORE THIS PART OF THE OUTPUT";
 run;
 data outeigc (replace=yes); set outeigc;
   if _type_='EIGENVAL';
   drop _type_ _name_;
 run;
 proc transpose data=outeigc prefix=eigc out=outeigc(replace=yes);
   var c1-c&nrecs;
 run;
 data wcorrel(type=corr); set iw;
   _type_ = 'corr';
 run;
 proc factor data=wcorrel method=prin n=0 outstat=outeigw;
```

```
    var w1-w&nrecs;
    TITLE1 "IGNORE THIS PART OF THE OUTPUT";
run;
data outeigw (replace=yes); set outeigw;
    if _type_='EIGENVAL';
    drop _type_ _name_;
run;
proc transpose data=outeigw prefix=eigw out=outeigw(replace=yes);
    var w1-w&nrecs;
run;
proc datasets lib=work;
    delete c w iw trsum / mt=data;
run;
data outeig; merge outeigc outeigw; * Correct eigenvalues ;
    id = _n_;                        * from PROC FACTOR.    ;
    eigenc=eigc1-1;
    eigenw=eigw1-1;
    LAMBDA=eigen&contype;
    drop eigc1 eigw1;
run;
proc sort data=outeig; by descending id; run;
proc print data=outeig; var eigenc eigenw;
    TITLE1 "Eigenvalues for MATRIX C and MATRIX W";
    TITLE2 "in file &dspath.\&flname..nei";
run;
proc univariate data=outeig noprint;
    var lambda;
    output out=extremel max=lmax min=lmin;
run;
data inveig; set extremel;
    if lmin = 0 then lmin = .00001; * Prevent division;
    if lmax = 0 then lmax = .00001; * by zero error.  ;
    inv_lmin = 1/lmin;
    inv_lmax = 1/lmax;
run;
proc print data=inveig; var lmin lmax inv_lmin inv_lmax;
    TITLE1 "Minimum and maximum eigenvalues, and the inverse of these";
    TITLE2 "for matrix &contype in file &dspath.\&flname..nei";
run;
data step1;                         * Sampled rho values;
    if _n_=1 then set extremel;     * for Jacobian       ;
    set outeig;                     * approximation      ;
    start  = 0.999/lmin;            * coefficients.      ;
    finish = 0.999/lmax;
    inc = (finish - start)/21;
    array jacob{22} jac1-jac22;
    rho=start;
    do i = 1 to 22 by 1;
        jacob{i} = -log(1 - rho*lambda);
        rho = rho + inc;
    end;
    drop i;
run;
proc means noprint data=step1;
    var jac1-jac22;
    output out=jacob1 mean=;
run;
proc transpose data=jacob1 out=jacob2;
    var jac1-jac22;
run;
data step1(replace=yes); set step1;
    if _n_=1 then set extremel;
    do rho = start to finish by inc;
```

```
      output;
    end;
  run;
  data step1(replace=yes);
    set jacob2; set step1;
    j = col1;
    drop col1;
  run;
  /*----------------------------------------|
  | Estimate parameters A1, A2, D1 and D2.  |
  | for Jacobian approximation method.      |
  |----------------------------------------*/
  proc nlin data=step1 maxiter=500 method=marquardt;
        &parmstr;
        &adbound;
        MODEL J = (RHO/(2*&nrecs))*(A2*(LMIN**2) - A1*(LMAX**2)) +
                  (1/&nrecs)*(A2*ABS(LMIN)*LOG(D2/ABS(LMIN)) +
                       A1*LMAX*LOG(D1/LMAX) -
                       A2*ABS(LMIN)*LOG(D2/ABS(LMIN) + RHO) -
                       A1*LMAX*LOG(D1/LMAX - RHO));
    TITLE1
       "Estimates for A1, A2, D1 and D2 from file &dspath.\&flname..nei";
    TITLE2
       "Variables A1, A2, D1, D2 are used as initial starting values "
       "when the";
    TITLE3
       "Jacobian approximation is used, rather than the exact solution.";
  RUN;

  data pltstep1; length horder$ 140;     * Make the Jacobian;
    set step1;                           * term vs. rho plot;
    rho=round(rho,0.01);                 * look pretty.      ;
    if _n_ = 1 then do; horder = trim(left(rho)); end;
    else do;
      horder = trim(left(horder))||','||trim(left(rho));
    end;
    retain horder;
    call symput('horder',horder);
    drop horder;
  run;
  symbol1 c=black v=dot i=none;
  proc gplot data=pltstep1;
    axis1 c=black w=2
          label = (c=black font=swissb h=1 'sampled values of '
                   font=cgreek h=1 'r')
          value = (c=black font=swissb h=0.7)
          major = (w=2 c=black)
          minor = none
          order = &horder;
    axis2 c=black w=2
          label = (a=90 r=0 font=swissb h=1 'Jacobian term')
          value = (c=black font=swissb h=0.7)
          major = (w=2 c=black)
          minor = none;
    plot j*rho / haxis=axis1
                 vaxis=axis2;
    title1 font=swissb h=1
           "Plot of the Jacobian term vs. selected values of rho";
    title2 font=swissb h=1
           "for matrix &contype file &dspath.\&flname..nei";
    title3 h=0.1 " ";
  run;
  quit; * Stop GPLOT running;
```

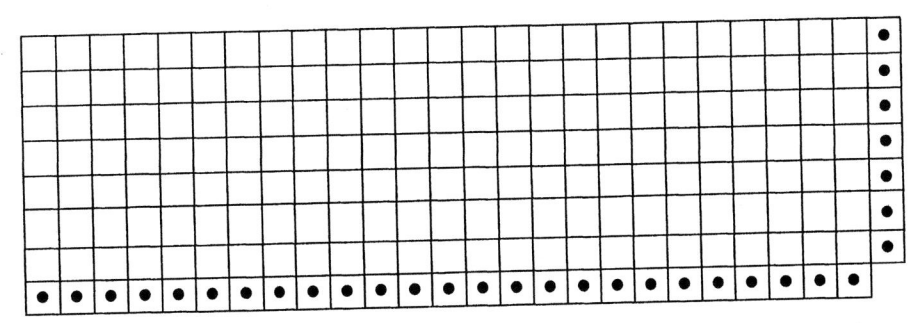

Map 3.5. The attached lattice column of missing values on the right-hand side and across the bottom of a K-by-L rectangular region, converting it to a (K+1)-by-(L+1) region, to trick SAS into correctly using the time series LAG function for spatial statistical analyses.

In addition to computing the eigenvalues for matrices **C** and **W**, this program also estimates the coefficients used in the approximation to the Jacobian term (§3.1.2), if this method will be employed instead of the exact solution, for any of the spatial autoregressive models. The variables A1, A2, D1 and D2 represent the coefficients α_1, α_2, γ_1 and γ_2, respectively, appearing in equation (3.10). The minimum and maximum eigenvalues for the selected connectivity matrix (**C** or **W**) computed prior to estimating these coefficients are automatically inserted into the appropriate places in (3.10). Starting values and bounds for each of the coefficients are explained in the first and second text boxes, and provided in the macro variables PARMSTR and INITAADD, after the number of observations in the neighbors file is determined. (This is because the number of observations in the neighbors file, i.e. the total number of areal units, is used as the upper bound for both A1 and A2). Estimation of these 4 coefficients is accomplished by dividing the feasible parameter space of ρ (given as the inverse of the minimum and maximum eigenvalues of the user-selected connectivity matrix) into 21 intervals, and then computing the Jacobian term for each of these 22 selected values of ρ. These data, along with the minimum and maximum eigenvalues, then are supplied to PROC NLIN for the coefficient estimations. (The user should note that if the exact solution for the Jacobian term is used in the spatial autoregressive routines presented in Appendices 3-A and 3-C, then this part of the output can be ignored.) Finally, a plot of the Jacobian term versus the selected values of ρ is displayed.

In the case of a regular rectangular tessellation, the LAG function, which is used for time series work, can be tricked into helping with spatial statistics computations by augmenting the data with two ordering variables, one indexing position in the horizontal direction and one in the vertical direction. In its time series specification, the LAG function incorrectly codes an areal unit at the end of one vertical row as being juxtaposed with an areal unit at the beginning of the next row of a rectangular region. Here the remedy is to introduce a lattice column of missing values on the right-hand side and across the bottom of the rectangular region (see Map 3.5); hence a K-by-L region becomes (K+1)-by-(L+1), but with the lower right-hand corner removed, and the data set has (L + K) additional lines of missing data inserted into it, one after each K original data records and L appended at the end of the data set (this method can be employed with MINITAB, too)—there is no need to insert the lower right-hand corner artificial missing value. Sample code for analyzing some variable Y distributed across a P-by-Q (P rows and Q columns) regular square tessellation is contained in the code found in Appendix 3-A, in the section denoted "regular" for the CAR model. The code assumes that the data initially are ordered from west to east (i.e., left to right) and north to south (i.e., top to bottom)—the northwest pixel is coordinate (1,1) and the southeast pixel is coordinate (P,Q). (The user should note, however, that arrangement of the data must be organized so that P is always the smaller of P and Q, and may result in the data values being rotated 90°. For data values already present in a data file, this step can be easily accomplished prior to utilizing the code in Appendix 3-A). This code initially sorts the data by the vertical coordinate in the sorted order of the horizontal coordinate, and then extracts the attribute value located north of each pixel with the LAG1 operator. This sorting rotates the original P-by-Q rectangular region 90° in a clockwise direction. The second sorting reorders the data to its original P-by-Q geometric arrangement, and uses the LAG1 operator to extract the attribute value located west of each pixel. The third sorting rotates the original P-by-Q rectangular region 180°, and uses the LAG1 operator to extract the attribute value located south of each pixel. Finally, the fourth sorting rotates the original P-by-Q rectangular region 270°, and uses the LAG1 operator to extract the attribute value located east of each pixel. The final data management manipulations ensure that all of the contrived missing data are handled correctly and then deleted from the data set, and computes **CY** as the sum of juxtaposed data (needed for a CAR model). Because attribute values for each of the four compass directions are extracted separately, only a simple data manipulation is needed in this step to retain directional sums and averages required for an anisotropic analysis. This same process may be applied for analysis using an AR or SAR model, and replacing **CY** with **WY**, but is not included in the SAS program in Appendix 3-C.

The two SAS programs presented in Appendices 3-A and 3-C provide the code necessary for implementing any of the three spatial autoregressive models presented in this chapter. CAR.SAS (Appendix 3-A) specifies a CAR model, is restricted to utilizing only the C version of the connectivity matrix, and provides options for analyzing data that are spatially arranged in an irregular manner, or regularly in the form of a regular square lattice (thus utilizing the LAG function described previously), and use of the exact solution for the Jacobian term, or its approximation, for an irregular arrangement. In SAR_AR.SAS (Appendix 3-C), the user selects the type of model to implement (AR or SAR) in the macro variable SPECTYPE, an option is provided to specify use of either the exact solution or the approximation of the Jacobian term, options are provided to allow regressor variables in the model, and the model specification is restricted to utilizing only the **W** version of the connectivity matrix. No option is provided in this program for distinguishing between data arranged spatially as regular or irregular. Again, explanations for macro variable input are given in the first text box for each of these programs.

For both of these programs, if the approximation to the Jacobian term is used (this option is specified in the macro variable JACOB), estimates for the coefficients A1, A2, D1 and D2 (from Program #6 above) are placed into the macro variable INITAADD. The bounds for the variable rho are the inverses of the minimum and maximum eigenvalues of either matrix **C** or **W** (the values of which, again, are taken from Program #6), and placed into the macro variable RHOBOUND. Likewise, the maximum and minimum eigenvalues computed from Program #6 are inserted into the macro variables MAXEIGEN and MINEIGEN, respectively. The Moran Coefficient can be supplied for the initial value of rho in the cases of the AR or SAR models. In the case of the CAR model, MC cannot be used as an initial value of rho as many times this value will exceed the upper bound of rho. For the CAR model, for positive spatial autocorrelation, an initial value of rho can be derived by dividing the upper rho bound value by 2. The user should note that if the data analysis variable is to be transformed and/or residuals from an OLS regression are to be analyzed, these steps need to be performed prior to computing the Moran Coefficient. For both CAR.SAS and SAR_AR.SAS, if the analysis variable is to be transformed, this step needs to be performed prior to using the programs, or alternatively, the user can modify the first data step (data indat) to include this information. If the residuals from an OLS regression are to be analyzed using a CAR model, an option within the code allows this step needs to be performed. To use the SAS code in SAR_AR.SAS where residuals from a regression will be analyzed, regressor variables are included directly in the model statement and the actual residuals from an OLS regression should not be used as data input values for the AR or SAR models. For the CAR model, $\bar{x}$ and s of the data analysis variable can be used as the initial values in the variables mu and sigma (specified in the macro variable PARMSTR), unless the analysis variable will be converted to z-scores first. Then the initial values for $\bar{x} = 0$ and s = 1. For the AR or SAR models, if regressor variables are included in a model specification, initial values can be taken from an OLS regression. In both the AR and SAR models, an initial value for B0 is required. In the case where at least one regressor variable is included in the model, the initial value for B0 is, obviously, the intercept of the regression. In the case where no regressor variables are included in the model, the initial value for B0 is simply the mean of the analysis variable. (Of special note is that the AR model in SAR_AR.SAS is implemented using the term b0*(1-rho) instead of only b0; different from its equational form. This implementation is utilized because the parameter estimates for b0 and any regressor variables will be more similar to their OLS counterparts than if only b0 is used in the model, and is the method used for most of the analyses in this book utilizing the AR model). For the CAR model, if the residuals from an OLS regression are to be analyzed, then, again, this step needs to be performed prior to computing $\bar{x}$ and s for the residuals. In the case of the AR or SAR models, if regressors are included in a model specification, then they are supplied in the macro variable

MIV, and the number of regressor variables specified in NMIV. Explicitly supplying the correct number of regressor variables is necessary for the program to write the model and derivative statements correctly. If no regressor variables are to be included, the macro variable MIV should be left blank and NMIV should equal 0. Caution must be used for regressor variable names in SAR_AR.SAS as they are limited to a maximum of 7 characters, minus the number digits involved in denoting the total number of areal units in the neighbors file, and the last character of the variable name cannot be a number. This is because any regressor variables included in the model are prefixed and suffixed at various places in the program. Also, regressor variable names are prefixed with a "b" in the PARMS statement (in the macro variable PARMSTR). The CAR.SAS program does not allow regressor variables in the CAR model. The form of the connectivity matrix (C or W) for both programs is required to be in the form of a neighbors file, with the same file name as the data analysis file. For the special case of a CAR model using a regular lattice, initial values only for $\bar{x}$ and s are specified in the PARMSTR macro variable and not for rho. Likewise, the bounds for rho are not required. This is because the eigen values of a regular square lattice are known analytically, are computed within the program, and the inverses of the minimum and maximum eigen values are supplied automatically by the program to define the upper and lower bounds of rho, respectively. The initial value of rho is computed as the upper rho bound x 0.90. If the user wishes to override these automatically supplied values, then the statements prior to the last PROC NLIN statement in the file need to be turned into comment statements (or deleted) and the values supplied explicitly at the beginning of the file.

Lastly, nonlinear estimation can sometimes be problematic as many local optima may be present, for instance, as visualized as a relatively flat curve of the Jacobian term vs. ρ plot. Thus, resulting parameter estimates can be highly sensitive to initial values of the parameters, particularly in the case of the CAR model. If a researcher suspects this is occurring, the model can be re-run several times, varying the starting values of, particularly, ρ until relatively consistent parameter estimates are achieved among different outputs. This strategy was required for some of the data sets analyzed in this book and, therefore, the reader may find differences in parameter estimates between some of the results presented in this book and when attempts are made to replicate our work using the suite of SAS code for spatial autoregressive modeling.

3.4.2. SAS CODE FOR SEMIVARIOGRAM MODELS

SAS code, in the form of macro programs, for the semivariogram models implemented throughout Chapters 4-8 of this book, in addition to the other models contained in Chapter 9, is presented in Appendix 3-E. Explicitly, the five semivariogram models most consistently implemented here are the exponential, Bessel, Gaussian, spherical, power and wave/hole, and are programmed separately as macro programs. In addition, Appendix 3-E contains a data preprocessing program (SEMIGRAM.SAS) and a program for plotting the modeling results (SEMIPLOT.SAS). To implement these SAS macro programs, the code from Appendix 3-E needs to be separated into specific files. Respective file names for the separate programs are contained in the text boxes at the beginning of each program. Thus, code for the file SEMIGRAM.SAS contains all of the lines of code from the line preceding the filename "SEMIGRAM.SAS" to the line preceding the comment line of the filename "SEMIEXPO.SAS". Consequently, a total of eight files could be created from the code in Appendix 3-E: SEMIGRAM.SAS, SEMIEXPO.SAS, SEMIBESS.SAS, SEMIGAUS.SAS, SEMISPHE.SAS, SEMIWAVE.SAS, SEMIPOWR.SAS and SEMIPLOT.SAS. These files must be placed together in a subdirectory, which can be separate from the data analysis files.

For semivariogram modeling, user inputs are supplied to macro variables in SEMIGRAM.SAS, and are described in the first text box in the program. The pathname where the programs are located is entered into the macro variable PROGPATH, and once defined, does not need to be changed. The pathname where the data analysis file is located (specifically the file with the "sam" file extension created from Program #3 in §1.7.1) is specified in the macro variable DSPATH. The same file name specified in Program #1 (§1.7.1) is specified in FLNAME and likewise for the name given to DSVAR in Program #1. This is the filename of the semivariogram data file to model and, again, required to have a file extension of "sam". A maximum distance (which is translated into a maximum lag distance) to model is given in the macro variable MAXLAG, and can be inferred from the semivariogram plot produced from Program #3.

For each of the semivariogram models implemented, initial starting parameter values are computed in the program SEMIGRAM.SAS. However, as seen for some of the semivariogram modeling results in this book, occasionally the initial starting values need to be supplied explicitly by the user. This user input is necessary because SAS will sometimes react poorly to the initial values supplied automatically by the program SEMIGRAM.SAS and fail to compute parameter estimates correctly. This problem is detectable when: the SASLOG contains an error statement like "A bug in SAS has been encountered..."; the plot containing the model fit(s) displays a horizontal line at or near 0; many times, the vertical axis [the axis for $\gamma(h)$] contains ridiculous values; and, values in the RSSE (and other estimated parameters) of the improperly estimated model are substantially different from those obtained for the other models. Although in these types of cases we have used the parameter estimates from WLSFIT as initial values, they may also be inferred by examination of the semivariogram plot from Program #3.

Finally, in order to refine parameter estimates for each of the different semivariogram models, we employ a two-step process of estimation that utilizes two PROC NLIN statements for each model. The first NLIN procedure employs the DUD method and initial parameter values supplied automatically by SEMIGRAM.SAS (or the user), and is less sensitive to initial parameter values than is the Marquardt method. The second NLIN procedure uses the Marquardt method and the parameter estimates taken from the first NLIN procedure as initial parameter estimates.

3.4.3. SPSS Code for Spatial Autoregressive and Semivariogram Models

SPSS code appearing in Appendices 3-B, 3-D, and 3-F closely, but does not exactly, replicate the SAS code discussed in § 3.4.1 and 3.4.2. Because of limitations in the macro facility of SPSS, Appendix 3-F does not replicate the macro structure of its counterpart, Appendix 3-E, and starting values in the SPSS code of Appendix 3-F must be specified manually—their positions are indicated as "_____" with comment lines. Because the ability of internally carrying forward estimated parameters from one command to another in SPSS is very limited, the portions of SAS code that execute this task are not translated into SPSS; the most conspicuous discrepancy of this type is for calculation of the R-squared value found in Appendix 3-E. And, for Appendix 3-D, the analytical derivatives for SAR (**W**) and AR (**C**) do not work.

Finally, because SPSS for windows (and probably for Unix [X-windows], too) presupposes that its users interactively specify graphical details, such as axis tick marks and axis ranges, after a chart with minimum details is produced, particular options of SAS PROC GPLOT do not translate (Appendix 3-E).

Nevertheless, basics of the necessary SPSS code for implementing spatial statistics are presented here for the interested reader.

APPENDIX 3-A
SAS code listing for the conditional autoregressive (CAR) model and the 20-by-25 regular square lattice Mercer-Hall data set

```
*THIS CODE IS COPYRIGHTED BY SYRACUSE UNIVERSITY, 1991*;
FILENAME ATTRIBUT 'H:\BOOK.OUP\MERCRHAL.DAT';
OPTIONS LINESIZE=72;
TITLE 'CAR MODEL FOR THE MERCER-HALL YIELD DATA';
************************************************************
*                                                          *
* SET P AND Q TO THE REGIONAL DIMENSIONS OF THE LATTICE    *
*                                                          *
************************************************************;
DATA STEP0;
P=20; Q=25;
RUN;
*----------------------------------------------------------*
*                                                          *
* DATA ARE FOR A SQRT(N)-BY-SQRT(N) REGULAR LATTICE, WHICH *
* HAS BEEN AUGMENTED TO A (SQRT(N)+1)-BY-(SQRT(N)+1) REGULAR*
* LATTICE.  THE ADDITIONAL LATTICE COLUMN IS ADDED TO THE  *
* EAST (OR RIGHT HAND) SIDE, AND CONTAINS MISSING VALUES.  *
* THE ADDITIONAL LATTICE ROW IS ADDED TO THE SOUTH (OR BOTTOM),*
* AND CONTAINS MISSING VALUES.  THESE ARE GENERATED IN DATA STEP2*
* DATA ARE ENTERED INTO "SQUARE DATA" BEGINNING WITH THE   *
* NORTHWEST CORNER OF THE LATTICE, AND MOVING FROM WEST TO *
* EAST.  THIS RESULTS IN A CONCATENATION OF SUCCESSIVE     *
* LATTICE ROW ENTRIES, WITH EACH SUCCESSIVE PAIR BEING     *
* SEPARATED BY A MISSING VALUE AREAL UNIT VALUE.           *
*                                                          *
*----------------------------------------------------------*;
DATA STEP1;
     INFILE ATTRIBUT;
     INPUT ORDER U V YIELD STRAW;
*********************************************
*                                           *
* EQUATE Y TO THE VARIABLE TO BE ANALYZED   *
*                                           *
*********************************************;
Y = YIELD;
UCO=U; VCO=V;
DROP U V YIELD STRAW;
RUN;
PROC UNIVARIATE NORMAL; VAR Y; RUN;
PROC STANDARD MEAN=0 STD=1 OUT=STEP1(REPLACE=YES); VAR Y; RUN;
DATA STEP1 (REPLACE=YES);
     SET STEP1;
Y2=Y*Y;
RUN;
DATA STEP2;
SET STEP0;
     DO I=1 TO Q;
          UCO = I;
```

```
              VCO = P+.1;
        OUTPUT;
        END;
        DO I=1 TO P;
              UCO = Q+.1;
              VCO = I;
        OUTPUT;
        END;
  DROP I P Q;
  RUN;
  DATA STEP1(REPLACE=YES);
        SET STEP1 STEP2;
  RUN;

  PROC SORT DATA=STEP1;
        BY VCO UCO;
  RUN;
  DATA STEP1(REPLACE=YES);
        SET STEP1;
  YN = LAG1(Y);
  RUN;

  PROC SORT DATA=STEP1;
        BY UCO VCO;
  RUN;
  DATA STEP1(REPLACE=YES);
        SET STEP1;
  YW = LAG1(Y);
  RUN;

  PROC SORT DATA=STEP1;
        BY DESCENDING UCO DESCENDING VCO;
  RUN;
  DATA STEP1(REPLACE=YES);
        SET STEP1;
  YS = LAG1(Y);
  RUN;

  PROC SORT DATA=STEP1;
        BY DESCENDING VCO DESCENDING UCO;
  RUN;
  DATA STEP1(REPLACE=YES);
        SET STEP1;
  YE = LAG1(Y);
  RUN;

  *----------------------------------------------------------------*
  *                                                                *
  * DETERMINATION OF THE NUMBER OF NEIGHBORS AND THE AVERAGE       *
  * NEIGHBORING VALUES FOR EACH AREAL UNIT.                        *
  *                                                                *
  *----------------------------------------------------------------*;

  DATA STEP1(REPLACE=YES);
        SET STEP1;
```

```
IF Y EQ '.' THEN DELETE;

IF YN EQ '.' THEN NBRN = 0; ELSE NBRN = 1;
IF YS EQ '.' THEN NBRS = 0; ELSE NBRS = 1;
IF YE EQ '.' THEN NBRE = 0; ELSE NBRE = 1;
IF YW EQ '.' THEN NBRW = 0; ELSE NBRW = 1;
CSUM = NBRN + NBRS + NBRE + NBRW;
IF YN EQ '.' THEN YN = 0;
IF YS EQ '.' THEN YS = 0;
IF YE EQ '.' THEN YE = 0;
IF YW EQ '.' THEN YW = 0;

CY = (YN + YS + YE + YW);

RUN;

DATA STEP2;
     SET STEP1;
YCY = Y*CY;
ITEMP = 1;
RUN;
PROC MEANS DATA=STEP2 NOPRINT;
     VAR Y Y2 YCY CY CSUM ITEMP;
     OUTPUT OUT=CYOUT3 SUM=YSUM Y2SUM YCYSUM CYSUM TCSUM IP;
RUN;

/*
PROC PRINT DATA=CYOUT3; RUN;
*/

DATA STEP3;
     INPUT IMU ISIGMA IRHO;
     CARDS;
   1 0 0
   0 1 0
   0 0 1
;
RUN;
*************************************************************
*                                                           *
* SET N TO THE NUMBER OF AREAL UNITS, P*Q.                  *
* P IS ALWAYS SET TO THE SMALLER OF THE TWO                 *
* RECTANGULAR DIMENSIONS.                                   *
*************************************************************;
DATA STEP0 (REPLACE=YES);
SET STEP0;
PI = ARCOS(-1);
ARRAY EVAL{500} LAM1-LAM500;
K=0;
DO I=1 TO P;
     DO J=1 TO Q;
          K=K+1;
          EVAL{K} = 2*(COS(I*PI/(P+1)) + COS(J*PI/(Q+1)));
     END;
  END;
```

```
RUN;

DATA CYOUT3 (REPLACE=YES);
     IF _N_=1 THEN SET CYOUT3;
IF _N_=1 THEN SET STEP0;
     IF _N_=1 THEN SET STEP1;
     SET STEP3;
RUN;

/*
PROC PRINT; RUN;
*/

*********************************************************
*                                                       *
* THE CAR SPATIAL STATISTICAL MODEL USING MATRIX C      *
*                                                       *
*********************************************************;
PROC NLIN METHOD=DUD MAXITER=500 DATA=CYOUT3;
     PARMS MU=0 SIGMA=0.21 RHO=0.245;
     BOUNDS 0<RHO<0.25232;

     ARRAY EVAL{500} LAM1-LAM500;
     JACOB = 0;
     DO I=1 TO 500;
          JACOB = JACOB + EVAL{I}/(1 - RHO*EVAL{I});
     END;
     Z = IMU*YSUM + ISIGMA*Y2SUM + IRHO*YCYSUM;
     MODEL Z = IMU*(MU*P*Q + RHO*(CYSUM - MU*TCSUM)) +
               ISIGMA*(P*Q*SIGMA + 2*MU*YSUM - (MU**2)*P*Q +
               RHO*(YCYSUM - 2*MU*CYSUM + (MU**2)*TCSUM)) +
               IRHO*(JACOB*SIGMA +2*MU*CYSUM - (MU**2)*TCSUM);
RUN;
```

APPENDIX 3-B
SPSS code[7] listing for the conditional autoregressive (CAR) model and the 20-by-25 regular square lattice Mercer-Hall data set[8]

```
/* SPSS code listing for the conditional autoregressive */.
/* (CAR) model and the 20-by-25 regular square lattice */.
/* Mercer-Hall data set */.

get file = 'c:\book.oup\Mercer-Hall.sav'
 /rename (U V = uc0 vc0) /drop straw.

descriptives variables= yield (y).

compute y2=y*y.
save outfile = 'c:\book.oup\step_1.sav'.

input program.
compute #p=20.
compute #q=25.
loop #i = 1 to #q.
  compute uc0 = #i.
  compute vc0 = #p + 0.1.
  compute y = 9999.
  compute y2 = 9999.
  end case.
end loop.
loop #i = 1 to #p.
  compute uc0 = #q +0.1.
  compute vc0 = #i.
  compute y = 9999.
  compute y2 = 9999.
  end case.
end loop.
end file.
end input program.
save outfile = 'c:\book.oup\step_2.sav'.

add files file = * / file = 'c:\book.oup\step_1.sav'.
save outfile = 'c:\book.oup\step_1.sav'.

sort cases by vc0 uc0.
compute yn = lag(y).
sort cases by uc0 vc0.
compute yw = lag(y).
sort cases by uc0 (d) vc0 (d).
compute ys = lag(y).
```

---

[7]Contributed by Dr. Akio Sone, Institute of Socioeconomic Planning, University of Tsukuba, Japan (asone@shako.sk.tsukuba.ac.jp).

[8]Note that the path name used here is 'c:\book.oup' and needs to be changed to that defining the drive and directory location of the user's data.

```
sort cases by vc0 (d) uc0 (d).
compute ye = lag(y).
save outfile = 'c:\book.oup\step_1.sav'
select if(y ~= 9999).
execute .

do if (yn = 9999) .
compute nbrn = 0.
else.
compute nbrn = 1.
end if.
do if (ys = 9999).
compute nbrs = 0.
else.
compute nbrs = 1.
end if.
do if (ye = 9999).
compute nbre = 0.
else.
compute nbre = 1.
end if.
do if (yw = 9999).
compute nbrw = 0.
else.
compute nbrw = 1.
end if.

compute csum = nbrn + nbrs + nbre + nbrw.

do if (yn = 9999).
compute yn = 0.
end if.
do if (ys = 9999).
compute ys = 0.
end if.
do if (ye = 9999).
compute ye = 0.
end if.
do if (yw = 9999).
compute yw = 0.
end if.

compute cy = (yn + ys + ye + yw).
compute ycy = y*cy.
compute itemp = 1.
execute.

matrix.
get cyout3 /file =* /
               variables = y y2 ycy cy csum itemp/missing=0.
compute csm = csum(cyout3).
release cyout3.
compute x = make(3,1,1).
compute csmm = x*csm.
save {csmm}/outfile= 'c:\book.oup\cyout3.sav'
 /variables = ysum y2sum ycysum cysum tcsum ip.
```

```
end matrix.
data list list /imu isigma irho.
begin data
1 0 0
0 1 0
0 0 1
end data.
list.
save outfile = 'c:\book.oup\step_3.sav'.

input program.
compute #p=20.
compute #q=25.
compute #pi = 3.14159265358979.
loop #i = 1 to #p.
loop #j = 1 to #q.
compute lmd1=2*(cos(#i*#pi/(#p+1))+cos(#j*#pi/(#q+1))).
end case.
end loop.
end loop.
end file.
end input program.
flip.
rename variables (var001 to var500 =lam1 to lam500).
save outfile = 'c:\book.oup\step01.sav'/drop = case_lbl.
save outfile = 'c:\book.oup\step02.sav'/drop = case_lbl.
add files file = 'c:\book.oup\step01.sav'/
                                 file = 'c:\book.oup\step02.sav'.
add files file = * / file = 'c:\book.oup\step02.sav'.
save outfile = 'c:\book.oup\step0.sav'.
erase file = 'c:\book.oup\step01.sav'.
erase file = 'c:\book.oup\step02.sav'.
execute.

match files /file = 'c:\book.oup\cyout3.sav' /
   file = 'c:\book.oup\step0.sav'/ file='c:\book.oup\step_3.sav'.
save outfile ='c:\book.oup\final.sav'.
execute.

/* THE CAR SPATIAL STATISTICAL MODEL USING MATRIX C */.

compute z =imu*ysum + isigma*y2sum + irho*ycysum.
model program mu=0 sigma=0.21 rho=0.245.
compute jacob=0.
vector eval = lam1 to lam500.
loop #i =1 to 500.
  compute jacob = jacob + eval(#i)/(1-rho*eval(#i)).
end loop.
compute p = 20.
compute q = 25.
compute pred = imu*(mu*p*q + rho*(cysum - mu*tcsum))+
              isigma*(p*q*sigma + 2*mu*ysum - (mu**2)*p*q* +
              rho*(ycysum -2*mu*cysum + (mu**2)*tcsum)) +
              irho*(jacob*sigma + 2*mu*cysum - (mu**2)*tcsum).
cnlr z / criteria = iter 500 /bounds 0.25232>rho>0.
```

APPENDIX 3-C
SAS code listing for the autoregressive response (AR) and the simultaneous autoregressive (SAR) models for the Wolfcamp aquifer data

```
FILENAME ATTRIBUT 'H:\BOOK.OUP\WOLFCAMP.DAT';
FILENAME CONN 'H:\BOOK.OUP\WOLFCAMP.CON';
OPTIONS LINESIZE=72;
TITLE 'SAR FOR WOLFCAMP AQUIFER DATA';
**********************************************************
*                                                        *
* SET N TO THE NUMBER OF AREAL UNITS,                    *
* INPUT THE EXTREME EIGENVALUES FOR MATRICES C AND W     *
*                                                        *
**********************************************************;
DATA STEP1;
    INFILE ATTRIBUT;
    INPUT ID Y U V LAMBDAC LAMBDAW;
LAMBDA=LAMBDAW;
RUN;
PROC STANDARD MEAN=0 OUT=STEP1(REPLACE=YES); VAR U V; RUN;
PROC STANDARD MEAN=0 STD=1 OUT=STEP1(REPLACE=YES);
                                            VAR Y; RUN;
PROC PRINT; VAR Y U V; RUN;
PROC REG;
    MODEL Y = U V;
RUN;
DATA STEP1 (REPLACE=YES);
    SET STEP1;
    INFILE CONN;
    INPUT C1-C85;
    ARRAY CONY{85} CY1-CY85;
    ARRAY CONU{85} CU1-CU85;
    ARRAY CONV{85} CV1-CV85;
    ARRAY CON{85} C1-C85;
    CSUM = 0;
    DO I=1 TO 85;
        CSUM = CSUM + CON{I};
        CONY{I} = Y*CON{I};
        CONU{I} = U*CON{I};
        CONV{I} = V*CON{I};
    END;
RUN;
PROC MEANS DATA=STEP1 NOPRINT;
    VAR CY1-CY85 CU1-CU85 CV1-CV85;
    OUTPUT OUT=CYOUT1 SUM=CY1-CY85 CU1-CU85 CV1-CV85;
RUN;
PROC TRANSPOSE DATA=CYOUT1 PREFIX=CY OUT=CYOUT2;
    VAR CY1-CY85;
RUN;
PROC TRANSPOSE DATA=CYOUT1 PREFIX=CU OUT=CUOUT;
    VAR CU1-CU85;
RUN;
```

```
PROC TRANSPOSE DATA=CYOUT1 PREFIX=CV OUT=CVOUT;
     VAR CV1-CV85;
RUN;
PROC TRANSPOSE DATA=STEP1 PREFIX=TLAM OUT=CYOUT3;
     VAR LAMBDA;
RUN;
DATA STEP1 (REPLACE=YES);
     IF _N_=1 THEN SET CYOUT3;
     SET STEP1;
     SET CYOUT2;
     SET CUOUT;
     SET CVOUT;
WU = CU1/CSUM;
WV = CV1/CSUM;
WY = CY1/CSUM;
RUN;

*******************************************************
*                                                     *
* THE AR SPATIAL STATISTICAL MODEL USING MATRIX W     *
*                                                     *
*******************************************************;

PROC NLIN DATA=STEP1 METHOD=DUD MAXITER=250;
     PARMS RHO=0.274 B0=0.0 B1=-0.01 B2=-0.01;
     BOUNDS 0<RHO<0.999999;
     ARRAY LAMBDAJ{85} TLAM1-TLAM85;
     JACOB = 0;
     DO I=1 TO 85;
          JACOB = JACOB + LOG(1 - RHO*LAMBDAJ{I});
     END;
     J=EXP(JACOB/85);

  ZY = Y/J;
  MODEL ZY = (RHO*WY + B0*(1 - RHO) + B1*U + B2*V)/J;
RUN;

*******************************************************
*                                                     *
* THE SAR SPATIAL STATISTICAL MODEL USING MATRIX W    *
*                                                     *
*******************************************************;

PROC NLIN DATA=STEP1 METHOD=MARQUARDT MAXITER=250;
     PARMS RHO=0.3 B0=0.0 B1=-0.01 B2=-0.01;
     BOUNDS 0<RHO<0.999999;
     ARRAY LAMBDAJ{85} TLAM1-TLAM85;
     JACOB = 0;
     DERJ = 0;
     DO I=1 TO 85;
          JACOB = JACOB + LOG(1 - RHO*LAMBDAJ{I});
          DERJ = DERJ + -LAMBDAJ{I}/(1 - RHO*LAMBDAJ{I});
     END;
     J=EXP(JACOB/85);
     DERJ = -DERJ/85;
```

```
        ZY = Y/J;

        MODEL ZY = (RHO*WY + B0*(1 - RHO) + B1*U - RHO*B1*WU
                  + B2*V - RHO*B2*WV)/J;
        DER.B0 = (1 - RHO)/J;
        DER.B1 = (U - RHO*WU)/J;
        DER.B2 = (V - RHO*WV)/J;
        DER.RHO = ((RHO*WY + B0*(1 - RHO) + B1*U - RHO*B1*WU
                  + B2*V - RHO*B2*WV - Y)*DERJ + WY - B0 -
                  B1*WU - B2*WV)/J;
RUN;

*****************************************************
*                                                   *
* THE AR SPATIAL STATISTICAL MODEL USING MATRIX C   *
*                                                   *
*****************************************************;

PROC NLIN DATA=STEP1 METHOD=DUD MAXITER=250;
     PARMS RHO=0.274 B0=0.0  B1=-0.01 B2=-0.01;
     BOUNDS 0<RHO<0.999999;
     ARRAY LAMBDAJ{85} TLAM1-TLAM85;
     JACOB = 0;
       DERJ = 0;
     DO I=1 TO 85;
            JACOB = JACOB + LOG(1 - RHO*LAMBDAJ{I});
            DERJ = DERJ + -LAMBDAJ{I}/(1 - RHO*LAMBDAJ{I});
            END;
     J=EXP(JACOB/85);
     DERJ = -DERJ/85;
     ZY = Y/J;
     MODEL ZY = (RHO*CY + B0*(1 - RHO*CSUM) +
                                           B1*U + B2*V)/J;
     DER.B0 = (1 - RHO*CSUM)/J;
     DER.B1 = U/J;
     DER.B2 = V/J;
DER.RHO = ((RHO*CY + B0*(1 - RHO*CSUM) + B1*U + B2*V -
                               Y)*DERJ + CY - B0*CSUM)/J;
RUN;
```

APPENDIX 3-D
SPSS code[9] listing for the autoregressive response (AR) and the simultaneous autoregressive (SAR) models for the Wolfcamp aquifer data[10]

```
/* spss code listing for the autoregressive response (ar) */.
/* and the simultaneous autoregressive (sar) models */.
/* for the wolfcamp aquifer data */.

get file='c:\book.oup\wolfcamp.sav'.

*selection of eigenvalues from a w or a c matrix.
rename variables (lambdaw = lambda).

descriptives variables =y (zy).
Rename variables (y = y0).
rename variables (zy = y).
save outfile='c:\book.oup\wlfcmp.sav'.

*matrix operations for preparing data for spatial models.
set mxloop = 90.
matrix.
get c /file = * /variables c1 to c85.
get y /file = * /variable = y.
get u /file = * /variable = u.
get v /file = * /variable = v.
get lambda /file = * /variable = lambda.
compute umean = csum(u)/nrow(u).
compute vmean = csum(v)/nrow(v).
compute u = u - make(nrow(u),1,umean).
compute v = v - make(nrow(v),1,vmean).
compute cy = c*y.
compute cu = c*u.
compute cv = c*v.
compute csm = rsum(c).
release c.
compute wy = cy&/csm.
compute wu = cu&/csm.
compute wv = cv&/csm.
compute tlam = make(85,85,0).
loop i = 1 to 85.
   compute tlam(:,i) = lambda.
end loop.
compute tlam = transpos(tlam).
save {y,u,v,cy,cu,cv,csm,wy,wu,wv,tlam} /outfile=*/
```

---

[9]Contributed by Dr. Akio Sone, Institute of Socioeconomic Planning, University of Tsukuba, Japan (asone@shako.sk.tsukuba.ac.jp).

[10]Note that the path name used here is 'c:\book.oup' and needs to be changed to that defining the drive and directory location of the user's data. Also note that the analytical derivatives for the SAR model, using watrix **W**, and the CAR model, using matrix **C**, do not work.

```
variables=y,u,v,cy,cu,cv,csum,wy,wu,wv,tlam1 to tlam85.
end matrix.
save outfile='c:\book.oup\wlfcmpx.sav'.

*ordinary linear regression model.
regression variables = y, u, v
 /dependent = y /method = enter u, v.


/* THE AR SPATIAL STATISTICAL MODEL USING MATRIX W */.

compute zy=0.
model program rho=0.274 b0=0.0 b1=-0.0 b2=-0.01.
    compute jacob = 0.
    vector lambdaj = tlam1 to tlam85.
    loop i =1 to 85.
       compute jacob = jacob + ln(1 - rho*lambdaj(i)).
    end loop.
    compute j= exp(jacob/85).
    compute pred =   (y-(rho*wy + b0*(1 - rho) + b1*u + b2*v))/j.
cnlr zy / criteria = iter 250 / bounds 0 < rho < 0.999999.


/* THE SAR SPATIAL STATISTICAL MODEL USING MATRIX W */.

compute zy=0.
model program rho= 0.3 b0=0.0 b1=-0.01 b2=-0.01.
    compute jacob = 0.
    compute derj = 0.
    vector lambdaj = tlam1 to tlam85.
    loop i=1 to 85.
       compute jacob = jacob + ln(1 - rho*lambdaj(i)).
       compute derj = derj -lambdaj(i)/(1 - rho*lambdaj(i)).
    end loop.
    compute j= exp(jacob/85).
    compute derj = -derj/85.
    compute zy = y/j.
    compute pred = (y- (rho*wy + b0*(1 - rho) +
                       b1*u - rho*b1*wu + b2*v - rho*b2*wv))/j.
    compute d.b0 = (1 - rho)/j.
    compute d.b1 = (u - rho*wu)/j.
    compute d.b2 = (v - rho*wv)/j.
    compute d.rho =
              ((rho*wy + b0*(1 - rho) + b1*u - rho*b1*wu + b2*v
                - rho*b2*wv - y)*derj + wy - b0 - b1*wu -
b2*wv)/j.
cnlr zy /criteria = iter 250 / bounds 0 < rho < 0.999999.


/* THE AR SPATIAL STATISTICAL MODEL USING MATRIX C */.

compute zy=0.
model program rho = 0.274 b0 = 0.0 b1 = -0.01 b2 = -0.01.
    compute jacob = 0.
    compute derj = 0.
    vector lambdaj = tlam1 to tlam85.
```

```
    loop i=1 to 85.
       compute jacob = jacob + ln(1 - rho*lambdaj(i)).
       compute derj = derj - lambdaj(i)/(1 - rho*lambdaj(i)).
    end loop.
    compute j= exp(jacob/85).
    compute derj = -derj/85.
    compute zy = y/j.
    compute pred =
               (y- (rho*cy + b0*(1 - csum*rho) + b1*u + b2*v))/j.
    compute d.b0 = (1 - rho*csum)/j.
    compute d.b1 = u/j.
    compute d.b2 = v/j.
    compute d.rho =
              ((rho*cy + b0*(1 - rho*csum) + b1*u + b2*v - y)*derj
                + cy - b0*csum)/j.
cnlr zy /criteria = iter 250 / bounds 0 < rho < 0.999999.
```

APPENDIX 3-E
SAS code listing for selected semi-variogram models: Bessel, exponential, Gaussian, spherical, wave/hole, power

```
/*----------------------------------------------------------|
|  SEMIGRAM.SAS                                              |
|  This program needs the files SEMIEXPO.SAS SEMIBESS.SAS    |
|  SEMIGAUS.SAS SEMISPHE.SAS SEMIWAVE.SAS SEMIPOWR.SAS       |
|  SEMIPLOT.SAS in the same subdirectory that is specified   |
|  in the macro variable "progpath".                         |
|  User inputs are:                                          |
|     progpath - path name for the program files.            |
|     dspath - path name for the data analysis file.         |
|     flname - file name of the data analysis file. Same     |
|              name as that specified in Program #1.         |
|     dsvar - name of analysis variable from Program #1.     |
|             This will be the same as the filename for      |
|             the "sam" file.                                |
|     maxlag - maximum lag distance of the points on the     |
|              semivariogram plot to model. Value is taken   |
|              from the semivariogram plot produced from     |
|              Program #3.                                   |
|     The next set of macro variables determine if the user  |
|        will explicitly input starting values for the       |
|        parameters C0, C1 and r or lambda. If not, all      |
|        choices should be set to "no".                      |
|     If starting parameters are to be explicitly supplied   |
|        by the user, the parameter values are placed in     |
|        the appropriate "PARMSTR" macro variables.          |
|----------------------------------------------------------*/

*libname plots "c:\nsf_geog\book";

%global nrecs dsvar dspath flname parmstr pparmstr goutsv
                                                          goutvm;

%let goutsv =   ; * %str(gout = plots.figschp1);
%let goutvm =   ; * %str(gout = plots.figschp2);

%let progpath = %str(c:\nsf_geog\programs);
%let dspath = %str(c:\nsf_geog\book\wolfcamp);
%let flname = wolfcamp;
%let dsvar = piezo;
%let maxlag = 0.9;

        * Explicitly specify initial parameter values for:
%let specexpo = no;    * exponential model? ;
%let specbess = no;    * Bessel model?       ;
%let specgaus = yes;   * Gaussian model?     ;
%let specsphe = no;    * spherical model?    ;
%let specwave = no;    * wave/hole model?    ;
%let specpowr = no;    * power model?        ;

*wave/hole;
```

```
  %let parmstrw=   %str(parm c0=0.0001   c1=0.013    r=0.005 );
*power model;
  %let pparmstr =  %str(parm c0=0.0001   c1=0.10     lambda=0.5 );
*all others;
  %let parmstr =   %str(parm c0=0.0001   c1=0.10     r=0.10 );

options mautosource sasautos="&progpath" nocenter nodate
                                                              pageno=1
         mprint symbolgen mlogic;

goptions device=win norotate;
                         *<--- default graphics is to monitor;
*goptions device=win norotate goutmode=append;
*goptions device=winprtg rotate;
*filename grafout1 "&dspath.\&dsvar..cgm";
*goptions device=cgmwpwa gsfname=grafout1 gsfmode=replace;

data _null_;
 infile "&dspath.\&dsvar..sam" missover;
 input c1;
 call symput('nrecs',left(_n_));
run;
%put number of observations = &nrecs;

DATA STEP2;  INFILE "&dspath.\&dsvar..sam";
   input dist npairs gamma;
RUN;
proc sort out=step2(replace=yes); by dist; run;

DATA STEP1;  set step2;
   if dist > &maxlag then delete;
RUN;
proc sort out=step1(replace=yes); by dist; run;

proc means mean var data=step1 noprint;
                             *** Compute starting value for C1 ;
   var gamma;                *** which is weighted mean of      ;
   freq npairs;              *** gamma values up to maxlag.     ;
   output out=ic1 mean=w_ic1 var=s2;
run;
data _null_; set ic1;
   sr= &maxlag / sqrt(3);
   call symput('smc1',w_ic1);
   call symput('s2',s2);
   call symput('sr',sr);
 run;

 data _null_; set step1;
    if _n_ = 2 then do;
       halfdist=dist/3;
       call symput('smr',halfdist);
       wavemr=&maxlag / 300;
                         /* Specify grid values for estimating */
       call symput('wmr',wavemr);
                         /* range (r) for the Wave/hole model. */
       wavemi=&maxlag / 20;
```

```
                                /* Specify grid values for estimating */
        call symput('wmi',wavemi);
                                /* range (r) for the Wave/hole model. */
        wavemc0=&smc1 / 30;
                                /* Specify grid values for estimating */
        call symput('wmc0',wavemc0);
                                /* nugget for the Wave/hole model.    */
    end;
run;
%put smc1 = &smc1 , smr = &smr , wmr = &wmr;
proc standard data=step1 out=step3 mean=0; var gamma; run;
                                        /* Compute CTSS to        */
proc univariate data=step3 noprint;     /* divide into error SS */
    var gamma;                          /* to compute RSSE.       */
    freq npairs;
    output out=totalss css=ctss;
run;

%let mexprss =   ;
%let mbesrss =   ;
%let mgaurss =   ;
%let msphrss =   ;
%let mwavrss =   ;
%let mpowrss =   ;
%let mexprpl =   ;
%let mbesrpl =   ;
%let mgaurpl =   ;
%let msphrpl =   ;
%let mwavrpl =   ;
%let mpowrpl =   ;

%semiexpo;                              /* exponential model */
    %let mexprss = %str(essetss);
    %let mexprpl = %str(expogrm);
%semibess;                              /* Bessel model      */
    %let mbesrss = %str(bssetss);
    %let mbesrpl = %str(bessgrm);
%semigaus;                              /* Gaussian model    */
    %let mgaurss = %str(gssetss);
    %let mgaurpl = %str(gausgrm);
%semisphe;                              /* spherical model   */
    %let msphrss = %str(sssetss);
    %let msphrpl = %str(sphegrm);
%semiwave;                              /* wave/hole model   */
    %let mwavrss = %str(wssetss);
    %let mwavrpl = %str(wavegrm);
%semipowr;                              /* power model       */
    %let mpowrss = %str(pssetss);
    %let mpowrpl = %str(powrgrm);

data plotgram; set step2 &mexprpl &mbesrpl &mgaurpl &msphrpl
                                            &mwavrpl &mpowrpl; run;
%semiplot;
data ssetss; set &mexprss &mbesrss &mgaurss &msphrss &mwavrss
                                                    &mpowrss; run;
```

```
proc print noobs data=ssetss;
  var svmodel rsse ctss errss c0 c1 r;
  title1 "Relative error sum of squares for &flname., variable
                                                    &dsvar.";
  title2 "Starting values: c0=0.000001 , c1=&smc1 , r=&smr";
  title3 "Maximum lag distance included = &maxlag.";
run;




/*----------------------------------------------|
|  SEMIEXPO.SAS                                 |
|  Exponential semi-variogram model macro file. |
|----------------------------------------------*/

%macro semiexpo;
  title "Exponential semi-variogram plot for data set "
        "&dsvar from file &flname";

  proc nlin method=dud maxiter=500 data=step1 noprint;
    %if &specexpo = yes %then %do;
       &parmstr;
    %end;
    %if &specexpo = no %then %do;
       parms c0=0.000001 c1=&smc1 r=&smr;
    %end;
    bounds c0>0, c1>0, 0.000001<r<&maxlag;
    if dist=0 then dist=0.00001;
    if r=0 then r=0.00001;
    model gamma = c0 + c1*(1 - exp(-(1/r)*dist));
    output out=expodud parms=c0 c1 r;
  run;
  data _null_; set expodud;
    if c0<0 then c0=0.0000001;
    call symput('mc0',c0);
    call symput('mc1',c1);
    call symput('mr',r);
  run;
  proc nlin method=marquardt maxiter=500 data=step1 noprint;
    parms c0=&mc0  c1=&mc1  r=&mr;
    bounds c0>0, c1>0, 0.000001<r<&maxlag;
    if dist=0 then dist=0.00001;
    if r=0 then r=0.00001;
    model gamma = c0 + c1*(1 - exp(-(1/r)*dist));
    _weight_=npairs;
    output out=expogram parms=c0 c1 r sse=errss;
  run;

  data expogrm; merge step2 expogram; by dist;
    gamma=c0 + c1*(1 - exp(-(1/r)*dist));
    svmodel="exponential";
    if (c0=. & c1=.) then delete;
  run;
```

```
   data essetss; set totalss; set expogram;
     rsse = errss/ctss;
     svmodel="exponential";
   run;
%mend semiexpo;




/*----------------------------------------------------------|
|   SEMIBESS.SAS                                             |
|   Bessel function semi-variogram model macro file.         |
|-----------------------------------------------------------*/

%macro semibess;
   title "Bessel semi-variogram plot for data set "
         "&dsvar from file &flname";

   proc nlin method=dud maxiter=500 data=step1 noprint;
      %if &specbess = yes %then %do;
         &parmstr;
      %end;
      %if &specbess = no %then %do;
         parms c0=0.000001 c1=&smc1 r=&smr;
      %end;
      bounds c0>0, c1>0, 0.000001<r<&maxlag;
      if r=0 then r=0.00001;
      distr = dist/r;
      if distr < 0.00001 then do; model gamma = c0; end;
      if distr > 88 then do; model gamma = c0 + c1; end;
      if (distr >= 0.00001 & distr <= 88) then do;
         p1 =   0.50000000;   p2 =   0.87890594;   p3 =   0.51498869;
         p4 =   0.15084934;   p5 =   0.02658733;   p6 =   0.00301532;
         p7 =   0.00032411;
         q1 =   0.39894228;   q2 =  -0.03988024;   q3 =  -0.00362018;
         q4 =   0.00163801;   q5 =  -0.01031555;   q6 =  -0.02282967;
         q7 =  -0.02895312;   q8 =   0.01787654;   q9 =  -0.00420059;
         if abs(distr) < 3.75 then do;
            c=(distr/3.75)**2;
            bessi1 =
 distr*(p1+c*(p2+c*(p3+c*(p4+c*(p5+c*(p6+c*p7))))));
         end;
         else do;
            ax = abs(distr);
            c = 3.75/ax;
            bessi1 =
                (exp(ax)/sqrt(ax))*(q1+c*(q2+c*(q3+c*(q4+c*(q5+c
                                   *(q6+c*(q7+c*(q8+c*q9))))))));
         end;
         p1 =   1.00000000;   p2 =   0.15443144;   p3 =  -0.67278579;
         p4 =  -0.18156897;   p5 =  -0.01919402;   p6 =  -0.00110404;
         p7 =  -0.00004686;
         q1 =   1.25331414;   q2 =   0.23498619;   q3 =  -0.03655620;
         q4 =   0.01504268;   q5 =  -0.00780353;   q6 =   0.00325614;
         q7 =  -0.00068245;
```

```
    if distr <= 2 then do;
      c=distr*distr/4;
      bessk1 =
             (log(distr/2)*bessi1)+(1/distr)*(p1+c*(p2+c*(p3+
                              c*(p4+c*(p5+c*(p6+c*p7))))));
    end;
    else do;
      c = 2/distr;
      bessk1 =
                  (exp(-distr)/sqrt(distr))*(q1+c*(q2+c*(q3+
                              c*(q4+c*(q5+c*(q6+c*q7))))));
    end;
      model gamma = c0 + c1*(1 - bessk1*(distr));
  end;
  output out=bessdud parms=c0 c1 r;
run;
data _null_; set bessdud;
  if c0<0 then c0=0.000001;
  call symput('mc0',c0);
  call symput('mc1',c1);
  call symput('mr',r);
run;
proc nlin method=dud maxiter=500 data=step1 noprint;
  parms   c0 = &mc0   c1 = &mc1   r = &mr;
  bounds c0>0, c1>0, 0.000001<r<&maxlag;
  if r=0 then r=0.00001;
  distr = dist/r;
  if distr < 0.00001 then do; model gamma = c0; end;
  if distr > 88 then do; model gamma = c0 + c1; end;
  if (distr >= 0.00001 & distr <= 88) then do;
    p1 =  0.50000000;  p2 =  0.87890594;  p3 =  0.51498869;
    p4 =  0.15084934;  p5 =  0.02658733;  p6 =  0.00301532;
    p7 =  0.00032411;
    q1 =  0.39894228;  q2 = -0.03988024;  q3 = -0.00362018;
    q4 =  0.00163801;  q5 = -0.01031555;  q6 = -0.02282967;
    q7 = -0.02895312;  q8 =  0.01787654;  q9 = -0.00420059;
    if abs(distr) < 3.75 then do;
      c=(distr/3.75)**2;
      bessi1 =
            distr*(p1+c*(p2+c*(p3+c*(p4+c*(p5+c*(p6+c*p7))))));
    end;
    else do;
      ax = abs(distr);
      c = 3.75/ax;
      bessi1 =
             (exp(ax)/sqrt(ax))*(q1+c*(q2+c*(q3+c*(q4+c*(q5+c
                              *(q6+c*(q7+c*(q8+c*q9))))))));
    end;
    p1 =  1.00000000;  p2 =  0.15443144;  p3 = -0.67278579;
    p4 = -0.18156897;  p5 = -0.01919402;  p6 = -0.00110404;
    p7 = -0.00004686;
    q1 =  1.25331414;  q2 =  0.23498619;  q3 = -0.03655620;
    q4 =  0.01504268;  q5 = -0.00780353;  q6 =  0.00325614;
    q7 = -0.00068245;
    if distr <= 2 then do;
      c=distr*distr/4;
```

```
      bessk1 =
           (log(distr/2)*bessi1)+(1/distr)*(p1+c*(p2+c*(p3+c*
                                  (p4+c*(p5+c*(p6+c*p7))))));
  end;
  else do;
    c = 2/distr;
    bessk1 =
              (exp(-distr)/sqrt(distr))*(q1+c*(q2+c*(q3+c*(q4+
                                   c*(q5+c*(q6+c*q7))))));
  end;
    model gamma = c0 + c1*(1 - bessk1*(distr));
  end;
  _weight_=npairs;
  output out=bessgram parms=c0 c1 r sse=errss;
run;

data bessgrm(replace=yes); merge step2 bessgram; by dist;
  distr = dist/r;
  if distr < 0.00001 then do; gamma = c0; svmodel="Bessel";
                                                          end;
  if distr > 88 then do; gamma = c0 + c1; svmodel="Bessel";
                                                          end;
  if (distr >= 0.00001 & distr <= 88) then do;
    p1 =  0.50000000;   p2 =  0.87890594;   p3 =  0.51498869;
    p4 =  0.15084934;   p5 =  0.02658733;   p6 =  0.00301532;
    p7 =  0.00032411;
    q1 =  0.39894228;   q2 = -0.03988024;   q3 = -0.00362018;
    q4 =  0.00163801;   q5 = -0.01031555;   q6 = -0.02282967;
    q7 = -0.02895312;   q8 =  0.01787654;   q9 = -0.00420059;
    if abs(distr) < 3.75 then do;
      c=(distr/3.75)**2;
      bessi1 =
            distr*(p1+c*(p2+c*(p3+c*(p4+c*(p5+c*(p6+c*p7))))));
    end;
    else do;
      ax = abs(distr);
      c = 3.75/ax;
      bessi1 =
              (exp(ax)/sqrt(ax))*(q1+c*(q2+c*(q3+c*(q4+c*(q5+c
                                  *(q6+c*(q7+c*(q8+c*q9))))))));
    end;
    p1 =  1.00000000;   p2 =  0.15443144;   p3 = -0.67278579;
    p4 = -0.18156897;   p5 = -0.01919402;   p6 = -0.00110404;
    p7 = -0.00004686;
    q1 =  1.25331414;   q2 =  0.23498619;   q3 = -0.03655620;
    q4 =  0.01504268;   q5 = -0.00780353;   q6 =  0.00325614;
    q7 = -0.00068245;
    if distr <= 2 then do;
      c=distr*distr/4;
      bessk1 =
              (log(distr/2)*bessi1)+(1/distr)*(p1+c*(p2+c*(p3+
                                   c*(p4+c*(p5+c*(p6+c*p7))))));
    end;
    else do;
      c = 2/distr;
      bessk1 =
```

POPULAR MODELS 185

```
                      (exp(-distr)/sqrt(distr))*(q1+c*(q2+c*(q3+
                                 c*(q4+c*(q5+c*(q6+c*q7))))));
     end;
     gamma = c0 + c1*(1 - bessk1*(distr));
     svmodel="Bessel";
  end;
  if (c0=. & c1=.) then delete;
  drop distr p1 p2 p3 p4 p5 p6 p7 q1 q2 q3 q4 q5 q6 q7 q8 q9
       bessi1 ax bessk1 c;
  run;

  data bssetss; set totalss; set bessgram;
    rsse = errss/ctss;
    svmodel="Bessel";
  run;
%mend semibess;




/*------------------------------------------|
|  SEMIGAUS.SAS                              |
|  Gaussian semi-variogram model macro file. |
|------------------------------------------*/

%macro semigaus;
  title "Gaussian semi-variogram plot for data set "
        "&dsvar from file &flname";

  proc nlin data=step1 method=marquardt maxiter=500 noprint;
    %if &specgaus = yes %then %do;
       &parmstr;
    %end;
    %if &specgaus = no %then %do;
       parms c0=0.000001 c1=&smc1 r=&smr;
    %end;
    bounds c0>0, c1>0, 0.000001<r<&maxlag;
    model gamma = c0 + c1*(1 - exp(-(dist**2)/(r**2)));
    output out=gaussdud parms=c0 c1 r;
  run;
  data _null_; set gaussdud;
    if c0<0 then c0=0.0000001;
    call symput('mc0',c0);
    call symput('mc1',c1);
    call symput('mr',r);
  run;
  proc nlin method=marquardt maxiter=500 data=step1 noprint;
    parms c0=&mc0   c1=&mc1   r=&mr;
    bounds c0>0, c1>0, 0.000001<r<&maxlag;
    model gamma = c0 + c1*(1 - exp(-(dist/r)**2));
    _weight_ = npairs;
    output out=gausgram parms=c0 c1 r sse=errss;
  run;

  data gausgrm; merge step2 gausgram;
```

```
    gamma=c0 + c1*(1 - exp(-(dist**2)/r**2));
    svmodel="Gaussian";
    if (c0=. & c1=.) then delete;
  run;

  data gssetss; set totalss; set gausgram;
    rsse = errss/ctss;
    svmodel="Gaussian";
  run;
%mend semigaus;




/*-------------------------------------------|
|   SEMISPHE.SAS                              |
|   Spherical semi-variogram model macro file.|
|---------------------------------------------*/

%macro semisphe;
  title "Spherical semi-variogram plot for data set "
        "&dsvar from file &flname";

 proc nlin  data=step1 method=dud maxiter=500 noprint;
    %if &specsphe = yes %then %do;
        &parmstr;
    %end;
    %if &specsphe = no %then %do;
        parms c0=&s2 c1=1 r=&sr;
    %end;
    bounds c0>0, c1>0, 0.000001<r<&maxlag;
    if dist = 0 then gamma=0;
    if 0 < dist < r then do;
     model gamma = c0 + c1*((3/2)*(dist/r) - (1/2)*(dist/r)**3);
    end;
    if dist >= r then do;
       model gamma = c0 + c1;
    end;
    output out=sphedud parms=c0 c1 r;
  run;
  data _null_; set sphedud;
    if c0<0 then c0=0.0000001;
    call symput('mc0',c0);
    call symput('mc1',c1);
    call symput('mr',r);
  run;
  proc nlin  data=step1 method=marquardt maxiter=500 noprint;
    parms c0=&mc0   c1=&mc1   r=&mr;
    bounds c0>0, c1>0, 0.000001<r<&maxlag;
    if dist = 0 then gamma = 0;
    if 0 < dist < r then do;
     model gamma = c0 + c1*((3/2)*(dist/r) - (1/2)*(dist/r)**3);
    end;
    if dist >= r then do;
      model gamma = c0 + c1;
```

```
      end;
      _weight_=npairs;
      der.c0=1;
      if dist <= r then der.c1=(3/2)*(dist/r) - (1/2)*(dist/r)**3;
         else der.c1=1;
      if dist <= r then der.r=c1*(-(3/2)*dist/r**2 -
                                  (1/2)*3*(-dist/r**2)*(dist/r)**2);
         else der.r=0;
      output out=sphegram parms=c0 c1 r sse=errss;
   run;

   data sphegrm; merge step2 sphegram; by dist;
      if dist <= r then do;
         gamma = c0 + c1*((3/2)*(dist/r) - (1/2)*(dist/r)**3);
         svmodel = "spherical";
      end;
      else do;
         gamma = c0 + c1;
         svmodel = "spherical";
      end;
      if (c0=. & c1=.) then delete;
   run;

   data sssetss; set totalss; set sphegram;
      rsse = errss/ctss;
      svmodel = "spherical";
   run;
%mend semisphe;




/*----------------------------------------------|
|   SEMIWAVE.SAS                                 |
|   Wave/Hole semi-variogram model macro file.   |
|-----------------------------------------------*/

%macro semiwave;
   title "Wave/hole semi-variogram plot for data set "
         "&dsvar from file &flname";

   proc nlin method=dud maxiter=500 data=step1 noprint;
      %if &specwave = yes %then %do;
         &parmstrw;
      %end;
      %if &specwave = no %then %do;
         parms c0=0.000001 to &smc1 by &wmc0
               c1=&smc1
               r= &wmi to &maxlag by &wmr;
      %end;
      bounds c0>0, c1>0, 0.00001<r<=&maxlag;
      if dist=0 then dist=0.00001;
      if r<=0 then r=0.00001;
      model gamma = c0 + c1*(1 - r*(sin(dist/r)/dist));
      output out=wavedud parms=c0 c1 r;
```

```
    run;
    data _null_; set wavedud;
      if c0<0 then c0=0.0000001;
      call symput('mc0',c0);
      call symput('mc1',c1);
      call symput('mr',r);
    run;

    proc nlin method=marquardt maxiter=500 data=step1 noprint;
      parms c0=&mc0   c1=&mc1   r=&mr;
      bounds c0>0, c1>0, 0.00001<r<&maxlag;
      if dist=0 then dist=0.00001;
      if r<=0 then r=0.00001;
      model gamma = c0 + c1*(1 - r*(sin(dist/r)/dist));
      _weight_=npairs;
      output out=wavegram parms=c0 c1 r sse=errss;
    run;

    data wavegrm; merge step2 wavegram; by dist;
      if dist=0 then dist=0.00001;
      gamma=c0 + c1*(1 - r*(sin(dist/r)/dist));
      svmodel = "Wave/hole";
      if (c0=. & c1=.) then delete;
    run;

    data wssetss; set totalss; set wavegram;
      rsse = errss/ctss;
      svmodel = "Wave/hole";
    run;
%mend semiwave;




/*-------------------------------------------|
|   SEMIPOWR.SAS                             |
|   Power semi-variogram model macro file.   |
|-------------------------------------------*/

%macro semipowr;
  title "Power semi-variogram plot for data set "
        "&dsvar from file &flname";

  proc nlin method=dud maxiter=500 data=step1 noprint;
    %if &specpowr = yes %then %do;
      &pparmstr;
    %end;
    %if &specpowr = no %then %do;
       parms c0=0.000001
             c1=&smc1
             lambda=0.1 to 2.0 by 0.1;
    %end;
    bounds c0>0.000000001, c1>0, 0<lambda<10;
    if dist=0 then dist=0.00001;
    model gamma = c0 + c1*dist**(lambda);
```

```
      output out=powrdud parms=c0 c1 lambda;
   run;
   data _null_; set powrdud;
      if c0<0 then c0=0.0000001;
      call symput('mc0',c0);
      call symput('mc1',c1);
      call symput('mr',lambda);
   run;
   proc nlin method=marquardt maxiter=500 data=step1 noprint;
      parms c0=&mc0   c1=&mc1   lambda=&mr;
      bounds c0>0.0000000001, c1>0, 0<lambda<10;
      model gamma = c0 + c1*dist**(lambda);
      _weight_=npairs;
      output out=powrgram parms=c0 c1 lambda sse=errss;
   run;

   data powrgrm; merge step2 powrgram; by dist;
      gamma = c0 + c1*dist**(lambda);
      svmodel = "power";
      if (c0=. & c1=.) then delete;
   run;

   data pssetss; set totalss; set powrgram;
      rsse = errss/ctss;
      svmodel = "power";
      r=lambda;
              /* set lambda to r only for printing convenience */
   run;
%mend semipowr;




/*---------------------------------------------------------------|
|  SEMIPLOT.SAS                                                  |
|  Takes input from semi-variogram models and creates:           |
|     (1) a semi-variogram plot with model fits overlayed on     |
|         the plot.                                              |
|     (2) ratio of variance-to-mean versus mean clustered        |
|         distance.                                              |
|----------------------------------------------------------*/

%macro semiplot;
   title1 h=1 c=black
                "Semi-variogram model plots for data set &dsvar "
                     "from file &flname";
   title2 h=1 c=black
             "Starting values: c0=0.000001 , c1=&smc1 , r=&smr";
   title3 h=1 c=black "Maximum lag distance included = &maxlag";
*title1 h=0.5 " ";

   symbol1 c=black v=dot  i=none;
   symbol2 c=black v=none w=3 l=14 i=spline;
   symbol3 c=black v=none w=3 l=8  i=spline;
   symbol4 c=black v=none w=3 l=3  i=spline;
```

```
  symbol5 c=black v=none w=3 l=29 i=spline;
  symbol6 c=black v=none w=3 l=2  i=spline;
  symbol7 c=black v=none w=3 l=1  i=spline;
  legend1 across = 2
    label = (position=left justify=center
             font=swissb h=1 'Semivariogram'
             justify=center font=swissb h=1 'Model')
    value = (c=black font=swissb h=1);
  proc gplot data=plotgram &goutsv;
    axis1 c=black w=2
      label =
         (c=black font=swissb h=1 'h (mean clustered distance)')
      value = (c=black font=swissb h=1)
      major = (w=2 c=black)
      minor = none;
    axis2 c=black w=2
      label =
            (a=90 r=0 font=cgreek h=1 'g' font=swissb h=1 '(h)')
      value = (c=black font=swissb h=1)
      major = (w=2 c=black)
      minor = none;
    plot gamma*dist = svmodel / haxis=axis1
                                vaxis=axis2
                                legend=legend1;
run;

  data modrvm; infile "&dspath.\&dsvar..rvm";
    input gammean gamvar rvm dist n;
    if dist <= &maxlag then do;
      lagdist = dist;
    end;
    else lagdist = . ;
    noacorr = 2;
  run;

  proc means mean stderr data=modrvm noprint;
    var rvm;
    where dist <= &maxlag;
    output out=mmrvm mean=mrvm stderr=srvm;
  run;
  data _null_; set mmrvm;
    call symput('lmrvm',mrvm);
    call symput('lsrvm',srvm);
  run;


  symbol1 c=black v=dot i=none;
  symbol2 c=black v=none w=3 i=join l=1;
  proc gplot data=modrvm &goutvm;
    axis1 c=black w=2
      label = (c=black font=swissb h=1
               'h (mean clustered distance)')
      value = (c=black font=swissb h=1)
      major = (w=2 c=black)
      minor = none;
    axis2 c=black w=2
```

```
        label = (a=90 r=0 font=swissb h=1
                                    'Variance-to-mean ratio of '
                 font=cgreek h=1 'g' font=swissb h=1 '(h)')
        value = (c=black font=swissb h=1)
        major = (w=2 c=black)
        minor = none;
     plot rvm*dist noacorr*lagdist / overlay
                                    haxis=axis1
                                    vaxis=axis2;
     title1 h=1
     "Ratio of the variance to the mean for each distance group "
               "up to distance &maxlag.";
     title2 h=1 "For data set &flname and variable &dsvar";
     title3 h=1 "mean rvm = &lmrvm , stderr rvm = &lsrvm";
*    title1 h=0.5 " ";
   run;
%mend semiplot;
```

SAS code for the additional semivariogram models discussed in Chapter 10 is as follows:

```
/* circular */
proc nlin maxiter=500 method=marquardt;
    parms c0=0.1 c1=0.1 r=0.02;
    bounds c0>0, c1>0, r>0;
pi=arcos(-1);
if mdist<=r then
    do;
        model mgamma = c0 + c1*((2/pi)*((mdist/r)*sqrt(1 -
                        (mdist/r)**2) + arsin(mdist/r)));
        _weight_ = _freq_;
    end;
else
    do;
      model mgamma = c0 + c1;
      _weight_ = _freq_;
    end;
run;

/* rational quadratic */
proc nlin maxiter=500 method=marquardt;
    parms c0=0.1 c1=0.1 r=0.01;
    bounds c0>0, 0<c1, r>0;
    model mgamma = c0 +
                   c1*( ((mdist/r)**2)/(1 + (mdist/r)**2) );
    _weight_ = _freq_;
run;

/* penta-spherical */
proc nlin method=marquardt;
    parms c0=0.1 c1=0.1 r=0.03;
    bounds c0>0, c1>0, r>0;
if mdist<=r then
    do;
        model mgamma = c0 + c1*((15/8)*(mdist/r) -
                 (5/4)*(mdist/r)**3 + (3/8)*(mdist/r)**5);
        _weight_ = _freq_;
    end;
else
    do;
      model mgamma = c0 + c1;
      _weight_ = _freq_;
    end;
run;

/* cubic */
proc nlin method=marquardt;
    parms c0=0.1 c1=0.1 r=0.02;
    bounds c0>0, c1>0, r>0;
```

```
if mdist<=r then
     do;
        model mgamma = c0 +
                  c1*(7*(mdist/r) - (35/4)*(mdist/r)**3 +
                  (7/2)*(mdist/r)**5 - (3/4)**(mdist/r)**7);
        _weight_ = _freq_;
     end;
else
     do;
        model mgamma = c0 + c1;
        _weight_ = _freq_;
     end;
run;

/* stable */
proc nlin method=marquardt;
    parms c0=0.03 c1=0.14 r=0.01 a=0.9;
    bounds c0>0, c1>0, r>0, 0<a<2;
    model mgamma = c0 + c1*(1 - exp(-(mdist/r)**a));
    _weight_ = _freq_;
run;

/* Cauchy */
proc nlin maxiter=500 method=marquardt;
    parms c0=0.02 c1=0.1 r=0.02 a=0.75;
    bounds c0>0, 0<c1, r>0.000001, 0<a<3;
    model mgamma = c0 + c1*(1 - 1/(1 + (mdist/r)**2)**a );
    _weight_ = _freq_;
run;
```

APPENDIX 3-F
SPSS code[11] listing for selected semi-variogram models:
Bessel, exponential, Gaussian, spherical, wave/hole[12]

```
/* SPSS code listing for selected semi-variogram models: */.
/* Bessel, exponential, Gaussian, spherical, wave/hole, */.
/* and power */.

/* Pre-processing program */.

set width = 90.

data list file = 'c:\book.oup\____.___' list
                                              /*specify data filename */
 / dist npairs gamma.
sort cases by dist.
save outfile = 'c:\book.oup\step2.sav'.

compute maxlag = ____                         /*specify maxlag value */.
select if (dist <= maxlag ).
sort cases by dist.
save outfile = 'c:\book.oup\step1.sav'.

weight by npairs.
descriptives variables = gamma
 / statistics = mean variance.

compute sr = maxlag / sqrt(3).
compute w_ic1 = _____    /* specify weighted mean of gamma */.
compute s2 =     _____   /* specify weighted variance of gamma */.

do if ($casenum = 2).
  compute halfdist = dist /3.
  compute smr = 3.
  compute wmr = maxlag /300.
  compute wmi = maxlag /20.
  compute wmc0 = smc1 /30.
end if.

*matrix operations for preparing data for spatial models.
```

---

[11]Contributed by Dr. Akio Sone, Institute of Socioeconomic Planning, University of Tsukuba, Japan (asone@shako.sk.tsukuba.ac.jp).

[12]Note that the path name used here is 'c:\book.oup' and needs to be changed to that defining the drive and directory location of the user's data. Also note that because of limitations in the macro facility of SPSS: (a) this code does not replicate exactly its SAS macro structure counterpart appearing in Appendix 3-E; (b) starting values in this code must be specified manually (their positions are indicated as "____" with comment lines); and, (c) the lack of a capacity to internally carry forward estimated parameters from one command to another prevents some portions of the SAS code from being translated into SPSS code (e.g., the calculation of R-squared). Further, because SPSS for windows presupposes that its users will interactively specify graphical details (e.g., axis tick marks and axis ranges) after a chart with minimum details is produced, detailed options of SAS's PROC GPLOT cannot be translated into SPSS code.

POPULAR MODELS 195

```
set mxloop = 100.
matrix.
get d /file = * /variable = dist.
get p /file = * /variable = npairs.
get g /file = * /variable = gamma.
compute gmean = csum(g)/nrow(g).
compute g = g - make(nrow(g),1,gmean).
save {d,p,g} /outfile = */variables = dist,npairs, gamma.
end matrix.
save outfile 'c:\book.oup\step3.sav'.

weight by npairs.
descriptives variables = gamma
 /statistics = vairance.


/* Bessel Model */.
title "Bessel semi-variogram plot".

get file = 'c:\book.oup\step1.sav'.

model program c0=0.000001
                    c1 = _____ r = _____  /* specify c1 and r */.
do if (r = 0).
  compute dist = 0.00001.
end if.
compute distr=dist/r.
do if (distr < 0.00001).
  compute pred = c0.
else if (distr > 88).
  compute pred = (c0 + c1).
else if (distr >= 0.00001 and distr <= 88).
  compute p1 =   0.50000000.
  compute p2 =   0.87890594.
  compute p3 =   0.51498869.
  compute p4 =   0.15084934.
  compute p5 =   0.02658733.
  compute p6 =   0.00301532.
  compute p7 =   0.00032411.
  compute q1 =   0.39894228.
  compute q2 =  -0.03988024.
  compute q3 =  -0.00362018.
  compute q4 =   0.00163801.
  compute q5 =  -0.01031555.
  compute q6 =  -0.02282967.
  compute q7 =  -0.02895312.
  compute q8 =   0.01787654.
  compute q9 =  -0.00420059.
    do if (abs(distr) < 3.75).
     compute c = (distr/3.75)**2.
     compute bessi1 =
             distr*(p1+c*(p2+c*(p3+c*(p4+c*(p5+c*(p6+c*p7)))))).
    else.
     compute ax = abs(distr).
     compute  c = 3.75/ax.
     compute bessi1 =
```

```
                    (exp(ax)/sqrt(ax))*(q1+c*(q2+c*(q3+c*(q4+c*(q5+c*(q6+
                                          c*(q7+c*(q8+c*q9))))))))).
   end if.
   compute    p1 =   1.00000000.
   compute    p2 =   0.15443144.
   compute    p3 =  -0.67278579.
   compute    p4 =  -0.18156897.
   compute    p5 =  -0.01919402.
   compute    p6 =  -0.00110404.
   compute    p7 =  -0.00004686.
   compute    q1 =   1.25331414.
   compute    q2 =   0.23498619.
   compute    q3 =  -0.03655620.
   compute    q4 =   0.01504268.
   compute    q5 =  -0.00780353.
   compute    q6 =   0.00325614.
   compute    q7 =  -0.00068245.
   do if (distr <= 2).
     compute c = distr*distr/4.
     compute besskl =
                 (ln(distr/2)*bessi1)+(1/distr)*(p1+c*(p2+c*
                            (p3+c*(p4+c*(p5+c*(p6+c*p7)))))).
   else.
      compute c = 2/distr.
      compute besskl = (exp(-distr)/sqrt(distr))*
                   (q1+c*(q2+c*(q3+c*(q4+c*(q5+c*(q6+c*q7)))))).
   end if.
  compute pred = c0 + c1*(1 - bessk1*(distr)).
end if.
cnlr gamma / criteria = iter 500 /
                           bounds c0>0; c1>0; 0.000001<r<  ___  /
outfile = param1 /save = pred /*specify the upper bound of r*/.

model program c0 c1 r
 /*If c0 < 0 in the above estimation, then let c0 = 0.0000001*/.
weight by npairs.
do if (r = 0).
  compute dist = 0.00001.
end if.
compute distr = dist/r.
do if (distr < 0.00001).
  compute pred = c0.
else if (distr > 88).
  compute pred = c0 + c1.
else if (distr >= 0.00001 and distr <= 88).
  compute p1 =   0.50000000.
  compute p2 =   0.87890594.
  compute p3 =   0.51498869.
  compute p4 =   0.15084934.
  compute p5 =   0.02658733.
  compute p6 =   0.00301532.
  compute p7 =   0.00032411.
  compute q1 =   0.39894228.
  compute q2 =  -0.03988024.
  compute q3 =  -0.00362018.
  compute q4 =   0.00163801.
```

```
  compute q5 = -0.01031555.
  compute q6 = -0.02282967.
  compute q7 = -0.02895312.
  compute q8 =  0.01787654.
  compute q9 = -0.00420059.
do if (abs(distr) < 3.75).
  compute c = (distr/3.75)**2.
  compute bessi1 =
             distr*(p1+c*(p2+c*(p3+c*(p4+c*(p5+c*(p6+c*p7)))))).
else.
  compute ax = abs(distr).
  compute  c = 3.75/ax.
  compute bessi1 =
          (exp(ax)/sqrt(ax))*(q1+c*(q2+c*(q3+c*(q4+c*(q5+c*(q6+
                                    c*(q7+c*(q8+c*q9)))))))).
end if.
    compute    p1 =  1.00000000.
    compute    p2 =  0.15443144.
    compute    p3 = -0.67278579.
    compute    p4 = -0.18156897.
    compute    p5 = -0.01919402.
    compute    p6 = -0.00110404.
    compute    p7 = -0.00004686.
    compute    q1 =  1.25331414.
    compute    q2 =  0.23498619.
    compute    q3 = -0.03655620.
    compute    q4 =  0.01504268.
    compute    q5 = -0.00780353.
    compute    q6 =  0.00325614.
    compute    q7 = -0.00068245.
do if (distr <= 2).
  compute c = distr*distr/4.
  compute bessk1 = (ln(distr/2)*bessi1)+(1/distr)*(p1+c*(p2+c*
                  (p3+c*(p4+c*(p5+c*(p6+c*p7)))))).
else.
  compute c = 2/distr.
  compute bessk1 = (exp(-distr)/sqrt(distr))*
                  (q1+c*(q2+c*(q3+c*(q4+c*(q5+c*(q6+c*q7)))))).
end if.
compute pred = c0 + c1*(1 - bessk1*(distr)).
end if.
cnlr gamma / criteria = iter 500 /
                             bounds c0>0; c1>0; 0.000001<r< ____
 / file = param1 /outfile = param2 /save = pred
                             /*specify the upper bound of r */.

match files / file = 'c:\book.oup\step2.sav'
 / file = 'c:\book.oup\bessgram.sav' / by dist.

compute distr=dist/r.
do if (distr < 0.00001).
  compute gamma = c0.
else if (distr > 88).
  compute gamma = (c0 + c1).
else if (distr >= 0.00001 and distr <= 88).
  compute p1 =  0.50000000.
```

```
compute p2 =   0.87890594.
compute p3 =   0.51498869.
compute p4 =   0.15084934.
compute p5 =   0.02658733.
compute p6 =   0.00301532.
compute p7 =   0.00032411.
compute q1 =   0.39894228.
compute q2 =  -0.03988024.
compute q3 =  -0.00362018.
compute q4 =   0.00163801.
compute q5 =  -0.01031555.
compute q6 =  -0.02282967.
compute q7 =  -0.02895312.
compute q8 =   0.01787654.
compute q9 =  -0.00420059.
  do if (abs(distr) < 3.75).
    compute c = (distr/3.75)**2.
    compute bessi1 =
              distr*(p1+c*(p2+c*(p3+c*(p4+c*(p5+c*(p6+c*p7)))))).
  else.
    compute ax = abs(distr).
    compute  c = 3.75/ax.
    compute bessi1 =
              (exp(ax)/sqrt(ax))*(q1+c*(q2+c*(q3+c*(q4+c*(q5+c*(q6+
                                     c*(q7+c*(q8+c*q9))))))))).
  end if.
  compute     p1 =   1.00000000.
  compute     p2 =   0.15443144.
  compute     p3 =  -0.67278579.
  compute     p4 =  -0.18156897.
  compute     p5 =  -0.01919402.
  compute     p6 =  -0.00110404.
  compute     p7 =  -0.00004686.
  compute     q1 =   1.25331414.
  compute     q2 =   0.23498619.
  compute     q3 =  -0.03655620.
  compute     q4 =   0.01504268.
  compute     q5 =  -0.00780353.
  compute     q6 =   0.00325614.
  compute     q7 =  -0.00068245.
  do if (distr <= 2).
    compute c = distr*distr/4.
    compute bessk1 =
                (ln(distr/2)*bessi1)+(1/distr)*(p1+c*(p2+c*
                                 (p3+c*(p4+c*(p5+c*(p6+c*p7)))))).
  else.
     compute c = 2/distr.
     compute bessk1 = (exp(-distr)/sqrt(distr))*
                   (q1+c*(q2+c*(q3+c*(q4+c*(q5+c*(q6+c*q7)))))).
  end if.
  compute pred = c0 + c1*(1 - bessk1*(distr)).
end if.

select if (missing (c0) = 0 and missing (c1) = 0).
save outfile = 'c:\book.oup\bessgrm.sav'.
```

POPULAR MODELS 199

```
/*Exponential semi-variogram function */.
title "Exponential semi-variogram plot".

get file = 'c:\book.oup\step1.sav'.

model program c0=0.000001 c1 = ____ r = ____ /* specify c1 and
r */.
do if (dist = 0).
compute dist = 0.00001.
end if.
do if (r = 0).
compute r = 0.00001.
end if.
compute pred  = c0 + c1*(1 - exp(-(1/r)*dist)).
cnlr gamma / criteria = iter 500 /bounds c0>0;c1>0;0.000001<r<
____ /
outfile = param1 /save = pred /*specify the upper bound of r*/.

model program c0 c1 r
 /*If c0 < 0 in the above estimation, then let c0 = 0.0000001*/.
weight by npairs.
do if (dist = 0).
compute dist = 0.00001.
end if.
do if (r = 0).
compute r = 0.00001.
end if.
compute pred = c0 + c1*(1 - exp(-(1/r)*dist)).
cnlr gamma / criteria = iter 500 /bounds c0>0;c1>0;0.000001<r<
____
 / file = param1 /outfile = param2 /save = pred
                            /*specify the upper bound of r */.

match files / file = 'c:\book.oup\step2.sav'
 / file = 'c:\book.oup\expogram.sav' / by dist.
compute gamma = c0 + c1*(1 - exp(-(1/r)*dist)).
select if (missing (c0) = 0 and missing (c1) = 0).
save outfile = 'c:\book.oup\expogrm.sav'.


/* Gaussian semi-variogram function */.

title "Gaussian semi-variogram plot ".

get file = 'c:\book.oup\step1.sav'.

model program c0=0.000001 c1 = ____ r = ____
                                       /* specify c1 and r */.
compute pred = c0 + c1*(1 - exp(-(dist**2)/(r**2))).
cnlr gamma / criteria = iter 500 /bounds c0>0;c1>0;
0.000001<r<0.9/
outfile = param1 /save = pred.

model program c0 c1 r
```

```
/*If c0 < 0 in the above estimation, then let c0 = 0.0000001*/.
weight by npairs.
compute pred = c0 + c1*(1 - exp(-(dist/r**2))).
cnlr gamma / criteria = iter 500 /
                        bounds c0>0;c1>0; 0.000001<r< ____
 / file = parm1 /outfile = param2 /save = pred
                        /*specify the upper bound of r */.

match files / file = 'c:\book.oup\step2.sav'
 / file = 'c:\book.oup\gausgram.sav' / by dist.
compute gamma = c0 + c1*(1 - exp(-(dist**2)/r**2)).
select if (missing (c0) = 0 and missing (c1) = 0).
save outfile = 'c:\book.oup\gausgrm.sav'.


/* Spherical semi-variogram function */.
title "Spherical semi-variogram plot ".

get file = 'c:\book.oup\step1.sav'.

model program c0 = ____   c1 = 1   r = ____
                                        /* specify c0 and r */.
do if (dist = 0).
  compute pred = 0.
else if (dist > 0 and dist < r).
  compute pred = c0 + c1*((3/2)*(dist/r)-(1/2)*(dist/r)**3).
else if (dist >= r).
  compute pred = c0 + c1.
end if.
cnlr gamma / criteria = iter 500 /
                        bounds c0>0;c1>0; 0.000001<r< ____ /
outfile = param1 /save = pred /*specify the upper bound of r */.

model program c0 c1 r
 /*If c0 < 0 in the above estimation, then let c0 = 0.0000001*/.
weight by npairs.
do if (dist = 0).
  Compute pred = 0.
else if (dist > 0 and dist < r).
  compute pred = c0 + c1*((3/2)*(dist/r)-(1/2)*(dist/r)**3).
else if (dist >= r).
  compute pred = c0 + c1.
end if.
  compute d.c0 = 1.
do if (dist <= r).
  compute d.c1 = (3/2)*(dist/r)-(1/2)*(dist/r)**3.
  compute d.r =
      c1*(-(3/2)*dist/r**2 - (1/2)*3*(-dist/r**2)*(dist/r)**2).
else.
  compute d.c1 = 1.
  compute d.r = 0.
end if.
cnlr gamma / criteria = iter 500 /
                        bounds c0>0;c1>0; 0.000001<r< ____
 /file = param1 /outfile = param2 /save = pred
```

```
                                      /*specify the upper bound of r */.
match files / file = 'c:\book.oup\step2.sav'
 / file = 'c:\book.oup\sphegram.sav' / by dist.
do if (dist <= r0.
   compute gamma = c0 + c1*((3/2)*(dist/r) - (1/2)*(dist/r)**3).
else.
   compute gamma = c0 + c1.
end if.
select if (missing (c0) = 0 and missing (c1) = 0).
save outfile = 'c:\book.oup\sphegrm.sav'.



/* wave/hole semi-variogram function */.
title "Wave/hole semi-variogram plot".

get file = 'c:\book.oup\step1.sav'.

model program c0 = 0.000001 c1 = ____  r = ____
                                              /* specify c1 and r */.
do if (dist = 0).
   compute dist = 0.00001.
end if.
do if (r <= 0).
   compute r = 0.00001.
end if.
compute pred = c0 + c1*(1 - r*(sin(dist/r)/dist)).
cnlr gamma / criteria = iter 500 /
                                  bounds c0>0;c1>0; 0.01<r< ____ /
 outfile = param1 /save = pred
                                  /*specify the upper bound of r */.

model program c0 c1 r
 /*If c0 < 0 in the above estimation, then let c0 = 0.0000001*/.
weight by npairs.
do if (dist = 0).
   compute dist = 0.00001.
end if.
do if (r <= 0).
   compute r = 0.00001.
end if.
compute pred = c0 + c1*(1 - r*(sin(dist/r)/dist)).
cnlr gamma / criteria = iter 500 /
                                  bounds 0.1>c0>0;c1>0; 0.01<r< ____
 / file = param1 /outfile = param2 /save = pred
                                  /*specify the upper bound of r */.

match files / file = 'c:\book.oup\step2.sav'
 / file = 'c:\book.oup\wavegram.sav' / by dist.
do if (dist = 0).
   compute dist = 0.00001.
end if.
   compute gamma = c0 + c1*(1 - r*(sin(dist/r)/dist)).
 select if (missing (c0) = 0 and missing (c1) = 0).
 save outfile = 'c:\book.oup\wavegrm.sav'.
```

```
/* Power semi-variogram function */.
title "Power semi-variogram plot".

get file = 'c:\book.oup\step1.sav'.

model program c0=0.000001 c1 = ____ lambda = 0.1 /* specify c1
*/.
do if (dist eq 0).
compute dist = 0.000001.
end if.
compute pred  = c0 + c1*dist**lambda.
cnlr gamma / criteria = iter 500 /bounds
c0>0.000000001;c1>0;10>lambda>0/
outfile = param1 /save = pred.

model program c0 = 0.0000001 c1 lambda
/*If c0 < 0 in the above estimation,  then let c0 = 0.0000001*/.
weight by npairs.
compute pred  = c0 + c1*dist**lambda.
cnlr gamma / criteria = iter 500
                      /bounds c0>0.000000001;c1>0;10>lambda>0
 / file = param1 /outfile param2/save = pred.

match files / file = 'c:\book.oup\step2.sav'
 / file = 'c:\book.oup\powrgram.sav' / by dist.
compute gamma = c0 + c1*dist**(lambda).
select if (missing (c0) = 0 and missing (c1) = 0).
save outfile = 'c:\book.oup\powrgrm.sav'.


/* Plotting part of programs */.

get file 'c:\book.oup\____.sav
                /* specify the data file name for ploting: */.
                /* bessgrm, expogrm, gausgrm, sphegrm,     */.
                /* wavegrm, powrgrm                        */.
variable labels dist "distance".
variable labels gamma "gamma".

graph
  /scatterplot(bivar)=dist with gamma.
  /title =
  'Semi-variogram model plots for data set ____ from file ____'
        'Starting values: c0=0.000001 , c1=____, r = ____'
  /subtitle = 'Maximum lag distance included = ____ '.


get file 'c:\book.oup\____.sav'
 / keep = gammean gamvar rvm dist n /* specify file name */.

do if (dist <= ____ )
         /* specify the maximum lag distance value (maxlag) */.
  compute lagdist = dist.
```

```
else.
  compute lagdist = 9999.
end if.
compute noacorr = 2.
missing values lagdist = 9999.

compute filter_$ = (dist <= ___ )
          /* specify the maximum lag distance value (maxlag) */.
filter by filter_$.
execute .

descriptives variables = rvm
  /statistics mean semean.

filter off.
variable labels dist "distance".
graph
  /scatterplot(overlay) = dist lagdist  with rvm noacorr (pair)
  /title =
          'Ratio of the variance to the mean '
          'for each distance group up to distance ___ '
          'For data set ___  and variable ___ '
  /subtitle = 'mean rvm = ___    , stderr rvm = ___    '.



/* SPSS code for the additional semivariogram models */.

/* circular */.
model program c0=0.1 c1=0.1 r=0.02.
compute pi = 3.1415926535.
do if (mdist <= r).
  compute pred =
            c0 + c1*((2/pi)*((mdist/r)*sqrt(1 - (mdist/r)**2) +
                    arsin(mdist/r))).
  weight by npairs.
else.
  compute pred = c0 + c1.
  weight by npairs.
end if.
clnr mgamma / criteria = iter 500 /
                                    bounds c0 > 0; c1 > 0;  r > 0.


/* rational quadratic */.
model program c0 = 0.01 c1 = 1.0 r = 0.01.
weight by npairs.
compute pred = c0 + c1*(((mdist/r)**2)/(1 + (mdist/r)**2)).
cnlr mgamma / criteria = iter 500 /bounds c0 > 0;c1 > 0;r > 0.


/* penta-spherical */.
model program c0 = 0.1 c1 = 0.1 r = 0.03.
do if (mdist<=r).
   compute pred = c0 + c1*((15/8)*(mdist/r) - (5/4)*(mdist/r)**3
+ (3/8)*(mdist/r)**5).
```

```
   weight by npairs.
else.
   compute pred = c0 + c1.
   weight by npairs.
end if.
cnlr mgamma / bounds c0 > 0; c1 > 0; r > 0.

/* cubic */.
model program c0 = 0.1 c1 = 0.1 r = 0.02.
weight by npairs.
do if (mdist <= r).
   compute pred = c0 + c1*(7*(mdist/r) - (35/4)*(mdist/r)**3 +
                  (7/2)*(mdist/r)**5 - (3/4)**(mdist/r)**7).
   weight by npairs.
else.
   compute pred = c0 + c1.
   weight by npairs.
end if.
clnr mgamma / bounds c0 > 0; c1 > 0; r > 0.

/* stable */.
model program c0 = 0.03 c1 = 0.14 r = 0.01 a = 0.9.
weight by npairs.
   compute pred = c0 + c1*(1 - exp(-(mdist/r)**a)).
Cnlr mgamma / bounds c0 > 0; c1 > 0; r > 0; 0 < a < 2.

/* Cauchy */.
model program parms c0=0.02 c1=0.1 r=0.02 a=0.75.
weight by npairs.
   compute  pred = c0 + c1*(1 - 1/(1 + (mdist/r)**2)**a ).
cnlr mgamma / criteria = iter 500 /bounds c0 > 0;c1 > 0;r >
0.00001; 0 < a < 3.
```

PART II

GEOREFERENCED DATA SET CASE STUDIES

4

ANALYSIS OF GEOREFERENCED SOCIOECONOMIC ATTRIBUTE VARIABLES

Georeferenced socioeconomic data sets are commonly used for example calculations in popular standard statistics books, such as Neter et al. (1996, pp.1367-69), Venables and Ripley (1994, 67, 74, 81, 127, 320), and Johnson and Wichern (1992, 392, 393, 604-7, 624, 626). Unfortunately, as Anselin and Griffith (1988) and Anselin and Hudak (1992) note, authors of this class of texts either gloss over (e.g., Johnson and Wichern 1992, 392) or acknowledge and then dismiss the importance of (e.g., Neter et al.1996, 1048) latent spatial dependencies. Notable exceptions are furnished by Bailey and Gatrell (1995, 249) and by Sen and Srivastava (1990), who not only recognize the presence and importance of spatial autocorrelation in georeferenced data, but also include two Chicago examples (pp. 94, 152) involving socioeconomic attribute variables, as well as a geographic connectivity matrix (p. 153). In practice, spatial effects latent in socioeconomic data increasingly are playing an explicitly important role in the everyday life of people. For example, the current act of transferring specific land to native people's in the province of Saskatchewan, Canada, is resulting in an increased tax burden to adjacent municipalities because of the loss of property tax revenues (a spatial spillover effect). Therefore, strong motivation exists for practitioners to adopt spatial statistical techniques when analyzing geographic distributions of socioeconomic attribute variables and for spatial scientists to provide spatial statistical applications based on this type of data.

  A preponderance of the small number of spatial autoregressive model applications to date have been to georeferenced socioeconomic attribute data sets. Thus far, few geostatistical analyses of this class of georeferenced data have been undertaken. Exceptions include Dubin (e.g., 1992), who furnishes kriging interpolations of the selling price of houses in Baltimore, Griffith et al. (1994), who use the semivariogram for sampled per capita income to evaluate the standard error formula for the battery of variables reported by the U.S. Census Bureau, Can and Megbolugbe (1997), who address the issue of spatial dependence when house price indices are constructed.

Anselin (1988, 206-10) presents a spatial econometrics analysis of a georeferenced socioeconomic attribute data set that relates to a spatial economics version of the classical economics Philips curve (i.e., a graph depicting the relationship between inflation and unemployment) that serves as a benchmark in this chapter. His study area comprises the 25 counties of southwest Ohio. The dependent variable in his analysis is the change in wage rates for 1983; his model specification and estimation are as follows:

$$\hat{Y} = 0.96650 + 0.96036 \times (1983 \text{ unemployment rate})$$
$$- 1.02679 \times (1983 \text{ net migration rate}) - 0.02609 \times (\text{SMSA indicator variable}),$$

where the binary SMSA indicator variable takes on values of 1 and −1. The S-W (Shapiro-Wilk) statistic[1] for this variables equals 0.74980[†††], the MC (Moran Coefficient) equals −0.13915 ($\hat{\sigma}_{MC} \approx 0.13484$), the GR (Geary Ratio) equals 1.06018, and Bartlett's homogeneity of variance[2] test statistic equals 8.937[††], whereas Levene's equals 0.906. These diagnostics imply that (1) the frequency distribution for the 1983 change in wage rates does not conform to a normal distribution, casting doubt on the Bartlett's diagnostic statistic implication, (2) the 1983 change in wage rates does not exhibit pronounced spatial autocorrelation, and (3) geographic variation across this landscape is reasonably constant. Next, Anselin reports results from fitting a linear regression model and then a spatial AR model to this variable, using the inverse unemployment rate, a net migration rate, and an SMSA indicator variable as predictors. The OLS (ordinary least squares) solution accounts for roughly 49% of the variation in the 1983 change in wage rates and is accompanied by MC = −0.27733 (GR = 1.29847), which suggests the presence of statistically significant negative spatial autocorrelation, significant Bartlett (6.987[†]) and Levene (2.543[†]) test statistics, and no evidence of a failure for the residual frequency distribution to conform to a normal one (S-W = 0.98071). This last finding is corroborated by other diagnostic statistics: A-D = 0.179, R-J = 0.9899, and K-S = 0.071 (§2.3). The AR model (using matrix **W**) yields a spatial autocorrelation parameter estimate of $\hat{\rho} = -0.623$. This model accounts for roughly 64% of the variation in the 1983 change in wage rates and generates residuals that display improved geographic homogeneity of variance (Bartlett = 2.628; Levene = 1.777), as well as deviate only slightly more from a normal distribution (S-W = 0.96826).

---

[1]Statistical significance is denoted throughout by [†] for a 10%, [††] for a 5%, and [†††] for a 1% level.

[2]The geographic landscape has been divided into four quadrants by centering the county centroids (which were obtained from U.S. Census Bureau TIGER files), and then by grouping counties together whose centroids are contained in each of the four standard quadrants of a plane.

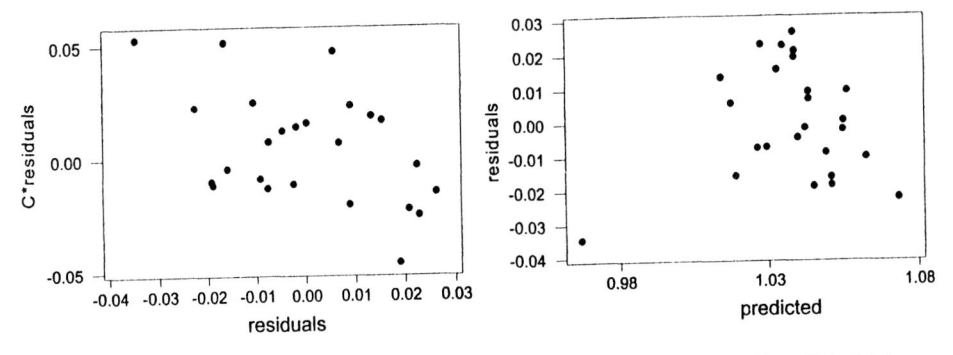

Figure 4.1. Left: (a) MC scatterplot of AR residuals for change in wage rates (Anselin's data). Right: (b) Traditional homogeneity of variance plot for AR residuals.

Griffith (1993b, 113-11) replicates this analysis, noting that one of the SMSA indicator variable values is coded incorrectly in Anselin's data table. The exact SAS solution does not differ from the one reported by Anselin; however, the solution reported in Griffith (1993b) uses the Jacobian approximation (§3.1.2) and renders estimates that differ just slightly from those reported by Anselin. Diagnostic statistics here concerning the normality assumption include S-W = 0.96826, A-D = 0.253, R-J = 0.9884, and K-S = 0.077. Hence, there is no evidence to indicate that the frequency distribution of the AR residuals does not conform to a normal distribution in the population.

Although the frequency distribution for the raw 1983 change in wage rates exhibits marked deviation from a normal distribution, because the ratio of its extreme values equals 1.14280 < 3, there is no reason to expect that the use of a Box-Cox power transformation would be helpful in this data analysis. The principal culprit appears to be the nonconstant mean response, which, when

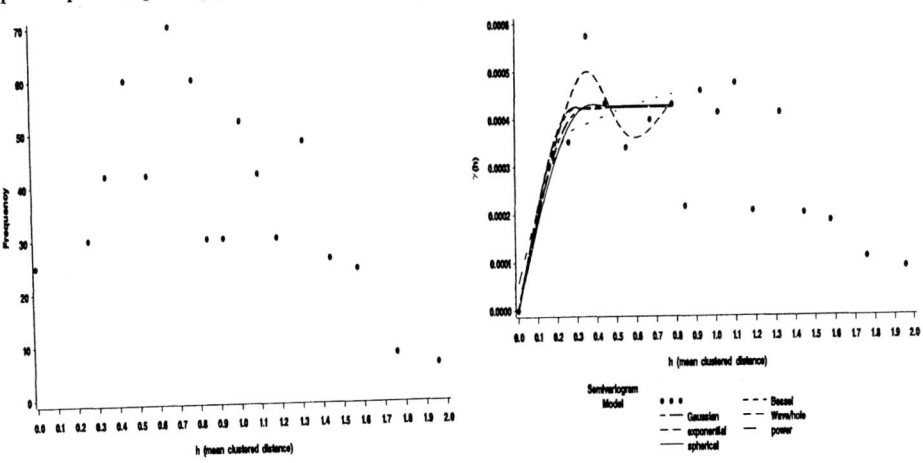

Figure 4.2. Left: (a) Semivariogram-type plot of the frequency of semivariance values in each distance class for wage rates (Anselin's data). Right: (b) Semivariogram plot of wage rates (Anselin's data), with selected semivariogram model curves superimposed.

removed from these data, renders residuals whose frequency distribution indeed does exhibit conformity to a normal distribution. In addition, the AR model specification results in an acceptable stabilization of the geographic variability across the southwestern Ohio county landscape.

But this data analysis continues to be afflicted by complications. First, the AR residuals yield MC = -0.15567 (GR = 1.09795), whose prominence is confirmed by inspection of a Moran scatterplot (Figure 4.1a), and a standard residuals-versus-predicted values plot (Figure 4.1b) suggests nonconstant variance in the more traditional sense of this concept. One of the problems with this analysis is, no doubt, the small sample size (n = 25), which places an increased emphasis on satisfying the normality assumption here. Nevertheless, the AR model increases the percent of variance accounted for in the 1983 change in wage rates by 15% and reveals that the standard errors for the regression coefficients estimated by OLS are inflated by roughly 12% (on average) for the predictor variables and 69% for the intercept term.

A geostatistical analysis also reveals the presence of pronounced negative spatial autocorrelation in the OLS residuals. Because there are only 25 data points in the data set, distances were divided into 17 classes. Using a maximum average distance class, including the rescaled[3] value of 0.8, the plot of the frequency of distances versus distance class (Figure 4.2a) contains an average of 47 distances per class. The scatter of points displayed in the semivariogram plot (Figure 4.2b) can be described as a dampening oscillation with increasing distance, a pattern indicative of negative spatial autocorrelation. Of the six semivariogram models fit to the data (Figure 4.2b), the RSSE of the wave/hole model—the only semivariogram model that effectively handles negative spatial autocorrelation (§3.2.1)—is markedly lower than that for the alternative models:

model	RSSE	$\underline{C_0}$	$\underline{C_1}$	$\underline{r}$
exponential	0.25940	0.00000	4.237×10^{-4}	0.06116
Bessel	0.25851	3.367×10^{-6}	4.205×10^{-4}	0.05396
Gaussian	0.25015	0.00000	4.264×10^{-4}	0.15913
spherical	0.23736	0.00000	4.275×10^{-4}	0.35846
wave/hole	0.09498	0.00000	4.105×10^{-4}	0.07740
power	0.35837	0.00000	4.727×10^{-4}	* $\lambda = 0.18355$.

Also visible in the semivariogram plot is the relatively steep ascent of the model curves over a short initial distance, indicative of relatively low levels of spatial autocorrelation, which is consistent with results reported for the spatial AR model ($\frac{\hat{\rho}}{|\rho_{min}|} = -0.366$). The wave/hole model is the semivariogram model of choice for describing the spatial autocorrelation present in these data.

---

[3] Distances are rescaled by dividing all (U, V) coordinate values by the maximum absolute U or V coordinate value in the set of georeferencings attached to the sample of areal units.

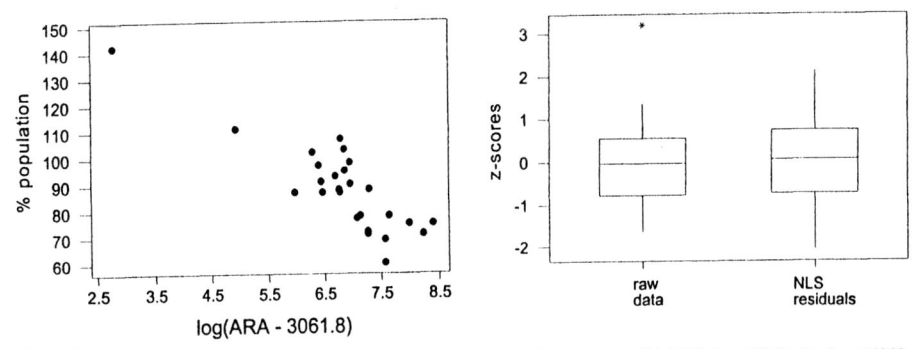

Figure 4.3. Left: (a) Scatterplot of percentage of population versus $LN(ARA - 3061.8)$ for Cliff and Ord's data. Right: (b) Boxplots of raw percentage of population and its associated regression residuals.

4.1. THE CLIFF-ORD EIRE POPULATION DATA

Griffith (1993b, 111-12) replicates the Cliff-Ord spatial autoregressive analysis of agricultural output data (Cliff and Ord 1981, 239). Cliff and Ord (pp. 208-10) also present initial spatial statistical results for O'Sullivan's regression of population percentage (for which S-W = 0.92528[†]) on an arterial road network index (ARA) for the 26 counties of Eire. This bivariate regression yields an R^2 of 0.404 when computed with raw data; the accompanying MC value reported is 0.19078, Bartlett = 9.029[††], and Levene = 2.951[†]. Cliff and Ord then find that taking logarithms of both the dependent and the predictor variables increases R^2 to 0.517, decreases MC to 0.13006, and decreases Bartlett to 7.129[†] and Levene to 2.394[†]. They also note that a nonlinear specification of this model may well be worth considering (p. 214).

Again, because $\frac{y_{max}}{y_{min}} = 2.37 < 3$, there is no reason to expect that the use of a Box-Cox power transformation would be helpful in this data analysis. Rather, following the Cliff-Ord suggestion, a more sophisticated exploration of the linearity assumption in this case replaces their log-log specification with the following model specification and estimation:

$$\hat{Y} = 178.67578 - 13.26754 \times LN(ARA - 3061.82631),$$

which renders an exemplary linearization scatter of points (Figure 4.3a) and increases R^2 to 0.735. Figure 4.3a reveals one problematic feature of these data, namely, the presence of two influential observations. Although one of these observations is a statistical distribution outlier with regard to the raw data, its regression residual loses this outlier status (Figure 4.3b). This estimation result is accompanied by the following diagnostic statistics: S-W = 0.98986 (A-D = 0.139, R-J = 0.9951, and K-S = 0.089), Bartlett = 3.185, Levene = 1.125, and MC = 0.04389 (GR = 0.95528). Because E(MC) = −0.05537 (which is almost identical

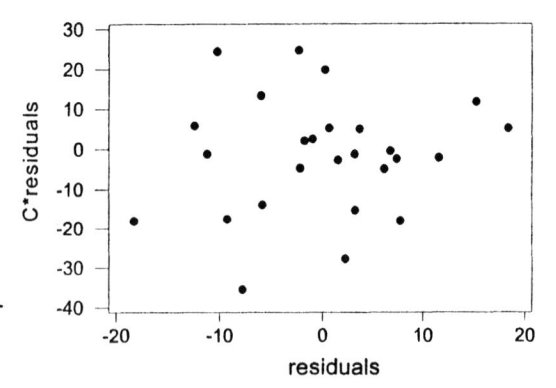

Figure 4.4. MC scatterplot of NLS residuals for percentage of population (Cliff and Ord's data).

to what Cliff and Ord report), and $\hat{\sigma}_{MC} \approx 0.13131$ (which also is almost identical to what Cliff and Ord report), findings described here support their contention that any spatial autocorrelation detected in the Eire population data is strictly attributable to a misspecified mean response function (the missing variable interpretation of spatial autocorrelation). This absence of residual spatial autocorrelation is further attested to by its corresponding Moran scatterplot (Figure 4.4).

The bothersome concentration of points in the lower left-hand quadrant of Figure 4.3a is accompanied by a lack of dispersion on the right-hand side of a standard residual-versus-predicted values scatterplot (Figure 4.5). Both of these graphs are indicative of a relative underrepresentation of extreme population percentages in the data set. Cliff and Ord suggest that a serious analysis of these data should include a comparison of estimation results with and without this influential observation (i.e., Dublin), which was done (1981, 214).

As in the case of the benchmark analysis, the number of distance classes[4] was reduced from the usual 50 (§1.6) to 17 due to a small sample size (n = 26) for the data set. Inspection of the frequency plot of distance pairs versus distance class

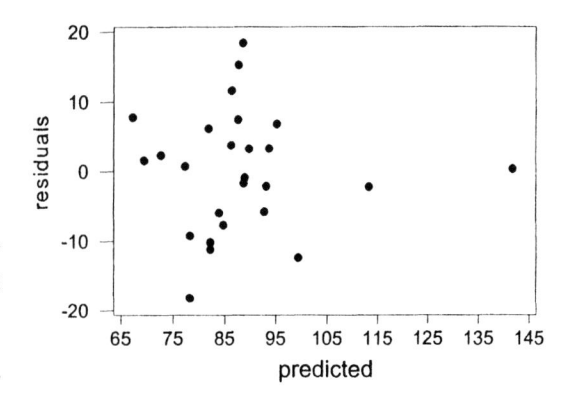

Figure 4.5. Traditional homogeneity of variance NLS residual plot for Cliff and Ord's percentage of population data.

---

[4]Centroids derived from a digitized version of Cliff and Ord's map (1981, 207) used in this analysis are reported in Appendix 4-A.

(Figure 4.6a) reveals that approximately 30 or more pairs of distances were present in each distance class up to the maximum average rescaled distance class of 0.8. A low level of spatial autocorrelation is indicated in the semivariogram scatterplot of the residuals (Figure 4.6b) from the model specification

$$\hat{Y} = 178.67578 - 13.26754 \times LN(ARA - 3061.82631),$$

corroborating the low level of estimated spatial autocorrelation reported for the spatial autoregressive analysis (MC = 0.04389). The scatter of points displayed in Figure 4.6b indicates a low level of spatial autocorrelation in the residuals, primarily due to the lack of gamma values at relatively small interpoint distances coupled with the presence of the 0 point (i.e., each attribute value covarying with itself). Semivariogram model descriptions of the scatter of points (Figure 4.6b) indicate moderately good fits; relatively low-range parameter values are expected due to the lack of spatial autocorrelation present in the residuals:

model	RSSE	C_0	C_1	r
exponential	0.32905	0.00000	90.851	0.10985
Bessel	0.32764	0.07950	90.379	0.07699
Gaussian	0.32151	0.00000	89.769	0.15544
spherical	0.31474	0.00000	89.545	0.32279
wave/hole	0.27641	0.47527	84.843	0.05575
power	0.36180	0.00000	101.469	* $\lambda = 0.18965$.

The r values seem artificially high in this case, partly because of the absence of small interpoint distances (Figure 4.6b)—a geographic discretization effect. In practice, none of these specifications is the preferred semivariogram model, given

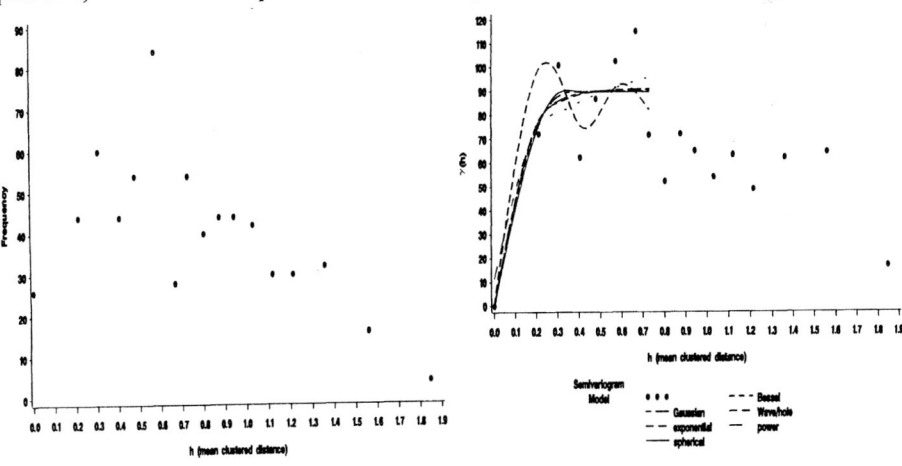

Figure 4.6. Left: (a) Semivariogram-type plot of the frequency of semivariance values in each distance class for Cliff and Ord's percentage of population data. Right: (b) Semivariogram plot of NLS residuals for Cliff and Ord's percentage of population data, with selected semivariogram model curves superimposed.

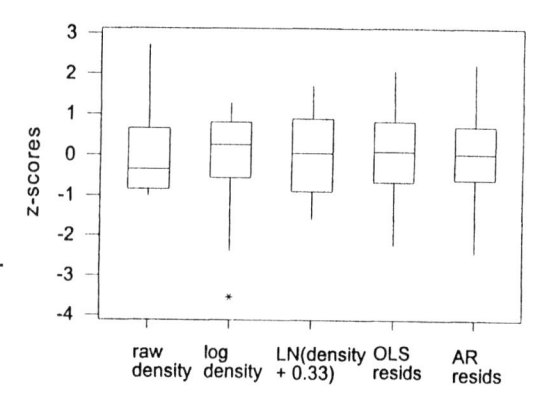

Figure 4.7. Boxplots of raw and power-transformed urban population density and their associated OLS and AR regression residuals (Griffith and Can's data).

the rather narrow RSSE spread, which should be the case when 0 spatial autocorrelation prevails.

4.2. URBAN POPULATION DENSITY

Population density is of interest to a wide range of spatial scientists (e.g., Carroll et al. 1997). Autoregressive versions of urban population density models have been posited for several decades now (e.g., Griffith 1981b; Anselin and Can 1986). Griffith and Can (1996, 231-49) present comparative estimation results for four data sets, namely 1980 and 1990 Syracuse 1986 Ottawa-Hull[5], and 1986 Toronto federal census data. They compare SAS and SpaceStat output, as well as SAS exact and approximate Jacobian term results. The succeeding analysis extends that by Griffith and Can for the 1980 Syracuse data set.

Several features of the 1980 Syracuse population density case merit underscoring here. First, the most suitable power transformation according to urban economic theory is the logarithm (i.e., $\gamma = 0$); estimating equation (2.1) yields $\hat{\gamma} \approx -0.00203$, which essentially is a logarithmic transformation. This finding also holds for the remaining four data sets analyzed by Griffith and Can (1996):

urban area	$\hat{\gamma}$	$\hat{\delta}$	raw data S-W	Ln transformed S-W
1990 Syracuse	-0.00495	> 0	0.85108	0.92000
1986 Ottawa-Hull	-0.00251	> 0	0.86581	0.93942
1986 Toronto	-0.00280	> 0	0.77388	0.97278

Second, the added constant, δ, improves symmetry of the frequency distribution of the dependent variable in every one of these four cases. While the raw data are positively skewed, simply applying a logarithmic transformation converts this

---

[5] Preliminary spatial analysis of this data set appears in Griffith (1993b).

positive skewness to negative skewness with an outlier (Figure 4.7). By adding δ to population density and then taking the log of the resulting sum, both the skewness and its accompanying outlier disappear. Third, a spatial autoregressive response model is employed because, conceptually speaking, a given magnitude of population density should directly encourage similar magnitudes of population density to accumulate around it; high densities inspire nearby high densities, and low densities inspire nearby low densities, especially considering the geography of zoning constraints. And, fourth, the added constant δ, of LN(population density + δ) can be interpreted in the following nonlinear model context:

$$LN(\text{population density} + \delta) \approx \mathbf{X}\boldsymbol{\beta} \rightarrow \text{population density} \approx \delta + EXP(\mathbf{X}\boldsymbol{\beta}).$$

In other words, although one needs to recognize that the error structure differs between these two model specifications, δ relates to the intercept term of a nonlinear least squares estimation of the model parameters.

Analysis of the 1980 Syracuse population data is guided by findings reported in Griffith and Can (1996). The set of X variables includes distance from the city center [i.e., the central business district (CBD)], the (u, v) pair of geo-referencing Cartesian coordinates (to account for a nonconstant geographic mean response), a low-density indicator variable (indexing census tracts #1, #11, and #12), and a local anomaly indicator variable (indexing census tracts #165.02 and #166; these are two towns, each completely surrounded by a single census tract). OLS and spatial autoregression estimation results together with diagnostic statistics for this data set include

raw			power transformation	
			$\hat{\delta} = 0$	
			$\hat{\gamma} = 0.32510$	
			RSSE = 5.1×10^{-2}	
S-W = 0.84826†††			S-W = 0.91660†††	
Bartlett = 32.545†††			Bartlett = 5.049	
Levene = 6.850†††			Levene = 2.202†	
MC = 0.63123 ($\hat{\sigma}_{MC} \approx 0.05074$)			MC = 0.67814	
GR = 0.34534			GR = 0.33616	

parameter estimates: LN(population density + 0.32510)

	OLS	s.e.	AR(W)	s.e.
ρ	*	*	0.55227	0.10448
b_0	0.87185	0.25886	0.94168× (1−0.55227)	0.49804
$b_{\text{distance}_{1,\text{CBD}}}$	−1.15838	0.07078	−0.58666	0.12411
$b_{u-\bar{u}}$	−0.14199	0.06614	−0.06502	0.05871
$b_{\text{low density}}$	−0.99792	0.15664	−0.87100	0.13682
$b_{\text{spatial anomalies}}$	0.52521	0.20063	0.57740	0.17281
(pseudo-)R^2	0.675		0.762	

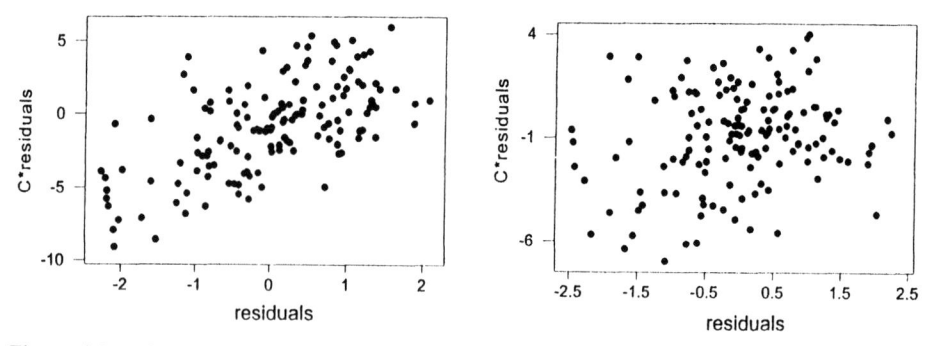

Figure 4.8. Left: (a) MC scatterplot of OLS residuals for urban population density (Griffith and Can's data). Right: (b) MC scatterplot of AR residuals for urban population density (Griffith and Can's data).

	residuals	
S-W	0.95732††	0.97420
A-D	0.786††	0.344
R-J	0.9893††	0.9958
K-S	0.053	0.048
Bartlett	1.376	1.365
Levene	0.628	0.434
MC	0.36271	0.09115
GR	0.71850	0.95683

These results highlight several commonly observed differences between OLS and spatial autoregressive results, namely, (1) a change in standard errors, (2) an increase of 5%-10% in the percent of variance accounted for (pseudo-R^2), and (3) improved compliance with underlying statistical model assumptions. Although the introduction of a nonconstant mean response reduces the MC value by nearly 47%, spatial autocorrelation remaining in the OLS residuals is still pronounced and conspicuous (Figure 4.8a).

These findings differ from those reported by Griffith and Can (1996) for two reasons. First, Griffith and Can used $\delta = 0$. Inspection of the preceding diagnostic statistics for the raw and power-transformed versions of population density reveals that letting δ deviate from 0 can have a dramatic impact on the diagnostic statistics (also note that Griffith and Can calculated a much smaller $\hat{\rho}$ = 0.26236). Second, Griffith and Can employed weighted nonlinear least squares to adjust for nonconstant variance. The AR model results reported here are accompanied by quite appealing diagnostic statistics (both the pseudo-R^2 and S-W statistics are improvements over the ones reported by Griffith and Can). The residuals appear to conform to a normal frequency distribution (see Figure 4.7, too), variance is constant across the four quadrants of the geographic landscape, and the residual spatial autocorrelation is relatively small. This spatial autocorrelation residue is not very pronounced (Figure 4.8b), but complications arising from the presence of nonconstant variance may well persist. The OLS residuals do not display constant variance at all levels of predicted population density (Figure

SOCIOECONOMIC VARIABLES 217

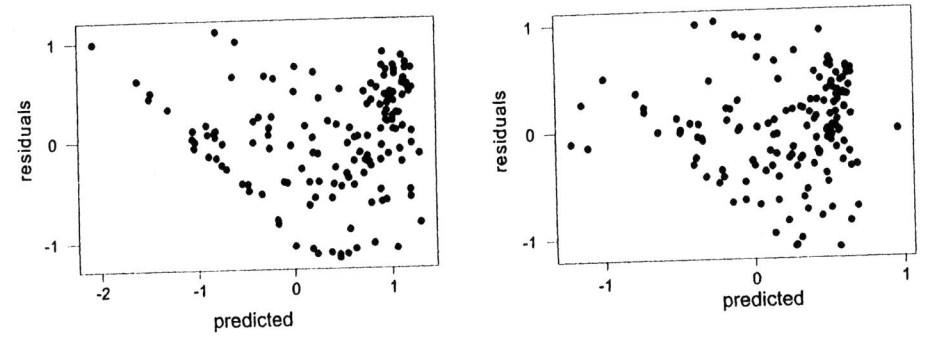

Figure 4.9. Left: (a) Traditional homogeneity of variance OLS residual plot for urban population density (Griffith and Can's data). Right: (b) Traditional homogeneity of variance AR residual plot for urban population density (Griffith and Can's data).

4.9a), as noted by Griffith and Can. Although much better behaved than their OLS counterparts, the AR residuals also display some heterogeneity of variance across the predicted value levels (Figure 4.9b). A more serious analysis of this data set would need to explore the weighting scheme proposed by Griffith and Can.

For this data set, the number of distance classes was set to 100 rather than to 50 because of a rather large sample size (n = 145) that yields 145^2 distance pairs. Inspection of Figure 4.10a indicates that the number of distance pairs per distance class far exceeds 50 for all classes through the maximum average rescaled distance class of 0.8. The accompanying scatter of points displayed in the semivariogram plot (Figure 4.10b) for the initial distance classes indicates a low to moderate level of spatial dependency in the residuals, which concurs with results reported for the spatial autoregression analysis ($\hat{\rho}$ = 0.55) and the OLS residuals (MC = 0.36271). Unfortunately, semivariogram model fits to the semivariogram plot (Figure 4.10b)

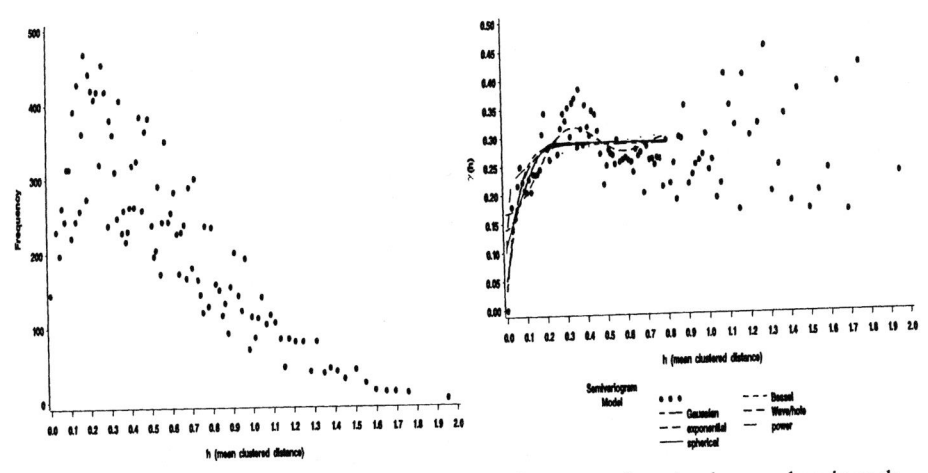

Figure 4.10. Left: (a) Semivariogram-type plot of the frequency of semivariance values in each distance class for Griffith and Can's OLS regression residuals. Right: (b) Semivariogram plot of OLS residuals for urban population density (Griffith and Can's data), with selected semivariogram model curves superimposed.

are relatively poor:

model	RSSE	C_0	C_1	r
exponential	0.55439	0.03391	0.25591	0.06554
Bessel	0.57416	0.05138	0.23611	0.04183
Gaussian	0.58623	0.14061	0.15010	0.12262
spherical	0.55973	0.10932	0.18115	0.24295
Wave/hole[†]	0.45820	0.16816	0.12212	0.07687
power	0.73460	0.00000	0.30871	* $\lambda = 0.09738$

[†]Initial parameter estimate values for the wave/hole semivariogram model were specified explicitly as $C_0 = 0.0001$, $C_1 = 0.3$, $r = 0.1$.

Although the wave/hole model provides the lowest RSSE for the competing model fits to the semivariogram plot, there is no reason to expect that negative spatial autocorrelation is present in these data, especially given the results of the preceding spatial autoregressive analysis (e.g., the MC is positive). And, while the power semivariogram model is expected to provide a good model fit because a spatial AR model specification provides a good fit to these data, the power model has the highest RSSE and provides a very poor fit at best to the semivariogram plot. These poor semivariogram model fits might best be explained by the "spikiness" of gamma values between distance classes 0.2-0.4 (inclusive), a reason that the wave/hole model is somewhat successful in tracking this scatterplot. A coarser resolution for distance classes further smooths the fluctuation of average gamma values at these distance class intervals; e.g., a decrease from 100 to 50 distance classes results in a decrease of approximately 12% in the RSSE for all models. Parameter estimation for the spherical model, in which r coincides with the range, renders an r value that indicates the presence of a moderately wide spatial dependency field, a finding that is consistent with $\hat{\rho} = 0.55$ for the spatial AR model.

4.3. RESIDENTIAL INSURANCE COVERAGE IN CHICAGO

In 1978, the U.S. Commission on Civil Rights compiled data by zip code area for the city of Chicago in an attempt to decide whether or not insurance companies were redlining Chicago neighborhoods. These data, available in Andrews and Herzberg (1985, 407-11), include a map on which zip code area 60622 is mislabeled as 60652 (which also appears on the map). MapInfo 4.0 was used to correct this mistake and to extract georeferencing coordinates for each of the 47 predominantly residential zip code areas (Map 4.1 displays their centroids) included in this data set. Birkes and Dodge (1993), Rasmussen (1992, 170-171), as well as others, use this as an example data set for illustrating aspects of, for instance, regression analysis. Birkes and Dodge recommend the application of a logarithmic transformation to the fire variable and the removal of zip code areas 60607 and 60611, which they identify as leverage points. The analysis reported in this section recommends retaining both of these areas and creating an indicator variable for zip code area 60607 to include in a regression model as an outlier.

SOCIOECONOMIC VARIABLES   219

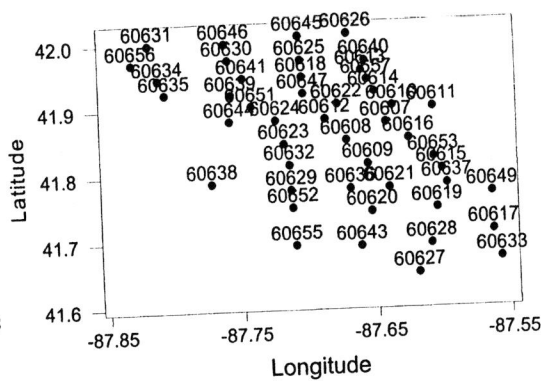

Map 4.1. Geographic distribution of zip code (longitude, latitude) centroids for the Chicago residential insurance coverage data.

One regression model that can be formulated with these data—prescribed to address the problem posed by the U.S. Commission on Civil Rights—describes "voluntary market activity" for insurance coverage (i.e., new homeowner policies plus renewals, minus cancellations and nonrenewals, per 100 housing units). Applying the following power transformation causes the frequency distribution for the insurance coverage (i.e., dependent) variable to better conform to that for a normal distribution:

raw	power transformation
	$\hat{\delta} = -0.5$
	$\hat{\gamma} = 0.62961$
	RSSE = 2.2×10^{-2}
S-W = 0.93650[††]	S-W = 0.95939
Bartlett = 6.202	Bartlett = 3.987
Levene = 2.064	Levene = 1.236
MC = 0.60020 ($\hat{\sigma}_{MC} \approx 0.10512$)	MC = 0.60038
GR = 0.31987	GR = 0.35658 .

These results indicate that the power transformation improves agreement between the empirical and normal frequency distributions, inducing a better homogenization of variance across the four quadrants of the Chicago geographic landscape. Marked positive spatial autocorrelation is displayed by this geographic distribution of insurance coverage.

Birkes and Dodge (1993) allude to the need to apply Box-Tidwell transformations [see equations (2.1) and (2.2)] to the individual predictor variables. Besides the fire variable, whose transformed version is LN(fire + 0.23395), the north-south Cartesian coordinate is converted to $LN[(v - \bar{v})^2 + 0.00101]$, theft to LN(theft + 10.37286), and income to LN(income − 3524.72476). The impact of these transformations on the pairwise scatterplots for these variables, with special attention to their pairings with the power-transformed insurance coverage variable, reveals improved linear alignments (Figure 4.11a-b). OLS estimation results together with diagnostic statistics for this data set include

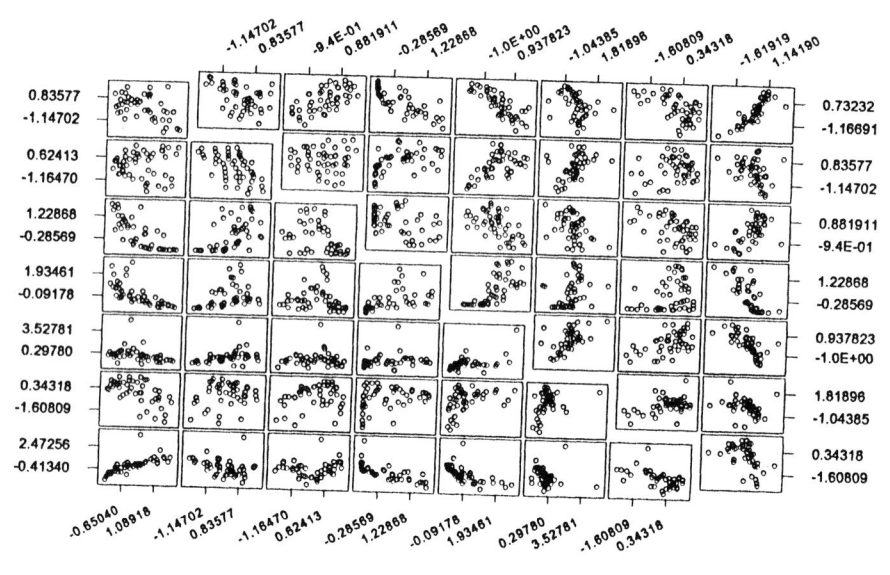

Figure 4.11. Left: (a) Scatterplot matrix for the raw residential insurance coverage data. Left: (b) Scatterplot matrix for the power-transformed residential insurance coverage data.

parameter estimates: (insurance coverage − 0.5)$^{0.62961}$

OLS		residuals	
b_0	−3.46793	S-W	0.97410
$b_{I_{outlier}}$	0.94421	A-D	0.308
$b_{LN[(v-\bar{v})^2 + 0.00101]}$	0.21796	R-J	0.9903
b_{race}	−0.00975	K-S	0.082
$b_{LN(fire + 0.23395)}$	−0.54510	Bartlett	2.784
b_{age}	−0.00871	Levene	0.802
$b_{LN(income - 3524.72476)}$	1.18812	MC	−0.03656
R^2	0.889	E(MC)	≈ −0.09992
		GR	0.84538

These results indicate that the detected spatial autocorrelation is nothing more than simple map pattern (which is accounted for by the conspicuous linear trend surface—captured by the v coordinate term—across the Chicago geographic landscape). The Moran scatterplot presented in Figure 4.12a corroborates this interpretation of the originally detected moderate positive spatial autocorrelation. The OLS results suggest that, after controlling for factors such as median family income, fires per 1,000 housing units, and percentage of housing units built in or before 1939, as the percentage of minority population increases (the race variable), voluntary market activity tends to decrease, which is consistent with the belief that redlining has occurred. The power-transformed theft variable is not a significant predictor in this OLS regression equation. In fact, the power-transformed income and fire variables are the only prominent predictor variables uncovered in this data set. The only disappointing feature of this analysis is that the OLS residuals do not

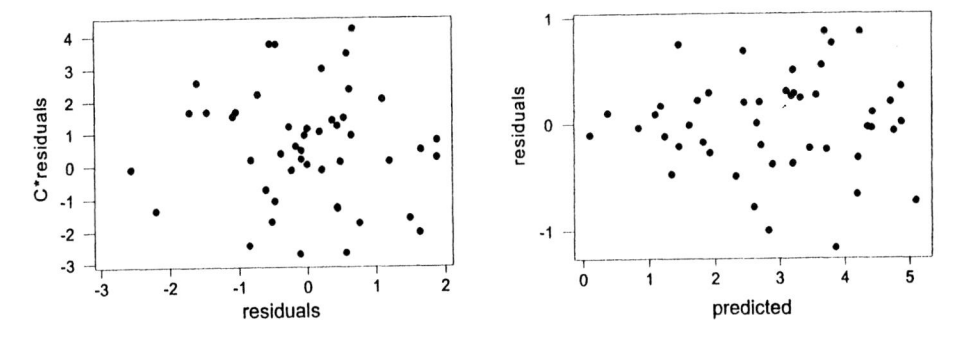

Figure 4.12. Left: (a) MC scatterplot of OLS residuals (residential insurance coverage data). Right: (b) Traditional homogeneity of variance OLS residual plot for the residential insurance coverage data.

display constant variance at all levels of predicted insurance coverage (Figure 4.12b), insinuating that a more serious analysis of these data may well require the use of weighted regression. As an aside, spatial autocorrelation detected in an OLS analysis of the raw, untransformed variables would partly be attributable to the presence of overlooked nonlinear relationships (i.e., dirty data), a result similar to the preceding finding for the Cliff-Ord Eire data.

Distance pairs were divided into 40 (instead of the usual 50) distance classes for the geostatistical analysis of these data because sample size is relatively small (n = 46, or 2,116 distance pairs total). The scatterplot of the frequency of distances versus distance classes indicates that approximately 50 distance pairs per distance class are available up to the maximum average rescaled distance class that includes 0.2 (Figure 4.13a). The scattering of points displayed in the semi-variogram plot (Figure 4.13b) indicates that virtually no spatial autocorrelation is pres-

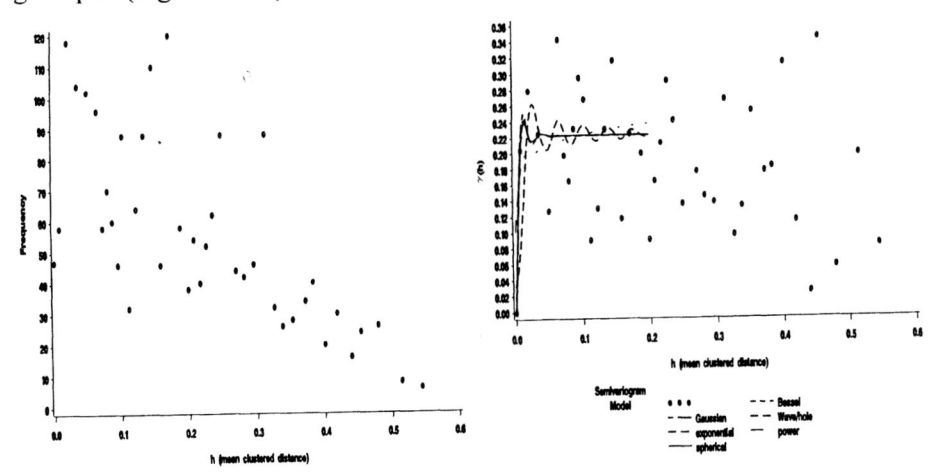

Figure 4.13. Left: (a) Semivariogram-type plot of the frequency of OLS residuals semivariance values in each distance class for the residential insurance coverage data. Right: (b) Semivariogram plot of OLS residuals for the residential insurance coverage data, with selected semivariogram model curves superimposed.

ent in the residuals, evident by (1) a rapid increase in gamma values at initial distance classes and (2) the dispersion of points (through the selected maximum average rescaled distance class), which essentially reveals a random pattern. In fact, if the 0 point were ignored, the nugget effect would essentially be equivalent to the sill, indicating an absence of spatial autocorrelation. All semivariogram model fits (Figure 4.13b) are poor, with all of the models having near-0 r values:

model	RSSE	C_0	C_1	r
exponential	0.74645	0.00000	0.22335	0.00304
Bessel	0.74744	0.00000	0.22288	0.00159
Gaussian	0.74591	0.00000	0.22354	0.00536
spherical	0.74591	0.00000	0.22354	0.01120
Wave/hole[†]	0.68465	0.03228	0.19277	0.00591
power	0.87419	0.00000	0.27339	* $\lambda = 0.08625$

[†]Initial parameter estimate values for the wave/hole semivariogram model were specified explicitly as $C_0 = 0.0001$, $C_1 = 0.22$, $r = 0.005$.

The spherical model r estimate suggests essentially a 0 width for the spatial dependency field. This finding is exactly what is expected given MC ≈ -0.1 for the OLS residuals.

    Two features of interest can be gleaned from these unautocorrelated OLS residuals that allow us to compare empirical results with results anticipated from random, unautocorrelated spatial data (§2.6.2), namely, the behavior of aspatial outliers and the ratio of the variance to the mean within each distance class in the semivariogram plot. The presence of an aspatial outlier in a georeferenced attribute is expected to be the dominant value (either high or low) for many, if not most, of the distance classes, depending on where the outlier value is located within a given landscape with respect to the edge of the map. When the indicator variable (containing presence/absence of the outlier for each observation in the data set) is removed from the Chicago insurance regression equation, the value 1.59 appears in 8 of the first 19 distance classes (Figure 4.14). Accounting for the presence of an aspatial outlier by including the indicator variable in the regression equation results in no single value dominating distance class extremes. Meanwhile, for unautocorrelated spatial data, the ratio of variance to mean within each distance class of the semivariogram plot is expected to be 2 (§2.6.2). Inspection of Figure 4.15 reveals that this is approximately the case once the presence of the aspatial outlier has been accounted for.

4.4. URBAN CRIME IN COLUMBUS, OHIO

The FBI announced in 1997 that crime nationwide in the United States had registered its fifth consecutive annual decline, and the steepest single decrement in the 35 years such records have been kept. Criminal behavior frequently has a salient spatial dimension. Recognizing this geographic feature of the phenomenon has encouraged city commissioners, police precinct captains, detectives, and others to sometimes assess crime patterns street by street. The displacement hypothesis

SOCIOECONOMIC VARIABLES 223

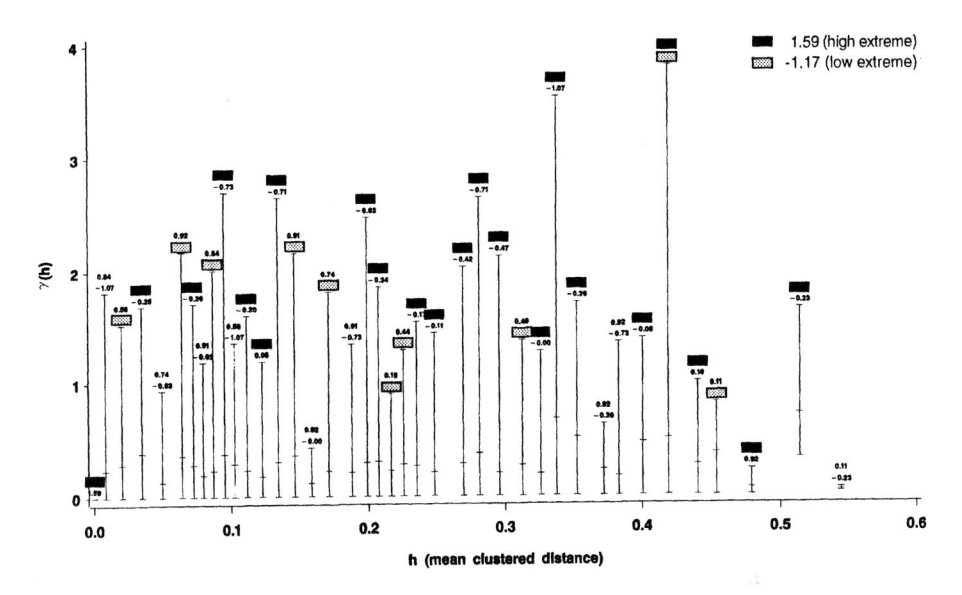

Figure 4.14. Semivariogram plot displaying the mean and extreme values for Figure 4.13b.

posits that as vigilance by police increases in an area, crime tends to spill over and relocate into surrounding areas. This geographic process, when coupled with the marked tendency for similar types of households and economic activities to cluster together in an urban area, has motivated researchers to study the geography of crime (e.g., Hubert et al. 1985; Griffith 1987, 31-43; Sen and Srivastava 1990, 15; Bailey and Gatrell 1995, 78). The analysis reported in this section reevaluates Anselin's (1988, 187-202) spatial model of determinants of neighborhood crime, which also has been reanalyzed by Getis (1995, 176-78). The geographic landscape under study comprises 49 contiguous planning neighborhoods in Columbus, Ohio; the data are for 1980. Because the recording of a crime at the "downtown" pre-

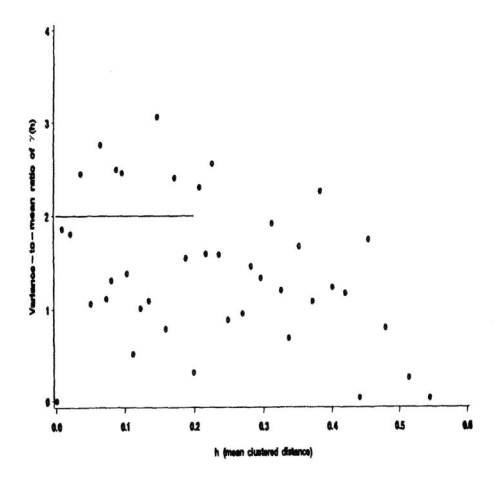

Figure 4.15. Semivariogram-type plot of the distance class variance-to-mean ratio for Figure 4.13b.

cinct, rather than in the neighborhood where it was committed, has been common practice, outlier areal units should be anticipated in these data.

Anselin (1988) proposes an initial regression model that describes crime (the combined total of residential burglaries and vehicle thefts per 1,000 households) as a function of income (in \$1,000s) and housing value (in \$1,000s); R^2 = 0.552. Searching for an appropriate power transformation as well as evaluation of Anselin's initial OLS results yields the following diagnostics:

```
raw                              OLS residuals
S-W = 0.96409                    S-W = 0.98303
Bartlett = 4.651                 Bartlett = 6.154
Levene = 1.185                   Levene = 0.746
MC = 0.52064 (ô_MC ≈ 0.09285)    MC = 0.24220
GR = 0.58367                     GR = 0.71114 .
```

They suggest that little would be gained by employing a power transformation; variation across the four quadrants of the Columbus geographic landscape displays a tolerable degree of heterogeneity, and both of the frequency distributions adequately conform to a normal frequency distribution. Marked positive spatial autocorrelation is displayed by the geographic distribution of OLS residuals, though.

A closer inspection of these data reveals two disturbing features of Anselin's initial OLS results. First, a Moran scatterplot reveals the presence of two spatial outliers, neighborhoods #4 and (marginally) #2—these are the two extreme left-hand points in Figure 4.16a. A traditional residuals-versus predicted values scatterplot also discloses that these two areal units have extreme negative residuals; in addition, this particular plot reveals that neighborhood #34 has an extreme positive residual (Figure 4.16b). The latter finding is in keeping with Anselin's determination of the presence of heterogeneous variance. Accordingly, a pair of indicator variables was created, one denoting the pair of spatial outliers and one denoting the single frequency distribution outlier. Introducing these variables into the OLS regression analysis results in housing value becoming insignificant and hence leaving the equation, and the R^2 increasing to 0.764.

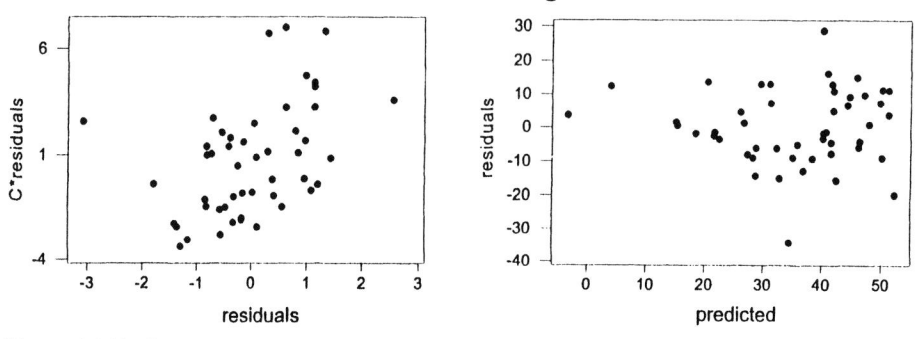

Figure 4.16 Left: (a) MC scatterplot of OLS residuals (Anselin's urban crime data). Right: (b) Traditional homogeneity of variance OLS residual plot for Anselin's urban crime data.

Because the detected latent positive spatial autocorrelation should be linked to a spatial process, such as crime displacement, and because variables more than likely are missing from the equation specification, spatial autocorrelation is viewed here as entering into the analysis through the error term; as such, an SAR model specification is posited. OLS and spatial SAR estimation results together with diagnostic statistics for this data set include

parameter estimates

	OLS	s.e.	SAR(W)	s.e.
ρ	*	*	0.54971	0.21010
b_0	67.29569	5.61134	58.84338	5.97432
			=(1-0.54971)	
b_{income}	- 2.42671	0.22182	- 1.98007	0.25613
$b_{spatial\ outliers}$	-19.50235	3.16536	-18.30916	2.74574
$b_{conventional\ outliers}$	15.83984	4.24046	13.05390	3.73039
(pseudo-)R^2	0.764		0.817	

residuals

S-W	0.96281	0.93503††
A-D	0.467	1.033††
R-J	0.9878	0.9757††
K-S	0.086	0.134††
Bartlett	1.295	2.526
Levene	0.294	1.192
MC	0.20265	0.03580
GR	0.74669	0.92754 .

These results once again reveal that the increase in percentage of variance accounted for in the dependent variable, namely crime, is only about 5% when a spatial autoregressive term is introduced into the regression model specification. Marked positive spatial autocorrelation is present in these crime data (beyond simple map pattern—given that the Cartesian coordinates were found to be trivial descriptors of the geographic distribution of crime) and essentially is completely accounted for by the spatial SAR model specification. The principal gain attributable to using the SAR model is that variance is homogeneous both across the four quadrants of the geographic region and across the range of predicted crime levels (Figure 4.17); the cost of using an SAR model is a noticeable increase in deviation by the frequency distribution of the residuals from a normal distribution. Given a sample size of only 49, this cost indicates the need for a more serious investigation of these data to involve additional remedial work; perhaps a power transformation of the crime variable would be justifiable after all.

For the geostatistical analysis, 40 distance classes were specified for these data. The plot of the number of distance pairs versus distance class (Figure 4.18a) indicates approximately 50 distance pairs per class through the maximum average rescaled distance class of 0.8, except for the second distance class. The scatter of points for initial distance classes in the semivariogram plot (Figure 4.18b) indicates

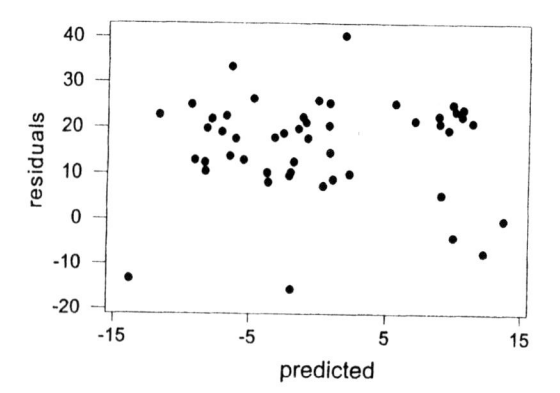

Figure 4.17. Traditional homogeneity of variance SAR residual plot for Anselin's urban crime data.

the presence of a moderate degree of spatial dependency in the residuals. The selected semivariogram models provide a relatively poor fit to the semivariogram plot based upon the OLS residuals:

```
model           RSSE        C₀        C₁        r
exponential     0.49368     3.43014   66.3137   0.13149
Bessel          0.50773     2.15725   65.9428   0.07414
Gaussian        0.52936     5.67704   61.3699   0.13173
spherical       0.51739     2.76986   64.2019   0.27537
Wave/hole       0.46574     9.37294   55.4250   0.06298
power           0.45109     0.00000   79.1659      *     λ = 0.24081 .
```

As with the population density data analysis (§4.2)—where $\hat{\rho} = 0.55$, too—the spherical model r estimate implies the presence of a moderately wide spatial dependency field. Because an SAR autoregressive model was fit to the OLS

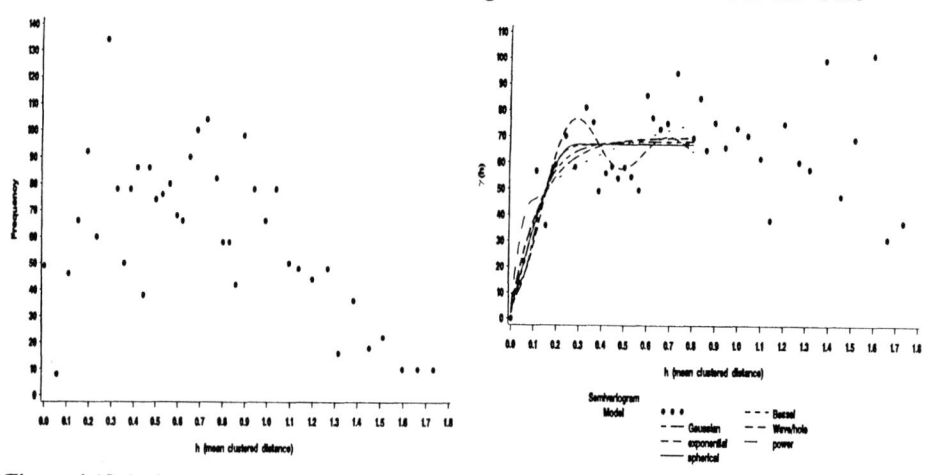

Figure 4.18. Left: (a) Semivariogram-type plot of the frequency of OLS residuals semivariance values in each distance class for Anselin's urban crime data. Right: (b) Semivariogram plot of OLS residuals for Anselin's urban crime data, with selected semivariogram model curves superimposed.

SOCIOECONOMIC VARIABLES   227

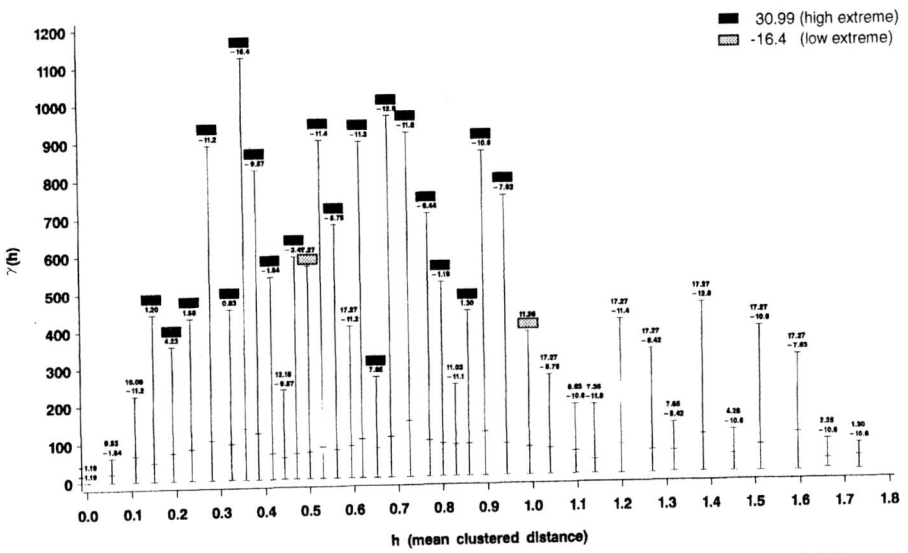

Figure 4.19. Semivariogram plot displaying the mean and extreme values for Figure 4.18b.

residuals, the best-fit semivariogram model is expected to be the Bessel function. The RSSE of the Bessel semivariogram model falls within the RSSE range found in the exponential, Gaussian, spherical, and power models, raising questions about whether it, or the power model (the lowest RSSE), should be the semivariogram model of choice here. A more serious analysis would further differentiate among these model fits using cross-validation.

As in §4.4, an aspatial outlier (ID = 34) was detected in the residuals of these data and is accounted for by including an indicator variable in the OLS regression equation. When this outlier is not accounted for there, the plot of extreme high and low gamma values contained in each distance class (Figure 4.19) is dominated by the outlier value of 30.99 in 16 of the first 23 classes, reinforcing the notion put forth in §2.6.2 that presence of an aspatial outlier value can be detected by examining extreme gamma values for each distance class.

4.5. GEOGRAPHIC DISTRIBUTION OF MINORITIES ACROSS SYRACUSE, NEW YORK

The geographic distribution of the percentage of people of color in U.S. cities frequently is studied when issues of segregation are addressed, as exemplified by the Nassau County public schools data set (Chatterjee et al. 1995, 72-77). Urban sociological theory argues, in part, that households with similar characteristics tend to cluster in geographic space. Reinforcing this tendency, urban economic theory argues that households with similar utility functions tend to cluster in geographic space. The analysis reported in this section explores the 1990 census blockgroup geographic landscape of Syracuse, New York, which comprises 193 coterminous areal units (Map 4.2); the number of people of color living in the city at that time

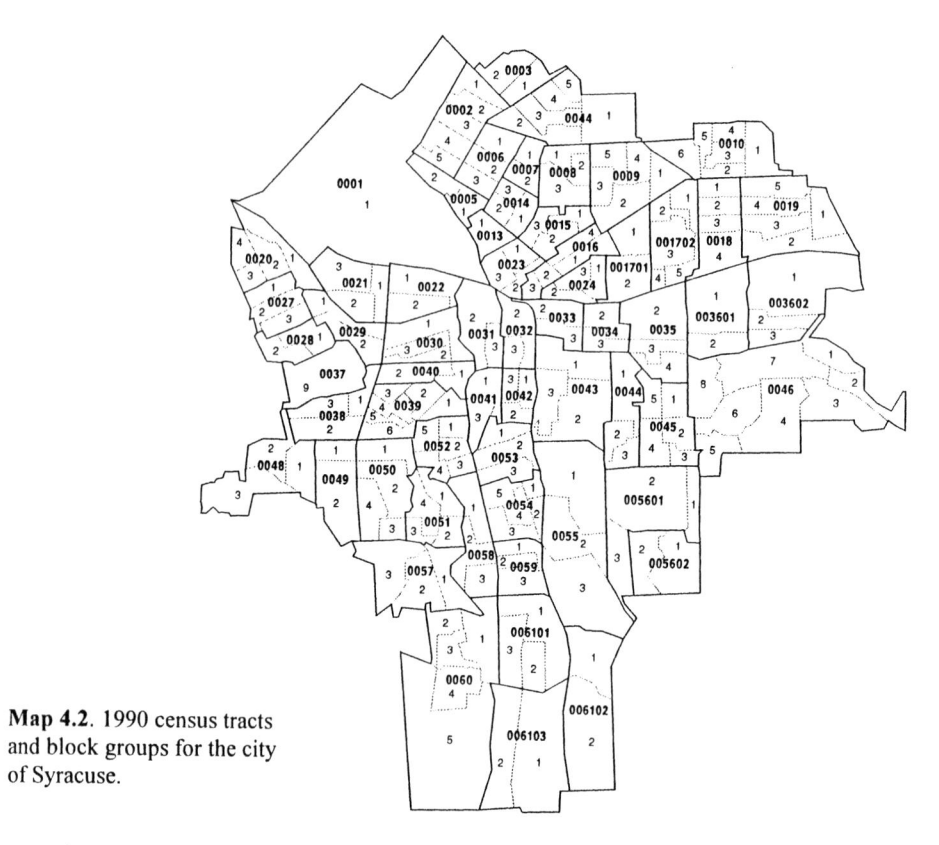

228 A CASEBOOK FOR SPATIAL STATISTICAL DATA ANALYSIS

Map 4.2. 1990 census tracts and block groups for the city of Syracuse.

constituted 25.02% of its population.

Searching for an appropriate power transformation of the measure

$$1 - \frac{\frac{\text{number of whites}}{\text{total population}}}{\frac{\text{number of whites}}{\text{total population}}}$$ yields the following diagnostics:

<table>
<tr><th>raw</th><th>power transformation</th></tr>
<tr><td></td><td>$\hat{\delta} = -0.00098$</td></tr>
<tr><td></td><td>$\hat{\gamma} = 0$</td></tr>
<tr><td></td><td>RSSE = 1.8×10^{-2}</td></tr>
<tr><td>S-W = 0.76469†††</td><td>S-W = 0.97481†</td></tr>
<tr><td>Bartlett = 35.711†††</td><td>Bartlett = 18.046</td></tr>
<tr><td>Levene = 11.791†††</td><td>Levene = 6.052</td></tr>
<tr><td>MC = 0.72125 ($\hat{\sigma}_{MC} \approx 0.04156$)</td><td>MC = 0.71340</td></tr>
<tr><td>GR = 0.32178</td><td>GR = 0.31437 .</td></tr>
</table>

These results suggest that some data analytic benefits would be gained by employing a power transformation: variation across the four quadrants of the Syracuse blockgroup geographic landscape displays a more tolerable degree of heterogeneity, and the frequency distribution better mimics a normal frequency distribution. Essentially, the same marked positive spatial autocorrelation is displayed by this geographic distribution, regardless of whether the raw or the power-transformed data are evaluated.

Theoretically, however, a binomial distribution does have a nonconstant variance, and using a normal approximation can introduce specification error into an analysis. Consequently, because considerable geographic variability and spatial autocorrelation remain after employment of a power transformation here, the binomial frequency distribution assumption is retained; accordingly, a logistic regression model has been estimated. Predictors entered into this model include 54 of the 63 eigenvectors, extracted from expression (1.10), displaying the most conspicuous levels of positive spatial autocorrelation. These eigenvectors—which are both orthogonal and uncorrelated —combined with population density (a surrogate for urban poverty) and the east-west Cartesian coordinate constitute the set of predictor variables. Maximum likelihood estimation results for this logistic regression equation include

parameter estimates

variable	parameter estimate	s.e.	Wald chi-square	Pr > chi-square
b_0	− 1.9381	0.0312	3865.6346	0.0001
$b_{v-\bar{v}}$	−25.7087	1.4414	318.1358	0.0001
$b_{population\ density}$	0.0008	0.0001	77.9368	0.0001
b_{EV1}	− 6.2296	0.3115	400.0662	0.0001
b_{EV2}	1.9535	0.1445	182.6763	0.0001
b_{EV3}	−10.8473	0.1432	5741.6280	0.0001
b_{EV4}	− 4.6425	0.1544	904.1436	0.0001
b_{EV5}	− 3.2160	0.1657	376.5268	0.0001
b_{EV6}	2.6109	0.1784	214.2725	0.0001
b_{EV7}	0.6228	0.2001	9.6905	0.0019
b_{EV8}	1.5875	0.1723	84.8429	0.0001
b_{EV9}	− 4.2889	0.1291	1103.9071	0.0001
b_{EV10}	− 7.5233	0.1355	3082.4562	0.0001
b_{EV11}	6.6223	0.1974	1125.8178	0.0001
b_{EV12}	5.5625	0.1262	1943.0933	0.0001
b_{EV14}	2.0050	0.1602	156.7016	0.0001
b_{EV15}	2.1112	0.1691	155.9578	0.0001
b_{EV16}	− 1.6674	0.1615	106.5438	0.0001
b_{EV17}	− 1.6055	0.1407	130.1775	0.0001
b_{EV19}	− 1.3403	0.1128	141.2125	0.0001
b_{EV20}	− 2.0073	0.1356	219.0790	0.0001
b_{EV21}	− 1.4960	0.1507	98.5375	0.0001
b_{EV22}	2.8115	0.1702	272.8508	0.0001
b_{EV23}	− 2.3140	0.1223	358.1028	0.0001
b_{EV24}	− 3.5639	0.1617	485.7341	0.0001
b_{EV25}	3.3563	0.1333	633.7008	0.0001
b_{EV26}	− 1.0977	0.1332	67.8953	0.0001
b_{EV27}	1.1923	0.1639	52.9124	0.0001
b_{EV28}	1.3069	0.1628	64.4728	0.0001

230   A CASEBOOK FOR SPATIAL STATISTICAL DATA ANALYSIS

b_{EV29}	1.0381	0.1420	53.4758	0.0001
b_{EV30}	− 1.5776	0.1437	120.5742	0.0001
b_{EV31}	− 1.6781	0.1581	112.5903	0.0001
b_{EV32}	− 0.7704	0.1346	32.7511	0.0001
b_{EV33}	− 3.1128	0.1458	455.5339	0.0001
b_{EV34}	− 0.2826	0.1270	4.9507	0.0261
b_{EV35}	3.4236	0.1222	785.1397	0.0001
b_{EV36}	− 1.9629	0.1324	219.6327	0.0001
b_{EV37}	− 1.0970	0.1152	90.6836	0.0001
b_{EV39}	− 2.4181	0.1350	320.6228	0.0001
b_{EV40}	− 1.4658	0.1335	120.5103	0.0001
b_{EV42}	− 3.0992	0.1305	563.9161	0.0001
b_{EV43}	− 0.9248	0.1128	67.2536	0.0001
b_{EV44}	2.5473	0.1069	568.3407	0.0001
b_{EV45}	0.8286	0.1294	40.9971	0.0001
b_{EV46}	3.5888	0.1357	699.6365	0.0001
b_{EV47}	− 1.1803	0.1028	131.7364	0.0001
b_{EV48}	− 0.9429	0.1564	36.3338	0.0001
b_{EV51}	1.6827	0.1228	187.7449	0.0001
b_{EV52}	1.2306	0.1192	106.6386	0.0001
b_{EV53}	− 1.7817	0.1526	136.2357	0.0001
b_{EV54}	1.5282	0.1358	126.6323	0.0001
b_{EV55}	− 0.5597	0.1377	16.5106	0.0001
b_{EV57}	− 0.6136	0.1127	29.6187	0.0001
b_{EV59}	0.2918	0.1145	6.4925	0.0108
b_{EV60}	− 0.3641	0.1242	8.5912	0.0034
b_{EV61}	0.6248	0.1116	31.3486	0.0001
b_{EV62}	1.2991	0.1285	102.2436	0.0001

The pseudo-R^2 associated with this logistic regression model is 0.892, indicating that a considerable amount of the variability in the percentage of people of color across the census blockgroups of Syracuse is accounted for by population density and various map patterns (i.e., v and the 54 eigenvectors).

The residuals from this logistic regression are quite symmetric, more so than are the original percentages (Figure 4.20). And, MC has been reduced from roughly 0.7 to −0.11050 (GR = 1.06533), which is consistent with the anticipated expected value of MC in this case, since for an OLS result it would be (§1.4.3) −0.21953. This expectation is corroborated by inspection of the MC scatterplot for

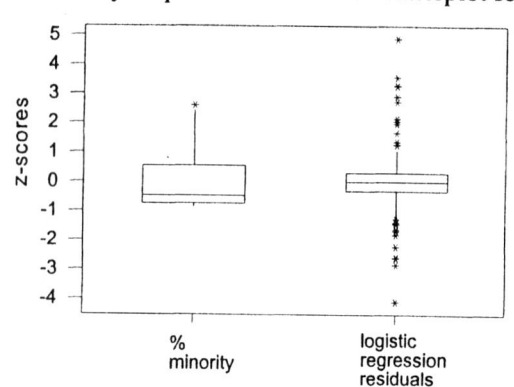

Figure 4.20. Boxplots of percentage of minorities and spatially filtered logistic regression residuals for percentage of minorities (U.S. Census Bureau data).

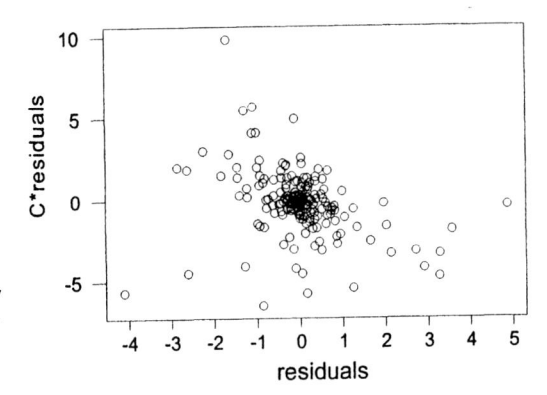

Figure 4.21. MC scatterplot of spatially filtered logistic regression residuals for percentage of minorities (U.S. Census Bureau data).

these residuals (Figure 4.21). Therefore, much of the spatial autocorrelation latent in the geographic distribution of 1990 percentage of minority population across the City of Syracuse is accounted for by various map patterns.

Using residuals from the power-transformed analysis variable and the OLS model specification of

$$\hat{Y} = -2.853165 - 2.894127 \times (v - \bar{v}) + 0.003257 \times (\text{population density})$$

for the geostatistical analysis, 100 distance classes were specified because of the relatively large number of distance pairs ($n^2 = 193^2$). The frequency plot of distance pairs per distance class (Figure 4.22a) shows that the number of distance pairs per class well exceeds 50. The relative gradual increase of gamma values for initial distance classes (through the maximum average rescaled distance class of 0.8) in the semivariogram plot appearing in Figure 4.22b indicates the presence of spatial dependency in the data, which also is indicated by the MC value (0.71340) for the

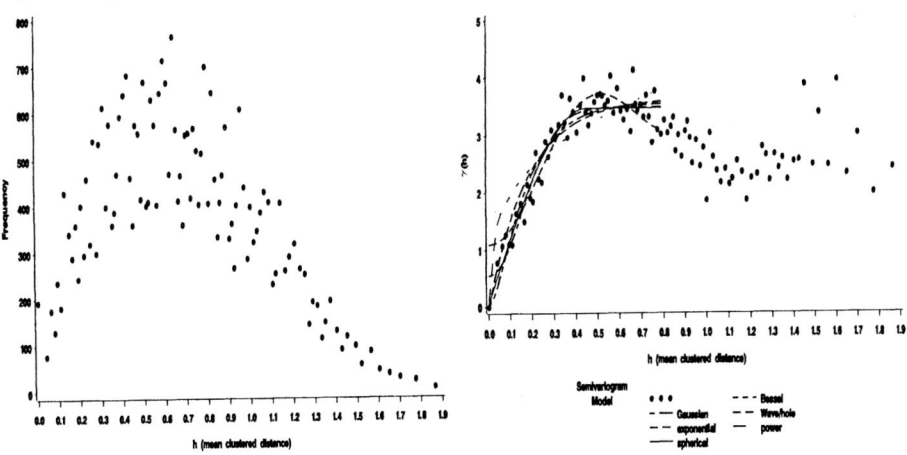

Figure 4.22. Left: (a) Semivariogram-type plot of the frequency of OLS residuals semivariance values in each distance class for percentage of minorities (U.S. Census Bureau data). Right: (b) Semivariogram plot of OLS residuals for percentage of minorities (U.S. Census Bureau data), with selected semivariogram model curves superimposed.

power-transformed data and by the relatively low RSSEs coupled with relatively large r values for the fitted semivariogram models:

```
model          RSSE      C₀        C₁        r
exponential    0.20435   0.00000   3.64009   0.18566
Bessel         0.17579   0.00000   3.55888   0.11523
Gaussian       0.15355   0.55035   2.94015   0.22229
spherical      0.14668   0.07709   3.40125   0.45005
Wave/hole      0.15939   1.09776   2.15762   0.11449
power          0.34561   0.00000   4.09369      *      λ = 0.32026 .
```

The spherical semivariogram provides the lowest RSSE, with RSSE values for the Gaussian and wave/hole models being very close to that for the spherical model. One should keep in mind that this analysis is exploratory in nature; the geo-referenced attribute studied in this case comes from an auto-binomial distribution. Accordingly, at present the theoretical background presented in §3.3 has not been developed to the point of supporting a preference for an anticipated appropriate semivariogram model.

4.6. CONCLUDING COMMENTS: SPATIAL AUTOCORRELATION AND SOCIO-ECONOMIC ATTRIBUTE VARIABLES

The geography, regional science, and statistics literatures are replete with geo-referenced socioeconomic data sets for which the presence of spatial autocorrelation is overlooked. Findings reported in this chapter should convincingly persuade researchers to alter this widespread malpractice. Conceptually illuminating results uncovered here may be summarized under three general headings. Foremost is the theme of articulating relationships between geostatistical and spatial autoregressive models (§3.3). Based largely upon estimated RSSEs, the following empirical linkages have been established in this chapter:

Table 4.1. Summary of linkage findings for Chapter 4

phenomenon	areal unit	autoregressive model	semivariogram model
unemployment	county	AR	wave/hole
population growth	county	NLS	none
urban population density	census tract	AR	exponential; spherical; Bessel; Gaussian
insurance coverage	zip code zone	OLS	none

	crime	planning neighborhood	SAR	power
percentage of minority population	census block group	filtered OLS	spherical; Gaussian; Bessel; exponential

The wave/hole semivariogram model is consistent with conceptual expectations given the presence of negative autocorrelation in the unemployment data, as is the failure to uncover a preferred semivariogram model for either the population growth or the insurance coverage data given an absence of spatial autocorrelation in these two cases. The anticipated linkage for the urban population density data is between the AR and power models, whereas the anticipated linkage for the crime data is between the SAR and Bessel function models. An inability to discriminate easily and simply between four of the competing models for both the urban population density and percentage of minority population data sets is disappointing and remains a topic for future research.

The final data analysis of this chapter involving the percentage of minority population revolves around the theme of the auto-binomial frequency distribution and supplies evidence that (1) spatial autocorrelation can be filtered out of georeferenced data with judiciously selected eigenvectors of the geographic weights matrix (§1.5), which summarizes locational information latent in georeferenced data, and (2) the use of these eigenvectors indicates that the number of dfs associated with georeferenced data are substantially smaller than would be the case with independent data. This data analysis also suggests that the auto-binomial model, in a modified form, can be easily and effectively used with georeferenced data.

The third major conceptual theme illuminated by results reported in this chapter concerns the meaning of spatial autocorrelation (§1.2). Spatial autocorrelation can arise from nonlinear model misspecification error, as well as from a missing variable having a conspicuous spatial expression. In many empirical cases, at least a part of detected spatial autocorrelation can be interpreted as simple map pattern (a linear gradient across the geographic landscape under study). But even when a simple map pattern is present, other sources of spatial autocorrelation (e.g., centripetal forces) can be operating. In addition, spatial autocorrelation can arise from quite complex map patterns.

Finally, taking spatial autocorrelation into account increases the percentage of variance accounted for in the response variables of every one of the data sets analyzed in this chapter. This increase is as little as 2% (in the case of introducing only simple map patterns into a specification) or as much as 63% (in the case of accounting for spatial autocorrelation with complex map patterns). Interestingly, nonlinear misspecification error confused with spatial autocorrelation is found to account for an additional 22% of variability in the response variate of the Eire data set.

APPENDIX 4-A
Centroids Derived from a Digitized Version of Cliff and Ord's Map of Eire (1981, 207) Using ArcInfo

Sequential Number	Cliff & Ord's Code	coordinates U	V	County Name
1	A	7.096813	4.092594	Carlow
2	B	6.514928	6.354832	Cavan
3	C	4.738920	4.272455	Clare
4	D	4.925203	2.567127	Cork
5	E	5.928473	8.002393	Donegal
6	F	7.638783	5.305707	Dublin
7	G	4.861348	5.185306	Galway
8	H	3.966842	2.931758	Kerry
9	I	7.082359	4.927536	Kildare
10	J	6.657759	3.795705	Kilkenny
11	K	6.515757	4.532493	Laoighis
12	L	5.814391	6.598311	Leitrim
13	M	5.042394	3.623049	Limerick
14	N	6.126397	5.846844	Longford
15	O	7.481897	6.246275	Louth
16	P	4.419265	6.196510	Mayo
17	Q	7.165770	5.731268	Meath
18	R	6.935245	6.663845	Monaghan
19	S	6.264349	4.933213	Offaly
20	T	5.576138	5.847149	Roscommon
21	U	5.188160	6.624058	Sligo
22	V	5.887674	3.888028	Tipperary
23	W	6.287020	3.055144	Waterford
24	X	6.418162	5.511110	Westmeath
25	Y	7.370715	3.612707	Wexford
26	Z	7.557760	4.565921	Wicklow

5

ANALYSIS OF GEOREFERENCED
NATURAL RESOURCES ATTRIBUTE VARIABLES

Perusing the *Proceedings* of the first and second international symposia on the spatial accuracy of natural resources and environmental sciences databases (Congalton 1994; Mowrer et al. 1996) reveals that one popular data analytic toolbox used by scientists working in this substantive area is geostatistics; geostatistical methodology for this applications area is summarized in Goovaerts (1997). Particularly within the context of geographic information systems (GISs), many natural resources data sets studied are sizeable and span the spectrum from acreage of forested land and bauxite reserves to zoological georeferenced data. Spatial autocorrelation materializes in these kinds of data mainly due to geographic processes: weather fronts move across regions, species invade a region as they reproduce and migrate, rainfall accumulates and then flows from one place to another, erosion forces tend to be similar in nearby locations, and certain chemicals and minerals percolate or leach through the ground and collect in geological substructure formations. Although a considerable amount of the data analytic work completed with such data has employed geostatistical techniques, this field of research also was the spawning ground for some of the first spatial autoregression model applications (Meade 1967). Geostatistical techniques are especially relevant to this category of georeferenced data, which may be characterized by a high degree of geographic continuity, and have been used to forecast natural resources potential in unsampled locations from an inventory of natural resources in nearby sampled locations. Continuous surfaces can be partitioned in meaningful ways to support spatial autoregression analysis through the use of Thiessen polygons—also known as Voronoi diagrams, which are the double of Delaunay tessellations (Okabe et al. 1992)—implemented here using Arc/Info 7.0 for UNIX. This approach is of special historical relevance to natural resources analysis, since

Thiessen (1911) devised his surface partitioning procedure as a way to improve the accuracy of average regional rainfall interpolation. One caution needs to be heeded here, though, namely that judiciously selected Thiessen polygon borders may need to be ignored; otherwise, unreasonably distant geographic points might be designated as being nearby. Such nonsensical connections usually occur along the edge of a region, where each peripheral Thiessen polygon has its borders extended to the boundary of the artificial study rectangle used for geometric closure in the polygon construction operation.

Cressie (1989; 1991, 212-24) presents a geostatistical analysis of a georeferenced piezometric-head water pressure data set, collected from drill stem tests conducted by the U.S. Department of Energy, in the deep brine Wolfcamp aquifer region of northern Texas and northeastern New Mexico; these data are reprinted in *A Handbook of Small Data Sets* (Hand et al. 1994). This study area, which essentially is that found in Harper and Furr (1986), comprises 85 suitable wells. Analysis of these data serves as a benchmark in this chapter. The dependent variable in this analysis is the number of feet above sea level to which water rises when a narrow pipe is drilled into the aquifer. The geographic distribution of well locations, together with their associated Thiessen polygon surface partitioning, appears in Map 5.1 (each ignored polygon border is denoted by a slash crossing it); Cressie (1991, 213) supplies a map of the aquifer upon which county boundaries have been superimposed. The S-W (Shapiro-Wilk) statistic[1] for this variable equals 0.93333[†††], the MC (Moran Coefficient) equals 0.62662 [E(MC) ≈ -0.01190; σ_{MC} ≈ 0.06594], the GR (Geary Ratio) equals 0.34225, Bartlett's homogeneity of variance[2] test statistic equals 30.295[†††], and Levene's equals 5.343[†††]. These diagnostics imply that (1) the frequency distribution for the piezometric-head water pressure does not conform to a normal distribution, casting doubt on the Bartlett's (but not Levene's) diagnostic statistic, (2) the piezometric-head water pressure ex-

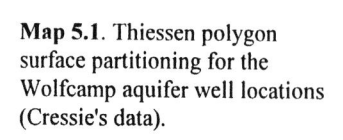

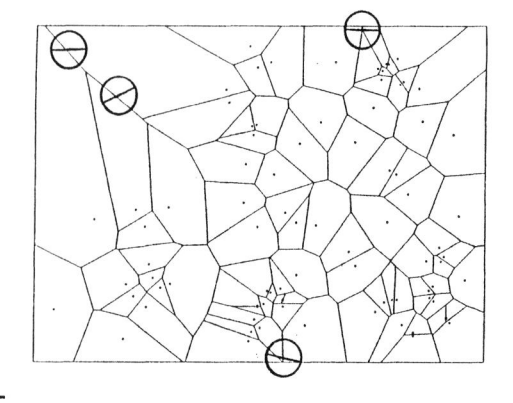

Map 5.1. Thiessen polygon surface partitioning for the Wolfcamp aquifer well locations (Cressie's data).

---

[1]Statistical significance is denoted throughout by † for a 10%, †† for a 5%, and ††† for a 1% level.

[2]The geographic landscape has been divided into four quadrants by centering the geographic coordinates of the wells and then grouping wells together whose locations are contained in each of the four standard quadrants of a plane.

hibits pronounced positive spatial autocorrelation, and (3) geographic variation across this landscape is heterogeneous.

The best power transformation uncovered for these data is the logarithm—$\hat{\delta} = 0.00246, \hat{\gamma} = 0.03007$, RSSE = 2.7×10^{-2}—yielding S-W = $0.95235^{\dagger\dagger}$ (implying non-normality), Bartlett = $26.368^{\dagger\dagger\dagger}$ and Levene = $7.313^{\dagger\dagger\dagger}$ (implying nonconstant variance), and MC = 0.60398 (GR = 0.35552; implying persistence of marked positive spatial autocorrelation). Consequently, a power transformation should not be used with these piezometric-head water pressure data, as it fails to dramatically improve upon the detected nonstationarity in spread and the non-symmetry (non-normality) of the empirical frequency distribution.

Cressie (1989, 200-201) furnishes a three-dimensional scatterplot and a three-dimensional interpolated surface for these data. Both of these graphics suggest, separately, that the piezometric-head water pressure contains a nonconstant geographic mean. Exploiting this feature, Layne et al. (1997) reanalyzed these data and discovered that one of the wells was omitted from the tabulated data listing in Cressie and Clark (1994, 4.1-4.2). OLS (ordinary least squares) results are based upon the standardized normal deviate for piezometric-head water pressure (z_y), coupled with the centering of the (u, v) Cartesian coordinates [i.e., $(u - \bar{u})$ and $(v - \bar{v})$], and include

$$\hat{z}_y = 0 - 0.01104(u - \bar{u}) - 0.00979(v - \bar{v}), R^2 = 0.892.$$

Residuals generated by this regression equation are reasonably well-behaved: S-W = 0.97093 (A-D = 0.475, R-J = 0.9881, K-S = 0.069); Bartlett = 3.538 and Levene = 0.718; and MC = 0.14234 [GR = 0.82612; E(MC) ≈ -0.029840]. The assumption violations initially detected in these data appear to arise from a nonconstant mean response; in fact, Cressie treats these data as anisotropic, a detected trait that also largely appears to be attributable to the overlooking of a nonconstant mean response. Furthermore, much of the apparent spatial autocorrelation in raw piezometric-head water pressure may be interpreted as map pattern.

Both a MC scatterplot and a traditional residuals-versus-predicted values plot insinuate that the drilled well located in the extreme northeastern New Mexico part of the Wolfcamp aquifer is an outlier (data value = 3571). Introducing an indicator variable into the regression model specification to account for this aberration yields the following set of estimation results:

parameter estimates

	OLS	AR(W)	SAR(W)
ρ	*	0.04917	0.37837
b_0	0.55665	0.57566×	0.59247×
		(1-0.04917)	(1-0.37837)
$b_{u-\bar{u}}$	-0.01060	-0.01023	-0.01053
$b_{v-\bar{v}}$	-0.00986	-0.00954	-0.01025
$b_{outlier}$	0.57007	0.56130	0.60987

(pseudo-)R²		0.907	0.907	0.916
residuals				
S-W		0.95842††	0.95929††	0.95643††
A-D		0.549	0.493	0.489
R-J		0.9903	0.9909	0.9902
K-S		0.073	0.057	0.074
Bartlett		0.652	0.532	2.056
Levene		0.816	0.582	1.604
MC		0.19807	0.16904	0.00517
	E(MC) ≈ -0.03008			
GR		0.78152	0.81336	0.96316

We can see several interesting aspects of these piezometric-head water pressure data. First, including a spatial autoregressive term adds at most 1% to the statistical explanatory power of the model. Second, all three model specifications generate residuals that display what can be considered homogeneous variability across the four quadrants of the Wolfcamp aquifer geographic landscape. Third, the respective frequency distributions of residuals come close to conforming to a normal distribution. Fourth, residuals rendered by both the OLS and the AR models exhibit conspicuous positive spatial autocorrelation, accompanied by near-perfect homogeneous variance. In contrast, residuals rendered by the SAR model exhibit essentially no spatial autocorrelation but are accompanied by noticeably more —although not significant—variance heterogeneity. Relatively speaking, when compared with geographic variability for the other two sets of residuals, variability for the SAR residuals is almost exactly the same in the western part of this aquifer, increases slightly in the northeast, and decreases markedly in the southeast. This trade-off between spatial autocorrelation and constancy of variance exposed by a comparison of these three models may well be indicative of the anisotropy identified and modeled by Cressie. From a conceptual point of view, a researcher should expect an AR model to best characterize piezometric-head water pressure, since pressure in nearby locations should directly impact upon pressure at any selected location.

Residuals from the OLS estimated regression equation,

$$\hat{Y} = 0.55665 - 0.01060 \times (u - \bar{u}) - 0.00986 \times (v - \bar{v}) + 0.57006 \times (I_{outlier}),$$

where the dependent variable has been transformed as just described, were used to analyze these data; this differs from the approach used by Cressie (1989). The importance of including an outlier indicator variable is seen in Figure 5.1, where the gamma value of 1.07 is present in 12 of the 32 distance classes having distances that do not exceed the maximum rescaled[3] distance value of 0.9. In fact, this same value is dominant in most of the larger distance classes beyond 0.9 and

---

[3]Distances are rescaled by dividing all (U, V) coordinate values by the maximum absolute U or V coordinate value in the set of georeferencings attached to the sample of areal units.

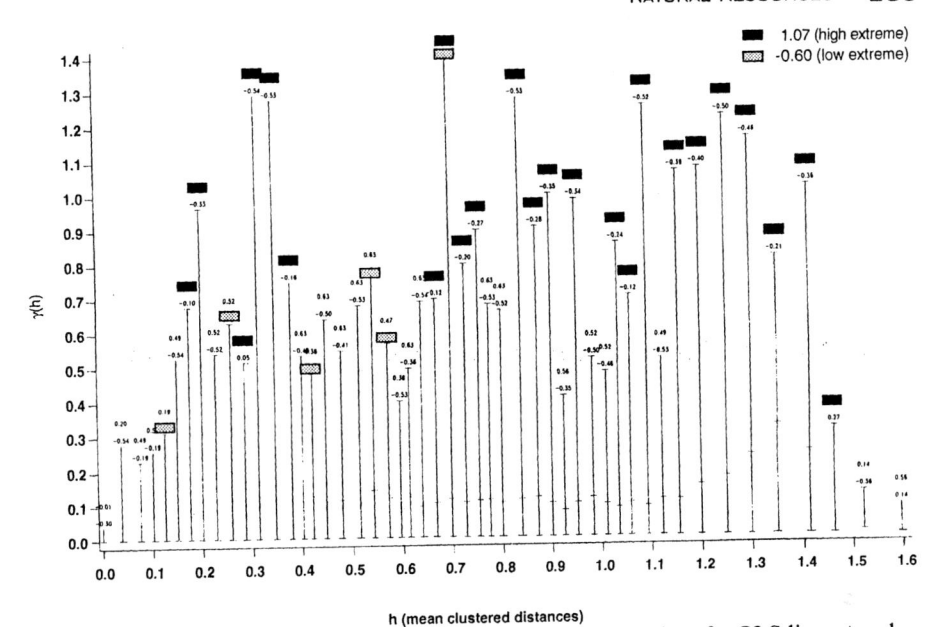

Figure 5.1. Semivariogram plot displaying the mean and extreme values for OLS linear trend model residuals for Cressie's Wolfcamp aquifer data.

is absent from the distance classes smaller than that class containing the data value 0.17, indicating that the outlier value lies toward the edge of the map (§2.6.2). This conclusion is further corroborated with the MC scatterplot described previously for these data (Layne et al. 1997).[4] The dominance of this outlier is removed when its presence is accounted for in the OLS regression.

Fifty distance classes were specified, which resulted in at least 90 distance pairs per class up to the class containing 0.9 (Figure 5.2a). The maximum distance class including 0.9 was selected because this value most closely corresponds to the value used in Cressie (1989, Figures 3a-b, 200). Unlike the exponentially increasing-like scatter of points analyzed by Cressie (1989), however, residuals from the OLS regression result in a well-behaved (§1.7) semivariogram plot (Figure 5.2b) that could be fitted with transition models. The selected semivariogram model fits to this scatter of points yield

model	RSSE	C_0	C_1	r
exponential	0.43679	0.00313	0.09765	0.10961
Bessel	0.43758	0.01660	0.08475	0.08224
Gaussian[‡]	0.43820	0.02857	0.07210	0.17328
spherical	0.41820	0.02030	0.08053	0.37107
wave/hole	0.31749	0.03570	0.06447	0.09706
power	0.58497	0.00000	0.10821	* $\lambda = 0.17682$.

[‡]Initial parameter estimate values for the Gaussian semivariogram model were specified explicitly as $C_0 = 0.0001$, $C_1 = 0.1$, $r = 0.1$.

---

[4]Analysis of these data for the *American Statistician* paper did not include the outlier indicator variable, because its results needed to be directly comparable to those presented in Cressie (1989).

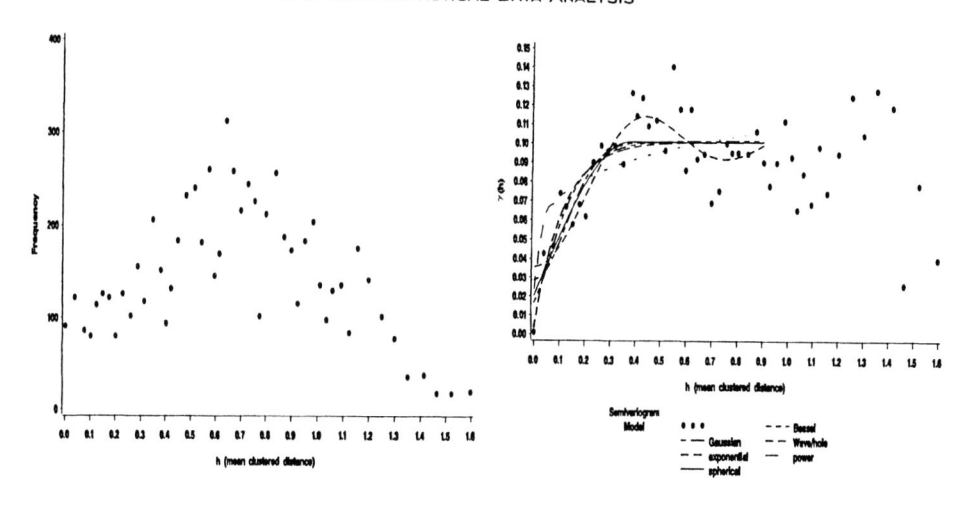

Figure 5.2. Left: (a) Semivariogram-type plot of the frequency of OLS residuals semivariance values in each distance class for Cressie's Wolfcamp aquifer data. Right: (b) Semivariogram plot of OLS residuals for Cressie's Wolfcamp aquifer data, with selected semivariogram model curves superimposed.

The wave/hole model has the best fit. The presence of negative spatial autocorrelation, however, is not indicated by the preceding spatial autoregressive analysis and its accompanying MC scatterplot, nor is it expected given the nature of the data. Inspection of Figure 5.2b suggests that the wave/hole tracks these data so well because of the conspicuous oscillation characterizing the latent pattern in the scatter of points. Instead, then, one of the first four models listed should be

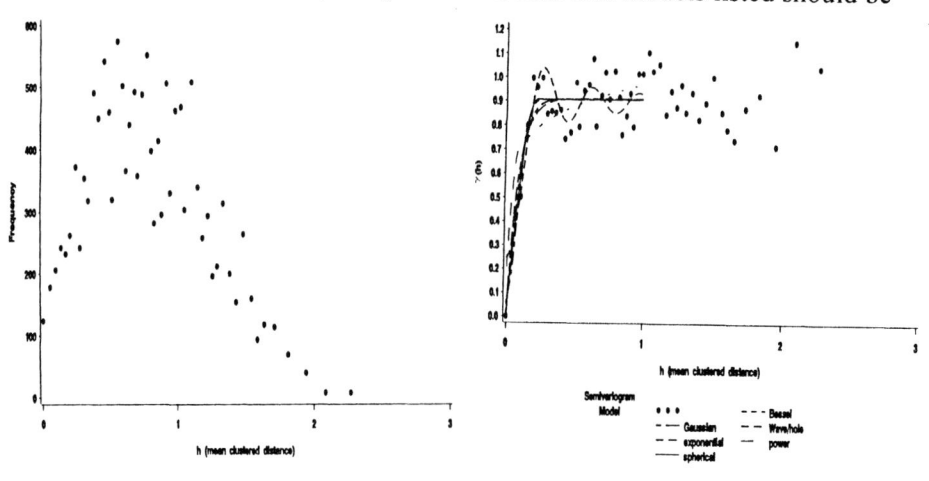

Figure 5.3. Left: (a) Semivariogram-type plot of the frequency of semivariance values in each distance class for thickness of the B division of the Mississippian Osage Series (Jones and Wrigley's Kansas oil wells data). Right: (b) Semivariogram plot of OLS residuals for thickness of the B division of the Mississippian Osage Series (Jones and Wrigley's data), with selected semivariogram model curves superimposed.

selected as the semivariogram model for describing the spatial autocorrelation in these data. The other selection criteria (e.g., cross-validation) may well need to be applied to single out a particular model. As an aside, other easily accessed natural resources data sets worth mentioning include average monthly water temperatures of 105 U.S. beaches (http://www.nodc.noaa.gov/NODC-Wnew/wtg.shtml) —pollution of beaches is mentioned in Chapter 7; Gleeson and McGilchrist's (1980) three replicates on a 7-by-7 square grid for a genotype of tall fescue, Cressie's (1986) n = 112 incomplete regular lattice percentage iron ore measurements (reprinted in Zimmerman and Zimmerman 1991); and Strauss's (1992) analysis of Bartlett's presence/absence of *Carex arenaria* data distributed over a 24-by-24 lattice (these last two data sets are reprinted in *A Handbook of Small Data Sets* [Hand et al. 1994]).

5.1. KANSAS OIL WELLS DATA

Exploration for subterranean natural resources inaugurated much of the seminal work in geostatistics. Mineral and oil explorations are like shooting at a target from behind it (i.e., the target is hidden), with oil, gas, and ore field discoveries being further impeded by intervening stratigraphy and geological structures. Hence, such exploration can become costly when the number of failures to locate deposits (i.e., a drilled well is dry) is large. Venables and Ripley (1994, 383) provide digital data sets for a pair of Russian and U.S. oil fields. Meanwhile, Jones and Wrigley (1995) list data for 124 oil wells, drilled between 1948 and 1963, in a 13-by-13-square-mile area of Stafford County, in south-central Kansas. They also provide a map of the locations of these wells; Map 5.2 (also see Map 3.1) displays the geographic distribution of these well locations together with their Thiessen polygon surface partitioning (each ignored polygon border is denoted by a slash crossing it). Part of the data tabulated by Jones and Wrigley are a reprint of data reported in Doveton (1973), and part were obtained from the Kansas Geological Survey. The two variables from this data set analyzed in this section are thickness of the B division of the Mississippian Osage Series and the percentage of shale in the Mississippian B division (an index of permeability).

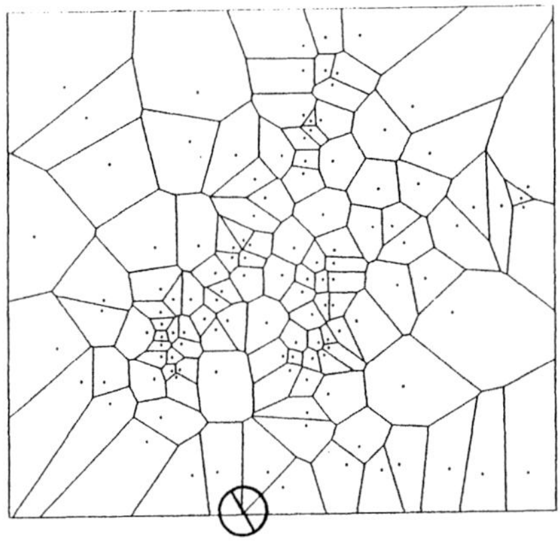

Map 5.2. Thiessen polygon surface partitioning for the oil well locations (Jones and Wrigley's data).

For these Kansas oil wells data, thickness appears to benefit from a power transformation:

raw	power transformation
	$\hat{\delta}$ = -0.00450
	$\hat{\gamma}$ = 0.44795
	RSSE = 6.1×10^{-3}
S-W = 0.95083†††	S-W = 0.97757
Bartlett = 2.752	Bartlett = 0.861
Levene = 0.273	Levene = 0.629
MC = 0.19464 ($\hat{\delta}_{MC} \approx$ 0.0.05384)	MC = 0.17815
GR = 0.79143	GR = 0.81247 .

The frequency distribution of the raw thickness measures fails to conform adequately to a normal frequency distribution, while its power-transformed counterpart does seem to conform to a normal distribution. Both versions of these data display acceptable fluctuation in variation across the four quadrants of the region, with the power transformation markedly reducing the degree of this fluctuation. Regardless of which measurement scale is employed, roughly the same degree of significant positive spatial autocorrelation is detected.

OLS results were computed based upon the standardized normal deviate for the power-transformed thickness (z_y) coupled with the centering of the u Cartesian coordinates [i.e., $(u - \bar{u})$]. Conceptually speaking, thickness of the Mississippian division at one location should not be directly determined by (although it should be correlated with) thicknesses at nearby locations, implying that the appropriate spatial autoregressive model in this case is the SAR. Positive spatial autocorrelation should be latent in these thickness data, because this phenomenon is analogous to the heterogeneous mixing of chocolate batter when making a marble cake—nearby thicknesses should tend to be similar. A poor statistical description of these data is furnished by standard OLS estimation, and the statistical description furnished by an SAR model is not much better:

parameter estimates

	OLS	s.e.	SAR(W)	s.e.
ρ	*		0.24291	0.17814
b_0	0	0.08572	-0.00351×	
			(1-0.24291)	0.11157
$b_{u-\bar{u}}$	-0.01283	0.00356	-0.01237	0.00456
(pseudo-)R²	0.096		0.131	
residuals				
S-W	0.98122		0.97752	
A-D	0.316		0.283	
R-J	0.9967		0.9966	
K-S	0.055		0.058	
Bartlett	6.880†		1.789	
Levene	1.161		1.239	

MC	0.09673	-0.01153
E(MC) ≈ -0.01527		
GR	0.88845	1.00101.

Although incorporating spatial autocorrelation into the model specification only modestly improves statistical description of these data, the accompanying residuals are much better behaved. Spatial autocorrelation is all but removed from the residuals, which conform quite closely to a normal frequency distribution. And, variance is constant across the four quadrants of the geographic landscape, while a standard residuals-versus-predicted-values plot suggests approximately constant variance across the range of predicted values.

Specification of 50 distance groups for thickness provides a minimum of 100 distance pairs per class up to the maximum distance class including the rescaled distance value of 1.0 (Figure 5.3a). The selected model fits to the scatter of points in the semivariogram plot (Figure 5.3b) yield

model	RSSE	c_0	c_1	r
exponential	0.40539	0.00000	0.90993	0.07420
Bessel	0.39039	0.00000	0.90816	0.04966
Gaussian	0.37502	0.05146	0.85462	0.09423
spherical	0.36473	0.00000	0.90579	0.19914
wave/hole	0.37179	0.24391	0.65288	0.05531
power	0.51940	0.00000	0.97074	*

$\lambda = 0.13509$.

The semivariogram plot produced with the OLS residuals implies the presence of weak spatial autocorrelation, which is indicated by the steep slope of the scatter of points up to the sill within the first five distance classes (Figure 5.3b), as well as the relatively small estimated values of the range parameter; this finding corroborates results from the preceding spatial autoregressive analysis. The spherical model has the best fit to the semivariogram plot; but, the RSSE associated with the spherical model is not very different from those for the other transition models. Perhaps except for the power model, any of these models might be selected as the semivariogram model for describing the trace amount of spatial autocorrelation. The other selection criteria (e.g., cross-validation) may well need to be applied to select a preferred model.

Percentage of shale can be analyzed in various ways. Based upon strictly empirical considerations, this variable appears to benefit from a power transformation:

raw	power transformation
	$\hat{\delta} = 0.14369$
	$\hat{\gamma} = 0.38139$
	RSSE = 7.4×10^{-3}
S-W = 0.96517[††]	S-W = 0.98145
Bartlett = 1.821	Bartlett = 1.971
Levene = 1.153	Levene = 1.136

MC = 0.02886 ($\hat{\delta}_{MC} \approx$ 0.0.05384)   MC = 0.02303
GR = 0.94650   GR = 0.94722 .

These results suggest that spatial autocorrelation is absent in the percentage of shale in this region, that a power transformation converts this variable to a measurement scale that better mimics a normal frequency distribution, and that variance is constant across the four quadrants of this geographic landscape.

But if these data actually come from a binomial distribution, theoretically speaking, the log-odds ratio [i.e., $LN(\frac{shale}{1-shale})$] should be approximately normally distributed, with nonconstant variance. Because two of the shale percentages are 0, a constant needs to be added to the value of shale when calculating this ratio. Therefore,

power transformation

$\hat{\delta} = 0$
$\hat{\gamma} = 0.14594$
RSSE = 7.6×10^{-3}

S-W = 0.98325
Bartlett = 2.019
Levene = 1.085
MC = 0.02202
GR = 0.94755

weighted power transformation

weight = $\sqrt{\frac{shale + \hat{\delta}}{1 - shale + \hat{\delta}}}$

S-W = 0.93607†††
Bartlett = 1.933
Levene = 1.138
MC = 0.03386
GR = 0.94481 .

These results corroborate the preceding ones, emphasizing that spatial autocorrelation is absent in the percentage of shale in this region. Because the sample size for each measure is unknown, and hence would need to be set to 1, and because two 0 values are present, estimation of the nonconstant variance term is problematic. Consequently, these data can be analyzed using standard maximum likelihood estimation techniques with logistic regression.

Geostatistical analysis also fails to detect any spatial autocorrelation for the power-transformed variable as evidenced by the lack of $\gamma(h)$ values lying between the origin (0 distance class) and the sill (Figure 5.4a), and by an average value of 2 for the variance-to-mean ratio (Figure 5.4b).

As an aside, the geographic distribution of the binary variable signifying dry/producing wells yields MC = 0.15438 (GR = 0.85385), implying that Jones and Wrigley (1995) should have employed an auto-logistic model specification. Their incorporation of Casetti's (e.g., Casetti and Jones 1992, 1-41) expansion method does not necessarily compensate for the presence of spatial autocorrelation.

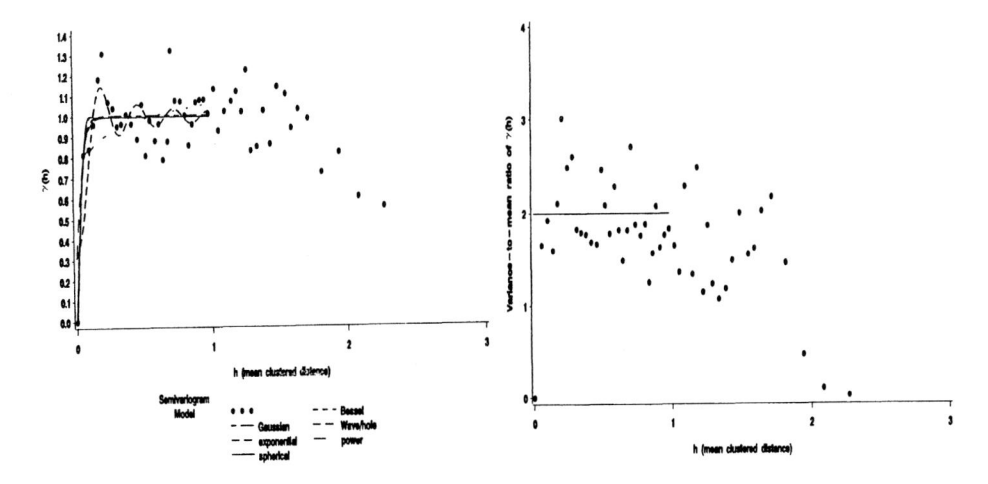

Figure 5.4. Left: (a) Semivariogram plot of OLS residuals for the percentage of shale (Jones and Wrigley's data), with selected semivariogram model curves superimpose. Right: (b) Semivariogram-type plot of the distance class variance-to-mean ratio for Figure 5.4a.

5.2. NATURAL RESOURCES INVENTORY DATA

The world's population reaps many valuable ecological and economic benefits from the natural environment and as such has a vested interest in maintaining and enhancing its quality and extent. Both the U.S. Environmental Protection Agency (EPA) and the U.S. Congress, as well as other government agencies and private organizations with environmental and natural resources interests, have long recognized the need to implement a monitoring scheme to collect data about the current extent, location, and geographic distribution of ecological resources with the goal of safeguarding them. To fulfill this need, the Environmental Monitoring and Assessment Program (EMAP[5]) was established (Messer et al. 1991). Two data sets that have been made available from this project thus far are for the northeastern forest and the Virginian near-coastal biogeographical province (Copeland et al. 1994). These data have been georeferenced to a hexagonal tessellation that has been superimposed upon the United States (Map 5.3). Selected natural resources inventoried in these two georeferenced data sets are analyzed in this section; pollution measures included are analyzed in Chapter 7. In addition, a set of forestry inventory data—comprising 513 forest stands for Township 153, Range 26—from the Minnesota forestry database also are analyzed in this section,[6] as is Cressie's reprint (1991, 581-85) of the longleaf pines (*Pinus palustris*) data set collected by

---

[5]Unfortunately, for all practical purposes, this program has been disbanded. Nevertheless, data sets generated by it are available, contain valuable information, and will and should be analyzed by spatial scientists.

[6]We thank Dr. Susanna McMaster, Department of Geography, Macalester College, for providing us with this data set.

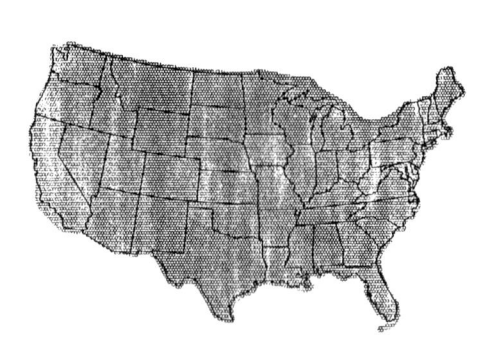

Map 5.3 (left). EMAP hexagonal tessellation superimposed upon the coterminous United States.

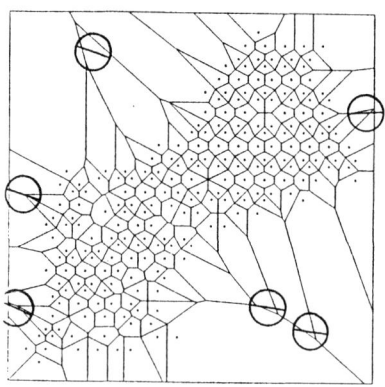

Map 5.4 (right). Thiessen polygon surface partitioning for the sampling grid (EMAP northeast forest data).

Platt and Rathbun. Additional available forestry data sets are furnished by Venables and Ripley (1994, 394, which consists of the Swedish pine data of Ripley 1981), Bailey and Gatrell (1995, 78; redwood seedlings in a forest), and Ogata and Tanemura (1985; Japanese black pines); these last data have been reprinted in *A Small Handbook of Data Sets* (Hand et al. 1994).

The northern forest sampling grid covers eastern Rhode Island (RI; $n_i = 2$), Massachusetts (MA; $n_i = 22$), Connecticut (CT; $n_i = 12$), Vermont (VT; $n_i = 22$), New Hampshire (NH; $n_i = 34$), Maine (ME; $n_i = 122$), and New Jersey ($n_i = 16$). Because New Jersey is a geographic outlier in this set of states, its data have been deleted from the analysis whose results are reported here; preliminary diagnostic analysis findings essentially are the same with or without these 16 data values. Map 5.4 displays the geographic distribution of the remaining 222 sampling locations together with their Thiessen polygon surface partitioning (each ignored polygon border is denoted by a slash crossing it). The remaining six states are treated as separate geographic groupings and, rather than the four quadrants of the plane employed elsewhere in this book, are used to evaluate homogeneity of variance across the geographic landscape. Location error is present in the georeferences, as latitude and longitude values have been taken from the center of the corresponding U.S. Geological Survey (USGS) quad sheets; actual hexagon centroids are unavailable. Only EMAP plots having some proportion of the sample area classified as forested land (using the U.S. Forestry Service's definition) are included in the data set. Two plots in the region were not measured again in 1993 because of denied access.

One variable of interest here is basal area (in square meters per hectare; a measure of the biomass present), which has been subjected to analysis of covariance (ANCOVA). In this context, basal area, for which there are no missing values in this particular data set, seems to benefit little from a power transformation:

raw	power transformation
	$\hat{\delta} = 0$
	$\hat{\gamma} = 0.77933$
	RSSE $= 7.0 \times 10^{-2}$
S-W $= 0.95805^{\dagger\dagger\dagger}$	S-W $= 0.96718^{\dagger\dagger\dagger}$
Bartlett $= 1.764$	Bartlett $= 1.767$
Levene $= 0.387$	Levene $= 0.489$
MC $= 0.02648$ ($\hat{\delta}_{MC} \approx 0.03962$)	MC $= 0.02538$
GR $= 0.93934$	GR $= 0.94188$.

Only slightly better conformity with a normal distribution is attained with the power transformation, but at the cost of slightly more variance heterogeneity; statistically significant deviation from a normal distribution on the new measurement scale persists, whereas the raw data display very minor variance heterogeneity across the states. The analysis of variance (ANOVA) assumption of normal distributions within groups, not simply across all groups, requires inspection of the following S-W diagnostic statistics:

state acronym	CT	ME	MA	NH	RI	VT
raw	0.94351	$0.93998^{\dagger\dagger\dagger}$	0.88318	0.96533	1	0.98482
transformed	0.96361	$0.95452^{\dagger\dagger\dagger}$	0.97604	0.97334	1	0.98861 .

These results also indicate that little is gained by employing a power transformation. All but one of the state subsets of data improve their mimicking of a normal distribution. Rhode Island has a sample of size 2, and so its S-W statistic actually is meaningless.

The preceding initial analysis of these data indicates that no spatial autocorrelation is present (Figure 5.5). Accordingly, a simple, conspicuous map pattern should not be detectable. ANCOVA output for these data includes

source	df	seq SS	adj SS	adj MS	F-ratio	Pr
$u - \bar{u}$	1	786.8	110.4	110.4	0.93	0.337
$v - \bar{v}$	1	323.8	0.9	0.9	0.01	0.930
state	5	436.1	436.1	87.2	0.73	0.601
Error	214	25529.4	25529.4	119.3		
Total	221	27076.1 .				

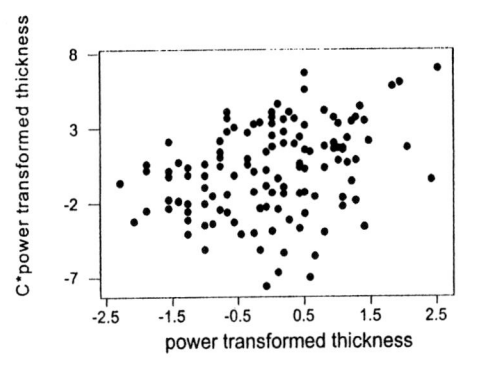

Figure 5.5. MC scatterplot of basal area (EMAP northeast forest data).

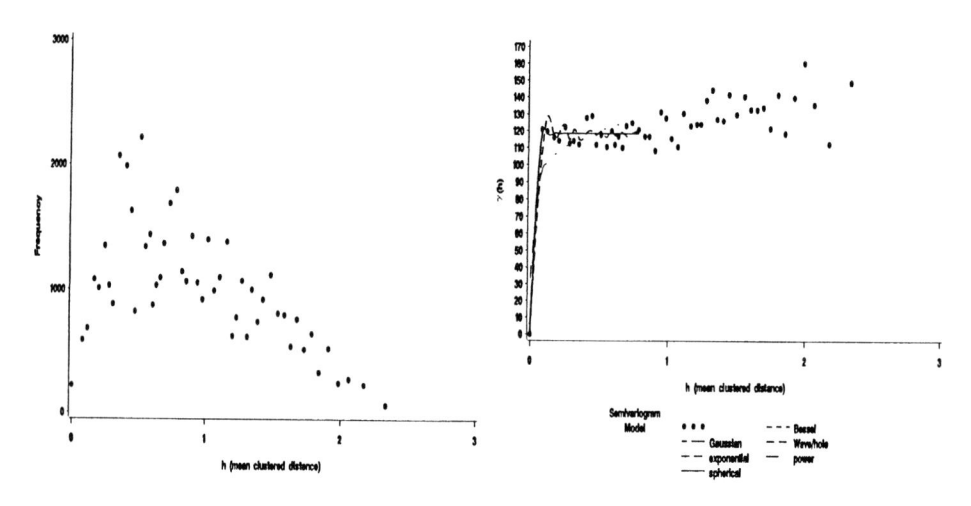

Figure 5.6. Left: (a) Semivariogram-type plot of the frequency of semivariance values in each distance class for basal area (EMAP northeast forest data). Right: (b) Semivariogram plot of OLS residuals for basal area (EMAP northeast forest data), with selected semivariogram model curves superimposed.

These results quantitatively corroborate the anticipated lack of a conspicuous basal area map pattern across this geographic region. They also suggest that the average basal area does not differ across state in this landscape, which must be viewed with some caution, given the failure to satisfy the normality assumption. However, because Maine is the sole state that contributes to this assumption violation, and its sample size is greater than 100, this caveat is not too bothersome. The negligible amount of spatial autocorrelation originally detected almost completely disappears from the ANCOVA residuals (MC = -0.02464). As an aside, two covariates that should be attached to this data set and explored are elevation and rainfall.

For the geostatistical analysis, the raw data values were used but only after the (u, v) coordinates were centered (Program #3; §1.7.1). Comparisons between use of the residuals from the ANCOVA model described and the raw data values indicate no difference in the semivariogram plots. Maintaining the standard number of distance classes (i.e., 50) yielded at least 200 distance pairs per class

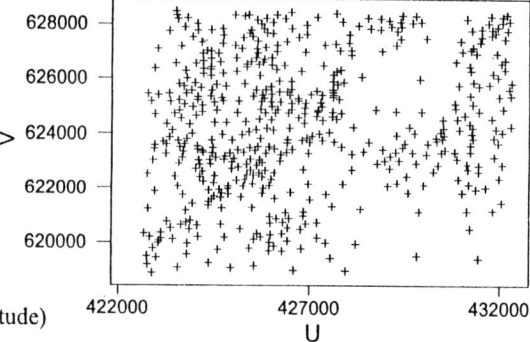

Map 5.5. Geographic distribution of Minnesota forest stand (longitude, latitude) centroids (McMaster's data).

(Figure 5.6a) up to the maximum distance class including the rescaled distance value of 0.8. As already indicated by the MC computed during the spatial autoregressive analysis, virtually no spatial autocorrelation is apparent in the semivariogram plot (Figure 5.6b); except for the first distance class [γ(0) = 0], all gamma values are approximately equal. The selected semivariogram model fits to this scatter of points yield

model	RSSE	C_0	C_1	r
exponential	0.22268	0.00000	118.497	0.00950
Bessel	0.22266	0.22346	118.272	0.00171
Gaussian	0.22267	0.00198	118.495	0.02577
spherical[‡]	0.22266	0.02967	118.467	0.07718
wave/hole[‡]	0.22946	5.77171	112.644	0.01058
power	0.57842	0.00000	130.786	* λ = 0.11950 .

[‡]Initial parameter estimate values for the spherical and wave/hole semivariogram models were specified explicitly as $C_0 = 0.0001$, $C_1 = 120$, $r = 0.01$.

Moderately low RSSE values are the result of the low degree of scatter among the points in the semivariogram plot. Lack of scatter also indicates that basal area does not differ among regions, corroborating the preceding ANCOVA analyses. An absence of spatial autocorrelation is indicated by the relatively small estimated range parameter values. The best semivariogram plot fit is given by the spherical model, although differences among this model and the exponential, Bessel, and Gaussian semivariogram models are trivial. Hence, there is no semivariogram model of choice for describing the trace amount of spatial autocorrelation in these data, which in fact should be the case.

The geographic distribution of the Minnesota stand centroids appears in Map 5.5. Analysis of basal area measures contained in this forestry inventory database yield

raw	power transformation
	$\hat{\delta}$ = 170.87647
	$\hat{\gamma}$ = 1.29908
	RSSE = 1.7×10^{-2}
S-W = 0.87047[†††]	S-W = 0.87059[†††]
Bartlett = 31.144[†††]	Bartlett = 35.319[†††]
Levene = 11.699[†††]	Levene = 12.928[†††]
MC = 0.32212 ($\hat{\sigma}_{MC} \approx 0.02850$)	MC = 0.32296
GR = 0.73972	GR = 0.73965 .

Applying a power transformation to these basal area data negligibly moves the empirical frequency distribution closer to a normal frequency distribution but only by increasing variance heterogeneity. Therefore, there are no apparent benefits to applying a power transformation here.

Basal area measures for the 513 forest stands range in value from 0 to 221, with 163 stands having a value of 0; the range for the 350 stands having positive basal area values is 2 to 221. One expectation is that stands having 0 basal area

250   A CASEBOOK FOR SPATIAL STATISTICAL DATA ANALYSIS

come from a different population, one involving clear-cutting, say. Creating an indicator variable, I_0, to differentiate between these two groups renders the following OLS and SAR estimation results:

<u>parameter estimates (z_y)</u>

	OLS	s.e.	SAR(W)	s.e.
ρ	*		0.36524	0.06954
b_0	-0.29105	0.03159	-0.28662	0.04557
$b_{v-\bar{v}}$	0.00003	0.00001	0.00003	0.00002
b_{I_0}	-0.79843	0.03159	-0.80714	0.03189
(pseudo-)R^2	0.558		0.604	
residuals				
S-W	0.95959†††		0.97317†††	
A-D	7.289†††		3.484†††	
R-J	0.9863†††		0.9921†††	
K-S	0.122†††		0.075†††	
Bartlett	3.846		5.786	
Levene	1.345		1.561	
MC	0.20878		0.01898	
GR	0.81402		0.99561	.

These results indicate, somewhat in contrast with the EMAP basal area measures, that the geographic distribution of the Minnesota forestry stand basal area measures are markedly positively spatially autocorrelated and fail to conform to a normal distribution. A portion of the spatial autocorrelation exhibited by the raw basal area data is attributable to simple map pattern—a north-south linear gradient—whereas separating basal area into the two populations of 0 and positive values largely accounts for the originally detected variance heterogeneity. The SAR model specification, which seems appropriate because autocorrelation may well arise from missing variables, almost completely accounts for the detected spatial autocorrelation in these basal area measures. In addition, this SAR specification increases the percentage of variance accounted for in the response variable (i.e., basal area) by roughly 5%. One conspicuous trade-off between results generated by the OLS and SAR specifications is that reduction in residual spatial autocorrelation and improvement of the residuals in conforming to a normal distribution are achieved with the SAR model at a cost of modestly increased variance heterogeneity; but this new level of variance heterogeneity still is not significant.

Residuals from the OLS estimated regression equation,

$$\hat{Y} = -0.29105 + .00003 \times (v - \bar{v}) - 0.79843 \times (I_0) ,$$

were subjected to geostatistical analysis. Specification of 200 distance classes resulted in at least 200 distance pairs per class up to the maximum distance class including the rescaled distance value of 0.75 (Figure 5.7a). The selected model fits to the scatter of points in the semivariogram plot (Figure 5.7b) yield

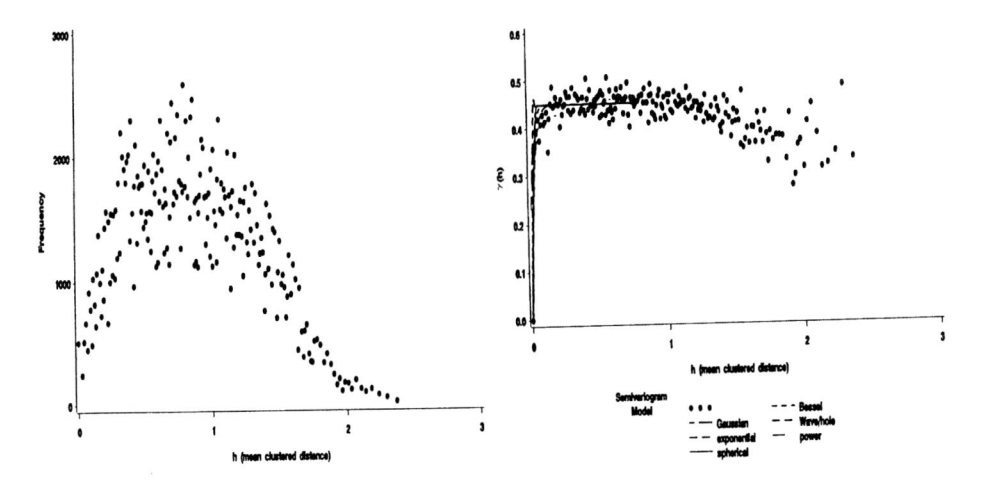

Figure 5.7. Left: (a) Semivariogram-type plot of the frequency of semivariance values in each distance class for basal area (McMaster's Minnesota forest stand data). Right: (b) Semivariogram plot of OLS residuals for basal area (McMaster's data), with selected semivariogram model curves superimposed.

model	RSSE	C_0	C_1	r
exponential	0.36211	0.01162	0.43998	0.02685
Bessel	0.37550	0.00493	0.44588	0.01386
Gaussian	0.38804	0.00347	0.44697	0.02494
spherical	0.38809	0.00502	0.44546	0.05539
wave/hole[†]	0.40457	0.00025	0.44983	0.00017
power	0.55833	0.00000	0.47463	* $\lambda = 0.06191$

[†]Initial parameter estimate values for the wave/hole semivariogram model were specified explicitly as $C_0 = 0.0001$, $C_1 = 0.45$ $r = 0.0002$.

Unlike the spatial autoregressive analysis finding of noticeable positive spatial autocorrelation in the residuals from the OLS regression, results from the semivariogram plot and model fits (Figure 5.7b) indicate that the presence of spatial autocorrelation is less obvious, evidenced by the relatively small estimated range parameter values, the intermediate RSSE values, the lack of very many points lying between the origin (0 distance class) and the sill, and the average value of 2 for the variance-to-mean ratios (Figure 5.8). Still, the exponential semivariogram model gives the best fit, albeit only moderate differences exist among all of the transition models and hence may be considered the semivariogram model of choice for describing the spatial autocorrelation in these data.

The 584 longleaf pine diameter at breast height (dbh) measures analyzed by Cressie are from a 4-hectare region of forest located in the Wade Tract, Thomas County, Georgia, collected on all longleaf pine trees at least 2 cm in diameter in 1979. This region has a relatively tame topography, is void of recent disturbances, and contains all sizes of trees. Cressie (1991, 580) provides a map showing the individual tree locations together with their relative diameters. Map 5.6 displays the geographic distribution of these locations together with their Thiessen polygon

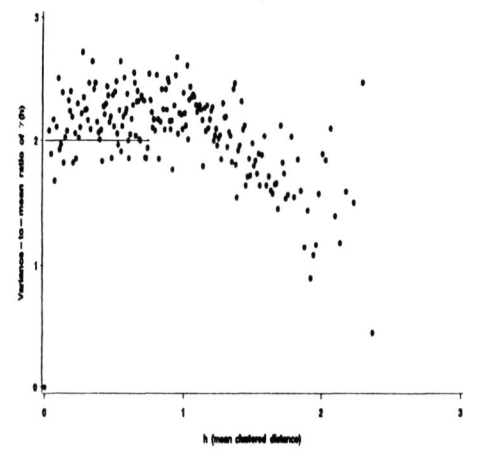

Figure 5.8 (left). Semivariogram-type plot of the distance class variance-to-mean ratio for Figure 5.7b.

Map 5.6 (right). Thiessen polygon surface partitioning for the locations of longleaf pine (Cressie's data).

surface partitioning (each ignored polygon border is denoted by a slash crossing it). Analysis of these dbh measures yields ($\frac{y_{min}}{y_{max}} \approx 38.0$):

<pre>
         raw   power transformation
                                                  δ̂   = -2
                                                  γ̂   = 0.68898
                                                  RSSE = 5.0×10⁻²
    S-W = 0.91574†††   S-W = 0.92343†††
    Bartlett = 30.260†††   Bartlett = 42.970†††
    Levene = 15.717†††   Levene = 26.173†††
    MC = 0.50718 (σ̂_MC ≈ 0.02452)   MC = 0.53439
    GR = 0.50471   GR = 0.47368 .
</pre>

Applying a power transformation to these dbh data moves the empirical frequency distribution just slightly closer to a normal frequency distribution but only by markedly increasing variance heterogeneity. Therefore, as with the preceding Minnesota basal area measures, there are no apparent benefits to applying a power transformation to these data.

Inspection of the graphical portrayal of this frequency distribution (Figure 5.9) discloses that two populations more than likely are being mixed here. Presumably one comprises the younger trees and is positively skewed with a mean less than 10 cm, whereas the other comprises the older trees and is approximately normally distributed with a mean around 40 cm. These two frequency distributions overlap, with their point of intersection appearing to be around 28 cm. An age covariate should discriminate between these two populations; Platt et al. (1988) report that a random sample of 400 tree cores revealed the presence of all ages of trees, up to 250 years old.

Dbh measures for the 584 pine trees range in value from 2 to 75.9. A boxplot of these measures fails to reveal any conspicuous outliers, whereas a histogram reveals potential bimodality (the previously mentioned two-parent-populations situation). Estimation results based on these data include

parameter estimates

	OLS	AR(W)
ρ	*	0.68203
b_0	23.23863	21.07069
$b_{u-\bar{u}}$	-0.08770	-0.03045
$b_{v-\bar{v}}$	-0.04512	*
$b_{(v-\bar{v})^2}$	0.00145	0.00058
(pseudo-)R^2	0.178	0.507

residuals

	OLS	AR(W)
S-W	0.94505†††	0.95882†††
A-D	6.531†††	4.267†††
R-J	0.9824†††	0.9855†††
K-S	0.095†††	0.070†††
Bartlett	13.007†††	23.684†††
Levene	4.576†††	3.534†††
MC	0.40969	-0.03667
GR	0.61273	1.12004

The geographic distribution of longleaf pine dbh measures, like the basal area for the Minnesota forestry stand, are markedly positively spatially autocorrelated and fail to conform to a normal distribution. A portion of the spatial autocorrelation exhibited by these dbh data is attributable to simple map pattern. Some of it also appears to relate to the missing variable that accounts for the visible bimodality in Figure 5.9; this bimodality persists in the OLS residuals (Figure 5.10) but not in the AR residuals (Figure 5.11). The AR model specification, which seems appropriate because autocorrelation may well arise from individual pairwise tree competitions for space, soil nutrients, moisture, and sunlight, almost completely accounts for the detected spatial autocorrelation in these dbh measures. In addition,

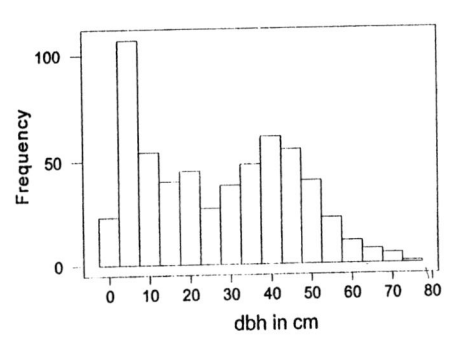

Figure 5.9. Histogram of dbh (Cressie's longleaf pine data).

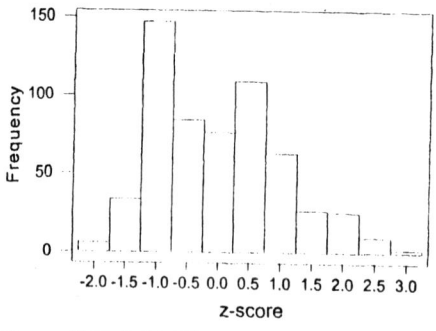

Figure 5.10 (left). Histogram of OLS residuals for dbh (Cressie's longleaf pine data).

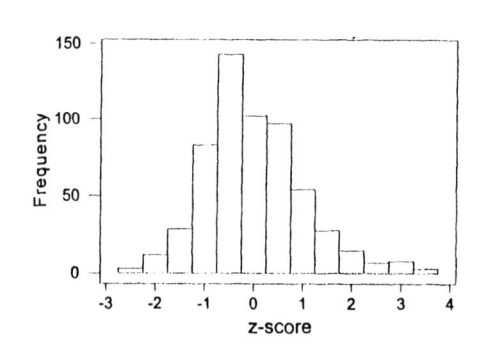

Figure 5.11 (right). Histogram of AR residuals for dbh (Cressie's longleaf pine data).

this AR specification increases the percentage of variance accounted for in the response variable (i.e., dbh) by more than 30%. Only trace spatial autocorrelation remains in the AR model residuals. And, given that these residuals do not conform to a normal frequency distribution, the more reliable Levene statistic implies that taking spatial autocorrelation into account helps mitigate the heterogeneous geographic variability situation, although variance heterogeneity remains significant.

A geostatistical analysis of the OLS residuals also suggests that relatively high levels of positive spatial autocorrelation are present in these data. Specification of 100 distance classes resulted in at least 900 distance pairs per class, including the maximum distance class containing the rescaled distance value of 0.75 (Figure 5.12a). The selected semivariogram model fits to this scatter of points yield

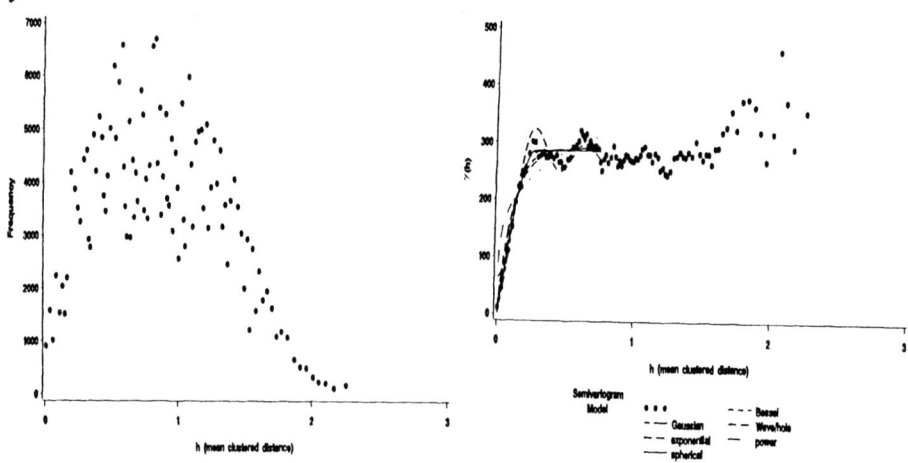

Figure 5.12. Left: (a) Semivariogram-type plot of the frequency of semivariance values in each distance class for dbh (Cressie's longleaf pine data). Right: (b) Semivariogram plot of OLS residuals for dbh (Cressie's longleaf pine data), with selected semivariogram model curves superimposed.

model	RSSE	C_0	C_1	r	
exponential‡	0.14722	0.00000	291.589	0.10666	
Bessel	0.12396	0.00000	288.573	0.06663	
Gaussian	0.11056	24.32310	261.625	0.12326	
spherical‡	0.10860	0.00000	285.778	0.25940	
wave/hole	0.15094	63.84530	213.939	0.05735	
power	0.33690	0.00000	336.697	*	$\lambda = 0.23818$

‡Initial parameter estimate values for the exponential and spherical semivariogram models were specified explicitly as $C_0 = 0.0001$, $C_1 = 300$ $r = 0.1$.

Although the range parameter estimates here are relatively small, the degree of spatial autocorrelation contained in the OLS residuals is quite high. Evidence of this strong level of spatial autocorrelation is furnished by the MC calculated for these OLS residuals, as well as by the relatively large number of $\gamma(h)$ values for the distance classes lying between the origin (0 distance class) and the sill. The spherical model provides the best fit to the semivariogram plot (Figure 5.12b) and hence may be considered the semivariogram model of choice for describing the spatial autocorrelation in these data; notably, all transition models have relatively good fits.

Turning attention to the second EMAP data set, benthic species' abundance has been measured for near-coastal waters at 70 sample sites in the following bays:

C: Chesapeake ($n_i = 26$)   LIGSN: Long Island ($n_i = 12$)
N: Nantucket ($n_i = 4$)   Great South ($n_i = 1$)
D: Delaware ($n_i = 25$)   Narragansett ($n_i = 1$)
            Chincoteague ($n_i = 1$)

Many of the pilot test studies of these data have been for Chesapeake Bay. Long Island, Great South, and Narragansett Bays are collected into a single group here because they are located around Long Island and off the nearby coast of Rhode Island. Chincoteague Bay is geographically isolated from the rest of these sample locations and so has been deleted from these spatial analyses whose results are reported in this section. The remaining four groups of sample sites constitute four separate geographic regions, and, rather than the four quadrants of the plane employed elsewhere in this book, are used to evaluate homogeneity of variance across the geographic landscape. Map 5.7 displays the geographic distribution of the remaining 69 sampling locations together with their Thiessen polygon surface partitioning (each ignored polygon border is denoted by a slash crossing it).

Because four geographic regions are of interest in this section, benthic species' abundance data are subjected to analysis of covariance (ANCOVA). In this context this abundance measure appears to benefit from a power transformation:

256   A CASEBOOK FOR SPATIAL STATISTICAL DATA ANALYSIS

Map 5.7. Thiessen polygon surface partitioning for the sampling grid (EMAP near-coastal waters bays data).

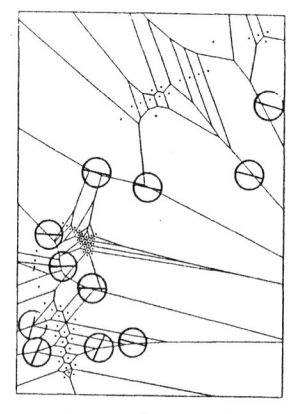

raw	power transformation
	$\hat{\delta} = 99.85505$
	$\hat{\gamma} = 0$
	RSSE = 1.6×10^{-2}
S-W = 0.83431†††	S-W = 0.97880
Bartlett = 16.893†††	Bartlett = 2.868
diff. of means F = 5.17†††	diff. of means F = 6.77††† .

These results indicate that the statistical behavior of the benthic species abundance variable benefits greatly from a logarithmic transformation: it conforms well to a normal distribution, and its variance stabilizes across the near-coastal bays. A pronounced difference of means is detected in the raw data and becomes more emphatic in the transformed data. And, conceptually, a logarithmic transformation seems sensible here, given that species counts should come from a Poisson distribution.

But, as mentioned in the preceding forestry data analysis discussion, ANOVA assumes normal distributions within groups not simply across all groups. Hence,

bay acronym	D	LIGSN	N	C
S-W: raw	0.92552	0.94039	0.83920	0.72504
S-W: power-transformed	0.92251	0.93381	0.80094	0.91392 .

The overall frequency distribution better mimics a normal distribution by the power transformation dramatically improving conformity to normality of data from the Chesapeake Bay; this improved behavior is gained at a cost of nominally increased deviations from normal distributions in the other three bays.

ANCOVA results for these data are as follows:

	ANCOVA sum of squares	parameter estimates
b_0	*	6.5322
u	12.0880	0.0102
v	0.1839	-0.0048

bay	14.3014		
$I_F - I_C$			0.3807
$I_{LIGSN} - I_C$			1.0382
$I_N - I_C$			-1.0513
error	39.5986		
TOTAL	66.1719		
R^2	0.402		

		MC	GR
residuals			
S-W = 0.96538	Chesapeake	-0.09682	0.95250
A-D = 0.530		($\hat{\sigma}_{MC} \approx 0.15617$)	
R-J = 0.9886	Delaware	0.14208	0.80956
K-S = 0.106†		($\hat{\sigma}_{MC} \approx 0.13608$)	
Bartlett = 5.000	Long Island-		
Levene = 0.987	Nantucket	-0.14050	1.01329
		($\hat{\sigma}_{MC} \approx 0.19245$)	.

These results insinuate that any latent spatial autocorrelation detected is attributable to simple map pattern—more specifically, to a conspicuous east-west trend. The bays themselves are another primary source of variation, with Delaware and Chesapeake Bays having very similar conditional amounts of benthic species' abundance, while Long Island, Great South, and Narragansett Bays have markedly more, and Nantucket Bay has markedly less. The ANCOVA results indicate that only the average level of benthic species' abundance varies across these bays, yielding parallel regression lines for them (Figure 5.13). In terms of residuals, heterogeneity of variance across the bays is tolerable, the residuals conform quite well to a normal frequency distribution (although K-S suggests that bothersome tail deviations may be present), and different levels of trivial, trace spatial autocorrelation are present in this set of bays.

For the geostatistical analysis, residuals from the OLS estimated regression equation

$$\hat{Y} = 6.5322 + 0.0102 \times (u - \bar{u}) - 0.0048 \times (v - \bar{v}) + 0.3807 \times (I_D - I_C) + 1.0382 \times (I_{LIGSN} - I_C) - 1.0513 \times (I_N - I_C)$$

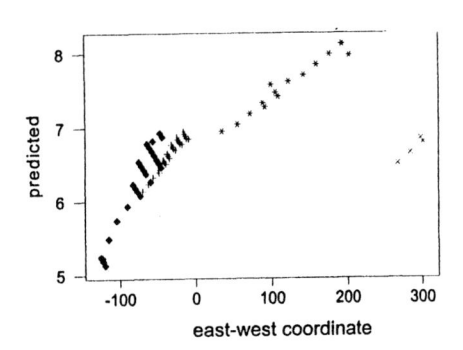

Figure 5.13. OLS regression lines for benthic species' abundance (EMAP near-coastal waters bays data).

were used and comprise a total of 69 observations. To conform to previously described ANCOVA analysis, however, the total data set was divided into three separate sub-data sets: Chesapeake Bay, Delaware Bay, and Long Island Coast, where data for Long Island, Great South, and Nantucket Bays were combined. Because of the relatively small sample sizes for each of the three sub-data sets, 10 distance classes were specified instead of the usual 50. Details for number of distance pairs, maximum distance used for modeling purposes, and minimum number of distance pairs per distance class for each of the three sub-data sets are as follows:

bay	number of distance pairs	minimum number per distance class	maximum distance
Chesapeake	26^2	54	0.40
Delaware	25^2	25	0.12
Long Island Coast	18^2	30	0.30

The selected model fits to the scatter of points in each of these three semivariogram plots (Figures 5.14a-c) yield

Chesapeake Bay

model	RSSE	C_0	C_1	r
exponential	0.44526	0.00000	0.39304	0.03286
Bessel[†]	0.43014	0.00000	0.39120	0.02095
Gaussian	0.42378	0.18436	0.20737	0.05722
spherical	0.41825	0.15160	0.24174	0.12585
wave/hole	0.20250	0.18461	0.20082	0.02762
power	0.66756	0.00000	0.48169	* $\lambda = 0.14581$

[†]Initial parameter estimate values for the Bessel semivariogram model were specified explicitly as $C_0 = 0.0001$, $C_1 = 0.4$, $r = 0.2$.

Delaware Bay

model	RSSE	C_0	C_1	r
exponential	0.01010	0.01393	0.82196	0.02697
Bessel	0.02507	0.02741	0.77818	0.01593
Gaussian	0.08366	0.03667	0.73010	0.02475
spherical	0.05634	0.09232	0.70043	0.06642
wave/hole	0.20453	0.30898	0.43546	0.01784
power	0.01886	0.00000	1.75946	* $\lambda = 0.31896$

Long Island Coast

model	RSSE	C_0	C_1	r
exponential	0.37792	0.00000	0.78755	0.03353
Bessel	0.35193	0.00000	0.78234	0.02036
Gaussian	0.32400	0.00000	0.78082	0.03389
spherical	0.32349	0.00000	0.78090	0.07718
wave/hole	0.15032	0.05239	0.72101	0.01617
power	0.44136	0.00000	1.14633	* $\lambda = 0.23547$

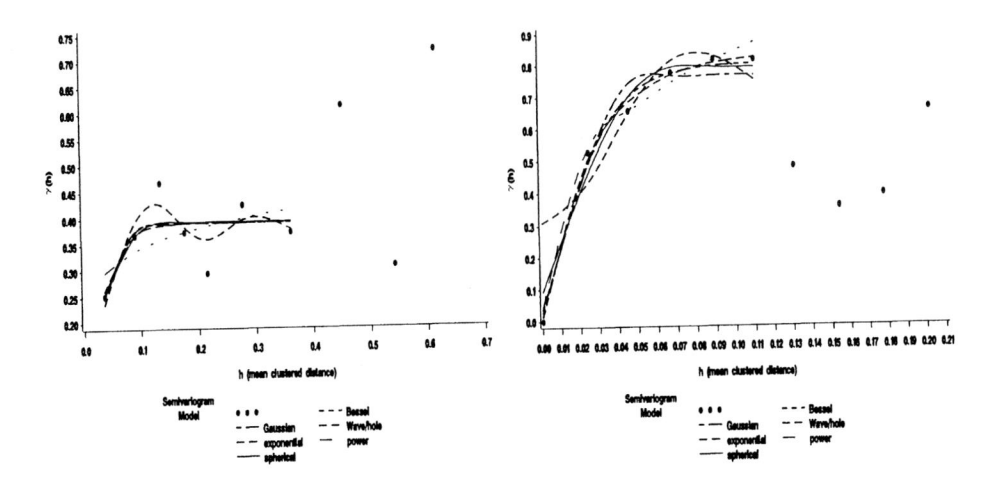

Figure 5.14. Top-left (a) Semivariogram plot of OLS residuals for benthic species' abundance for the Chesapeake Bay (EMAP near-coastal waters bays data), with selected semivariogram model curves superimposed. Top-right: (b) Semivariogram plot of OLS residuals for benthic species' abundance for the Delaware Bay (EMAP near-coastal waters bays data), with selected semivariogram model curves superimposed. Bottom-right: (c) Semivariogram plot of OLS residuals for benthic species' abundance for the Long Island Coast (EMAP near-coastal waters bays data), with selected semivariogram model curves superimposed.

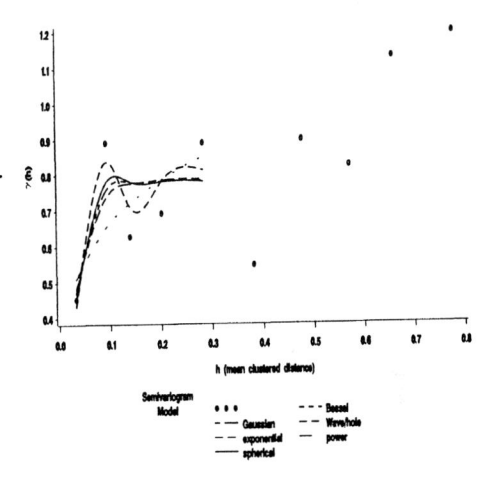

Results from the geostatistical analysis parallel those from the preceding ANCOVA analysis. The scatter of points in the semivariogram plot for Chesapeake Bay (Figure 5.14a) is reminiscent of semivariogram plots where low to no spatial autocorrelation is present. The MC for Chesapeake Bay indicates slightly negative spatial autocorrelation in the residuals, and, in fact, the wave/hole semivariogram model has the lowest RSSE, corroborating presence of a trace amount of negative spatial autocorrelation. The MC for Delaware Bay indicates positive spatial autocorrelation in the residuals, and inspection of the scatter of points in Figure 5.14b reveals a pattern typical of positive spatial autocorrelation in a semivariogram plot. For this sub-data set the exponential model achieves the best fit (lowest RSSE) among the set of semivariogram models, although the power model also appears to provide an adequate description of spatial dependency for these data. The MC for the Long Island Coast sub-data set indicates the presence of a low to moderate level of negative spatial autocorrelation. Inspection of Figure 5.14c suggests that the presence of negative spatial autocorrelation is not obvious;

however, the lowest RSSE value for this sub-data set is for the wave/hole model, indicating negative spatial autocorrelation in the residuals, which corroborates results of the preceding ANCOVA analysis.

Salinity—the total amount of dissolved salts derived from sodium, potassium, magnesium, calcium, carbonate, silicate, and halide ions present in seawater—also has been measured for the 69 near-coastal water sample sites in the set of four bays. Naturally occurring waters vary in salinity content, from the almost pure water (devoid of salts) in snowmelt to the saturated solutions in salt lakes, such as the Dead Sea. Salinity throughout the open oceans averages roughly 35 parts per thousand, being more variable in near-coastal waters, where seawater is diluted with fresh water from runoff or from the emptying of rivers. This brackish near-coastal water forms a barrier separating marine and freshwater organisms. Hence, salinity may be viewed in terms of pollution (i.e., is the fresh water in an estuary being displaced by salt water?), or it can be viewed in terms of a natural resource (i.e., are the habitats of stenohaline organisms, which have a low tolerance to salinity changes, being displaced or destroyed?). In this section, salinity is viewed and analyzed as a natural resource.

As with the preceding benthic species' abundance measures, because four geographic regions are of interest in this section salinity data are subjected to ANCOVA. In this context, salinity measures appear to benefit from a power transformation (although $\frac{y_{max}}{y_{min}} > 3$, these data frustrate optimal power transformation identification):

raw	power transformation
	$\hat{\delta} = 1.84107$
	$\hat{\gamma} = 2.13088$
	RSSE = 3.3×10^{-2}
	NOTE: convergence not attained
S-W = 0.87360†††	S-W = 0.93260†††
Bartlett = 9.267††	Bartlett = 2.234
diff. of means F = 2.13	diff. of means F = 5.06††† .

The statistical behavior of the salinity variable benefits greatly from a power transformation: it better conforms to a normal distribution, and its variance stabilizes across the near-coastal bays. A pronounced difference of means is obscured in the raw data, becoming quite pronounced in the transformed data.

But, as mentioned in the preceding benthic species' abundance analysis, ANOVA assumes normal distributions within groups not simply across all groups. Hence,

bay acronym	D	LIGSN	N	C
S-W: raw	0.81901†††	0.58276†††	0.63162†††	0.94406
S-W: power-transformed	0.93260	0.78263††	0.63362†††	0.90109† .

The overall frequency distribution better mimics a normal distribution by the power transformation dramatically improving conformity to normality of data from the Delaware Bay and Long Island near-coastal waters; this improved behavior is gained at a cost of increased deviations from normal distributions in the Chesapeake Bay.

ANCOVA results for these data[7] are as follows:

	ANCOVA sum of squares	parameter estimates
b_0	*	1260.81436
u	4755750	5.98631
v	995101	-2.84975
bay:	3213026	
$I_D - I_C$		-147.98319
$I_{LIGSN} - I_C$		310.09587
$I_N - I_C$		-198.43841
bay×temp:	6623917	
$(I_D - I_C) \times temp$		40.99610
$(I_{LIGSN} - I_C) \times temp$		41.87001
$(I_N - I_C) \times temp$		70.73727
error	5118587	
TOTAL	20706381	
R^2	0.742	

residuals			MC	GR
S-W = 0.97512	Chesapeake	0.27603	0.65574
A-D = 0.429	($\hat{\sigma}_{MC} \approx 0.15617$)
R-J = 0.9871	Delaware	0.47895	0.45391
K-S = 0.075	($\hat{\sigma}_{MC} \approx 0.13608$)
Bartlett = .170	Long Island–
Levene = .335	Nantucket	-0.29663	1.61767
	($\hat{\sigma}_{MC} \approx 0.19245$) .

These results imply that some of the latent spatial autocorrelation detected in these data is attributable to simple map pattern. Another primary source of variation is interaction between the bays and water temperature. A third principal source of variation is the bays themselves. In terms of residuals, heterogeneity of variance across the bays is tolerable, the residuals conform quite well to a normal frequency distribution, and significant, positive spatial autocorrelation is present in one, and perhaps two, of these bays (Figure 5.15a-c).

Because salinity at one location directly influences the magnitude of salinity at nearby locations, a spatial AR model is the most appropriate autoregressive specification for describing these data. Because the 69 EMAP sample

---

[7]OLS results included the Chincoteague Bay sample point, which was not included in the geostatistical analyses, because this point did not contribute to the observed spatial dependency. OLS parameter estimates are sensitive to this single data point's inclusion or exclusion.

points comprise three distinct groups (i.e., the connectivity matrix **C** is block-diagonal) and autocorrelation appears to vary from bay to bay, three autoregressive parameters need to be estimated here. This can be introduced into the nonlinear regression by partitioning the spatial lag information (i.e., **WY**) with indicator variables for the different bays. This estimation procedure yields

$$\hat{\rho}_{Chesapeake\,Bay} = 0.03569 \;,\; \hat{\rho}_{Delaware\,Bay} = 0.48154 \;,\; \text{and}\; \hat{\rho}_{Long\,Island\,Coast} = -0.01096\;,$$

with and accompanying S-W = 0.95041. These estimation results corroborate the aforementioned MC results, suggesting that only Delaware Bay salinity measures contain significant spatial autocorrelation. A repartitioning of the sums of squares in the presence of this autocorrelation is less straightforward.

Residuals from the OLS estimated regression equation

$$\hat{Y} = 1123.8368 + 2449.5403\times(u - \bar{u}) - 790.4838\times(v - \bar{v}) + \\ 19.8753\times(I_D - I_C) + 183.3942\times(I_{LIGSN} - I_C) - 544.6450\times(I_N - I_C) + \\ 41.5818\times[(I_D - I_C)\times(temp - \mu_{temp})] + 42.5778\times[(I_{LIGSN} - I_C)\times(temp - \mu_{temp})] + \\ 68.8553\times[(I_N - I_C)\times(temp - \mu_{temp})]$$

were used for the geostatistical analyses. As in the case of the benthic species were used for the geostatistical analyses. As in the case of the benthic species geostatistical analyses, the total data set containing 69 observations was divided into the same three sub-data sets: Chesapeake Bay, Delaware Bay, and Long Island

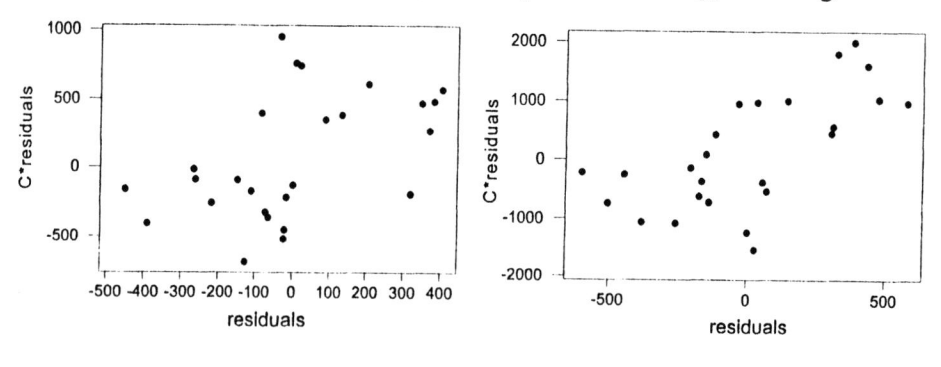

Figure 5.15. Top-left (a) MC scatterplot of salinity for the Chesapeake Bay (EMAP near-coastal waters bays data). Top-right: (b) MC scatterplot of salinity for the Delaware Bay (EMAP near-coastal waters bays data). Bottom-right (c) MC scatterplot of salinity for the Long Island Coast (EMAP near-coastal waters bays data).

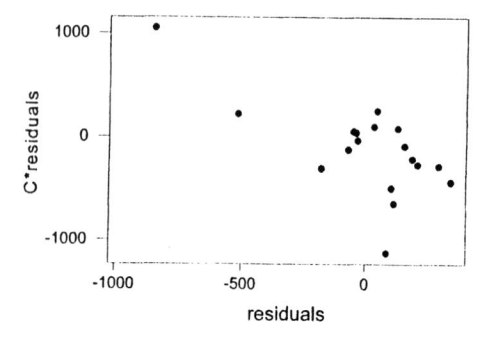

Coast. Again, because of the relatively small sample sizes, 10 distance classes were specified instead of the usual 50. Details for number of distance pairs, maximum distance used for modeling purposes, and minimum number of distance pairs per distance class for each of the three sub-data sets are as follows:

	distance pairs		
bay	number of	minimum number per distance class	maximum distance
Chesapeake	26^2	54	0.40
Delaware	25^2	18	0.18
Long Island Coast	18^2	20	0.70

The selected semivariogram model fits to the scatter of points in each of these three semivariogram plots (Figures 5.16a-c) yield

Chesapeake

model	RSSE	C_0	C_1	r	
exponential	0.15467	23883.0	42050.0	0.12466	
Bessel	0.14708	29458.9	35673.0	0.08279	
Gaussian	0.12733	32800.5	29490.3	0.13342	
spherical	0.13306	27181.6	34524.7	0.25762	
wave/hole	0.14090	34587.0	23945.5	0.05941	
power	0.17939	0.00000	85707.1	*	$\lambda = 0.26323$

Delaware

model	RSSE	C_0	C_1	r	
exponential	0.01357	3077.8	290116	0.18000	
Bessel	0.02377	17858.0	213360	0.07837	
Gaussian	0.03742	33994.7	157198	0.10387	
spherical	0.02586	5207.6	166308	0.18000	
wave/hole	0.04432	37208.7	119342	0.04005	
power	0.00731	3836.6	767321	*	$\lambda = 0.80399$

Long Island Coast

model	RSSE	C_0	C_1	r	
exponential	0.80550	0.00000	86966.2	0.03077	
Bessel	0.75934	0.01127	89266.6	0.02428	
Gaussian	0.69230	0.00000	90050.1	0.04554	
spherical[‡]	0.70375	0.00000	89284.8	0.10000	
wave/hole[‡]	0.53637	0.00000	90675.4	0.02734	
power	0.98324	0.00000	89371.1	*	$\lambda = 0.06969$

[‡]Initial parameter estimate values for the spherical and wave/hole semivariogram models were specified explicitly as $C_0 = 0.0001$, $C_1 = 90000$, and, respectively, $r = 0.1$ and $r = 0.05$.

Unlike results reported for the benthic species analysis, comparisons between the geostatistical and spatial autoregressive analyses are less straightforward in the cases of the Delaware Bay and Long Island Coast sub-data sets. However, for the Chesapeake Bay sub-data set, inspection of the scatter of points in Figure 5.16a reveals the typical case of a semivariogram plot in which a moderate level of

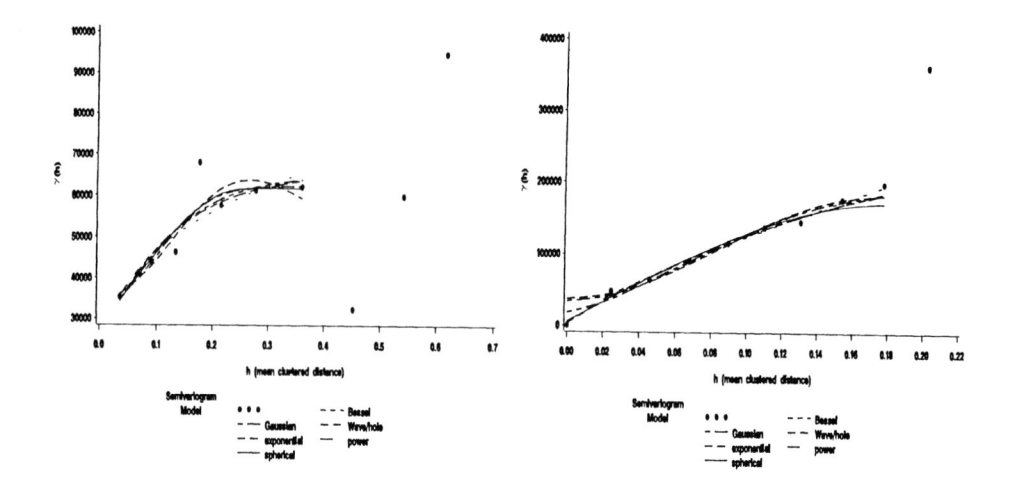

Figure 5.16. Top-left: (a) Semivariogram plot of OLS residuals for salinity for the Chesapeake Bay (EMAP near-coastal waters bays data), with selected semivariogram model curves superimposed. Top-right: (b) Semivariogram plot of OLS residuals for salinity for the Delaware Bay (EMAP near-coastal waters bays data), with selected semivariogram model curves superimposed. Bottom-left: (c) Semivariogram plot of OLS residuals for salinity for the Long Island Coast (EMAP near-coastal waters bays data), with selected semivariogram model curves superimposed.

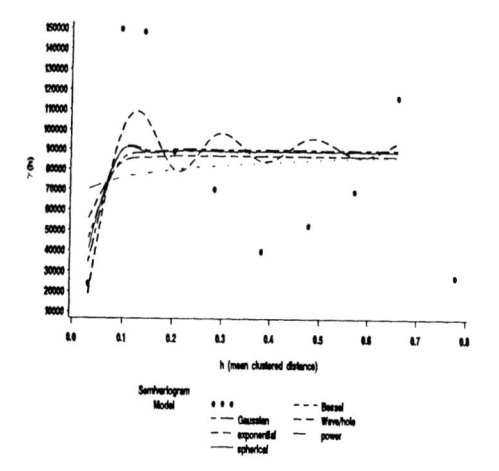

positive spatial autocorrelation is present. Here the Gaussian model achieves the best fit (lowest RSSE) among the set of semivariogram models. The MC for the Delaware Bay sub-data set indicates the presence of a moderate to high level of positive spatial autocorrelation in the residuals. Inspection of the scatter of points in the semivariogram plot (Figure 5.16b), however, could be interpreted in two alternative ways: high positive spatial autocorrelation or omission of an explanatory variable in the OLS regression leading to a nonconstant mean across the surface, the latter typical when a power semivariogram model should be employed. In fact, the power semivariogram model has the lowest RSSE, although the exponential model actually provides an acceptable alternative transition semivariogram model for these data. Interpreting the scatter of points in the semivariogram plot for the Long Island Coast (Figure 5.16c) presents a challenge for any statistician, as the pattern observed does not readily align with that for either positive or negative spatial autocorrelation or even for a nonconstant mean surface (§1.6, 2.6.2, 2.6.3). The MC from the spatial autoregressive analysis indicates that a low to moderate level of negative spatial autocorrelation is present in the residuals from the OLS regression. For this sub-data set, semivariogram model selection based on lowest RSSE does corroborate the MC value estimated (the wave/hole model achieves the best fit); but, again, the scatter of points appearing in the semivariogram plot does not necessarily indicate that spatial dependency is negative.

5.3. ISLAND BIOGEOGRAPHY: PLANT SPECIES DATA

Identification of determinants of the survival and evolution of different species of life is of considerable scientific interest. One of the most famous data sets used to address this issue inventories the number of plant species of the Galapagos Islands (Map 5.8), of which 30 have been surveyed. Initially analyzed using traditional OLS regression techniques by Johnson and Raven (1973), these data have been reprinted by Upton and Fingleton (1985), who reanalyze them using spatial autoregressive techniques. These data also have been analyzed by Hamilton et al. (1963) and Grant et al. (1980) and reprinted by Andrews and Herzberg (1985), who provide a map of this archipelago. A map more suitable for geocoding purposes appears in Grant (1986). A second set of similar data for 26 California-Mexico coastal locations (Map 5.9), also analyzed by Johnson and Raven (1970), is reprinted in Upton and Fingleton (1985), who reanalyze them using spatial autoregressive techniques and provide a map of their locations. Some of the areal unit labeling appearing in Upton and Fingleton is difficult to decipher. Other such data—such as those for bird species in the northern Andes of South America, analyzed by Manly (1991) and reprinted in *A Handbook of Small Data Sets* (Hand et al. 1994), and those for *Euphydryas editha* butterfly colonies in California and Oregon analyzed by McKechnie et al. (1975) and reprinted together with a map in Manly (1994)—also are readily available.

Analysis results reported by Upton and Fingleton (1985, 295-97) for the Galapagos Islands data are based upon 23 of the 30 islands; they deleted 7 islands

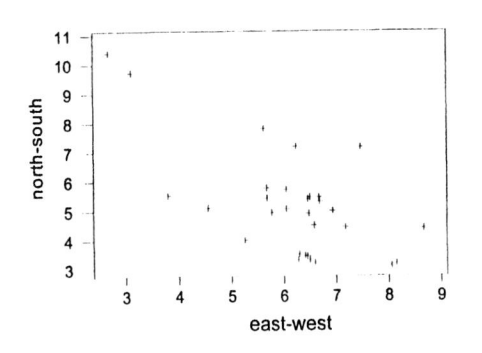

Map 5.8 (left). Geographic distribution of the centroids of the Galapagos Islands.

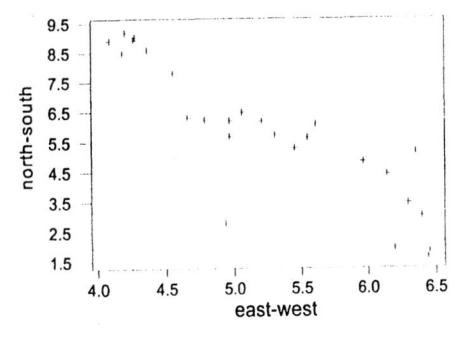

Map 5.9 (right). Geographic distribution of the centroids of the California-Mexico coastal localities.

for which some of the data are missing from the database. Because the dependent variable is number of plant species, a counts variable, these data should be analyzed using Poisson regression. But a good normal approximation may be, and in fact is, furnished by employing a logarithmic transformation:

```
raw                                    power transformation
                                        δ̂ = 2.64201
                                        γ̂ = 0
                                        RSSE = 1.4×10⁻²
S-W = 0.71426†††                        S-W = 0.95774
Bartlett = 17.303†††                    Bartlett = 9.077††
Levene = 1.116                          Levene = 5.141†††
MC = -0.08715 (σ̂_MC ≈ 0.17408)          MC = -0.03800
GR = 2.57135                            GR = 1.46702 .
```

The logarithmic transformation successfully converts the frequency distribution for the original counts to one that adequately conforms to a normal distribution and suggests the expected link between a log-Gaussian and Poisson model specification. In addition, any spatial autocorrelation that may be latent in these species counts data is negative, a finding that is consistent with conceptual expectations. The power transformation also reduces the degree of heterogeneous geographic variability.

Upton and Fingleton (1985, 297) conclude that the principal predictor variable is elevation and that the accompanying OLS regression residuals contain virtually no spatial autocorrelation. But by moving from the raw to the power-transformed data space, area replaces elevation as the principal predictor, which makes more intuitive sense. Presumably, this switch in predictor variable importance is attributable to specification error. A Box-Tidwell transformation was applied to the area variable for analysis purposes in this section; the resulting transformed variable is $LN(\text{area} + 0.00584)$. Scatter diagrams for both the raw and the transformed variable pairs appear in Figure 5.17. Clearly, application of the twopower transformations in this case makes the linearity assumption more tenable. Bivariate OLS regression results based upon these transformed variables include $b_0 = 3.14375$ and $b_{f(\text{area})} = 0.34426$, and $R^2 = 0.783$. This statistical description is substantially better than that achieved by Upton and Fingleton, who report an R^2 value of roughly 0.50. Diagnostic statistics calculated with the residuals from this regression include

```
Bartlett = 1.392              S-W = 0.96060
Levene = 0.497                A-D = 0.266
MC = 0.00103                  R-J = 0.9884
GR = 0.96396                  K-S = 0.101 .
```

An absence of spatial autocorrelation in the residuals also is reported by Upton and Fingleton, and in this case is confirmed by inspection of the associated MC

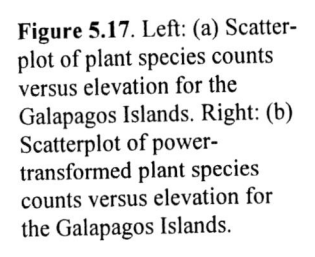

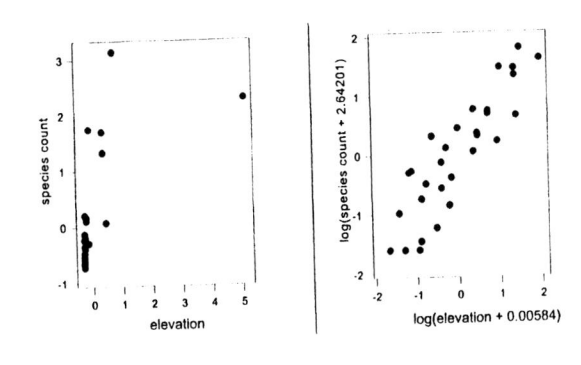

Figure 5.17. Left: (a) Scatterplot of plant species counts versus elevation for the Galapagos Islands. Right: (b) Scatterplot of power-transformed plant species counts versus elevation for the Galapagos Islands.

scatterplot (Figure 5.18). Furthermore, constant geographic variability across the four quadrants of the region is exhibited by these OLS residuals, whereas the traditional residuals-versus-predicted-values scatterplot (Figure 5.19) is consistent with that for constant variance across all levels of the mean response.

The geostatistical analyses involved constructing 25 distance classes instead of the usual 50, owing to the small sample size of 30 island locations. Division of the 900 distance pairs into 25 classes resulted in a minimum of 24 distance pairs per class (Figure 5.20a). Fitting the selected models to the semi-variogram plot (Figure 5.20b), where a maximum rescaled distance of 0.8 was used, results in the following goodness-of-fit and parameter estimates:

model	RSSE	C_0	C_1	r
exponential	0.76809	0.00000	0.37455	0.00320
Bessel	0.76809	0.00032	0.37422	0.00171
Gaussian	0.76812	0.00387	0.37067	0.01169
spherical[‡]	0.76809	0.00040	0.37412	0.03901
wave/hole	0.64038	0.00000	0.38155	0.01225

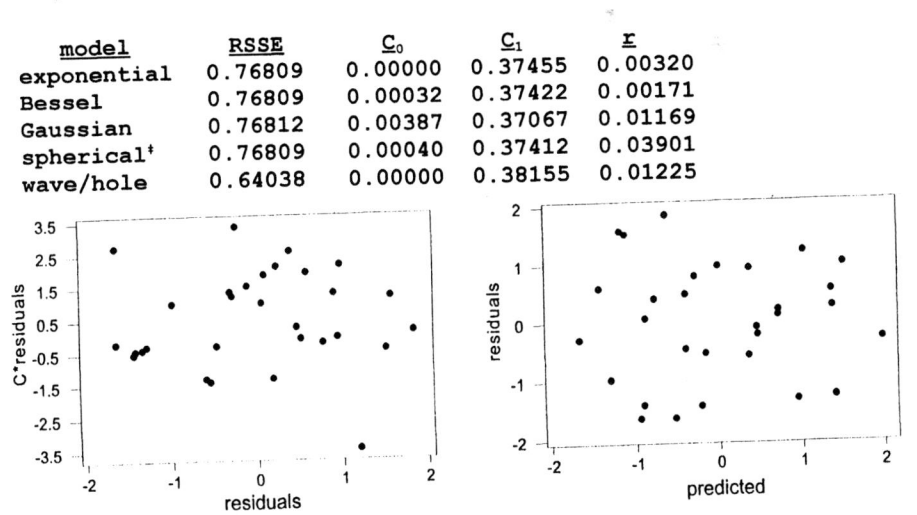

Figure 5.18 (left) MC scatterplot of OLS residuals for plant species counts (Galapagos Islands data).

Figure 5.19 (right). Traditional homogeneity of variance OLS residual plot for plant species counts (Galapagos Islands data).

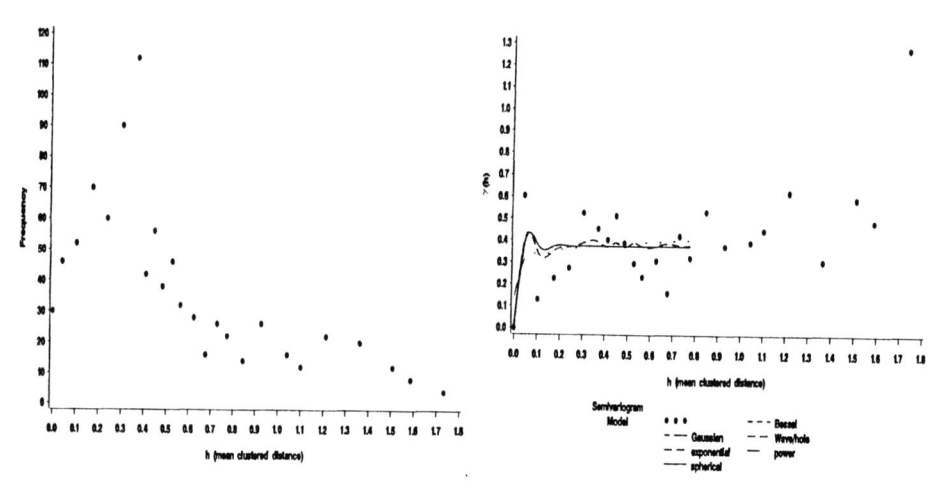

268   A CASEBOOK FOR SPATIAL STATISTICAL DATA ANALYSIS

Figure 5.20. Left: (a) Semivariogram-type plot of the frequency of semivariance values in each distance class for plant species counts (Galapagos Islands data). Right: (b) Semivariogram plot of OLS residuals for plant species counts (Galapagos Islands data), with selected semivariogram model curves superimposed.

```
power        0.84006   0.00000   0.41317     *      λ = 0.09236 .
```
‡Initial parameter estimate values for the spherical semivariogram model were specified explicitly as $C_0 = 0.0001$, $C_1 = 0.04$, and $r = 0.1$.

Although RSSEs are relatively high, fits are virtually identical for all the transition models except for the wave/hole semivariogram model. High RSSE values and low range parameter values indicate a lack of spatial autocorrelation in the residuals from the OLS regression, a result that is similar to findings reported for the spatial autoregressive analysis, although the variance-to-mean ratio plot (Figure 5.21) has an average less than the expected value of 2 (§2.6.2). Presumably, much of the variation contained in the RSSE values is the result of $\gamma(1)$ (≈ 0.6), suggesting that islands at the shortest distances are more dissimilar than islands whose separation distances are only slightly greater.

Upton and Fingleton (1985, 311-12) also report estimation results for Poisson regression—a more appealing specification from a theoretical point of view. Utilizing a Poisson regression model (implemented with STATA) to describe these data results in $b_0 = 3.2732$ and $b_{f(area)} = 0.33774$, and pseudo-$R^2 = 0.737$. This output is very similar to that based upon the normal approximation. The accompanying spatial autocorrelation diagnostics are MC = -0.10603 and GR = 2.46892 (this pair of values is worrisomely inconsistent), which suggest that latent negative spatial autocorrelation may well be present in the Poisson regression residuals. Following the prescriptions outlined in §2.8.1 (only eigenvectors affiliated with |MC| ≥ 0.38 are considered here) yields

```
        variable          parameter estimate      s.e.
        intercept                3.10318          0.04925
```

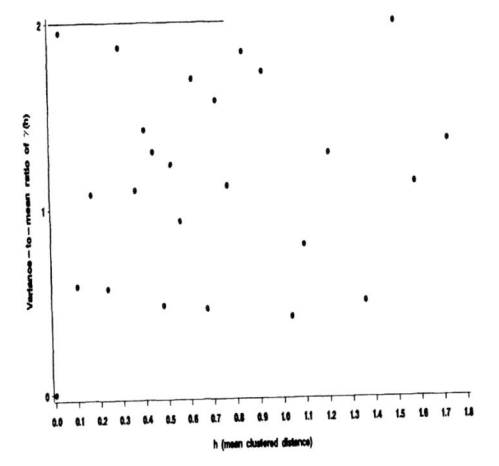

Figure 5.21 (left). Semivariogram-type plot of the distance class variance-to-mean ratio for Figure 5.20b.

Figure 5.22 (right). MC scatterplot of Poisson regression residuals for plant species counts (Galapagos Islands data).

LN(area)	0.36676	0.00983
$I_{Fernandina}$	-1.22948	0.11377
E_2 (-)	0.80803	0.09823
E_4 (+)	0.20964	0.14271
E_5 (-)	-0.55080	0.08259
E_6 (-)	-0.84677	0.08487
E_{13} (+)	-1.14907	0.15157
E_{14} (-)	-0.84978	0.11771 ,

for which the pseudo-R^2 increases to 0.976; the nature of the spatial autocorrelation associated with eigenvector E_j is denoted in the parentheses following it. Although other potential outliers are detectable in these data, the island of Fernandina is a particularly problematic plant species count outlier in this analysis and, accordingly, has been labeled with an indicator variable as coming from a different population. These Poisson regression results include an intercept term and coefficient for LN(area) that are very similar to the previously estimated ones and render MC = -0.05990 and GR = 0.91811, which constitute a more compatible pair; the MC scatterplot (Figure 5.22) for the Poisson regression residuals corroborates this conclusion. The importance especially of the negative autocorrelation terms in this regression equation relates island species counts to some competitive spatial process; the importance of a number of complex map patterns contradicts the prevailing notion of island insularity.

Upton and Fingleton (1985, 275, 293) report results for the California coastal localities data set, too, and detect marked positive spatial autocorrelation in the plant species counts variable. Because the geographic distribution of these localities involves a wide range of latitudes but a rather narrow longitude range geographic variability is evaluated only along the north-south axis. Again, because the number of different plant species is a count, assuming a Poisson distribution of errors conceptually is more appropriate. Nevertheless, as in the Galapagos Islands data analysis, a good normal approximation is furnished by employing a logarithmic transformation of this dependent variable:

```
raw                                    power transformation
                                              δ̂ = 10.09855
                                              γ̂ = 0
                                       RSSE = 2.6×10⁻²
S-W = 0.78760†††                       S-W = 0.96018
Bartlett = 0.301                       Bartlett = 0.134
Levene = 1.847                         Levene = 0.087
MC = 0.05604 (δ̂_MC ≈ 0.17150)          MC = 0.00155
GR = 0.75365                           GR = 0.87977 .
```

The logarithmic transformation successfully converts the frequency distribution for the original counts to one that adequately conforms to a normal distribution. In addition, any spatial autocorrelation that may be latent in these species counts data is positive in nature, a finding also reported by Upton and Fingleton (1985, 293). In contrast with the Galapagos Islands case, here the pair of MC and GR statistics is consistent.

Upton and Fingleton (1985, 275) conclude that the principal predictor variables are elevation, area, and latitude. A Box-Tidwell transformation was applied to both the elevation and the area variables for analysis purposes in this section; the respective transformed variables are LN(elevation + 164.56246) and LN(area + 0.13596). OLS regression results based upon these transformed variables reveal that only LN(area + 0.13596) and latitude are important predictor variables. Trivariate OLS regression results based upon the transformed variables include $b_0 = 0.36284$, $b_{f(area)} = 0.38169$, and $b_{latitude} = 0.11330$, and $R^2 = 0.917$. This statistical description is equivalent to that achieved by Upton and Fingleton with three predictor variables (their R^2 value is roughly 0.88); presumably, elevation appeared in their set of pronounced predictor variables due to specification error. Diagnostic statistics calculated with the residuals from this regression tender evidence failing to contradict the validity of the underlying model assumptions and include

```
Bartlett = 0.067            S-W = 0.95622
Levene = 0.020              A-D = 0.401
MC = 0.01695                R-J = 0.9769
GR = 0.90116                K-S = 0.120 ,
```

which suggest that spatial autocorrelation is primarily attributable to simple map pattern, variance is constant across the geographic landscape, and the residuals conform to a normal distribution. Unfortunately, the traditional residuals-versus-predicted-values scatterplot reveals the possibility of nonconstant variance across the range of mean response. Consequently, the marked positive spatial autocorrelation uncovered by Upton and Fingleton appears to arise from specification error, much like that found by Cliff and Ord for the Eire data (§4.1). Such complications are commonly encountered pitfalls during the analysis of dirty data.

Another option here is to posit a Poisson regression specification, as was done with the preceding Galapagos Islands data. Utilizing this model (implemented with STATA) to describe these data results in $b_0 = 3.22527$, $b_{LN(area)} = 0.26826$, $b_{latitude} = 0.13416$, and $b_{LN(elevation)} = 0.18956$, and pseudo-$R^2 = 0.941$. This output differs noticeably from that based upon the normal approximation; elevation now is an important predictor variable, and coefficients for both the intercept term and the area variable are discernibly smaller. The accompanying spatial autocorrelation diagnostics are MC = -0.00592 and GR = 0.94028, which imply an absence of spatial autocorrelation in the Poisson regression residuals. In turn, this finding once more suggests that any spatial autocorrelation detectable in the geographic distribution of the number of plant species along the California-Mexico coastline is attributable to simple map pattern (i.e., a latitude trend) and insinuates that the spatial autocorrelation uncovered by Upton and Fingleton presumably is attributable to specification error.

Residuals from the OLS estimated regression equation

$$\hat{Y} = 4.1554 + 0.38169 \times f(area) + 0.11330 \times (latitude) ,$$

where the number of species and area respectively were transformed using the power transformations given above, were used for the geostatistical analysis. Division of the 676 distance pairs (n = 26) into 20 distance classes yields a minimum of 20 pairs per class (Figure 5.23a). Fitting the selected models to the semivariogram plot (Figure 5.23b), using a maximum rescaled distance of 0.8, results in the following goodness-of-fit and parameter estimates:

model	RSSE	C_0	C_1	r
exponential	0.89932	0.10687	0.03710	0.38205
Bessel	0.88633	0.10823	0.03314	0.20296
Gaussian[†]	0.85471	0.10722	0.03162	0.28333
spherical	0.87599	0.10374	0.03312	0.50993
wave/hole	0.82045	0.10466	0.03140	0.10914
power	0.91087	0.10809	0.03830	* $\lambda = 0.65208$.

[†]Initial parameter estimate values for the Gaussian semivariogram model were specified explicitly as $C_0 = 0.0001$, $C_1 = 0.05$, $r = 0.2$.

Although the wave/hole semivariogram model has the lowest RSSE, the high RSSE of all models and the scattering of γ(h) values for the distance classes suggest low to no spatial autocorrelation in the residuals, similar to findings reported for the spatial autoregressive analysis. The relatively high range parameter values may suggest otherwise, except that for these data the average gamma value is much greater than the 0.0 value normally associated with the first distance class, giving a shallower slope to the model curves than expected when this gamma value

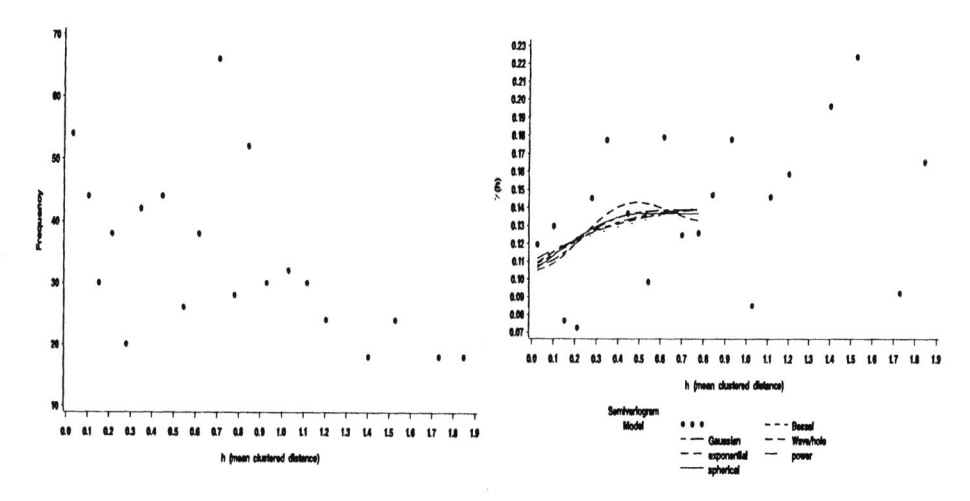

Figure 5.23. Left: (a) Semivariogram-type plot of the frequency of semivariance values in each distance class for plant species counts (Upton and Fingleton's California-Mexico coastline data). Right: (b) Semivariogram plot of OLS residuals for plant species counts (Upton and Fingleton's California-Mexico coastline data), with selected semivariogram model curves superimposed.

is 0. Much smaller range parameter values are expected if the number of distance classes is increased to retain the class of $\gamma(0) = 0$; but the already small number of distance pairs per class (Figure 5.23a) strongly discourages further disaggregation of the groups of distance pairs.

As an aside, various results reported here deviate from those reported by Upton and Fingleton because of the use of power transformations and because of differences in specifications of the geographic weights matrices. The fact that counts conform to a Poisson distribution furnishes a persuasive argument for employing a power transformation when analyzing numbers of species, as has been done here. Deriving geographic connectivity matrices is somewhat subjective, an art rather than a science; impacts of this subjectivity are critiqued by Griffith and Lagona (1997, 1998).

5.4. WEATHER STATION RAINFALL DATA

Water is an important natural resource, an essential priceless one in arid regions of the world. One principal source of water is rainfall, motivating considerable interest in studying its geographic distribution (e.g., Bailey and Gatrell 1995, 9, 146; Carr 1995, Chapters 9 and 11). Using a set of weather stations in the Kansas-Nebraska border region, Haining et al. (1984) focus on residuals from a CAR model for which a linear trend surface was removed and then the latent spatial autocorrelation added back in (all parameter estimates were simultaneously computed). A complete set of these residuals is listed in Griffith (1988a, 141). The original article includes both a map of the study area and a map of the Thiessen polygon surface partitioning used to construct the geographic connectivity matrix.

Estimation results were obtained with FORTRAN code that utilized IMSL subroutines. Of interest here is how the following SAS estimation results based upon equations (3.7a)-(3.7c) compare with those report by Haining et al.:

<u>CAR estimation results using matrix C</u>

$\hat{\mu}$ = 0.66983 - 0.89083 = -0.22100
(because original data are centered first)
$\hat{\sigma}^2$ = 0.58276×1.74095 = 1.01456
(because original data are converted to z-scores first)
$\hat{\rho}$ = 0.18125/(1/5.40234) → 0.97917
(because λ_{max} = 5.40234).

For the listed residuals, $\bar{x}$ = -0.89083, s^2 = 1.74095, and MC = 0.39452 ($\hat{\sigma}_{MC}$ ≈ 0.10050; GR = 0.57538). The foregoing estimate of $\hat{\rho}$ is exactly what these researchers report. Because OLS residuals have a mean of 0 by construction, with this value being deviated from when nonlinear estimation is involved, the result of -0.221 is consistent with this expectation. And, because adjusting for spatial autocorrelation will reduce the variance estimate, the result of 1.01456 also is consistent.

Other rainfall data sets are readily available for analysis (e.g., Bailey and Gatrell 1995, 9). Dingman et al. (1988) present an analysis of mean annual precipitation for 102 northeastern rain gauge sites. These data have been reprinted in Hamilton (1992); the gauge cite named Fitzwater should be Fitzwilliam W., and a number of the place name abbreviations commonly used in the national database are not presented as such in Hamilton. Georeferencing has been attached to the individual rain gauge sites by consulting the National Climatic Data Center's U.S. information system (Monthly Precipitation Data for All Cooperative and NWS Stations) through the Internet (searching through the "Station List with longitude & latitude" option). The geographic distribution of rain gauge locations, together with their associated Thiessen polygon surface partitioning, appears in Map 5.10 (each ignored polygon border is denoted by a slash crossing it). The New England region is partitioned into five states, each treated as separate geographic groupings,

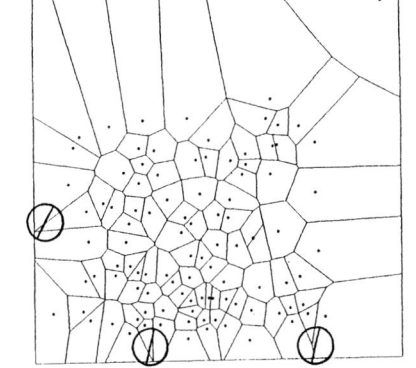

Map 5.10. Thiessen polygon surface partitioning for the northeast rain gauge locations (Hamilton's data).

rather than using the convention of four quadrants of the plane employed elsewhere in this book, for the purpose of evaluating homogeneity of variance across this geographic landscape.

For these New England rain gauge locations, rainfall data appear to benefit considerably from a power transformation:

```
raw                                          power transformation
                                             δ̂ = -58.52701
                                             γ̂ = 0
                                             RSSE = 6.6×10⁻³
S-W = 0.78773†††                             S-W = 0.95996†††
Bartlett = 25.163†††                         Bartlett = 17.428†††
Levene = 0.792                               Levene = 2.075†
MC = 0.09609  (σ̂_MC ≈ 0.06052)               MC = 0.20798
GR = 0.85352                                 GR = 0.70521 .
```

The frequency distribution of the raw rainfall measures fails to conform to a normal frequency distribution; these data display marked fluctuation in variation across the five states. Significant, marked positive spatial dependency is obscured in the raw data by their dirtiness but is successfully detected in their power-transformed version. Two rain gauges are located at excessively high elevations relative to the remaining 100 rain gauges contained in this data set. The elevation variable essentially serves as an indicator variable that designates these 2 locations as being from a different population.

Because missing variables are inherent to rainfall data (e.g., unrecorded weather systems), spatial autocorrelation latent in these data materializes in the error term of a regression model, and hence these data should be described with an SAR model. A respectable statistical description of these data is furnished by standard OLS estimation, with the description furnished by an SAR model being the usual roughly 5% better:

parameter estimates

	OLS	s.e.	SAR(W)	s.e.
ρ	*		0.45572	0.21598
b_0	-0.69730	0.10013	-0.72976×	0.13802
			(1-0.45572)	
$b_{u-\bar{u}}$	-0.31981	0.08197	-0.34877	0.13293
$b_{u-\bar{u}}$	-0.63650	0.08715	-0.69924	0.13293
$b_{elevation}$	0.00270	0.00029	0.00273	0.00028
(pseudo-)R^2	0.562		0.618	
residuals				
S-W	0.98024		0.97411	
A-D	0.282		0.393	
R-J	0.9936		0.9902	
K-S	0.047		0.059	
Bartlett	5.062		4.205	

NATURAL RESOURCES 275

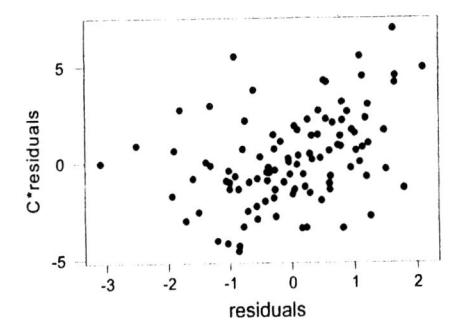

Figure 5.24 (left). MC scatterplot of OLS residuals for rainfall (Hamilton's New England rain gauge data).

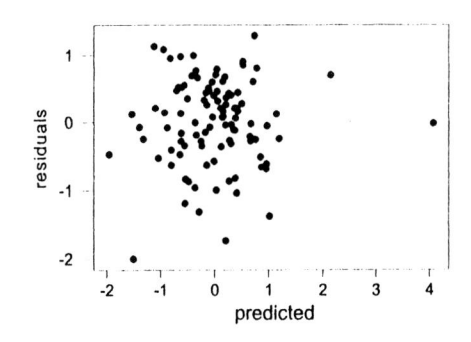

Figure 5.25 (right). Traditional homogeneity of variance OLS residual plot for Hamilton's New England rainfall data.

Levene	1.128	0.996
MC	0.16839	0.02022
GR	0.81514	0.98169

These results indicate that although incorporating spatial autocorrelation into the model specification only modestly improves statistical description, the accompanying residuals are better behaved. Spatial autocorrelation is all but removed from the residuals (Figure 5.24), which conform quite closely to a normal frequency distribution. Variance is constant across the five states of the geographic landscape. Regardless of whether constructed with the OLS or the SAR residuals, a standard residuals-versus-predicted-values plot suggests that nonconstant variance remains across the range of predicted values (Figure 5.25); the rain gauge located on Mt. Washington, New Hampshire, (one of the elevation outliers) persists in being somewhat anomalous here.

The frequency distribution of rainfall values undergoes an interesting transition through these four analyses (Figure 5.26). For the raw data, the 2 high-elevation rain gauge locations are especially conspicuous. The logarithmic transformation of these measures shrinks the rainfall distance separating recordings

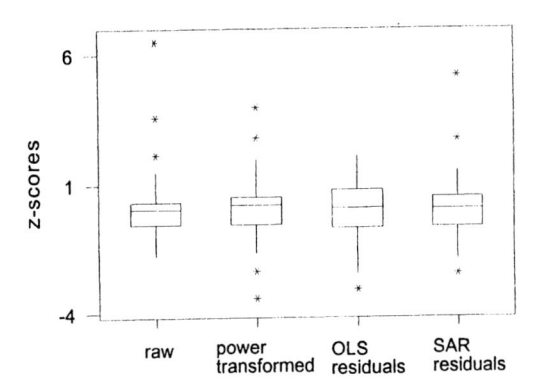

Figure 5.26. Boxplot metamorphosis of Hamilton's New England rainfall data: raw measures to SAR residuals.

at these 2 sites from those logged at the remaining 100 sites but in doing so causes the two lowest rainfall recordings to become atypical. The three variables included in the OLS regression adjust for this power transformation artifact in all but the lowest rainfall case (Peru, New York). The SAR model modestly improves the statistical description but causes all three of these extreme cases to have abnormal residuals.

Interpretation of the findings indicate that spatial autocorrelation latent in rainfall data arises from both simple map pattern and more complex spatial processes. Once latitude, longitude, and elevation have been taken into account, similar values of rainfall still have a marked tendency to cluster in the New England region. Rainfall tends to increase as elevation increases, and as the position of a rain gauge moves south in the region and toward the east coast.

Geostatistical analyses also uncover spatial dependency in the residuals, a finding evidenced by the semivariogram plot (Figure 5.27a), which reveals the presence of several gamma values located between the 0 distance class and the approximate location of the sill. Here residuals from the power-transformed New England rain gauge measures and the OLS estimated regression equation

$$\hat{Y} = -0.69730 - 0.31981 \times (u - \bar{u}) - 0.63650 \times (v - \bar{v}) + 0.00270 \times (\text{elevation}),$$

were analyzed. Division of the 10,404 distance pairs (n = 102) into 50 distance classes yielded a minimum of approximately 90 pairs per class (Figure 5.27b). Fitting the selected models to the semivariogram plot (Figure 5.27a), using a maximum rescaled distance of 0.6, yields the following results:

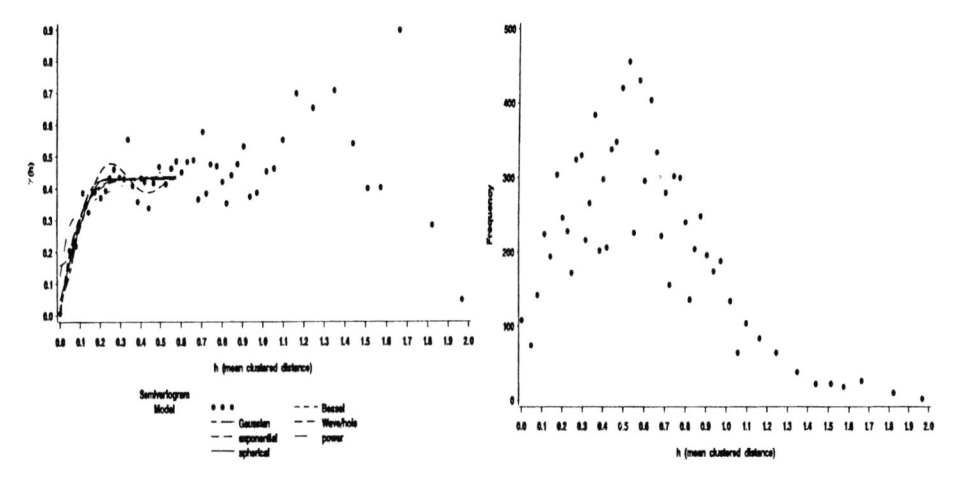

Figure 5.27. Left: (a) Semivariogram plot of OLS residuals for Hamilton's New England rainfall data. Right: (b) Semivariogram-type plot of the frequency of semivariance values in each distance class for Hamilton's New England rainfall data.

model	RSSE	C_0	C_1	r
exponential	0.31062	0.00000	0.43773	0.08070
Bessel	0.31146	0.01988	0.41438	0.05326
Gaussian‡	0.33854	0.05017	0.37913	0.09703
spherical	0.33244	0.04642	0.38489	0.22754
wave/hole‡	0.41717	0.15757	0.26412	0.05639
power	0.40802	0.00000	0.51685	* $\lambda = 0.19791$

‡Initial parameter estimate values for the Gaussian and wave/hole semivariogram models were specified explicitly as $C_0 = 0.0001$, $C_1 = 0.4$, and, respectively, $r = 0.1$ and $r = 0.2$.

The exponential model is seen as the best-fitting semivariogram model and hence is the preferred model, although the difference in RSSE between the exponential and Bessel models is quite small. While the range parameter values are relatively small, given the detected presence of moderate levels of positive spatial autocorrelation by the spatial autoregressive analysis, these small range parameter values suggest that spatial dependencies in the data are present only for relatively close rain gauge locations.

5.5. DRAINAGE BASIN RUNOFF DATA

River basin analyses are of interest in various contexts. Haith (1976) provides a tabulation of the various major land use types occupying 20 New York river basins. These data have been reprinted in Allen and Cady (1982) and in Hamilton (1992) and reanalyzed by Simpson et al. (1989). Hicks et al. (1990) furnish a tabulation of annual precipitation and runoff for 19 mountain basins in New Zealand. These data also have been reprinted in Hamilton (1992). Carr (1995) reprints runoff data for Cave Creek, Kentucky, appearing in Haan (1977). In the first of these two cases, georeferencing can be attached to the individual basins by consulting the USGS's Geographic Names Information System through the Internet (using "stream" as the feature type). In addition, Griffith and Amrhein (1991, 455) report rainfall, runoff, snowfall, and precipitation for 28 Ontario drainage basins, data explored in this section. This data set contains the following two recording errors: for Big Otter, LAT should be 42.9; and, for Fawn River, LAT should be 54.7 and LONG should be 89.6. Griffith and Amrhein (1991, 28) supply a map displaying the geographic distribution of these drainage basins. Map 5.11 displays the Thiessen polygon counterpart to their map (each ignored polygon border is denoted by a slash crossing it).

Precipitation, the phenomenon spotlighted in the preceding section of this chapter in the form of rainfall, is the precursor to runoff. For these Ontario river basins, precipitation data appear to benefit little from a power transformation, which is not surprising because $\frac{y_{min}}{y_{max}} < 3$:

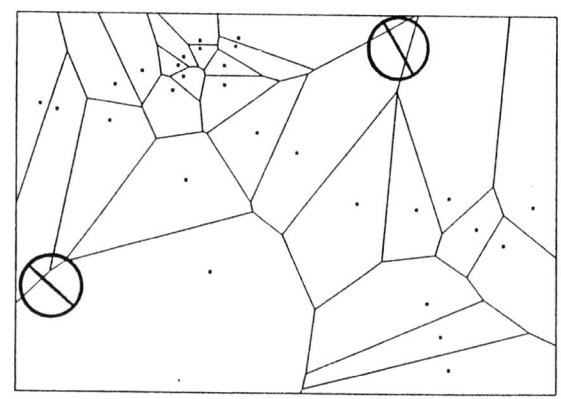

Map 5.11. Thiessen polygon surface partitioning for the Ontario river basins (Griffith and Amrhein's data).

raw

S-W = 0.965211
Bartlett = 2.892
Levene = 0.907
MC = 0.46824 ($\hat{\sigma}_{MC} \approx 0.13608$)
GR = 0.44864 .

power transformation

$\hat{\delta}$ = 0.01699
$\hat{\gamma}$ = 2.64575
RSSE = 1.8×10^{-2}
S-W = 0.95948

The frequency distribution of the raw precipitation measures seems to conform adequately to a normal frequency distribution, whereas these data display acceptable fluctuation in variation across the four quadrants of the province. Significant, marked positive spatial dependency is detected.

Positive spatial autocorrelation should be latent in these precipitation data, given the prevailing meteorological and climatological nature of the atmosphere hovering over this section of North America. A respectable statistical description of these data is furnished by standard OLS estimation, whose results include

parameter estimates (for z_y-scores)

b_0 0
b_{LAT} -0.19512
R^2 0.444

residuals

W-S = 0.95211
A-D = 0.317
R-J = 0.9844
K-S = 0.107
Bartlett = 2.662
Levene = 0.667
MC = 0.05391
GR = 0.75345 .

These diagnostic statistics indicate that virtually all of the spatial autocorrelation detected in the geographic distribution of Ontario river basin precipitation is accounted for by simple map pattern; there is a pronounced north-south trend, with precipitation tending to decrease as the location of a river basin is positioned further north. All diagnostic statistics, which already fall within their respective

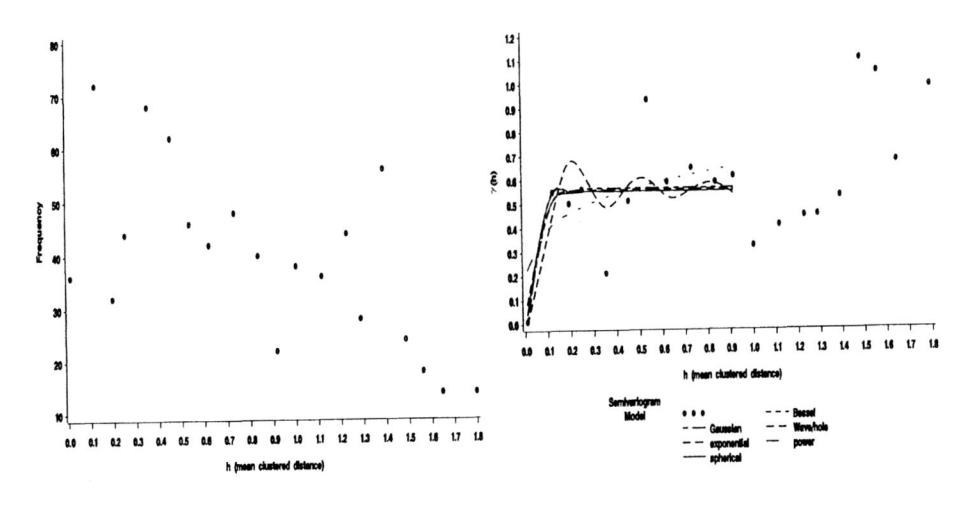

Figure 5.28. Left: (a) Semivariogram-type plot of the frequency of semivariance values in each distance class for Griffith and Amrhein's river basins precipitation data. Right: (b) Semivariogram plot of OLS residuals for Griffith and Amrhein's river basins precipitation data, with selected semivariogram model curves superimposed.

null hypothesis confidence intervals, marginally improve. In addition, these residuals display constant variance across the range of mean response values. The only disappointing feature of this georeferenced data analysis, then, is that it does not produce a more complete statistical description of the geographic distribution of precipitation across river basins located in the province of Ontario.

For the geostatistical analysis of precipitation, 20 distance classes for the 784 distance pairs (n = 28) yielded a minimum of approximately 22 pairs per class, up to the maximum rescaled distance of 1.0 (Figure 5.28a). The semivariogram plot (Figure 5.28b) derived using the residuals from the OLS estimated regression equation

$$\hat{Y} = -0.19512 \times (LAT) ,$$

yield the following model estimation results:

model	RSSE	$\underline{C}_0$	$\underline{C}_1$	$\underline{r}$
exponential	0.62389	0.00000	0.55908	0.05002
Bessel[‡]	0.62040	0.01431	0.55137	0.03795
Gaussian	0.61376	0.00000	0.55216	0.05204
spherical[‡]	0.61578	0.00000	0.55287	0.13690
wave/hole	0.51990	0.00000	0.55845	0.04660
power	0.64167	0.00000	0.66377	* λ = 0.22147 .

[‡]Initial parameter estimate values for the Bessel and spherical semivariogram models were specified explicitly as $C_0 = 0.0001$, $C_1 = 0.6$, $r = 0.2$.

The wave/hole model has the lowest RSSE, although positive spatial autocorrelation is expected given the preceding MC value. Inspection of the semivariogram plot, however, reveals that the scatter of points is more like the pattern seen when spatial autocorrelation is absent; excessive variability of gamma values in the semivariogram plot results in the wave/hole model achieving a slightly better fit than do any of the other transition models. The variance-to-mean ratio plot (Figure 5.29) also suggests the absence of spatial autocorrelation, with an average value, up to the maximum rescaled distance of 1.0, that is slightly below the expected value of 2 (§2.6.2).

Besides rainfall, precipitation also consists of snowfall (depending upon conditions, 4 to 15 inches of snow melt to 1 inch of water; a useful rule of thumb is that 10 inches of snow melt to 1 inch of water), which can be a critical source of reservoir water replenishment during the spring months. Approximately a cube-root power transformation appears to be appropriate for the Ontario river basin snowfall data:

```
      raw                                    power transformation
                                             δ̂ = 21.13470
                                             γ̂ = 0.37173

                                             RSSE = 2.5×10⁻²
   S-W = 0.96914                             S-W = 0.97562
   Bartlett = 0.060                          Bartlett = 0.316
   Levene = 0.165                            Levene = 0.363
   MC = 0.34843  (σ̂_MC ≈ 0.13608)
   GR = 0.61515 .
```

But this power transformation improves the normality diagnostic statistic only slightly, while modestly deteriorating the homogeneity of variance diagnostic statistics. Because the original diagnostic statistics are not significant and little is gained by employing this power transformation—which again is not surprising because $\frac{y_{min}}{y_{max}} < 3$—statistical analyses should be conducted with the raw data.

These raw snowfall data contain significant positive spatial autocorrelation, which should be expected given the range of latitudes involved, the manner in which winter storms move across North America, and the presence of lake effect-localized weather systems in the southern part of this region. Interestingly, these snowfall data do not correlate strongly with their georeferencing Cartesian coordinates (latitude and longitude). Conceptual differentiation between the spatial AR and SAR models when only a constant mean response is included is moot. Estimation of an AR model in this case yields

```
   parameter estimates (for z-scores)              residuals
                                             W-S = 0.95628
      ρ              0.60723                 A-D = 0.418
      b₀             0.05334                 R-J = 0.9768
      (Pseudo-)R²    0.417                   K-S = 0.118
                                             Bartlett = 0.119
```

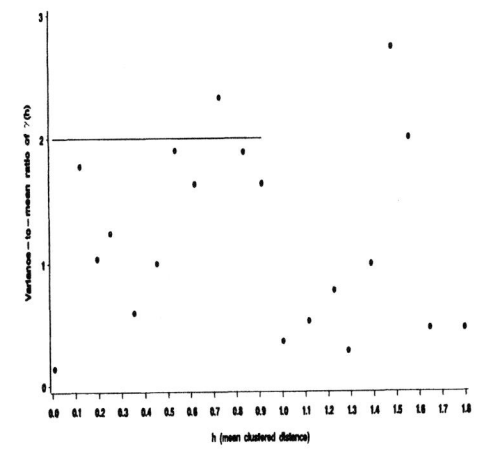

Figure 5.29 (left). Semivariogram-type plot of the distance class variance-to-mean ratio for Figure 5.28b.

Figure 5.30 (right). Traditional homogeneity of variance OLS residual plot for Griffith and Amrhein's river basins snowfall data.

                                  Levene = 0.125
                                  MC = 0.07179
                                  GR = 0.88029 .

The AR/SAR model generates well-behaved residuals: they adequately conform to a normal frequency distribution, and they display essentially homogeneous variability across the four quadrants of the Ontario geographic landscape. Unfortunately, these residuals do not display constant variance across the range of mean response values (Figure 5.30), indicating that a serious analysis of these data would need to explore the use of weighted generalized least squares regression.

As found with the spatial autoregressive analysis, virtually no spatial autocorrelation is detectable from the semivariogram plot and model fits (Figure 5.31). Model fits for these data, using a maximum rescaled distance of 1.0, yield the following goodness-of-fit and parameter estimates:

model	RSSE	C_0	C_1	r
exponential	0.71746	0.49003	2.00937	1.00000
Bessel[‡]	0.72585	0.68775	3.08893	0.99790
Gaussian[‡]	0.72792	0.73937	1.52645	0.82951
spherical	0.72387	0.50886	1.14182	1.00000
wave/hole	0.67728	0.00000	1.24946	0.04670
power	0.69879	0.00000	1.68497	* λ = 0.37496 .

[‡]Initial parameter estimate values for the Bessel and Gaussian semivariogram models were specified explicitly as $C_0 = 0.0001$, $C_1 = 1.0$, $r = 1.0$.

Similar to results obtained for the precipitation data, the wave/hole model has the lowest RSSE. Even lower RSSE values for other transition models might be attainable given that each of the range parameter values for the exponential and

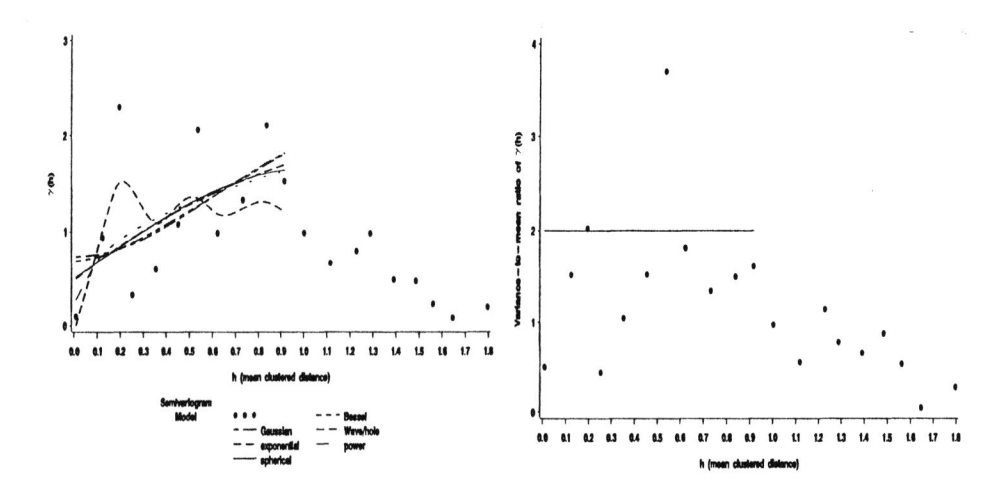

Figure 5.31 (left). Semivariogram plot of OLS residuals for Griffith and Amrhein's river basins snowfall data, with selected semivariogram model curves superimposed.

Figure 5.32 (right). Semivariogram-type plot of the distance class variance-to-mean ratio for Figure 5.31.

spherical models hits its upper-limit constraint value of 1; again, however, the scatter of gamma values in the semivariogram plot suggests the absence of spatial autocorrelation, which also is suggested by the variance-to-mean ratio plot (Figure 5.32), where average values, up to the maximum rescaled distance of 1, are near the expected value of 2 (§2.6.2).

As with the Ontario precipitation measures, the Ontario runoff data appear not to benefit from a power transformation. Consider the following results:

```
raw                              power transformation
                                 δ̂ = -96.87978
                                 γ̂ =   0.91686
                                 RSSE = 1.7×10⁻²
S-W = 0.96222                    S-W = 0.96413
Bartlett = 1.074
Levene = 0.424
MC = 0.45485  (δ̂_MC ≈ 0.13608)
GR = 0.35385 .
```

The estimate of a power transformation exponent that nearly equals 1 implies that little would be gained by employing such a transformation. The raw runoff data already exhibit a tolerable degree of heterogeneity across the four quadrants of the region, while its frequency distribution adequately conforms to a normal frequency distribution. Meanwhile, significant positive spatial autocorrelation is detectable in these data.

Conceptually speaking, autoregressive model specification in this phenomenon should cast the presence of spatial autocorrelation in the runoff measures as arising in the error term. Runoff in one of a set of geographically dispersed drainage basins should not directly impact upon runoff in nearby basins. Rather, geographic commonalities underlying runoff should be governed, in part, by similar basin topography and geological substructures—missing geographic variables in this analysis. Estimation results based on these Ontario data include

	parameter estimates (for z_y-scores)					
	OLS		AR(W)		SAR(W)	
	parameter	s.e.	parameter	s.e.	parameter	s.e.
ρ	*		0.59869	0.30671	0.59795	0.30223
b_0	0	0.15008	-0.07171	0.32792	-0.06640	0.32492
b_{LONG}	-0.12061	0.02942	-0.05229	0.04288	-0.11932	0.05605
(pseudo-)R^2	0.393		0.594		0.595	

residuals			
W-S	0.98260		0.97506
A-D	0.234		0.214
R-J	0.9924		0.9917
K-S	0.098		0.100
Bartlett	0.246		0.366
Levene	0.086		0.227
MC	0.35071 ($\hat{\sigma}_{MC} \approx 0.13608$)		0.0470
GR	0.57137		0.85887

These results indicate that spatial autocorrelation beyond simple map pattern accounts for roughly 20% of the variation in runoff across Ontario river basins. The simple map pattern portion accounts for nearly 40% of this variation and reveals an east-west trend across the province. In contrast to the spatial SAR model, the spatial AR model is unable to separate successfully this trend from other manifestations of spatial autocorrelation. The frequency distributions of the residuals for both the OLS and the SAR cases adequately conform to a normal frequency distribution. Each of these two sets of residuals also displays what can be considered homogeneous variability across the four quadrants of the Ontario geographic landscape. The OLS residuals contain significant positive spatial autocorrelation, which appears to be absent from the SAR residuals. The standard error of $\hat{\rho}$ for the SAR model seems too large (§2.7); given that the MC for the OLS residuals is significant, this parameter estimate also should be significant. Finally, the standard residual-versus-predicted-values scatterplot (Figure 5.33) does not indicate the presence of nonconstant variance across the range of mean response values. These conclusions are consistent with conceptual expectations about runoff measures.

Unlike results reported for precipitation and snow, positive spatial autocorrelation is observed in the pattern of gamma values in the semivariogram plot (Figure 5.34) using residuals from the OLS estimated regression equation $\hat{Y}$ = -0.12061×(LONG). Semivariogram model parameter estimation with these data, using a maximum rescaled distance of 1.0, yields the following results:

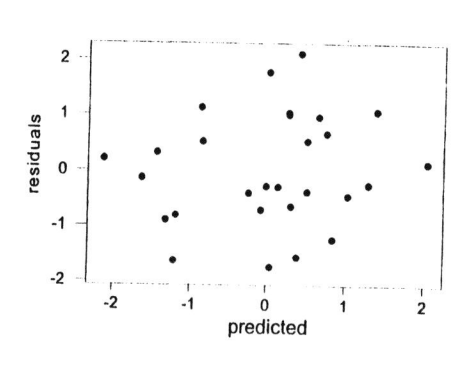

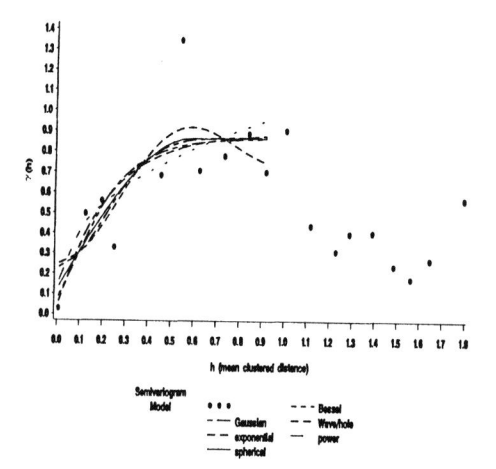

Figure 5.33 (left). Traditional homogeneity of variance OLS residual plot for Griffith and Amrhein's river basins runoff data.

Figure 5.34 (right). Semivariogram plot of OLS residuals for Griffith and Amrhein's river basins runoff data, with selected semivariogram model curves superimposed.

model	RSSE	c_0	c_1	r
exponential	0.42401	0.02973	0.85892	0.21850
Bessel	0.43298	0.08315	0.78106	0.13432
Gaussian	0.44494	0.23031	0.63725	0.29491
spherica	0.40423	0.12328	0.73955	0.58235
wave/hole	0.41486	0.24972	0.54867	0.13018
power	0.46100	0.00000	0.98468	* $\lambda = 0.37197$

Here the spherical model has the lowest RSSE and hence is the preferred geostatistical model, although the other transition models have only slightly poorer fits.

5.6. DIGITAL ELEVATION DATA

Digital elevation models (DEMs)—which are closely related digital terrain models (DTMs), and digital terrain elevation data (DTED)—are digital representations of terrain elevations for ground positions at regularly spaced horizontal intervals on the earth's surface. Each data value of a DEM computer image represents a height above sea level value. The Earth Science Information Center (ESIC), through the USGS) as part of the National Mapping Program, produces and distributes five different digital elevation products, each varying only in sampling interval, geographic reference system, areas of coverage, and accuracy. DEMs may be used in the generation of three-dimensional visualizations displaying terrain slope, aspect (direction of slope), and terrain profiles between selected points—critical components for undertaking meaningful land process studies. These data often are attached to georeferenced natural resources data (see §5.2, §5.3, §5.4), especially when searching for energy resources, locating watersheds for analysis, determining optimal reservoir and dam locations, calculating the volume of proposed reservoirs, estimating landslide probabilities, and assessing site suitability for potential corridors and proposed facilities locations. Besides quantifying topography and aiding in the understanding of hydrology, DEMs support performing spatial autocorrelation analysis via automated photogrammetric systems. A current additional goal of USGS is to promote incorporation of existing DEM data into global change modeling activities.

Although a wealth of public domain topographic data is both available and easily accessible, especially through the Internet, both Venables and Ripley (1994, 383) and Carr (1995, 551-68) provide small topographic height data sets for illustrative purposes, while Isaaks and Srivastava (1989) report two tabulated physical geography variables, derived from DEM data publicly available from the National Cartographic Information Center, for the Walker Lake region in Nevada, near the California-Nevada border. These two variables, which relate to topographic features of the region, are calculated from 1.95 million data points, with adjacent points separated by approximate 200 feet, aggregated into a regular square tessellation surface partitioning comprising 78,000 points that form a 260-by-300 rectangular grid. The first of these two variables is the variance of 25 aggregated elevation values in each 5-by-5 subsection of this region; it indexes the roughness of the topography. The second is the mean elevation multiplied by the log-variance of each subsection; it is a locally weighted average elevation index. A detailed map of this region appears in Griffith (1993b). A 10-by-10 regular lattice of aggregate sample locations was selected from this region by Isaaks and Srivastava which they then used to compile the data set.

The index of roughness variable, namely the within subsection elevation variance data, appears to benefit adequately from a power transformation:

raw	power transformation
	$\hat{\delta}$ = 7.54323
	$\hat{\gamma}$ = 1.77695
	RSSE = 7.7×10^{-3}
S-W = 0.95807[†††]	S-W = 0.97729
Bartlett = 18.283[†††]	Bartlett = 15.125[†††]
Levene = 2.950[††]	Levene = 4.521[†††]
MC = 0.42226 ($\hat{\sigma}_{MC} \approx$ 0.07454)	MC = 0.50531
GR = 0.50809	GR = 0.43502 .

This power transformation brings the frequency distribution of the roughness index variable into better compliance with a normal frequency distribution assumption and modestly diminishes the degree of variance fluctuation across the four quadrants of this region. Significant, marked positive spatial dependency is detected.

An SAR model is selected here to describe these DEM data because one would expect that spatial autocorrelation enters them through a measurement error process: elevation measurements for nearby locations should contain a similar

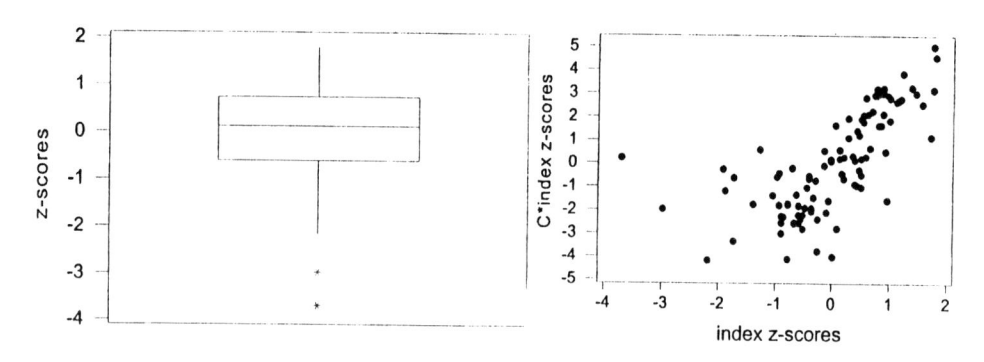

Figure 5.35. Left: (a) Boxplot of the DEM local elevation variance (Isaaks and Srivastava's data). Right: (b) MC scatterplot of the DEM local elevation variance (Isaaks and Srivastava's data).

nature and degree of error. Not surprisingly, a conspicuous systematic pattern is discernible across this region for the variance of elevation values aggregated by subsection. In addition, the presence of mountain ranges hints at the possibility of multiple elevation populations for this geographic region. Given this context, a number of spatial/aspatial outliers appear to be present in these data (Figures 5.35a-b); 1 is a relative peak, 11 others occur in six scattered pockets of low values. Estimation results based on these data include

parameter estimates (for z_y-scores)

	OLS parameter	s.e.	SAR(W) parameter	s.e.
ρ	*		0.77897	0.12695
b_0	0.03272	0.28413	0.24195	0.25613
$b_{u-\bar{u}}$	0.20686	0.01913	0.20061	0.05395
$b_{outlier-low}$	-1.04716	0.08786	-0.79062	0.06634
$b_{outlier-high}$	0.86684	0.27319	0.86000	0.18143
(pseudo-)R^2	0.714		0.859	
residuals				
W-S	0.96724[†]		0.95270[†††]	
A-D	0.576		1.268[†††]	
R-J	0.9925		0.9795[†††]	
K-S	0.066		0.111[†††]	
Bartlett	7.062[†]		5.033	
Levene	3.195[††]		1.687	
MC	0.38636		0.04168	
GR	0.61612		0.97306 .	

These results indicate that the roughness-of-topography index variable contains a sizeable amount of positive spatial autocorrelation, a portion of which is attributable to simple map pattern, which in turn destabilizes the geographic variability displayed by the index. But gaining a more constant level of variability across this

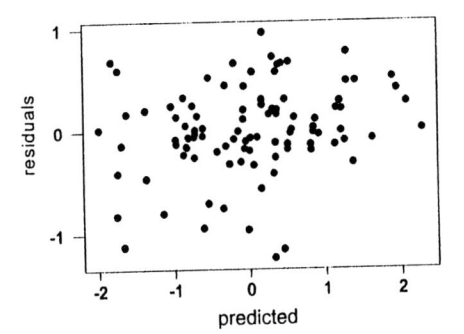

Figure 5.36. Traditional homogeneity of variance SAR residual plot for Isaaks and Srivastava's DEM local elevation variance data.

geographic landscape occurs at the cost of normality; the SAR residuals deviate more from a normal frequency distribution than do the OLS residuals, especially in the tail portions of the empirical frequency distribution. This trade-off seems worthwhile, because nonconstant variance interjects far more serious complications into a statistical analysis than does non-normality (§2.1). Further variance stability is corroborated by the traditional residuals-versus-predicted-values plot (Figure 5.36), which depicts constant variance across all levels of the mean response.

Likewise, residuals from the OLS estimated regression equation for the index-of-roughness variable

$$\hat{Y} = 0.3272 + 0.20686 \times (u - \bar{u}) - 1.04716 \times (I_{outlier-low}) + 0.86684 \times (I_{outlier-high})$$

were subjected to geostatistical analysis. Effects of the high and low outlier values can be seen in plots for the maximum and minimum gamma values for each distance class (Figures 5.37a-b). Omission of the single indicator variable for the low outlier values, for example, results in a sole value of -2.74 dominating the minimum value in the max/min plot (17 of 23 cases, up to the maximum rescaled distance of 1.5; Figure 5.37a). Likewise, omission of the indicator variable for the high outlier values results in the single value of 1.71 dominating the maximum value (13 of 23 cases, up to the maximum rescaled distance of 1.5; Figure 5.37b). Inclusion of the pair of indicator variables for high and low outlier values in the OLS regression removes both of these patterns.

Division of the 10,000 distance pairs (n = 100) into 50 distance classes gives more than 100 distance pairs per distance class (Figure 5.38a). Fitting selected models to these semivariogram plots (Figure 5.38b), using a maximum rescaled distance of 1.5, yields the following results:

model	RSSE	C_0	C_1	r
exponential	0.41093	0.00000	0.30734	0.18912
Bessel	0.38271	0.00000	0.30680	0.13153
Gaussian	0.35872	0.00902	0.29618	0.25424
spherical	0.35330	0.00752	0.29830	0.57048
wave/hole	0.30882	0.09664	0.20456	0.15309
power	0.58506	0.00000	0.30183	*

$\lambda = 0.13780$.

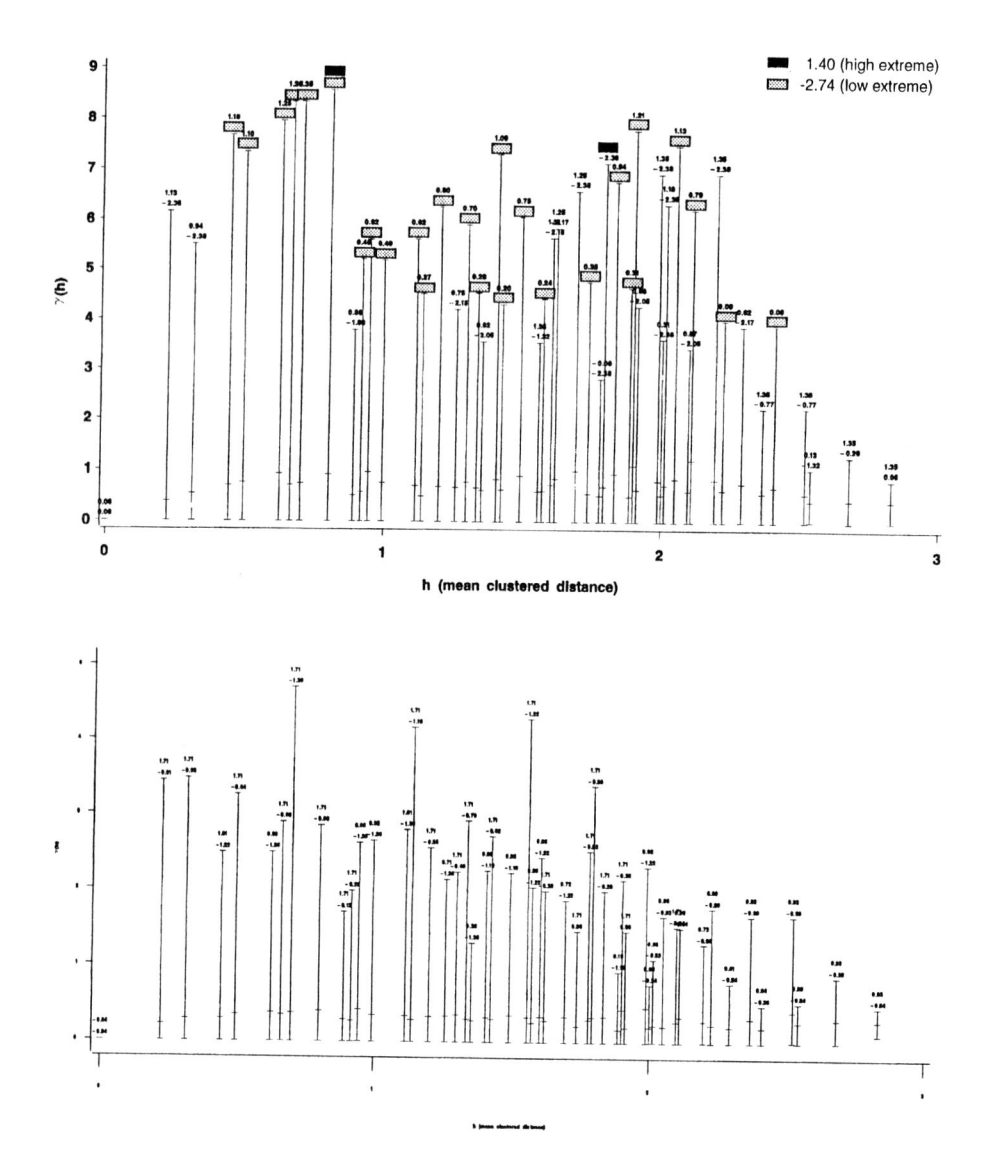

288   A CASEBOOK FOR SPATIAL STATISTICAL DATA ANALYSIS

Figure 5.37. Top: (a) Semivariogram plot displaying the mean and extreme values for OLS high outlier-adjusted linear trend model residuals for the DEM index-of-roughness (Isaaks and Srivastava's data). Middle: (b) Semivariogram plot displaying the mean and extreme values for OLS low outlier-adjusted linear trend model residuals for the DEM index-of-roughness (Isaaks and Srivastava's data).

A typical pattern in the scatter of points of the semivariogram plot indicative of positive spatial autocorrelation can be seen in Figure 5.38b. Values for the range parameter are relatively high, corroborating results obtained with spatial autoregressive analysis. The relatively high RSSE values presumably are due to the drop in gamma values beginning at roughly a distance of 1.0. This variability in

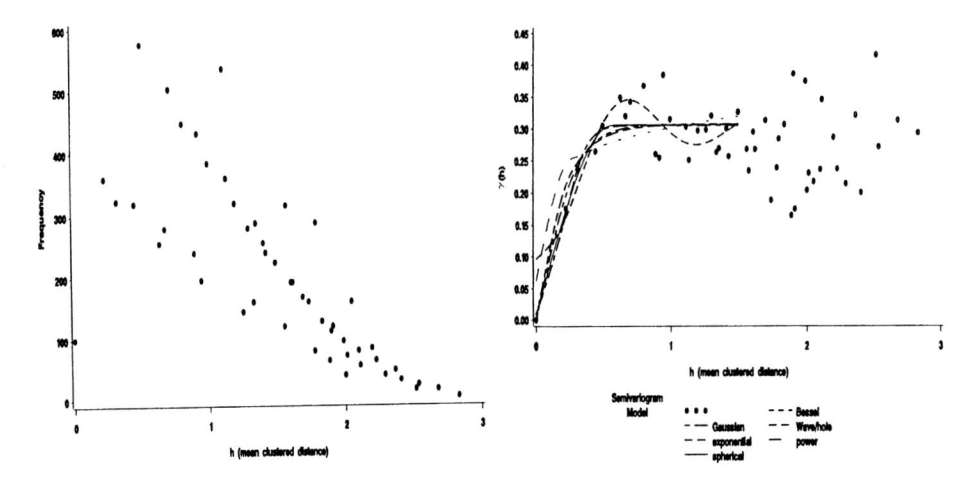

Figure 5.38. Bottom-left: (a) Semivariogram-type plot of the frequency of DEM index-of-roughness semivariance values in each distance class (Isaaks and Srivastava's data). Bottom-right: (b) Semivariogram plot of OLS residuals for Isaaks and Srivastava's DEM index-of-roughness data, with selected semivariogram model curves superimposed

gamma values suggests that the wave/hole model is the preferred transition model, although the Gaussian or spherical semivariogram models would provide a more conceptually appealing description of spatial dependency, given the preceding discussion of the nature of these data.

The locally weighted average elevation index variable also appears to benefit sufficiently from a power transformation:

```
raw                          power transformation
                             δ̂ = 43.07364
                             γ̂ = 0
                             RSSE = 1.3×10⁻²
S-W = 0.96911                S-W = 0.97880
Bartlett = 14.201†††         Bartlett = 10.537††
Levene = 3.738††             Levene = 3.181††
MC = 0.42671                 MC = 0.42456
GR = 0.57633                 GR = 0.57207 .
```

The respective frequency distributions of both the raw and power-transformed versions of this index seem to conform adequately to a normal frequency distribution, with somewhat less but still marked variance fluctuation displayed by the power-transformed measures across the four quadrants of the region. This latter finding is not surprising, given the five mountain ranges and the single valley that occupy the Walker Lake region. Significant, marked positive spatial dependency is detected in these data.

Although a systematic pattern across this region is not conspicuous for the mean multiplied by the log-variance of elevation, one value (i.e., 55) appears to be a distinct large-value outlier, towering over all the others and presumably coinciding with a mountain peak; a potential small-value outlier is detectable, too, which presumably coincides with the valley subregion. Estimation results based on these data include

parameter estimates (for z_y-scores)

	OLS parameter	s.e.	SAR(W) parameter	s.e.
ρ	*		0.74780	0.13065
b_0	1.47384	0.48169	1.26122	0.42191
$b_{outlier}$	1.50391	0.48169	1.28193	0.32354
(Pseudo-)R^2	0.091		0.584	
residuals				
W-S	0.96453[†]		0.93686[†††]	
A-D	0.687[†]		2.245[†††]	
R-J	0.9927		0.9707[†††]	
K-S	0.089[†]		0.160[†††]	
Bartlett	6.340[†]		3.552	
Levene	2.013		0.530	
MC	0.44784		0.06881	
GR	0.53731		0.91593	

The locally weighted average elevation index variable contains a sizeable amount of positive spatial autocorrelation, which in turn destabilizes the geographic variability displayed by the index. As is the case with the roughness index, gaining a more constant level of variability across this geographic landscape occurs at the cost of normality; the SAR residuals deviate modestly more from a normal frequency distribution than do the OLS residuals, especially in the tail portions of the empirical frequency distribution. Again, this trade-off seems worthwhile, because nonconstant variance interjects far more serious complications into a statistical analysis than does non-normality (§2.1). Meanwhile, the traditional residuals-versus-predicted-values plot suggests that a serious analysis of these data should involve weighted generalized least squares estimation procedures.

Similar to the geostatistical analysis for the index-of-roughness variable, outlier values were identified for the locally weighted average elevation index variable. When the indicator variable for outlier values is removed from the OLS estimated regression equation

$$\hat{Y} = 1.47384 + 1.50391 \times (I_{outlier}),$$

a single value of 2.98 dominates the max/min plot of gamma values in 13 of 18 cases, using a maximum rescaled distance of 1.3 (Figure 5.39). Inclusion of the outlier indicator variable removes the dominance of this value. The same number of distance classes (50) as was used in the index-of-roughness variable analysis is used here, resulting in more than 100 distance pairs per class (Figure 5.38a). Semivariogram model parameter estimation for these data (Figure 5.40), using a maximum rescaled distance of 1.3, yields

NATURAL RESOURCES 291

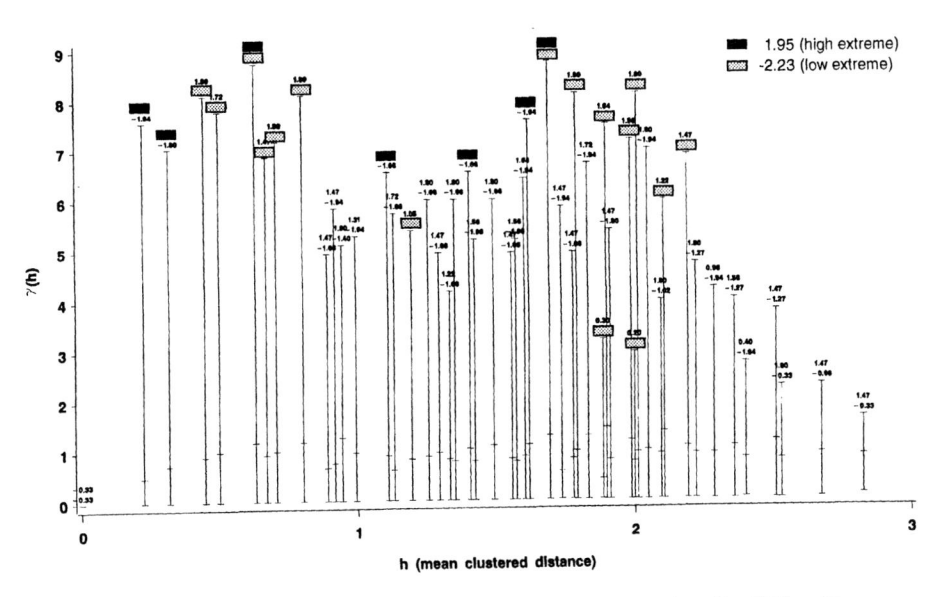

Figure 5.39. Semivariogram plot displaying the mean and extreme values for OLS outlier-adjusted residuals for the DEM locally weighted average elevation index (Isaaks and Srivastava's data).

model	RSSE	C_0	C_1	r
exponential	0.56635	0.00000	0.93530	0.18151
Bessel	0.53690	0.00000	0.93493	0.12808
Gaussian	0.49239	0.00000	0.93175	0.25039
spherical	0.49155	0.00000	0.93043	0.54126
wave/hole	0.38000	0.18891	0.72956	0.14236
power	0.70688	0.00000	0.93109	*

$\lambda = 0.15091$.

The scatter of points in the semivariogram plot (Figure 5.40) is indicative of and forms a pattern typical when spatial autocorrelation is present, corroborating results

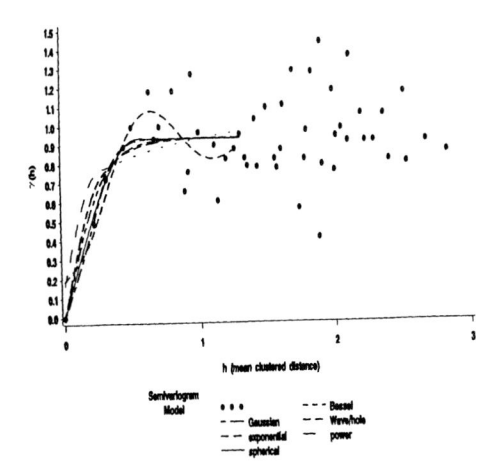

Figure 5.40. Semivariogram plot of OLS residuals for Isaaks and Srivastava's DEM index-of-roughness data, with selected semivariogram model curves superimposed.

from the spatial autoregressive analysis. The relatively high large parameter values also point to the presence of positive spatial autocorrelation. As in the case of the index-of-roughness variable, however, the wave/hole semivariogram model (indicating negative spatial autocorrelation) has the best fit among the transition models. Again, the relatively high RSSE values and the selection of the wave/hole model as the best-fit one presumably are due to a drop in gamma values at roughly a distance of 1.0. Unlike the index-of-roughness variable, though, differences between the RSSE of the wave/hole model and those of the other semivariogram models are at least about twice that found for the locally weighted average elevation index variable. These differences may well be attributable to the shorter maximum rescaled distance used for this analysis (1.5 versus 1.3). Despite these RSSE differences between the wave/hole model and the other transition models, selection of the Gaussian or spherical semivariogram model for describing the spatial autocorrelation in these data is warranted here. As an aside, Oliver and Webster (1986) find that spatial dependency in topographic height may be described with a linear combination of the linear, spherical, and wave/hole semivariogram models.

5.7. PRECLASSIFIED REMOTELY SENSED IMAGE REFLECTANCE DATA

The final type of data treated in this chapter is that produced by sensors mounted on satellites; these data have been publicly available since 1972 and comprise reflectance values constructed for bands of the electromagnetic spectrum. Otherwise known as remotely sensed data, they are massive and voluminous[1], are organized on a regular grid of pixels, and range in value from 0 (no reflectance) to 255 (maximum reflectance). The LANDSAT 5 "Thematic Mapper" (TM) sensor generates data for seven bands, coming from the visible and infrared segments of the spectrum. These data must undergo considerable preprocessing before they are released for them to be usable for analysis purposes. Understanding these data has been a taxing task for scientists over the more recent decades; one small data set available for quantitatively exploring the attachment of meaning to these spectral bands is provided by Campbell and Kiiveri (1988) and reprinted in *A Handbook of Small Data Sets* (Hand et al. 1994). Meanwhile, for illustrative purposes, Bailey and Gatrell (1995, 254) analyze data from a one-square-kilometer, 30-by-30 pixels portion of the High Peak District in England; in practice a serious spatial scientist would rarely analyze a LANDSAT image this small. This region includes a mix of land uses, from a reservoir located in the southwest part, through mixed deciduous and coniferous woodland inhabiting the south-central area, to rough grazing and moorland covering the northern part of this region. The High Peak varies in elevation from 700 meters in its southwestern part, to over 1,200 meters in its northern part.

---

[1] More than two dozen new data-gathering satellites are planned to be launched worldwide by 2000 (Tomerlin 1998).

The focus of the analysis here is on the ratio of Band #4 to Band #3, bands that, respectively, represent the *near-infrared* and *red* wavelengths of the electromagnetic spectrum. This ratio provides a good picture of spatial variation in biomass, with healthy green vegetation reflecting strongly in Band #4, which measures the vigor of vegetation, whereas its energy absorption is strongly sensed in Band #3, which aids in the identification of plant species. For the High Peak study area, the values for Band #4 range from 16 to 106; those for Band #3 range from 18 to 51. In addition, the ratio of the extreme values of $\frac{\text{Band \#4}}{\text{Band \#3}}$ for this geographic landscape equals 4.53 > 3, suggesting that these data may well benefit from the application of a power transformation. Unfortunately, little appears to be gained from such a data manipulation:

raw

S-W = 0.94904[†††]
Bartlett = 91.561[†††]
Levene = 9.640[†††]
MC = 0.88047
GR = 0.10820

power transformation
$\hat{\delta}$ = 0.42491
$\hat{\gamma}$ = 0
RSSE = 1.3×10^{-2}
S-W = 0.97298[†††]
Bartlett = 91.595[†††]
Levene = 12.876[†††]
MC = 0.87525
GR = 0.10132 .

Each of the two frequency distributions seems to deviate markedly from a normal frequency distribution, with roughly the same level of very marked variance fluctuation across the four quadrants of the region. The raw data set contains outliers in the high mean response segment of its measurement scale, whereas the power-transformed data set contains outliers in the low mean response segment. The variation contrast is between the northern and the southern subregions of the High Peak, which is not surprising given the aforementioned land use differences in them. Either version of these data contains pronounced positive spatial autocorrelation. But the corrupting nature of the power transformation becomes visibly apparent by comparing a MC scatterplot for these two data sets (Figures 5.41a-b). Overall, then, few benefits are gained by reverting to the power-transformed values, and hence the raw data should be retained for analysis purposes.

Of particular note here is the excessive degree of geographic heterogeneity. This is often a feature of remotely sensed data and suggests that weighted analyses should be the rule when undertaking a serious exploration of this type of data. Positive spatial autocorrelation detected in this ratio of Band #4 to Band #3 displays no simple map pattern across the High Peak (i.e., the Cartesian coordinates are not significant covariates). Rather, a few selected eigenvectors (after removal of the principal eigenvector; see §1.5), approximately equivalent to those extracted from expression (1.10), account for a sizeable portion of the variation exhibited in this ratio, while a spatial SAR model describes these data reasonably well:

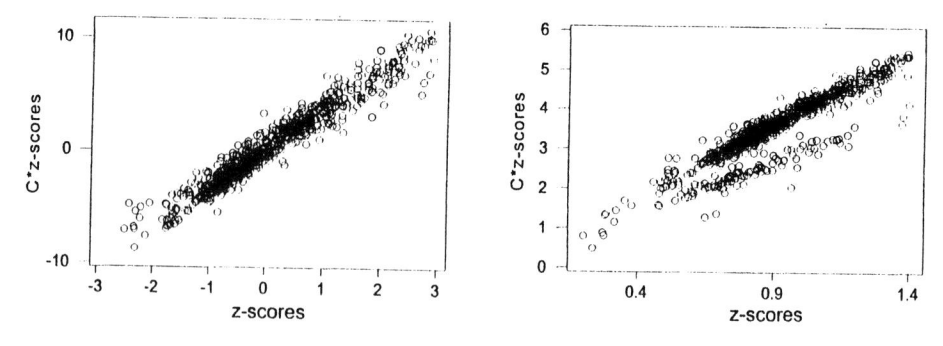

Figure 5.41. Left: (a) MC scatterplot of the raw $\frac{Band\#4}{Band\#3}$ LANDSAT TM remotely sensed spectral ratio (Bailey and Gatrell's data). Right: (b) MC scatterplot of the power-transformed $\frac{Band\#4}{Band\#3}$ LANDSAT TM remotely sensed spectral ratio (Bailey and Gatrell's data).

	parameter estimates (for z_y-scores)		
	SAR (W)		OLS
	parameter	eigenvector	% variance accounted
ρ	0.98812	$E_{1,3}$	25.6
b_0	-0.01340	$E_{2,3}$	11.2
		$E_{1,7}$	5.2
		$E_{3,2}$	3.3
		$E_{2,6}$	2.2
		$E_{1,2}$	2.2
(pseudo-)R^2	0.953		0.504
residuals			
W-S	0.95641†††		0.98150††
Bartlett	39.201†††		40.659†††
Levene	5.558†††		8.755†††
MC	0.16838		0.75390
GR	0.82489		0.20424 .

We see that the latent spatial autocorrelation is extremely strong, that it contributes substantially to the variance heterogeneity across this geographic landscape, and that even after roughly half the variation in this ratio is accounted for by rather complex map patterns (i.e., the six selected eigenvectors), marked positive spatial autocorrelation remains.

As an aside, the 30-by-30 regular square tessellation of pixels to which the High Peak spectral bands are locationally tagged render a 900-by-900 geographic connectivity matrix from which 900 eigenvectors can be extracted. To avoid the accompanying massive computational demands, an analytical solution to this eigenvector problem was derived. Combining results reported in Gasim (1988) with standard ones reported in Usmani (1987, 137) yields, for a P-by-Q regular square tessellation,

$$E_{i,j} = \frac{2 \times 2}{\sqrt{(P+1) \times (Q+1)}} SIN\left(\frac{iU\pi}{P+1}\right) \times SIN\left(\frac{jV\pi}{Q+1}\right), \qquad (5.1)$$

$$i = 1, 2, ..., P \ \& \ j = 1, 2, ..., Q.$$

As n=P×Q increases, these eigenvectors converge upon those for expression (1.10), after, of course, $E_{1,1}$ (at least for relatively small n) is removed. Prominent eigenvectors can be identified here simply by sequentially executing a nonlinear regression with the following steps:

STEP 1: estimate i and j by regressing equation (5.1) on the remotely sensed measure;
STEP 2: round off i and j to their respective nearest integers, and use them to compute $E_{i,j}$ using equation (5.1);
STEP 3: enter the result from Step 2 into the nonlinear regression equation as an additional predictor variable, retaining the right-hand side of equation (5.1);
STEP 4: return to Step 1.

Steps 1-4 should be repeated until only trivial increases in the resulting R^2 value occurs. This procedure is implemented to search for the six selected eigenvectors for the High Peak data; findings have been corroborated here by inspecting the first 225 eigenvectors rendered by equation (5.1).

For the geostatistical analysis, the mean response of values from the ratio of Band #4 to Band #3 were used. Specification of 50 distance classes gives a minimum of 900 (where n = 900) distance pairs per class (Figure 5.42a). Using a maximum rescaled distance of 1.1, semivariogram model fits to the scatter of points (Figure 5.42b) yield

model	RSSE	C_0	C_1	r
exponential	0.00676	0.00000	0.42206	0.55748
Bessel	0.00596	0.02767	0.35044	0.29634
Gaussian	0.01720	0.07698	0.27328	0.50055
spherical	0.00569	0.02585	0.32125	0.99744
wave/hole	0.02820	0.09676	0.20513	0.21117
power	0.03232	0.00000	0.35990	*	$\lambda = 0.57197$.

The pattern exhibited by the scatter of points in this semivariogram plot (Figure 5.42b), combined with the relatively high range parameter values, is typical of the presence of strong positive spatial autocorrelation in these data, corroborating results from the spatial autoregressive analysis. Relatively low RSSE values for these model fits apparently are from a lack of variation in the gamma values relative to the model curves (Figure 5.42b). For these data, the spherical model has the best fit and hence is the model of choice for describing spatial autocorrelation, although there is little distinction between estimation results for this semivariogram model and for the Bessel model.

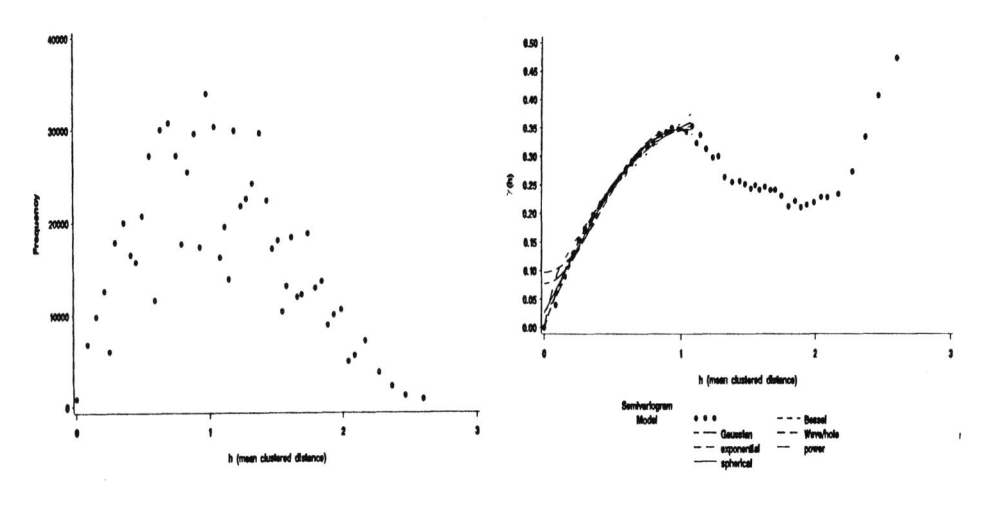

Figure 5.42. Left: (a) Semivariogram-type plot of the frequency of LANDSAT TM remotely sensed spectral ratio semivariance values in each distance class (Bailey and Gatrell's data). Right: (b) Semivariogram plot of the $\frac{Band\#4}{Band\#3}$ LANDSAT TM remotely sensed spectral ratio (Bailey and Gatrell's data), with selected semivariogram model curves superimposed.

But as previously mentioned, most remotely sensed scenes that are studied are much larger than the 30-by-30 one for the High Peak. Accordingly, one for the Adirondack State Park involving a 511-by-503 grid of pixels is explored here. This case, in which n=257,033, is a more common scene size and is nearly 300 times the size of the High Peak image just analyzed. This particular LANDSAT image is from the northwest corner of the park, just southwest of Star Lake, and is roughly 75% forested, with the remainder covered by wetlands, cultural areas, and logged zones. The scene is dotted with lakes and has a considerable amount of undulation because of hills and valleys.

The Adirondack sample[2] size here is extremely large. Its values for Band #4 range from 6 to 139; the values for Band #3 range from 9 to 43. In addition, the ratio $\frac{Band\ \#4}{Band\ \#3}$ ranges from 0.55 to 9.42; the ratio of extreme values for this geographic landscape equals 17.26 > 3, suggesting that these data may well benefit from the application of a power transformation. But there are only 4,690 possible distinct ratios that could be computed with these two ranges of values; and, in fact, there are only 963 distinct ratios that occur in this data set. Therefore, considerable computational time can be saved by aggregating these data—using, say, PROC MEANS in SAS—and using statistical formulae that involve frequencies (e.g., Griffith and Amrhein 1991, 81). The distribution of frequencies in this case resembles a Poisson distribution, with the magnitude of frequencies ranging from

---

[2] We thank Dr. William T. Elberty, Department of Geography, St. Lawrence University, for providing us with this EOSAT TM image subset, and Mr. Peter L. Fellows, Department of Geography, Syracuse University, for helping us acquire it in digital form.

1 to 8,137. In contrast with the High Peak example, this index of spatial variation in biomass appears to benefit from a power transformation:

raw	power transformation
	$\hat{\delta}$ = 9.35301
	$\hat{\gamma}$ = 6.04722
	RSSE = 6.6×10^{-3}
K-S = 0.12190[†††]	K-S = 0.04468[†††]
Bartlett = 23441.0[†††]	Bartlett = 15280.5[†††]
MC = 0.88221	MC = 0.84549
GR = 0.11774	GR = 0.15429.

This power transformation conspicuously improves conformity of the ratio with a normal frequency distribution, and it dramatically reduces variance heterogeneity but without removing it (the image was divided into 49 quadrats, and then Bartlett's statistic was calculated using the formula listed in Griffith and Amrhein 1991, 292). As with the High Peak District, positive spatial autocorrelation detected in this ratio of Band #4 to Band #3 is pronounced[3] and displays no simple map pattern across this section of the Adirondack landscape (i.e., the Cartesian coordinates are not significant covariates). In contrast to the High Peak case, the MC scatterplots (Figure 5.43a-b) do not exhibit a degeneration of the prevailing spatial dependency link when moving from the raw to the power-transformed measurement scale, although increased nonconstant variance across the range of attribute values seems to be more visible in the power-transformed scatterplot. In addition, the level of spatial autocorrelation detected is roughly the same for both measurement scales. Therefore, a more comprehensive spatial analysis of this LANDSAT scene should be pursued employing the power transformation.

To fit a spatial autoregressive model to these remotely sensed data, which involves a 257,033-by-257,033 geographic connectivity matrix, the Jacobian approximation given by equation (3.10) must be used. Calibration of this equation may well need to be based upon fewer systematically selected sample points across the feasible parameter space due to, say SAS, memory limitations; here 15 were used, rather than the 22 utilized by Griffith (1993b). The resulting approximation calibrated for this surface partitioning is given by

$$0.15278 \times [2LN(1.14545) - LN(1.14545 + \rho) - LN(1.14545 - \rho)],$$
$$RSSE = 2.9 \times 10^{-4},$$

where the eigenvalues of matrix **W** have been approximated, by extending the

---

[3] Previously developed SAS code (e.g., Griffith 1993b) has been modified to use only the LAG1 time series operator. To do this, an artificial set of missing values pixels must be concatenated to both the eastern and the southern edges of a remotely sensed image. By sequentially sorting by the U and the V coordinates, first in ascending and then in descending order, the LAG1 function can be used to identify all four of the neighboring values. This procedure circumvents the upper limit on the lag function but requires considerably more time to execute.

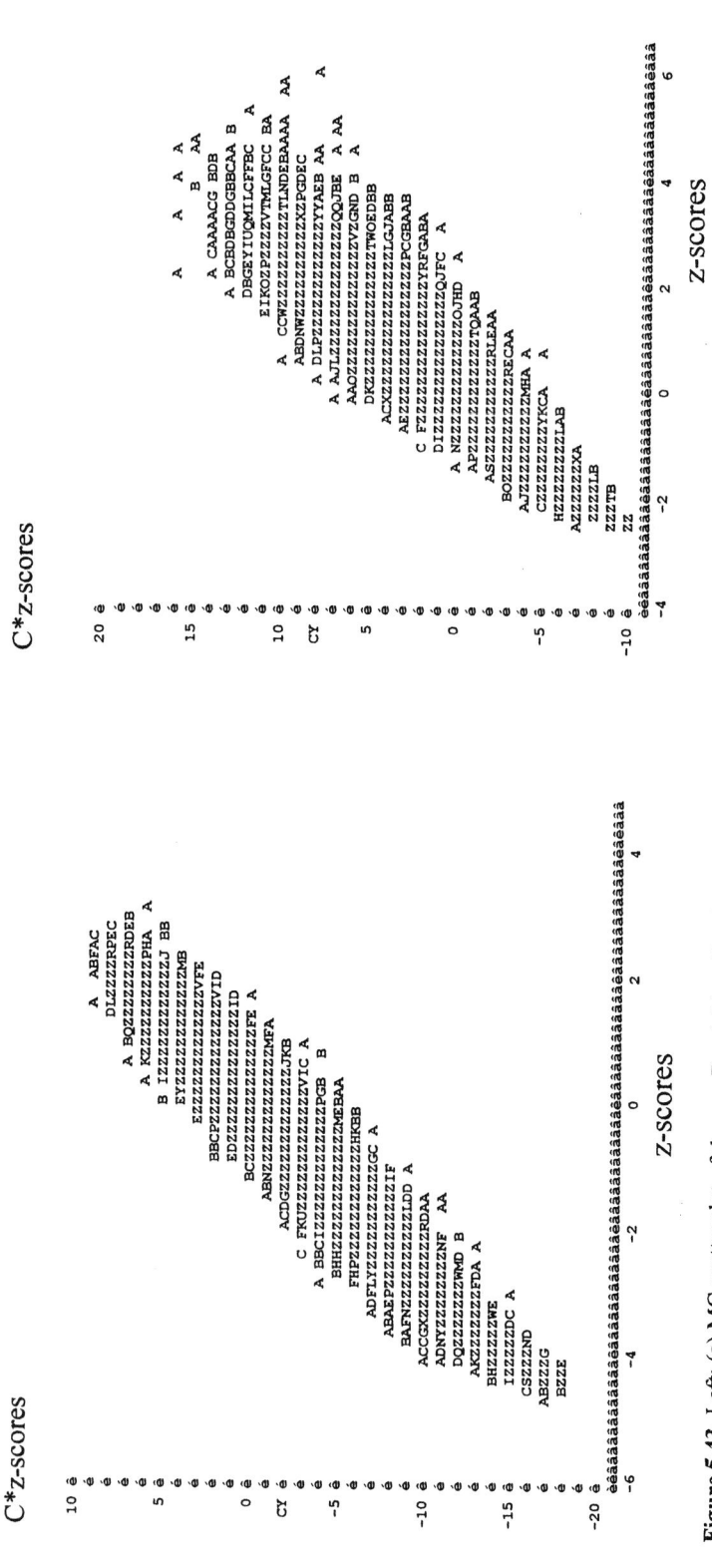

Figure 5.43. Left: (a) MC scatterplot of the raw Band #4-to-Band #3 LANDSAT TM remotely sensed spectral ratio (Elberty's data). Right: (b) MC scatterplot of the power-transformed Band #4-to-Band #3 LANDSAT TM remotely sensed spectral ratio (Elberty's data).

exact solution for the linear geographic landscape case (see Berman and Plemmons 1979; Griffith 1980; Martin and Griffith 1998), with

$$[COS(\frac{i-1}{P-1}\pi) + COS(\frac{j-1}{Q-1}\pi)]/2 \ , \ i = 1, 2, ..., P \ \& \ j = 1, 2, ..., P \ .$$

These eigenvalues are asymptotically correct[4], and are the approximation counterparts to those for matrix **C** reported in §3.3. Similar to the High Peak results, an SAR specification based upon matrix **W** yields the following parameter estimates: $\hat{\rho}$ = 0.94190 and $\hat{\mu}$ = -0.00006, R^2 = 0.885. But unlike the High Peak case, the SAR residuals in this case deviate further from normality, yielding K-S = 0.06274[†††]. Nevertheless, the important computational feature of this spatial analysis is that it can be done; the spatial autoregressive model can be implemented, even though n is so large. And, as in the High Peak case, the aforementioned stepwise eigenvector identification procedure using a nonlinear routine is a feasible method for uncovering salient complex map pattern components of spatial autocorrelation. This procedure identifies $E_{2,1}$ as a salient eigenvector, accounting for 2.02% of the variability in the power-transformed biomass index. Relatively speaking, this is a sizeable amount, given that there are 257,033 eigenvectors from which to select.

For the geostatistical analysis, the image is far too large for a comprehensive examination like the one done for the High Peak. The 257,003 pixels generate more than 3.3×10^{10} pairs of distances. Even restricting this set to those distances having a rescaled value that does not exceed 1.1 results in roughly 5.5×10^8 pairs of distances. In practice, sampling is used to circumvent this massive data set complication; this is exactly the solution to the problem implemented in GEO-EAS. This situation naturally lends itself to a resampling evaluation of semivariogram model parameter estimates, a research endeavor meriting far more attention (see §3.2.2). The sampling design developed for the EMAP data collection task, which produced those data analyzed in §5.2 and §7.2, involves geographically stratified random sampling. It provides better geographic coverage of a region, avoiding the patchiness typical of the Poisson point distribution generated by a simple random sample of pixel coordinates, and the possibility of

---

[4]The approximation here is extremely good, as the following sequence of mean squared differences implies:

P = Q =	5	10	15	20	25	30	40	50
MSE×10⁻²	0.1326	0.0291	0.0109	0.0049	0.0032	0.0021	0.0011	0.0007

The trend exhibited by these values can be described with the following equation:

$$\frac{M\hat{S}E}{100} = 0.00041 + \frac{75.18938}{(P + 2.80897)^{3.08647}},$$

RSSE = 2.2×10^{-7}. Clearly, by the time a geographic landscape is roughly 500-by-500 in size, the approximate eigenvalues are indistinguishable from the actual ones.

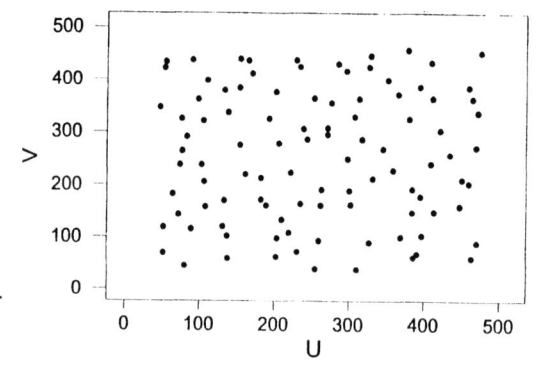

Map 5.12. Geographic distribution of the 25 central pixels for each of the sampled 65-by-65 clusters of pixels for the 503-by-511 Adirondack Park remotely sensed image.

alignment with periodicities across a region that could materialize by use of a systematic sample. But stratified random sampling alone allows for the selection of few pairs of nearby pixels, diminishing the amount of information contained in a sample concerning low-order spatial autocorrelation, spatial dependency being of most interest. Although GEO-EAS example data sets were constructed with a mixed simple random/systematic geographic sample, which helped to alleviate this problem, a risk persists of obtaining insufficient information about low-order spatial autocorrelation. The solution implemented here is a stratified random sample of clusters: (1) a 438-by-446 central region of the remotely sensed image was partitioned into 100 roughly 44-by-44 sets of pixels; (2) a pixel was selected at random from each of these 100 geographic strata; and (3) each randomly selected pixel constituted the central pixel of a 65-by-65 cluster of pixels. The size of clusters was based upon the semivariogram plot for the High Peak remotely sensed image previously analyzed. Because of the massive amounts of data involved here, only 25 replications were drawn using this sampling design. The geographic distribution of central sample pixels constituting the set of samples drawn appears in Map 5.12.

The power-transformed version of the ratio $\frac{\text{Band \#4}}{\text{Band \#3}}$ is the subject of analysis here, as it is in the preceding spatial autoregressive analysis. The 25 pooled samples render the semivariogram plot appearing in Figure 5.44. As with the High Peak image, this scatterplot is very smooth. The three candidate models suggested by the High Peak analysis yield the following parameter estimates for model fits to the scatter of points (Figure 5.44), using a maximum rescaled distance of 1.1:

model	RSSE	$\underline{C_0}$	$\underline{C_1}$	$\underline{r}$
exponential	0.014	3.94	19.6	0.218
Bessel	0.029	7.57	15.8	0.159
spherical	0.081	10.00	13.1	0.702 .

In contrast with the High Peak data, the exponential semivariogram model best describes these Adirondack data; the Bessel function and spherical models deviate

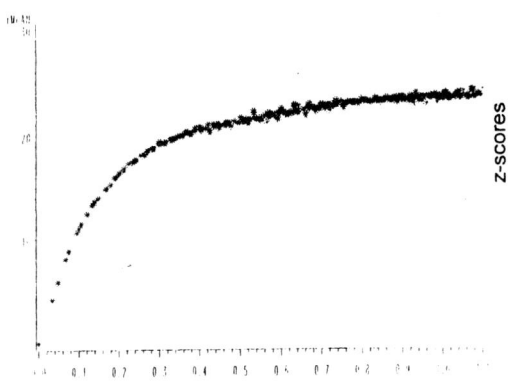

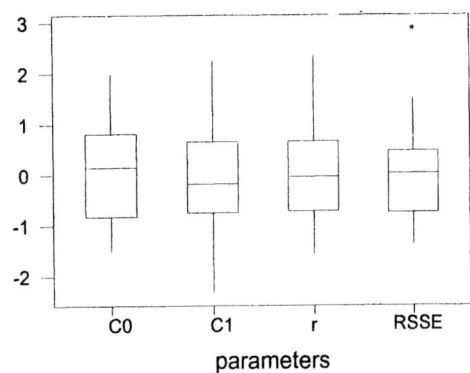

Figure 5.44 (left). Semivariogram plot of the power-transformed $\frac{Band\#4}{Band\#3}$ LANDSAT TM remotely sensed spectral ratio constructed by pooling 25 stratified random cluster samples drawn (Elberty's data).

Figure 5.45 (right). Boxplots of exponential semivariogram model parameter estimates for 25 stratified random cluster samples drawn (Elberty's data).

most near the semivariogram plot origin, furnishing a poor description of low-order spatial autocorrelation. Hence, the exponential model was fit to each of the 25 replications of the resampling exercise. Sample-by-sample estimates obtained are

C_0	C_1	r	RSSE	C_0	C_1	r	RSSE
3.03	21.0	0.203	0.084	5.54	19.2	0.224	0.050
5.71	18.1	0.242	0.057	0.17	21.1	0.166	0.043
6.06	19.4	0.266	0.038	2.38	19.5	0.190	0.026
7.17	16.9	0.349	0.068	4.14	20.9	0.164	0.073
5.07	18.3	0.262	0.028	4.07	21.4	0.264	0.048
0.74	23.5	0.154	0.041	6.24	17.7	0.233	0.059
1.69	21.4	0.209	0.023	1.88	19.6	0.188	0.078
1.05	22.0	0.168	0.119	5.95	17.8	0.269	0.047
0.31	21.7	0.140	0.049	2.00	23.4	0.205	0.011
5.22	18.1	0.260	0.027	5.51	18.6	0.258	0.030
8.41	14.2	0.355	0.049	4.77	18.0	0.232	0.054
2.35	18.9	0.205	0.029	0.65	25.5	0.186	0.011
2.84	22.2	0.263	0.013				

$\bar{x}$	=	3.72	19.94	0.226	*
s	=	2.34	2.46	0.054	*
S-W	=	0.95025	0.07658	0.93587	0.93815 .

The S-W statistics combined with boxplots (Figure 5.45) suggest that each of these four sampling distributions conforms to a normal distribution. The means of the sample parameter estimates are not very close to the pooled samples parameter estimates, a result largely due to the nonlinear estimation required. Whether or not the estimated standard errors are precise remains to be established; they at least are sensible.

Interpretation of these parameter estimates is consistent with that for the High Peak image. The pattern exhibited by the scatter of points in the semivariogram plot for the pooled sample (Figure 5.44), and the moderately high range parameter values are typical of the presence of strong positive spatial autocorrelation in these data, corroborating results from the spatial autoregressive analysis. Relatively low RSSE values for these model fits (one outlier is revealed in the boxplots) apparently are due to a lack of variation in the gamma values relative to the model curves. For these data, the exponential model has the best fit and hence is the model of choice for describing spatial autocorrelation. The inference here is that the exponential model, with essentially a 0 nugget effect, a sill of roughly 20, and a rescaled range parameter value of approximately 0.25, will best describe the power-transformed ratio $\frac{\text{Band \#4}}{\text{Band \#3}}$ for this 503-by-511 pixels Adirondack remotely sensed image.

These two satellite image explorations reveal that with the current state of computing power, software sophistication, and amount of easily accessible remotely sensed data, spatial statistics stands ready to deliver exciting new georeferenced data analyses. Besides vegetation health monitoring afforded by the index studied in this section, the insurance industry, for example, is interested in using remotely sensed data for disaster damage assessment, whereas the real estate industry is interested in using remotely sensed data for local development decision making (Tomerlin 1998).

5.8. Concluding Comments: Spatial Autocorrelation and Natural Resources Attribute Variables

By their very georeferenced nature, natural resources data almost always should benefit from being spatial statistical analyses, a contention attested to especially by the number of geostatistical analyses published in such tomes as the *Proceedings* of the first and second international symposia on the spatial accuracy of natural resources and environmental sciences databases (Congalton 1994; Mowrer et al. 1996). Findings reported in this chapter expand upon these empirical applications by initiating establishment of empirical links between the semivariogram models commonly used by natural resources scientists and the set of popular spatial autoregression specifications (§3.3) that are available. Linkages reported in this chapter include

Table 5.1. Summary of linkage findings for Chapter 5

phenomenon	areal unit	autoregressive model	semivariogram model
piezometric-head pressure	point	SAR	exponential; Bessel; spherical; Gaussian
B division soil thickness	point	SAR	spherical
basal area (#1)	hexagon	OLS	none
basal area (#2)	stand	SAR	exponential
dbh	point	AR	spherical; Gaussian
benthic species	hexagon	OLS	wave/hole & exponential; power
salinity	hexagon	AR	wave/hole & Gaussian & power
# plant species (#1)	island	OLS; filtered Poisson regression	none
# plant species (#2)	island & county	OLS	none
rainfall	point	SAR	exponential; Bessel
precipitation	point	OLS	none
snowfall	point	AR/SAR	none
runoff	point	SAR	spherical
elevation index (#1)	aggregated pixels	SAR	spherical; Gaussian
elevation index (#2)	aggregated pixels	SAR	spherical; Gaussian
remotely sensed data (#1)	pixel	SAR; filtered OLS	exponential; Bessel; spherical
remotely sensed data (#2)	pixel	SAR; filtered OLS	exponential

These articulations of relationships between geostatistical and spatial autoregressive models continue to reflect findings of Chapter 4 (§4.6), namely a frequent inability to relate simply and easily a single semivariogram model to a particular spatial autoregressive specification. The spherical and Gaussian specifications often qualify as the preferred semivariogram models. Because the spherical covariance specification is not twice differentiable with respect to the range parameter, it has a tendency to produce multimodal likelihoods, complicating likelihood-based parameter estimation techniques (Mardia and Watkins 1989). Because the Gaussian covariance specification is infinitely differentiable with respect to the range parameter, its tendency essentially is the other extreme, causing it to be too regular or smooth to be physically realistic for most georeferenced data (Stein and Handcock 1989). And, while the practice of using at least a small nugget with a Gaussian specification to avoid an ill-conditioned kriging coefficient matrix is well known (Ababou et al. 1994), obtaining non-0 nugget estimates by default for roughly 80% of the Gaussian semivariogram models fitted in this chapter is reassuring. Consequently, Zimmerman and Zimmerman (1991) note that the use of these two semivariogram models is being discouraged; but these two models are widely used in practice, for reasons like those uncovered here, with experience showing that they are useful specifications.

A modified auto-binomial specification is presented in Chapter 4. Its modified auto-Poisson specification counterpart is presented in this chapter. The spatial statistical analysis of the Galapagos data set (§5.3) supplies evidence that spatial autocorrelation can be filtered out of this type of georeferenced data by also including judiciously selected eigenvectors of the geographic weights matrix (§1.5) in the mean response function specification.

The final georeferenced data analysis of this chapter, involving a sizeable remotely sensed image, reveals that spatial autoregressive models can be estimated for enormous remotely sensed images using current computer software (e.g., SAS) and desktop PC resources, coupled with an excellent normalizing factor approximation. These massively large data sets also can be effectively analyzed with OLS procedures that incorporate eigenvectors from the geographic weights matrix. These eigenvectors, as well as their corresponding eigenvalues, can be computed analytically, avoiding the need to perform numerically intensive operations on mammoth n-by-n matrices.

6

ANALYSIS OF GEOREFERENCED AGRICULTURAL YIELD VARIABLES

Much of modern analytical statistics developed hand in hand with advances in agricultural science at the Rothamsted Experimental Station, near Harpenden, Hertfordshire, England, roughly 25 miles from London. This facility, which is the oldest agricultural research station in the world, was established in 1843 by Lawes, who had inherited it as an estate from his father in 1822. Lawes ensured the continuation of the Rothamsted Experimental Station by setting up the Lawes Agricultural Trust in 1889. In 1919, Fisher, who was hired to bring modern statistical methods to the task of analyzing data collected on the station's Broadbalk field since 1844 (and recorded since 1852), went to Rothamsted, where he obtained much of his statistical inspiration (and data). There Fisher pioneered the application of statistical procedures to the design of scientific experiments, seeking to provide more amounts of useful information with less investments of time, effort, and money, and in doing so devised the analysis of variance technique.

The principles of experimental design promoted by Fisher included replication (in space and sometimes in time), blocking (i.e., restricting the randomization of treatments to field plots to gain statistical efficiency), and randomization. He went to great lengths to remove impacts of spatial dependence from agricultural field experiments (Cressie 1991, 7), in which the union of spatially contiguous experimental plots are called a field, through the use of randomization, whose purpose in well-designed agricultural field experiments at least partly is to neutralize (but does not actually erase) spatial or temporal dependence (Grondona and Cressie 1991, 381). By making the adjacency of any pair of treatments equally likely, spatial and temporal correlations theoretically converge on 0 as the number of replications increases. Accordingly, classical experimental design of agricultural trials ignores the spatial position of treatments in the design. But, notably, randomization fails to neutralize the spatial correlation at geographic scales larger or smaller than the plot dimensions utilized in a study (Cressie 1991, 7).

One possible objective of an agricultural experiment is to determine conditions under which a crop will achieve high yields. Therefore, obtaining precise estimators of treatment effects (in crop production, treatments might be fertilizers on a particular crop variety or different varieties of the same crop) is essential, emphasizing the importance of reducing or controlling for residual variation not attributable to treatment effects. Moreover, the goal is efficient estimation of treatment effects and contrasts. Recently, there has been much interest in examining the effects of possibly correlated errors among plots in agricultural field trials, particularly with regard to obtaining more efficient treatment-contrast estimators by exploiting spatial variation contained in yield data (Cressie 1991, 249). After all, the classical statistics approach is sufficient only when factors influencing the outcome of an experiment exhibit random variability over a field with little or no spatial correlation.

To the chagrin of agricultural scientists who carefully design an experiment, a number of external factors may well give rise to the presence of spatial autocorrelation. Fisher recognized the influence of rainfall early on. Bhatti et al. (1991, 1523) stress soil type variability, slope position, and topsoil depth. Other influential factors include disease and insect damage. Consequently, data from a select number of uniformity trials—experiments in which all of the field plots are treated with a control treatment, and the data consist of plot yields for one specific crop—commonly are analyzed by statisticians to detect systematic patterns and establish cause-and-effect relationships. Federer and Schlottfildt (1954) report tobacco plant height data for an 8-by-7 grid, while Gleeson and McGilchrist (1980) report plant weights for a 7-by-7 square grid. Venables and Ripley (1994, 177) include a digital file of data from an oats and manure experiment compiled by Yates (1935) and subsequently analyzed by many authors (e.g., John 1971). Lyons (1980) furnishes data for annual yields of winter wheat in each of the years from 1970 to 1973 at 12 different sites in England, while Box et al. (1978) provide data from a randomized experiment on tomato yields; these latter two data sets are reprinted in *A Handbook of Small Data Sets* (Hand et al. 1994). Haining (1978) prints 1964 and 1969 corn and wheat yield data on a map of 45 Nebraska and Kansas counties. Andrews and Herzberg (1985) report the wheat data from the uniformity trial analyzed by Wiebe, the annual Broadbalk data from Rothamsted, and the wheat and straw data from a uniformity trial carried out at Rothamsted in 1910 and first analyzed by Mercer and Hall (1911). Spatial statistical analyses of both the Broadbalk and Wiebe data are presented in this chapter, as is a spatial statistical analysis of observational (rather than experimental) Puerto Rican sugar cane and milk production data. Haining (1978, 194-95) also analyzes these type of data, reporting tabulations for counties straddling the Kansas-Nebraska border; although Haining provides a map (pp. 190-91), he does not report county centroids. Analysis of the well-known Mercer-Hall wheat yield data—also analyzed by

Fairfield Smith (1938), Whittle (1954), Besag (1974), Ripley (1981), McBratney and Webster (1981), Wilkinson et al. (1983), Kuensch (1985), and Cressie (1991)—serves as a benchmark in this chapter.

Cressie (1991, 453-57), who reprints the Mercer-Hall wheat yield data collected on a 20-by-25 lattice of field plots, presents a geostatistical analysis of these data. These yield measures are considered heterogeneous, displaying a conspicuous trend across the field (Cressie 1991, 456). When analyzing these data, Whittle (1954) overlooks the conspicuous trend, acknowledges the presence of anisotropy (§3.2.4), and employs an SAR model based upon matrix C coupled with a Jacobian approximation (§3.1.2). In this context, his spatial autocorrelation parameter estimation results are $\hat{\rho}_1 = 0.213$ and $\hat{\rho}_2 = 0.102$. Reanalyzing these data, but including a linear geographic trend term and employing a CAR model based upon matrix C coupled with Whittle's Jacobian approximation, Besag (1974) reports spatial autocorrelation parameter estimates of $\hat{\rho}_1 = 0.368$ and $\hat{\rho}_2 = 0.107$. By utilizing a more complicated detrending procedure, Kuensch (1985) also finds that the spatial CAR model furnishes a good description of these data. Meanwhile, Cressie (1991, 248-59) presents a geostatistical analysis of these data, identifying the exponential model as the appropriate descriptor of their semivariogram.

None of these earlier analyses report subjecting the Mercer-Hall wheat yield data to a power transformation. These data do not appear to benefit from the application of a power transformation, supporting that decision, which is not surprising since $\frac{y_{max}}{y_{min}} < 3$:

```
raw                                    power transformation
                                         δ̂   = -0.27290
                                         γ̂   =  0.94821
                                         RSSE = 1.7×10⁻³
S-W      = 0.98041†                    S-W     = 0.98112
Bartlett = 4.888                       Bartlett = 4.742
Levene   = 1.430                       Levene   = 1.389
MC = 0.40553 (σ̂_MC ≈ 0.03236)
GR = 0.58819 ,
```

where S-W is the Shapiro-Wilk statistic, MC is the Moran Coefficient, and GR is the Geary Ratio. More specifically, the optimal exponent for a power transformation is very close to unity (implying a power transformation is unnecessary), and improvement in conformity to a normal frequency distribution is slight (although S-W goes from being significant to not being significant[1]), as is improvement in variance homogeneity.

Estimation results based upon the CAR code presented in §3.1.1, using an exact Jacobian term—which is far more difficult to specify in its reduced eigen-

---

[1] Statistical significance is denoted throughout by † for a 10%, †† for a 5%, and ††† for a 1% level.

value form when two spatial dependency parameters are involved—for these georeferenced data include

	CAR(C) parameter estimates isotropic	anisotropic
$\hat{\rho}$	0.23853	*
	($\rightarrow$ 0.23853/0.25233 = 0.94531)	
$\hat{\rho}_{EW}$	*	0.11617 (ρ_{max} = 0.50367)
$\hat{\rho}_{NS}$	*	0.36092 (ρ_{max} = 0.50564)
$\hat{\mu}$	$-0.02541 \times 0.21002 + 3.94864$	$-0.02905 \times 0.21002 + 3.94864$
$\hat{\sigma}^2$	0.62916×0.21002	0.57687×0.21002 .

The isotropic estimate $\hat{\rho}$ roughly equals the weighted average of the two anisotropic estimates, $\hat{\rho}_{EW}$ measuring horizontal axis spatial dependence and $\hat{\rho}_{NS}$ measuring vertical axis spatial dependence. Bivariate linear regression results include $b_{u-\bar{u}} = -0.01904$, $R^2 = 0.090$, S-W = 0.97998†, and MC = 0.34074 (GR = 0.64687). Now estimation results based upon the CAR code presented in §3.1.1 include

	CAR(C) parameter estimates isotropic	anisotropic
$\hat{\rho}$	0.22583	*
	($\rightarrow$ 0.22583/0.25233 = 0.89498)	
$\hat{\rho}_{EW}$	*	0.08989
$\hat{\rho}_{NS}$	*	0.36273
$\hat{\mu}$	$-0.01631 \times 0.19113 + 0$	$-0.01956 \times 0.19113 + 0$
$\hat{\sigma}^2$	0.70461×0.19113	0.63340×0.19113 .

These results compare favorably with those reported by Besag (1974), with 0.08989 being roughly equal to 0.107 and 0.36273 being roughly equal to 0.368. Again, the isotropic estimate $\hat{\rho}$ roughly equals the weighted average of the two anisotropic estimates, $\hat{\rho}_{EW}$ and $\hat{\rho}_{NS}$. In addition, the linear gradient across the geographic landscape accounts for only a modest amount of variability in wheat yield and very little of the detected latent spatial dependency.

A geostatistical analysis of these grain yield data, assuming isotropy, renders the following goodness-of-fit measures, using disaggregated distances up to the maximum rescaled[2] distance of 1.0:

model	RSSE: raw data	RSSE: OLS residuals
exponential	0.276	0.345
Bessel	0.314	0.376
Gaussian	0.366	0.446
spherical	0.338	0.436

---

[2] Distances are rescaled by dividing all (U, V) coordinate values by the maximum absolute U or V coordinate values in the set of georeferencings attached to the sample of areal units.

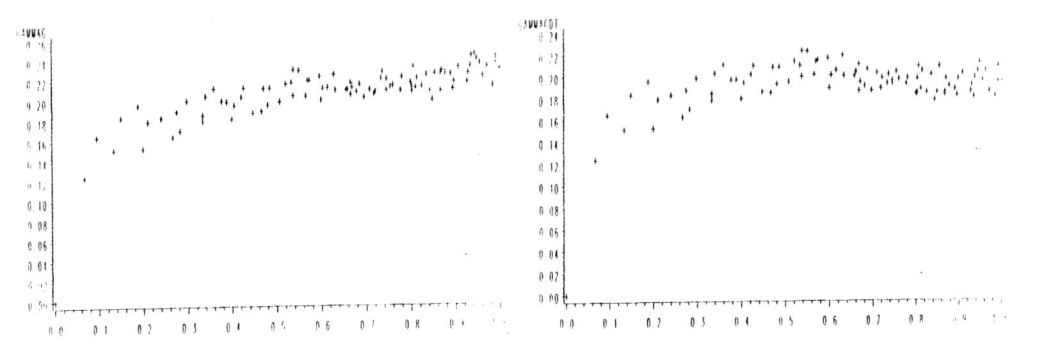

Figure 6.1. Left: (a) sotropic semivariogram plot of the Mercer-Hall raw grain yield agricultural field plot data. Right: (b) Isotropic semivariogram plot of OLS residuals for the Mercer-Hall grain yield agricultural field plot data.

```
wave/hole                 0.421                0.506
power                     0.205                0.493 .
```

where OLS is the acronym for ordinary least squares. The scatter of points in the respective aggregate semivariogram plots (Figure 6.1a-b) reveals a reasonable amount of variability at neighboring distances, and it is consistent with an exponential semivariogram model specification, which overall furnishes the best description of these two sets of data. Disaggregated semivariogram plots (Figures 6.2a-d) suggest quite a different spatial dependency in the two orthogonal coordinate axes directions in the agricultural field, a finding that certainly is consistent with the foregoing CAR results. A two-dimensional isopleth description of the semivariogram (Figure 6.3) visually bolsters this interpretation and further suggests that the arbitrary coordinate system superimposed on the Mercer-Hall agricultural field may not be coincident with the anisotropy axes; the angle of rotation, denoted here by θ, required for alignment appears to be modest in magnitude. It is helpful to recall that for geometric anisotropy (see §3.2.4), this rotation is similar to an axis rotation in conventional multivariate statistical analysis. The angle of rotation will allow the major and minor anisotropic ranges of the two-dimensional semivariogram to be jointly maximized.

The parameter estimates from selected exponential model specification fits to the scatter of points in the anisotropic semivariogram plot are

```
model                RSSE    C₀       C₁       rᵤ       rᵥ       θ
isotropic            0.469  0.10771  0.12856  0.30343  0.30343   *
horizontal axis (U)  0.461  0.14759  0.08700  0.28785     *      *
vertical axis (V)    0.999  0.19342  0.02852     *     0.41842   *
two-dimensional      0.455  0.09974  0.12816  0.18168  0.39235   *
rotated two-
  dimensional        0.373  0.09443  0.13249  0.15848  0.42831
                                                              -16.245° .
```

These results reveal that the underlying anisotropy may be somewhat zonal

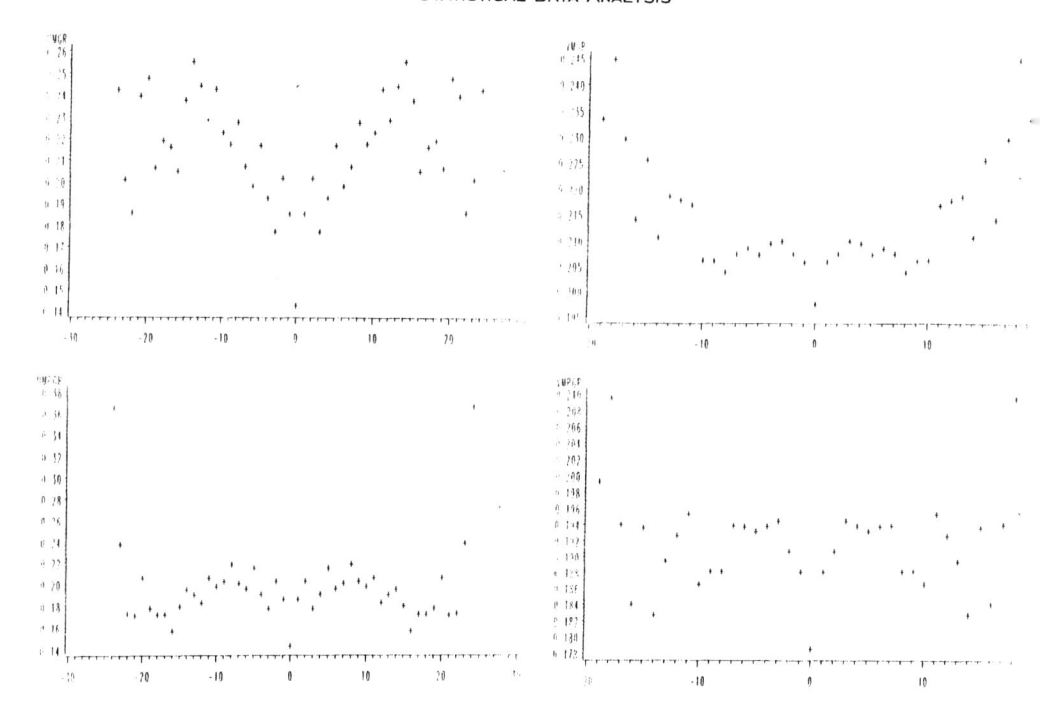

Figure 6.2. Top left: (a) Anisotropic semivariogram plot of the Mercer-Hall raw grain yield agricultural field plot data along the horizontal georeferenced coordinates axis. Top right: (b) Anisotropic semivariogram plot of the Mercer-Hall raw grain yield agricultural field plot data along the vertical georeferenced coordinates axis. Bottom left: (c) Anisotropic semivariogram plot of OLS residuals for the Mercer-Hall grain yield agricultural field plot data along the horizontal georeferenced coordinates axis. Bottom right: (d) Anisotropic semivariogram plot of OLS residuals for the Mercer-Hall grain yield agricultural field plot data along the vertical georeferenced coordinates axis.

(differences in the directional sills) and is markedly geometric (differences in the directional range parameters). The maximum horizontal and vertical lags used here for parameter estimation purposes were, respectively, 15 and 7 (see Figures 6.2a-b). While the angle of rotation for the anisotropic axes is modest, employing it further accents the difference in the directional spatial dependencies. In addition, the anisotropic model supplies a substantially better description of these data.

Because part of this anisotropy might be attributable to the linear horizontal geographic trend detected in these grain yield data, a geostatistical analysis was conducted on the residuals from the OLS regression model. Results from this analysis more strongly signify that the exponential semivariogram model specification is the appropriate one to describe these data; this analysis produces the following parameter estimates:

model	RSSE	c_0	c_1	r_U	r_V	θ
isotropic	0.558	0.03241	0.16938	0.09414	0.09414	*
horizontal axis (U)	0.582	0.12791	0.07470	0.07056	*	*
vertical axis (V)	0.959	0.17890	0.02019	*	0.12432	*

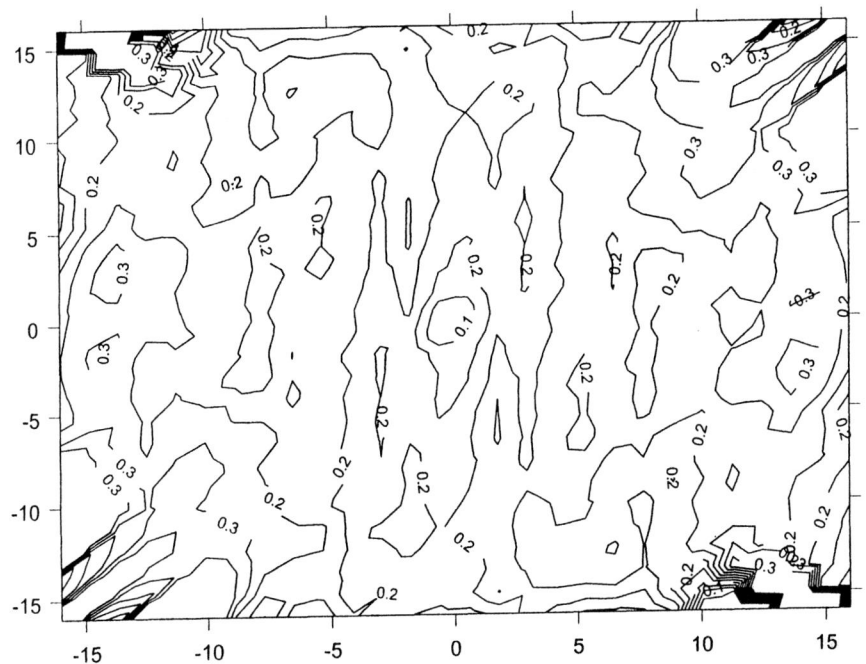

Figure 6.3. Isosemivariance contours for a 15-by-15 spatial lag grid (Mercer-Hall raw grain yield agricultural field plot data).

```
two-dimensional          0.455 0.03108 0.17052 0.04787 0.20698    *
rotated two-
     dimensional         0.433 0.04009 0.16314 0.05834 0.22794
                                                             -8.204° .
```

These results are consistent with the preceding ones for the raw grain yield data and for the foregoing CAR findings. The maximum horizontal and vertical lags used here for parameter estimation purposes were, respectively, 11 and 7 (see Figures 6.2c-d). Adjusting for the linear geographic trend reduces the detected level of spatial autocorrelation (range parameters decrease and RSSE values increase). The same nature and degree of anisotropy remains; notably the angle of rotation is considerably smaller. Therefore, the anisotropic exponential semivariogram specification is the preferred model for describing these data.

6.1. THE MERCER-HALL STRAW YIELD DATA

Many of the field plot data sets include both yield and straw harvest measures. Exploration of these straw data can be of interest to researchers who are looking for some sort of replications of the yield results. In this instance, the correlation between these two variables is 0.730, with a conspicuous trend displayed by the accompanying scatterplot (Figure 6.4).

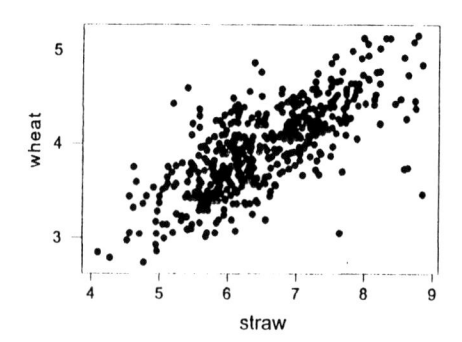

Figure 6.4. Scatterplot of the wheat yield versus straw yield by agricultural field plot for the Mercer-Hall data.

As with the corresponding wheat harvest measures, these straw data have $\frac{y_{max}}{y_{min}}$ < 3 and do not appear to benefit from a power transformation:

```
raw                              power transformation
                                       δ̂ = -3.09516
                                       γ̂ =  0.64642
                                       RSSE = 5.3×10⁻³
S-W = 0.97314†††                 S-W = 0.98643
Bartlett = 5.627                 Bartlett = 9.235††
Levene = 4.380††                 Levene = 4.432†††
MC = 0.52059
    (δ̂_MC ≈ 0.03236)
GR = 0.47900 .
```

More specifically, improvement in conformity to a normal frequency distribution is modest (although S-W goes from being marked to not being significant), while geographic variance homogeneity deteriorates. Therefore, a power transformation should not be applied to these data.

Bivariate linear regression results using the georeferenced coordinates include $b_{u-\bar{u}} = -0.05914$, $b_{v-\bar{v}} = 0.03360$, $R^2 = 0.272$, S-W = 0.98549, and MC = 0.32774 (GR = 0.65895). Not only does this linear gradient across the field account for a sizeable amount of geographic variability in straw yield, but it also accounts for a large decrease in the MC value; in order words, positive spatial autocorrelation detected in these straw data is partly attributable to simple map pattern. Estimation results based upon the CAR code presented in §3.1.1 include

```
CAR(C) parameter estimates based on regression residuals
                isotropy                  anisotropic
ρ̂               0.22253                       *
ρ̂_EW              *                         0.07555
ρ̂_NS              *                         0.37120
μ̂           -0.02384×0.58712 + 0      -0.03095×0.58712 + 0
σ̂²           0.71986×0.58712           0.63326×0.58712 .
```

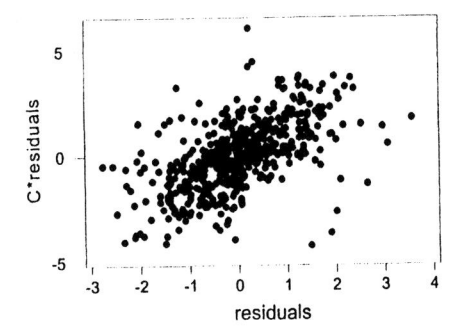

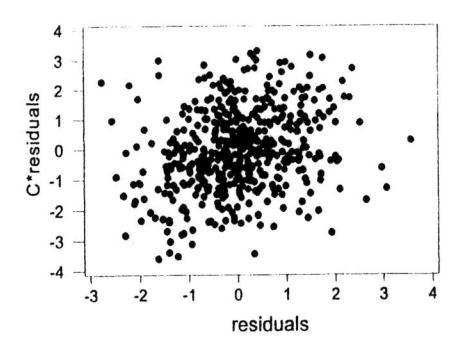

Figure 6.5. Top-left: (a) Isotropic MC scatterplot of OLS residuals for the Mercer-Hall straw yield agricultural field plot data. Top right: (b) Anisotropic MC scatterplot of OLS residuals for the Mercer-Hall straw yield agricultural field plot data along the horizontal georeferenced coordinates axis. Bottom right: (c) Anisotropic MC scatterplot of OLS residuals for the Mercer-Hall straw yield agricultural field plot data along the vertical georeferenced coordinates axis.

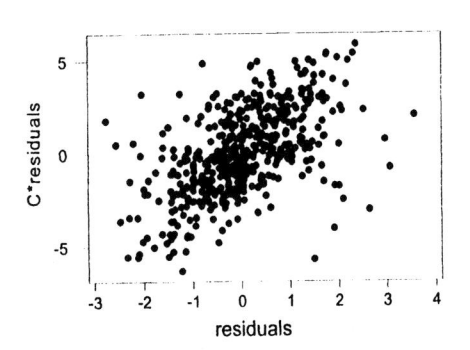

Once more, the isotropic estimate $\hat{\rho}$ roughly equals the weighted average of the two anisotropic estimates, $\hat{\rho}_{EW}$ and $\hat{\rho}_{NS}$. The anisotropy contained in straw yield (Figure 6.5a-c) exhibits approximately the same nature, degree, and directionality as that for wheat yield.

A geostatistical analysis of these straw yield data, assuming isotropy, renders the following goodness-of-fit measures, using disaggregated distances up to the maximum rescaled distance of 1.2:

model	RSSE: raw data	RSSE: OLS residuals
exponential	0.099	0.469
Bessel	0.105	0.483
Gaussian	0.110	0.520
spherical	0.099	0.519
wave/hole	0.110	0.650
power	0.097	0.686 .

As with the grain yield case, the scatter of points in the respective aggregate semivariogram plots (Figure 6.6a-b) reveals a reasonable amount of variability at neighboring distances. The plot based upon the raw data exhibit a pronounced drift, which is not apparent in the plot based upon the OLS residuals; this drift is attributable to the marked linear trend across the agricultural field that accounts for roughly one-fourth of the variance in straw yield. The OLS residual plot is consistent with an exponential semivariogram model specification, which overall

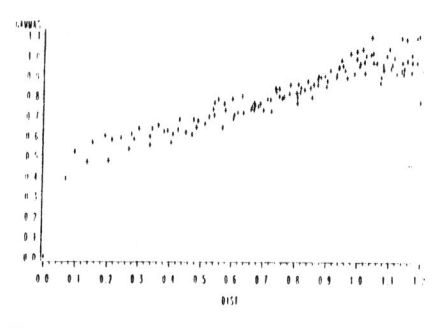

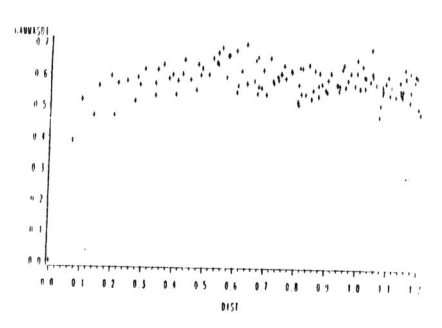

Figure 6.6. Left: (a)Isotropic semivariogram plot of the Mercer-Hall raw straw yield agricultural field plot data. Right: (b) Isotropic semivariogram plot of OLS residuals for the Mercer-Hall straw yield agricultural field plot data.

furnishes the best description of these two sets of data. Disaggregated semivariogram plots (Figures 6.7a-d) suggest quite a different spatial dependency in the two orthogonal directions in the agricultural field, a finding that, as with the grain yield case, is consistent with the foregoing CAR results. A two-dimensional isopleth description of the semivariogram (Figure 6.8) visually bolsters this interpretation and further suggests that the arbitrary coordinate system superimposed on the Mercer-Hall agricultural field may not be coincident with the anisotropy axes—the angle of rotation, θ, required for alignment appears to be rather pronounced.

The parameter estimates from selected exponential model specification fits to the scatter of points in the anisotropic semivariogram plot are

```
model              RSSE    C₀       C₁       rᵤ       rᵥ        θ
isotropic          0.242  0.35942  2.30749   3‡       3‡        *
horizontal axis (U) 0.240  0.46364  1.99855   3‡       *         *
vertical axis (V)   0.985  0.69688  0.38612   *       2.72654    *
two-dimensional    0.220  0.38034  1.71214  2.10347   3‡        *
rotated two-
  dimensional      0.169  0.35890  1.88025  2.10504   3‡
                                                             -31.484°.
```

‡upper limit suggested by Heuvelink's WLSFIT program that is encountered as a constraint on this parameter estimate.

These results reveal that the underlying anisotropy may be somewhat zonal and is markedly geometric; the drift detectable in Figure 6.6a, combined with a repeated encountering of the upper-bound range parameter value during estimation, signifies that these data are nonstationary. The maximum horizontal and vertical lags used here for parameter estimation purposes were, respectively, 15 and 7 (see Figures 6.7a-b). As in the case of the grain yield, the anisotropic model supplies a substantially better description of these data.

Because part of this anisotropy might be attributable to the conspicuous linear geographic trend across the field detected in these straw yield data, a

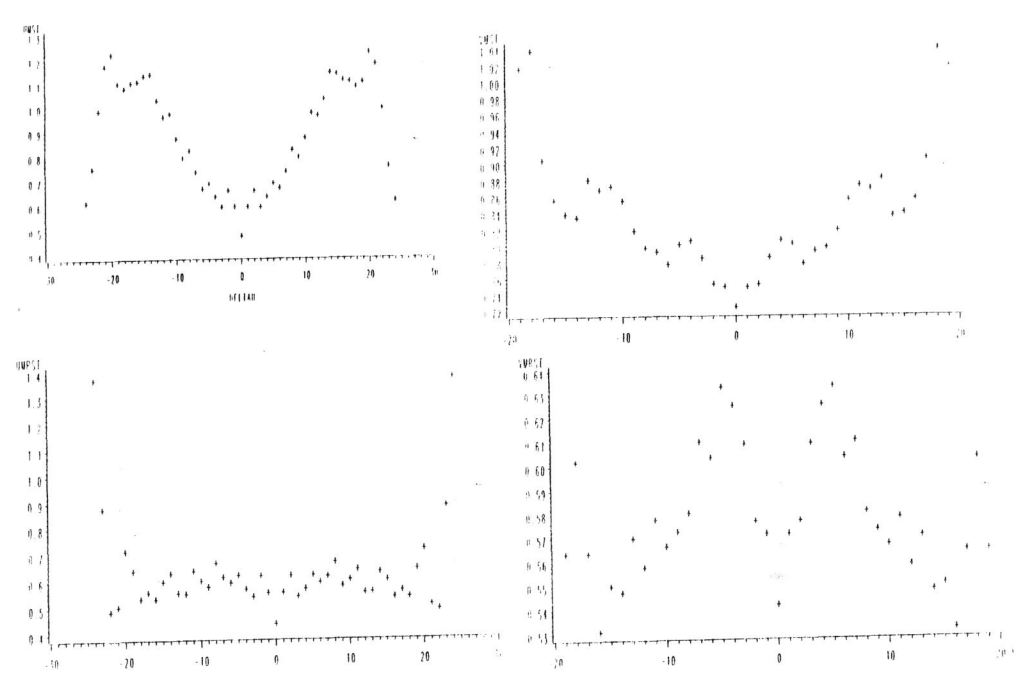

Figure 6.7. Top left: (a)Anisotropic semivariogram plot of the Mercer-Hall raw straw yield agricultural field plot data along the horizontal georeferenced coordinates axis. Top right: (b) Anisotropic semivariogram plot of the Mercer-Hall raw straw yield agricultural field plot data along the vertical georeferenced coordinates axis. Bottom left: (c) Anisotropic semivariogram plot of OLS residuals for the Mercer-Hall straw yield agricultural field plot data along the horizontal georeferenced coordinates axis. Bottom right: (d) Anisotropic semivariogram plot of OLS residuals for the Mercer-Hall straw yield agricultural field plot data along the vertical georeferenced coordinates axis.

geostatistical analysis is conducted on the residuals from the OLS regression model. Results from this analysis strongly signify that the exponential semivariogram model specification is most appropriate to describe these data (decisively removing both the spherical and power specifications as candidate models); this analysis produces the following parameter estimates:

model	RSSE	C_0	C_1	r_U	r_V	θ
isotropic	0.569	0.07159	0.54567	0.08255	0.08255	*
horizontal axis (U)	0.618	0.39346	0.22356	0.05034	*	*
vertical axis (V)	0.907	0.53352	0.09213	*	0.14829	*
two-dimensional	0.467	0.06633	0.55000	0.03867	0.18341	*
rotated two-dimensional	0.455	0.06628	0.55137	0.04041	0.19343	-5.157°

These results are consistent with the preceding ones for the raw straw yield data, and for the foregoing CAR findings. The maximum horizontal and vertical lags used here for parameter estimation purposes were, respectively, 11 and 7 (see

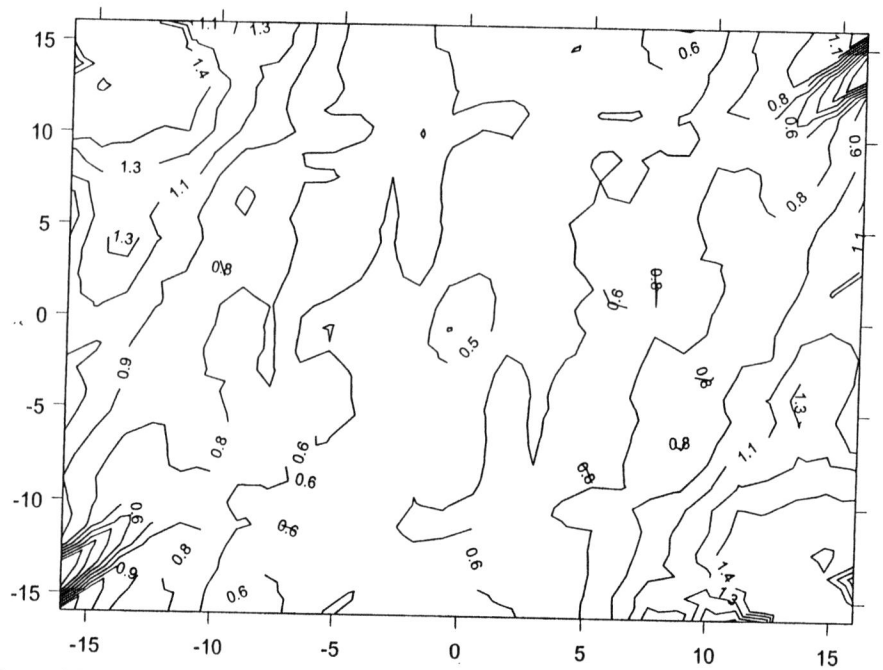

Figure 6.8. Isosemivariance contours for a 15-by-15 spatial lag grid (Mercer-Hall raw straw yield agricultural field plot data).

Figures 6.7c-d). Adjusting for the linear geographic trend not only reduces the detected level of spatial autocorrelation (range parameters decrease and RSSE values increase) but also makes the parameter estimation exercise easier (e.g., the semivariogram scatter of points is better behaved—no excessively large range parameters are obtained). The OLS results suggest that the geometric anisotropy is very similar to that detected for the corresponding grain yield data. Therefore, the anisotropic exponential semivariogram specification is the preferred model for describing these data, too.

6.2. THE WIEBE WHEAT YIELD DATA

A second readily available wheat yield data set (Andrews and Herzberg 1985) was compiled from a 1927 agricultural uniformity trial by Wiebe and comprises 1,500 nursery plots arranged in a 125-by-12 lattice. This field was located in the Aberdeen Substation, Idaho, where the average rainfall was roughly seven inches. The average yield per acre for the Federation wheat grown there in 1927 was 63 bushels, which was above average for that region at that time. Wiebe (1935) also records the quantity of straw harvested and the height of plants, and furnishes a contour map of the wheat yields.

In contrast with the Mercer-Hall wheat yield data, these Wiebe wheat yield data appear to benefit from application of a power transformation, even though

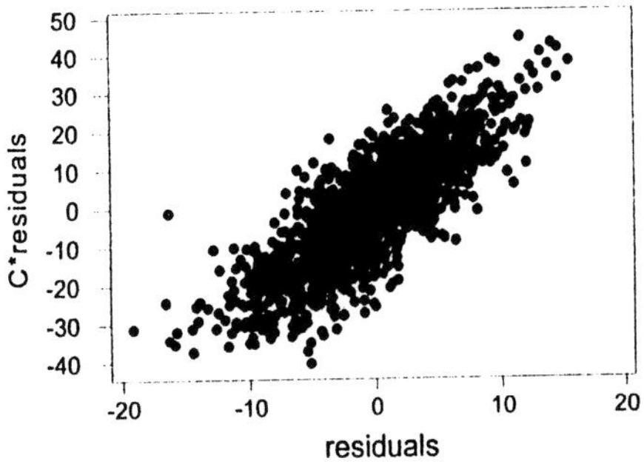

AGRICULTURAL YIELD 317

Figure 6.9. Isotropic MC scatterplot of the Wiebe raw wheat yield agricultural field plot data.

$\frac{y_{max}}{y_{min}} \approx 2.7 < 3$:

<u>raw</u>

S-W = 0.96610†††
Bartlett = 88.423†††
Levene = 21.409†††
MC = 0.71010
 ($\hat{\sigma}_{MC} \approx$ 0.01869)
GR = 0.30343

<u>power transformation</u>
 $\hat{\delta}$ = -367.59598
 $\hat{\gamma}$ =   0.57006
 RSSE = 1.6×10^{-3}
S-W = 0.98689
Bartlett = 22.997†††
Levene = 6.218†††
MC = 0.70113
GR = 0.31310

What is essentially a square-root transformation modestly improves conformity with a normal frequency distribution (increasing S-W so that it no longer is a significant statistic) while dramatically decreasing variance homogeneity (which remains significant). In addition, these data display a much higher level of positive spatial autocorrelation than do the Mercer-Hall data, suggesting the need to employ a second-order, rather than a first-order, spatial autoregressive model here (Figure 6.9). Because spatial autocorrelation is expected to be at least partly attributable to missing variables, the SAR model specification is selected for estimation purposes.

An OLS linear trend surface model furnishes a poor description of these data, whereas an SAR model that includes the same linear trend surface terms furnishes a reasonably good description:

	parameter estimates (for z_y-scores)			
	OLS	(se)	SAR(W)	(se)
ρ	*		0.83935	(0.02233)
b_0	0.0	(0.02310)	0.01284	(0.08675)
b_u-$\bar{u}$	0.00688	(0.00064)	0.00649	(0.00240)
b_v-$\bar{v}$	-0.10790	(0.00669)	-0.06151	(0.02259)
(pseudo-) R^2	0.200		0.732	

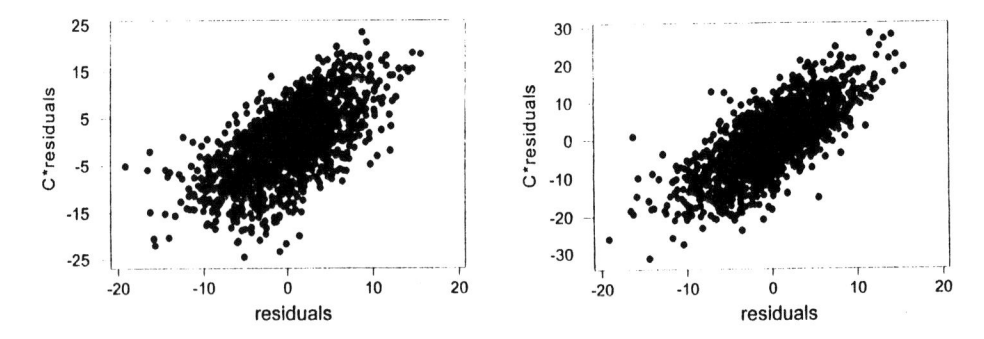

Figure 6.10. Left: (a) Anisotropic MC scatterplot of OLS residuals for the Wiebe wheat yield agricultural field plot data along the horizontal georeferenced coordinates axis. Right: (b) Anisotropic MC scatterplot of OLS residuals for the Wiebe wheat yield agricultural field plot data along the vertical georeferenced coordinates axis.

residuals		
S-W	0.98693	0.98835
A-D	0.671†	1.378†††
R-J	0.9989†	0.9971†††
K-S	0.027†††	0.034†††
Bartlett	11.878†††	2.426
Levene	1.670	0.909
MC	0.59958	0.03599
GR	0.39252	0.96689

These results reveal that simple map pattern accounts for a substantial portion of the positive spatial autocorrelation detected in these wheat yield data, although a very sizeable amount of positive spatial autocorrelation is attributable to other sources. This spatial autocorrelation appears to contribute substantially to the variance heterogeneity across the four planar quadrants of the geographic landscape. The SAR model residuals are virtually void of spatial autocorrelation, exhibit little heterogeneity of variance across the geographic landscape, and more or less conform to a normal frequency distribution, although some conspicuous deviations seem to occur in the tails of their distribution. These results differ substantially from those for the Mercer-Hall data.

The directional MC scatterplots (Figure 6.10a-b) suggest that a two-parameter SAR model is not needed here, another prominent difference between the Mercer-Hall and Wiebe wheat yield data. In addition, a 45° rotation of the field plot surface partitioning (which is equivalent to identifying neighbors with the "bishop's" adjacency definition), reduces $\hat{\rho}$ to roughly 0.47676 but without truly changing the other parameter estimates ($b_0 = 0.01203$, $b_{u-\bar{u}} = 0.00644$, $b_{v-\bar{v}} = -0.07797$, and S-W = 0.98816). In fact, for an isotropic surface, the theoretical relationship between these two estimates of ρ is $\rho_{bishop} = \dfrac{\rho_{rook}^2}{\sqrt{2}}$, which yields a value of 0.49816 here; again, there is no evidence to indicate that isotropy does not prevail with these Wiebe data.

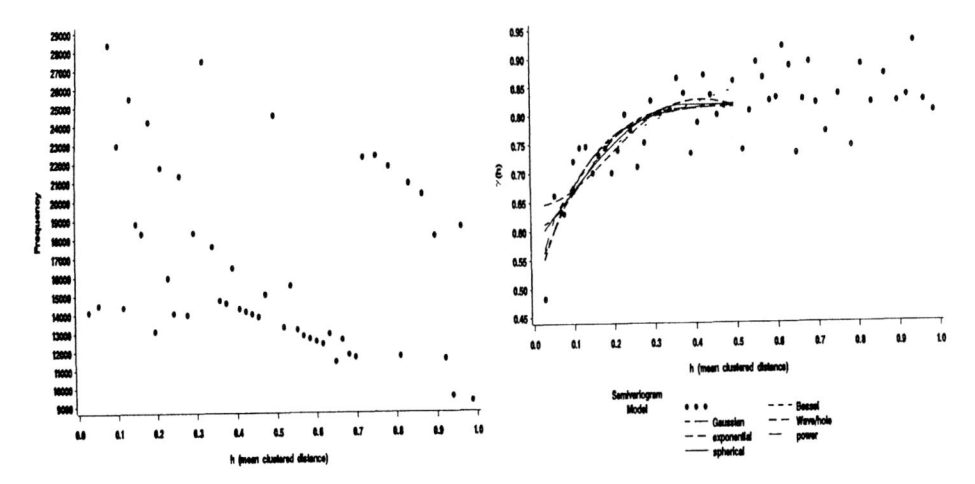

Figure 6.11. Left: (a) Semivariogram-type plot of the frequency of semivariance values in each distance class for the Wiebe wheat yield data. Right: (b) Semivariogram plot of OLS residuals for the Wiebe wheat yield data, with selected semivariogram model curves superimposed.

Geostatistical analysis of the residuals from the OLS regression model is consistent with the results from the spatial autoregressive analyses. Since the data set is relatively large ($1,500^2$ distance pairs), the program SEMIBVAR.SAS was used to compute distances and γ values, rather than using SEMI-VAR.SAS (§1.7.1), and two modifications of default values were made in the construction of the semivariogram plot. The first was that all distances greater than the rescaled distance of 1.0 were truncated; this truncation value being specified using the macro variable TRUNCDIS (§1.7.1). Secondly, the amount of rounding error specified in the distance rounding before clustering was changed from the usual 0.001 (§1.7.1) to 0.01, specified using the macro variable ROUNDTOL (§1.7.1). These two deviations from the default values used for almost all of the other data sets analyzed in this book were necessary in order to significantly reduce the amount of disk space and time required for computing the semivariogram values. Specification of 50 distance groups for wheat yield provides a minimum of 13,108 distance pairs per class up to the maximum rescaled distance of 0.5 (Figure 6.11a). The parameter estimates from selected model fits to the scatter of points in the semivariogram plot (Figure 6.11b) are

model	RSSE	C_0	C_1	r
exponential	0.24993	0.48216	0.34326	0.11928
Bessel	0.26977	0.53545	0.28241	0.08096
Gaussian	0.32738	0.60723	0.20682	0.17412
spherical	0.29622	0.57629	0.24257	0.38088
wave/hole	0.35974	0.64386	0.15079	0.09230
power[‡]	0.23687	0.00000	0.93003	* $\lambda = 0.13629$.

[‡] Initial parameter estimate values for the power semivariogram model were specified explicitly as $C_0 = 0.0001$, $C_1 = 0.8$, and $r = 0.5$.

The scatter of points in the semivariogram plot (Figure 6.11b) conforms to a pattern portraying moderate to high positive spatial autocorrelation, where several intermediate gamma values exist between the first distance class and the range. The selected model fits to the semivariogram plot also underscore this assertion with their relatively low RSSEs and values of the range parameter. RSSE and C_0 values are somewhat higher than expected presumably from the variation in the scatter of points and the lack of a 0 distance class. Based strictly on RSSE values, the power model achieves the best fit, although among the transition models, the exponential model has the better fit, followed by the Bessel semivariogram model.

Two effects of changing the ROUNDTOL macro variable value and using the TRUNCDIS macro variable are seen Figure 6.11b. Changing the ROUNDTOL value from the default of 0.001 to 0.01 results in the loss of the 0 distance class along with its accompanying $\gamma(0)$ value of 0. $\gamma(0)$ values are included with non-zero γ values resulting in a higher than expected mean γ value (0.48) for the first distance class. Dividing the same set of data used to construct Figure 6.11b into 100 distance classes will produce $\gamma(0) = 0$ but there is little gained in this increase of distance classes as the RSSE's increase substantially and C_0 values change very little, as seen in the following semivariogram model fits:

```
model         RSSE        C₀          C₁          r
exponential   0.42169     0.42156     0.39073     0.09073
Bessel        0.45023     0.47666     0.32394     0.05929
Gaussian      0.50452     0.59346     0.21520     0.15713
spherical     0.47509     0.57160     0.24672     0.37325
wave/hole     0.53180     0.64223     0.15161     0.09118
power         0.37363     0.00000     0.93312     *          λ = 0.13735 .
```

Finally, truncating rescaled distances greater than 1.0 reduces the overall amount of data used to construct the semivariogram plot but this is trivial for this data set as the $\gamma(h)$ values show no overall pattern of increase or decrease after a rescaled distance of approximately 0.4 (Figure 6.11b). Again, most of the spatial autocorrelation depicted by a semivariogram plot is contained within the first few distance classes (§1.7.1) and there is no loss of information resulting from the truncation of rescaled distance classes greater than 1.0.

6.3. THE BROADBALK WHEAT AND STRAW YIELD DATA

Annual wheat yields and straw harvests for the 17 linearly arranged field plots constituting the Broadbalk fields at Rothamsted are available for the period 1852-1925 (Andrews and Herzberg 1985). During 1843-51, many plot treatments varied from year to year; after 1925 a portion of each plot lay fallow each year. The Broadbalk data are of scientific interest for a number of reasons. First, because these data constitute a space-time series, if the experiments conducted by Fisher indeed neutralized spatial autocorrelation, this result should become apparent through analysis of these data. Second, two different questions can be explored here: whether or not the quantity of straw harvested from a plot in time t can help predict the quantity to be harvested in time t+1, and whether or not quantities harvested from a given plot can help predict those quantities to be harvested in adjacent plots. Third, because both wheat yield and straw quantities have been recorded, these data can be explored for consistency of findings. Because each geographic series has such a small sample size (i.e., n = 17), the space-time series has been pooled here to address the need to employ a power transformation.

6.3.1. SPATIAL AUTOREGRESSIVE ANALYSIS OF BROADBALK WHEAT YIELDS

Sample size alone suggests that these data may well benefit from application of a power transformation; the estimation results indicate that for the entire 17-by-74 data set only modest gains are achieved by applying a power transformation:

raw	power transformation
	$\hat{\delta}$ = -0.17
	$\hat{\gamma}$ = 0.75681
	RSSE = 4.8×10^{-3}
S-W = 0.97246	S-W = 0.97867
$\hat{\delta}_{MC}$ ≈ 0.25 .	

Meanwhile, selected (ten-year intervals) annual results based upon this power transformation are presented and described in the ensuing paragraphs.

The first point in time for which a previous set of data are available is 1853. Results for this year are as follows:

	1853 parameter estimates (CAR from z-scores)		
	raw	residuals	CAR(C)
ρ	*	*	0.19078
a	1.00008	-0.00210	0.00944×0.09264
			+ 1.00008
b	*	0.77681	0.87427×0.09264
R²	*	0.290	
S-W	0.89123††	0.95048	
A-D	0.785††	0.434	
R-J	0.9496††	0.9725	
K-S	0.186	0.161	
Bartlett	0.514	0.775	
Levene	2.088	0.005	
MC	0.19780	-0.11104, E(MC) ≈ -0.07985	
GR	0.80914	1.09958 .	

These results suggest that the geographic distribution of wheat yield in 1853 is modestly predictable from the geographic distribution of wheat yield in 1852 (Figure 6.12) and that this regression generates residuals that are well behaved and essentially void of spatial autocorrelation—although variance may not be constant

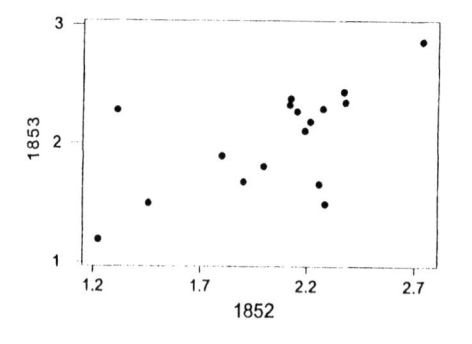

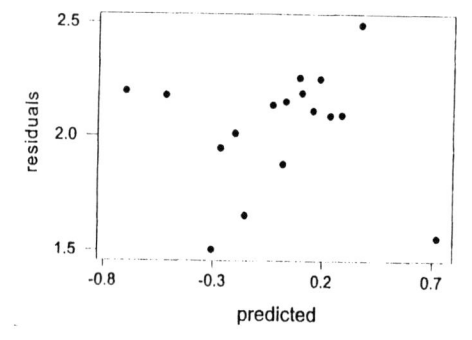

Figure 6.12 (left). Scatterplot of the power-transformed 1853 versus 1852 wheat yield by field plot (Broadbalk data).

Figure 6.13 (right). Traditional homogeneity of variance OLS residual plot for the power-transformed 1853 wheat yields regressed on the 1852 yields (Broadbalk data).

across the range of mean response levels (Figure 6.13). Considering only the 1853 geographic distribution, there is a very weak tendency for similar yields of wheat to cluster together across the linear arrangement of Broadbalk field plots (Figure 6.14a).

Both spatial and temporal trends are stronger by 1863, with the average yield having increased substantially when compared with the 1853 results:

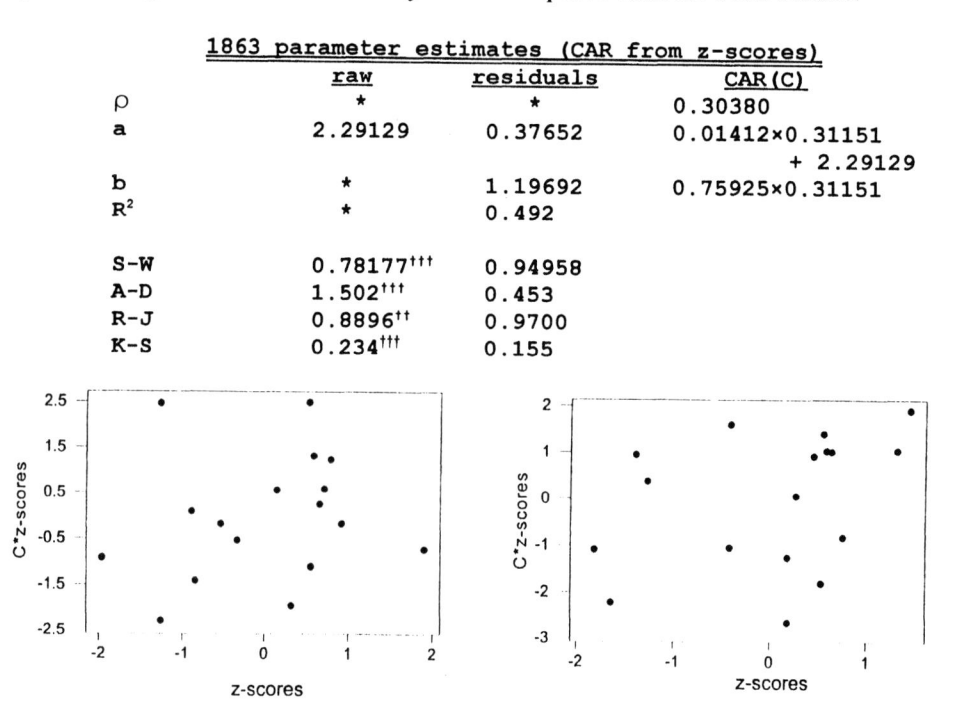

```
        1863 parameter estimates (CAR from z-scores)
                  raw          residuals        CAR(C)
ρ                  *                *           0.30380
a               2.29129          0.37652        0.01412×0.31151
                                                    + 2.29129
b                  *             1.19692        0.75925×0.31151
R²                 *             0.492

S-W             0.78177†††       0.94958
A-D             1.502†††         0.453
R-J             0.8896††         0.9700
K-S             0.234†††         0.155
```

Figure 6.14. Left: (a) MC scatterplot of the 1853 wheat yield (Broadbalk data). Right: (b) MC scatterplot of the 1863 wheat yield (Broadbalk data).

Bartlett	0.714	1.972
Levene	0.998	0.059
MC	0.33786	0.08230
GR	0.67964	0.83876 .

The geographic distribution of wheat yield in 1863 is predictable from that in 1862; this regression generates residuals that are well behaved and contain little spatial autocorrelation. Considering only the 1863 geographic distribution, there is a weak to moderate tendency for similar yields of wheat to cluster together across the linear arrangement of Broadbalk field plots (Figure 6.14b).

In 1873, spatial and temporal trends return to their 1852-53 levels, although the geographic distribution of wheat yield in 1872 accounts for roughly 10% more of the 1873 geographic variability than it does in 1862-63:

1873 parameter estimates (CAR from z-scores)

	raw	residuals	CAR(C)
ρ	*	*	0.23663
a	1.15723	0.34533	0.02360×0.09320 + 1.15723
b	*	0.57225	0.83625×0.09320
R^2	*	0.585	
S-W	0.94533	0.91470	
A-D	0.351	0.643†	
R-J	0.9790	0.9482†	
K-S	0.141	0.190†	
Bartlett	0.054	3.894††	
Levene	0.439	1.091	
MC	0.24957	-0.29099	
GR	0.73412	1.25477 .	

These results suggest that the geographic distribution of wheat yield in 1873 is quite predictable from the geographic distribution of wheat yield in 1872; but this regression generates residuals that are less well behaved than in earlier periods, ones that have replaced conspicuous positive spatial autocorrelation with weak negative spatial autocorrelation. Considering only the 1873 geographic distribution, there is a weak tendency for similar yields of wheat to cluster together across the linear arrangement of Broadbalk field plots.

From 1880 to 1945, the drilling of seed was switched from November to October, which could have impacted upon overall yield amounts. Average yield increased between 1873 and 1883, but the relationship between successive geographic distributions appears to have weakened:

1883 parameter estimates (CAR from z-scores)

	raw	residuals	CAR(C)
ρ	*	*	0.27387
a	1.62150	0.63464	0.02931×0.19987 + 1.62150
b	*	0.66167	0.79731×0.19987

		0.437
R²	*	
S-W	0.89323†	0.94959
A-D	0.803†	0.402
R-J	0.9513†	0.9623
K-S	0.182	0.145
Bartlett	0.587	1.629
Levene	1.345	0.230
MC	0.29565	-0.25910
GR	0.70703	1.23944 .

The geographic distribution of wheat yield in 1883 is modestly predictable from that in 1882, and this regression generates residuals that are well behaved but that again have replaced moderate positive with weak negative spatial autocorrelation. Considering the 1883 geographic distribution alone, there is a weak tendency for similar yields of wheat to cluster together across the linear arrangement of Broadbalk field plots.

In the autumn of 1893, the size of Broadbalk field plots was decreased. In addition, during 1891-99, plots were plowed both immediately after harvest and again just before fertilizers were applied. Both of these practices may partially account for a detected deterioration of spatial and temporal relationships:

1893 parameter estimates (CAR from z-scores)

	raw	residuals	CAR(C)
ρ	*	*	0.10466
a	1.00019	0.40512	0.02183×0.11930
			+ 1.00019
b	*	0.45019	0.92202×0.11930
R²	*	0.342	
S-W	0.93714	0.90979	
A-D	0.349	0.585	
R-J	0.9622	0.9527†	
K-S	0.136	0.215††	
Bartlett	1.830	0.003	
Levene	1.489	0.099	
MC	0.10125	0.14291	
GR	0.68685	0.67749 .	

These results suggest that the geographic distribution of wheat yield in 1893 is somewhat predictable from the distribution in 1892 and that this regression generates residuals whose behavior is a mixture of improvement and degeneration, displaying an increase in positive spatial autocorrelation. Considering only the 1893 geographic distribution, there is a very weak tendency for similar yields of wheat to cluster together across the linear arrangement of Broadbalk field plots.

Much of the characterization of space-time results for 1892-93 persists in 1902-03:

1903 parameter estimates (CAR from z-scores)

	raw	residuals	CAR(C)
ρ	*	*	-0.04923
a	1.11347	0.06301	-0.00293×0.17805 + 1.11347
b	*	0.64715	0.93688×0.17805
R²	*	0.381	
S-W	0.96347	0.89189††	
A-D	0.266	0.658†	
R-J	0.9862	0.9313††	
K-S	0.124	0.233††	
Bartlett	0.775	4.336††	
Levene	1.993	1.367	
MC	-0.04914	-0.58434	
GR	0.99988	1.52749 .	

The geographic distribution of wheat yield in 1903 remains weakly predictable from that in 1902, and this regression generates residuals whose behavior is more ill behaved than before, containing quite marked negative spatial autocorrelation. Considering the 1903 geographic distribution by itself, there is virtually no tendency for similar or dissimilar yields of wheat to cluster together across the linear arrangement of Broadbalk field plots.

1913 results are more similar to the 1893 results:

1913 parameter estimates (CAR from z-scores)

	raw	residuals	CAR(C)
ρ	*	*	0.10251
a	1.01210	0.67542	0.01210×0.12915 + 1.01210
b	*	0.75690	0.92251×0.12915
R²	*	0.329	
S-W	0.94069	0.88360††	
A-D	0.464	0.748††	
R-J	0.9752	0.9462†	
K-S	0.167	0.177	
Bartlett	1.393	0.210	
Levene	2.454	0.776	
MC	0.10216	0.28855	
GR	0.86002	0.70822 .	

These results suggest that the geographic distribution of wheat yield in 1913 is somewhat predictable from the distribution in 1912 and that this regression generates residuals whose frequency distribution conspicuously deviates from a normal distribution; the residuals display a marked increase in positive spatial autocorrelation. In the 1893 geographic distribution, there is a very weak tendency for similar yields of wheat to cluster together across the linear arrangement of Broadbalk field plots.

During the time interval 1844-1914, previous crop stubble on the

Broadbalk field plots was either shallow plowed or harrowed, with any considerable weed and plant debris being carted off the plots. Since 1915, the plots have been plowed only once, after which manure is applied, and since 1920, tractors have been used for cultivation. These changes in farming practices may well have detectable impacts of the spatial and temporal relationships being explored here.

	1923 parameter estimates (CAR from z-scores)		
	raw	residuals	CAR(C)
ρ	*	*	0.09834
a	0.77030	0.32865	0.01573×0.10404 + 0.77030
b	*	0.42171	0.92412×0.10404
R^2	*	0.332	
S-W	0.92982	0.92251	
A-D	0.440	0.533	
R-J	0.9715	0.9615	
K-S	0.167	0.187	
Bartlett	1.219	0.765	
Levene	1.817	0.011	
MC	0.09675	0.17472	
GR	0.82351	.80564 .	

The geographic distribution of wheat yield in 1923 is somewhat predictable from that in 1922. This regression generates residuals that are reasonably well behaved but that display an increase in positive spatial autocorrelation. In the 1923 geographic distribution, there is a very weak tendency for similar yields of wheat to cluster together across the linear arrangement of Broadbalk field plots.

To summarize, these eight sets of results reveal that (1) latent spatial autocorrelation is positive and was stronger in the earlier years, (2) time-series residuals are better behaved than are the raw data, (3) the geographic distribution of the previous year's yield accounts for roughly one-third of the geographic variance displayed by yield in a given year, and (4) regressing yield at time t on yield at time t-1 does not filter out the latent spatial autocorrelation. Based on the last year alone, it is unclear whether or not spatial autocorrelation is neutralized by Fisher's randomization practices.

6.3.2. SPATIAL AUTOREGRESSIVE ANALYSIS OF BROADBALK STRAW YIELDS

The quantity of straw harvested, which involves the same small sample size n = 17, also appears to benefit slightly from application of a power transformation:

raw	power transformation
	$\hat{\delta}$ = -0.41
	$\hat{\gamma}$ = 0.64091
	RSSE = 6.1×10^{-3}

S-W = 0.96390   S-W = 0.97897
$\hat{\sigma}_{MC} \approx 0.25$.

In this instance, analysis of the annual power-transformed data renders the following eight sets of results, which essentially parallel those already reported for wheat yields:

1853 parameter estimates (CAR from z-scores)

	raw	residuals	CAR(C)
ρ	*	*	0.12658
a	2.03957	0.69962	0.00460×0.18518 + 2.03957
b	*	0.65479	0.91235×0.18518
R^2	*	0.371	
S-W	0.95261	0.98237	
A-D	0.463	0.204	
R-J	0.9747	0.9868	
K-S	0.170	0.111	
Bartlett	0.119	1.742	
Levene	0.751	0.493	
MC	0.12849	−0.29305	
GR	0.87512	1.27534	

1863 parameter estimates (CAR from z-scores)

	raw	residuals	CAR(C)
ρ	*	*	0.27727
a	2.63079	0.08276	0.00004×0.50515 + 2.63079
b	*	1.20058	0.79261×0.50515
R^2	*	0.571	
S-W	0.88505††	0.93646	
A-D	0.772††	0.578	
R-J	0.9454†	0.9603	
K-S	0.178	0.179	
Bartlett	0.730	2.344	
Levene	1.274	0.021	
MC	0.30244	−0.07588	
GR	0.71504	1.00329	

1873 parameter estimates (CAR from z-scores)

	raw	residuals	CAR(C)
ρ	*	*	0.22203
a	1.38643	0.29158	0.01946×0.15856 + 1.38643
b	*	0.53813	0.84943×0.15856
R^2	*	0.655	
S-W	0.95931	0.88037††	
A-D	0.240	1.056†††	
R-J	0.9857	0.9295††	
K-S	0.114	0.260††	
Bartlett	0.619	4.928††	

Levene	1.612	1.874
MC	0.23267	−0.30947
GR	0.75787	1.27307

1883 parameter estimates (CAR from z-scores)

	raw	residuals	CAR(C)
ρ	*	*	0.26548
a	1.75984	0.64190	0.01733×0.34632 + 1.75984
b	*	0.49835	0.80635×0.34632
R^2	*	0.476	
S-W	0.90053[†]	0.88321[††]	
A-D	0.681[†]	0.816[††]	
R-J	0.9574	0.9221[††]	
K-S	0.177	0.181	
Bartlett	0.441	4.268[††]	
Levene	1.004	1.195	
MC	0.28635	−0.27988	
GR	0.72536	1.25870	

1893 parameter estimates (CAR from z-scores)

	raw	residuals	CAR(C)
ρ	*	*	−0.03655
a	0.84755	0.26147	−0.00577×0.08686 + 0.84755
b	*	0.36105	0.93884×0.08686
R^2	*	0.447	
S-W	0.92820	0.89514[†]	
A-D	0.355	0.716[†]	
R-J	0.9612	0.9392[††]	
K-S	0.135	0.197[†]	
Bartlett	3.817[†]	0.002	
Levene	3.235[†]	0.016	
MC	−0.03554	−0.00494	
GR	0.81758	0.79522	

1903 parameter estimates (CAR from z-scores)

	raw	residuals	CAR(C)
ρ	*	*	−0.03621
a	1.56102	0.32609	−0.00214×0.43981 + 1.56102
b	*	0.58977	0.93885×0.43981
R^2	*	0.351	
S-W	0.95623	0.91176	
A-D	0.311	0.549	
R-J	0.9804	0.9424[††]	
K-S	0.135	0.163	
Bartlett	1.117	2.979[†]	
Levene	2.075	0.970	
MC	−0.03611	−0.50736	
GR	1.00532	1.47486	

1913 parameter estimates (CAR from z-scores)

	raw	residuals	CAR(C)
ρ	*	*	0.05884
a	1.53236	0.79471	0.00513×0.38379 + 1.53236
b	*	1.04127	0.93505×0.38379
R^2	*	0.427	
S-W	0.91874	0.84341†††	
A-D	0.521	1.180†††	
R-J	0.9640	0.9209††	
K-S	0.177	0.294††	
Bartlett	0.783	0.180	
Levene	2.064	1.041	
MC	0.05856	0.22602	
GR	0.90975	0.77499	

1923 parameter estimates (CAR from z-scores)

	raw	residuals	CAR(C)
ρ	*	*	0.16340
a	1.54927	-0.19052	0.02217×0.47246 + 1.54927
b	*	1.08661	0.89301×0.47246
R^2	*	0.620	
S-W	0.88515††	0.92597	
A-D	0.839††	0.474	
R-J	0.9448†	0.9543†	
K-S	0.226††	0.174	
Bartlett	3.473†	2.558	
Levene	5.395††	0.418	
MC	0.16519	0.06276	
GR	0.78927	0.93775	

These results reveal that (1) latent spatial autocorrelation is positive and was stronger in the earlier years, (2) time-series residuals often are not better behaved than are the raw data, (3) when compared with wheat yield, the geographic distribution of the previous year's straw harvest accounts for more of the geographic variance displayed by harvest in a given year, and (4) regressing straw harvest at time t on harvest at time t-1 frequently induces negative spatial autocorrelation rather than filters out any latent spatial autocorrelation. Consequently, some, but not all, of the previous findings for wheat yield are corroborated.

As an aside, a more serious analysis of these data would employ a space-time model specification. In addition, with only 153 distance pairs, undertaking a geostatistical analysis with these data is ill advised, because a minimum number of pairs per distance class (§1.6) cannot be accomplished while at the same time safeguarding that there will be a sufficient number of semivariogram plot points to support legitimate estimation of model parameters. This obstacle could be circumvented by devising a procedure for pooling observations over time; such a procedure is not obvious at present, though, especially given the spatial autoregressive analysis results reported in this section.

6.4. SUGAR CANE PRODUCTION IN PUERTO RICO

The island of Puerto Rico is partitioned into 72 areal units (*municipios*), which aggregate into five agricultural administrative regions (AAR; n_j = 19, 11, 16, 14, 13). The Commonwealth of Puerto Rico (CPR) Department of Agriculture, with annual field surveys (involving thousands of personal interviews with farmers and tens of thousands of interviewers with truckers at markets and supermarkets), collects agricultural production data for the island and tabulates much of it by municipios and AARs. One of the traditional triad of products that has been monitored closely for a 16-year period (fiscal years 1958/59 to 73/74) is sugar cane. Griffith and Amrhein (1991) report tabulations and analyses for the 1974 CPR data. In this section, area of sugar cane harvested is analyzed, paralleling those analyses presented in §6.3, for the selected years of 1959/60, 1966/67 and 1973/74. The space-time data series of *cuerdas* (0.9712 acres) of sugar cane harvested differs from the two Broadbalk space-time data series in that Puerto Rico has fewer consecutive years of data but considerably more areal units for which data have been collected, and records observational rather than experimental data. Hence, one relevant question associated with these data replicates the question addressed with the Broadbalk wheat yields and straw harvests, and asks whether or not the geographic distribution of sugar cane production in time t can be used to predict its geographic distribution in time t+1 when more areal units are present and when inertia in farming practices prevails rather than agricultural field experiments.

Historically, the sugar industry in Puerto Rico was closely regulated by both U.S. and CPR legislation, which imposed certain restrictions on the production and marketing of various forms of sugar cane and sugar. Sugar cane production peaked in 1951/52. In 1959, sugar cane production covered roughly 340,000 cuerdas and accounted for 40% of Puerto Rico's gross farm income, being the island's single most important export item at that time. The best soils, especially in the coastal lowland areas of the island, and virtually all irrigated lands in 1959 were devoted to sugar cane production. By 1974, considerable sugar cultivation in the coastal lowlands had been displaced by dairy farming, with sugar cane production then covering only about 122,000 cuerdas (down nearly two-thirds from 1959). Concomitantly, during this period the number of sugar mills had decreased from 29 to 11. The SAR spatial autoregressive model is employed for analysis of these data because autocorrelation should arise from missing variables, such as rainfall, soil type, and steepness of terrain. All sugar cane production has been converted to density of sugar cane production to adjust for differences in municipio sizes. Because the AARs partition the island into five reasonably compact regions, these AARs are used to evaluate homogeneity of geographic variance. And because the AARs are mutually exclusive, Bonferroni adjustments are utilized when separate statistics are evaluated for them.

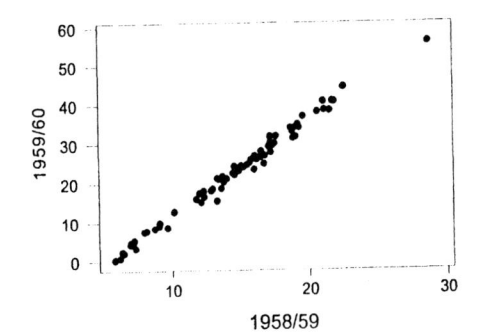

Figure 6.15. Scatterplot of the power-transformed 1959/60 versus 1958/59 density of cultivated sugar cane by municipio (CPR data).

6.4.1. SPATIAL AUTOREGRESSIVE ANALYSES OF PUERTO RICAN SUGAR CANE PRODUCTION

The 1959/60 CPR density of sugar cane cultivated land data benefit noticeably from the application of a power transformation:

```
raw                                power transformation
                                    δ̂ = 0
                                    γ̂ = 0.68312
                                    RSSE = 2.2×10⁻²
S-W = 0.94829††                    S-W = 0.96800
Bartlett = 9.901†                  Bartlett = 5.125
Levene = 1.190                     Levene = 0.441
MC = 0.33959 (σ̂_MC ≈ 0.07625)      MC = 0.34478
GR = 0.62330                       GR = 0.69100
```

AAR:	San Juan	Arecibo	Mayaguez	Ponce	Caguas
S-W(raw):	0.94318	0.90203	0.95789	0.88669	0.95217
S-W(p-t):	0.94307	0.86606	0.94180	0.91230	0.93977

These results suggest that the power-transformed measures conform to a normal distribution both across the island and within each of the five AARs, that the geographic variability of density of cuerdas of cultivated sugar cane production is reasonably constant across Puerto Rico, and that this variable contains weak to moderate positive spatial autocorrelation.

The first point in time for which a previous set of data are available is 1959/60. A very strong linear relationship exists between these consecutive annual measures (Figure 6.15) once the 1958/59 figures are subjected to a Box-Tidwell transformation ($\hat{\delta} = 23.69031$, $\hat{\gamma} = 0.55530$). Results for this year are as follows:

1959/60 parameter estimates (spatial results from z-scores)			
	OLS temporal	OLS spatial	SAR(W)
ρ	*	*	0.53825
b_0	-13.56800	0.0	0.11996
b_{t-1}	2.47163	*	*
$bu-\bar{u}$	*	-0.10362	-0.10056
(pseudo-) R^2	0.990	0.077	0.303

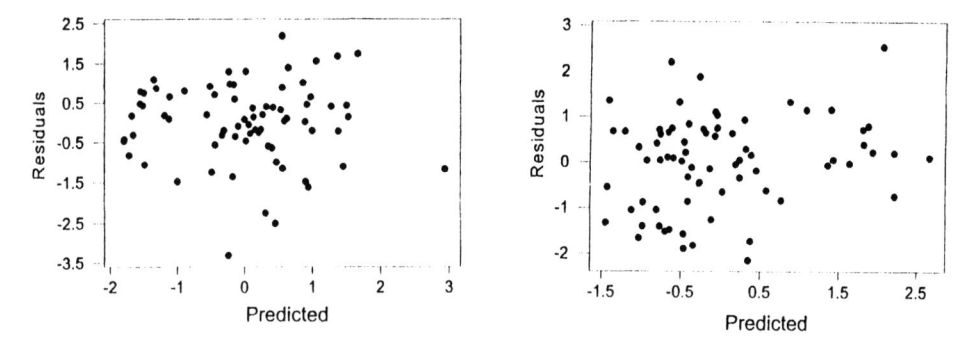

332   A CASEBOOK FOR SPATIAL STATISTICAL DATA ANALYSIS

Figure 6.16. Left: (a) Traditional homogeneity of variance OLS residual plot for the 1959/60 density of cultivated sugar cane regressed on the 1958/59 density (CPR data). Right: (b) Traditional homogeneity of variance SAR residual plot for the 1959/60 density of cultivated sugar cane regressed on the 1958/59 density (CPR data).

```
residuals
S-W         0.97266      0.95618†        0.97203
A-D         0.556        1.141†††        0.651†
R-J         0.9842†      0.9824††        0.9903
K-S         0.086        0.122†††        0.108††
Bartlett    6.916        3.698           4.535
Levene      1.099        0.265           0.410
MC          0.06742      0.30816         0.09031
GR          0.76604      0.74820         0.95483  .
```

Simple map pattern accounts for a small part of the spatial autocorrelation contained in these georeferenced data; 1959/60 density of cultivated cuerdas can be predicted from 1958/59 with a reasonable degree of accuracy and precision; and the spatial SAR model furnishes a modestly good description of the 1959/60 geographic distribution. Bothersome results include a discrepancy between MC and GR values for the OLS temporal model residuals (in contrast, the corresponding pair of results for the SAR residuals is compatible) and somewhat of a deviation from normality—in the tails of the frequency distribution—by the SAR residuals. According to the MC value, spatial autocorrelation latent in the 1958/59 figures effectively filters out spatial autocorrelation in the 1959/60 figures. Both the OLS temporal specification and the SAR specification render residuals that display relatively constant variance across the range of mean response levels (Figure 6.16a-b). And although MC is somewhat large for the SAR residuals, the corresponding MC scatterplot does not reveal a conspicuous trend (Figure 6.17).

The 1966/67 CPR density of sugar cane cultivated land data also markedly benefit from the application of a power transformation:

```
raw                         power transformation
                              δ̂ = 0.00702
                              γ̂ = 0.51079
                              RSSE = 2.0×10⁻²
S-W = 0.89518†††              S-W = 0.94998†
```

AGRICULTURAL YIELD 333

```
Bartlett = 23.533ttt          Bartlett = 5.524
Levene  = 2.808tt             Levene  = 0.328
MC      = 0.49404             MC      = 0.49557
GR      = 0.45732             GR      = 0.53421
```

AAR:	San Juan	Arecibo	Mayaguez	Ponce	Caguas
S-W(raw):	0.87189[t]	0.94219	0.97405	0.89426	0.93948
S-W(p-t):	0.93898	0.91616	0.88144	0.90344	0.97495

These results suggest that the power-transformed measures only marginally deviate from a normal distribution for the island as a whole but conform to one within each of the five AARs, that the geographic variability of density of cuerdas of cultivated sugar cane production is reasonably constant across Puerto Rico, and that this variable contains moderate positive spatial autocorrelation.

Once more a very strong linear relationship exists, now between the 1965/66 and 1966/67 measures, once the 1965/66 figures have been subjected to a Box-Tidwell transformation ($\hat{\delta} = 0.77706$, $\hat{\gamma} = 0.48961$). Estimation results for the 1966/67 geographic distribution of density of cultivated sugar cane include

1966/67 parameter estimates (spatial results from z-scores)

	OLS temporal	OLS spatial	SAR(W)
ρ	*	*	0.68578
b_0	-0.90823	0.0	0.20534
b_{t-1}	1.17203	*	*
$b_{u-\bar{u}}$	*	-0.14040	-0.13066
(pseudo-) R^2	0.992	0.142	0.501
residuals			
S-W	0.96829	0.96084[t]	0.98281
A-D	0.727[t]	0.834[tt]	0.327
R-J	0.9881	0.9877	0.9945
K-S	0.136[ttt]	0.101[t]	0.085
Bartlett	2.058	3.253	8.828[t]
Levene	0.564	0.346	1.020
MC	0.16226	0.42440	0.09936
GR	0.85452	0.61875	0.95927

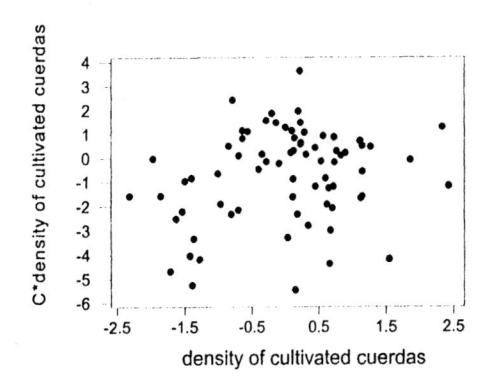

Figure 6.17. MC scatterplot of the SAR residual of Figure 6.16b.

Simple map pattern accounts for a small part of the spatial autocorrelation contained in these georeferenced data; 1966/67 density of cultivated cuerdas can be predicted from 1965/66 with a reasonable degree of accuracy and precision; and the spatial SAR model furnishes a respectable description of the 1965/66 geographic distribution. These are the same conclusions reached with the 1959/60 data set. Bothersome results include non-negligible spatial autocorrelation in residuals from the temporal OLS model specification and somewhat of a deviation from a normal frequency distribution—again in the tails—by these same residuals. Since 1959/60, both the simple map pattern trend and the degree of positive spatial autocorrelation have increased.

The 1973/74 CPR density of sugar cane cultivated land data dramatically benefits from the application of a power transformation:

```
raw                              power transformation
                                 δ̂ = 0
                                 γ̂ = 0.47091
                                 RSSE = 1.5×10⁻²
S-W = 0.79217†††                 S-W = 0.90923†††
Bartlett = 26.450†††             Bartlett = 2.336
Levene = 2.998††                 Levene = 0.401
MC = 0.39725                     MC = 0.52938
GR = 0.48217                     GR = 0.47372


AAR:          San Juan    Arecibo    Mayaguez    Ponce       Caguas
S-W(raw):     0.74433†††  0.86053    0.91013     0.82232††   0.82483†
S-W(p-t):     0.82040†    0.87826    0.91365     0.94616     0.91051 .
```

These results suggest that although the power-transformed measures more closely resemble a normal frequency distribution than do their raw data counterparts, they still deviate from a normal distribution for the island as a whole, while essentially conforming to one within each of the five AARs (in every one of the five cases the power transformation has increased S-W) that the geographic variability of density of cuerdas of cultivated sugar cane production is reasonably constant across Puerto Rico, and that this variable contains moderate positive spatial autocorrelation.

Similar to the two preceding cases, a very strong linear relationship exists between the 1972/73 and 1973/74 measures, once the 1972/73 figures have been subjected to a Box-Tidwell transformation ($\hat{\delta} = 0$, $\hat{\gamma} = 0.47532$). Estimation results for the 1973/74 geographic distribution of density of cultivated sugar cane include

```
1973/74 parameter estimates (spatial results from z-scores)
                    OLS temporal     OLS spatial     SAR(W)
ρ                   *                *               0.70812
b₀                  0.02398          0.0             0.23560
b_{t-1}             0.94961          *               *
bu-ū                *                -0.12013        -0.09617
(pseudo-) R²        0.954            0.104           0.518
```

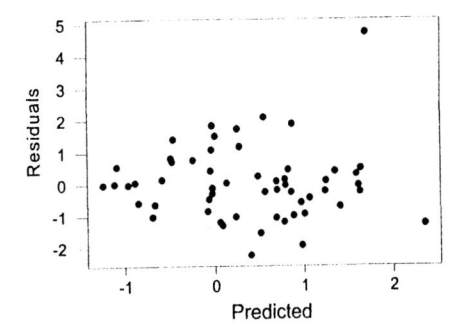

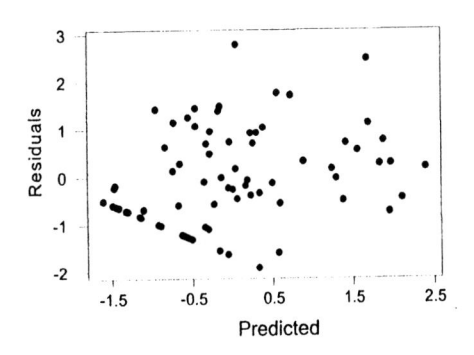

Figure 6.18. Left: (a) Traditional homogeneity of variance OLS residual plot for the 1973/74 density of cultivated sugar cane regressed on the 1972/73 density (CPR data). Right: (b) Traditional homogeneity of variance SAR residual plot for the 1973/74 density of cultivated sugar cane regressed on the 1972/73 density (CPR data).

residuals			
S-W	0.89347†††	0.93768†††	0.97144
A-D	2.740†††	1.274†††	0.445
R-J	0.9328††	0.9783††	0.9902
K-S	0.186†††	0.126†††	0.078
Bartlett	12.928††	1.630	7.695
Levene	0.253	0.376	2.427†
MC	0.19510	0.46340	0.10997
GR	0.66361	0.52536	0.84637

Simple map pattern accounts for a small part of the spatial autocorrelation contained in these georeferenced data; 1973/74 density of cultivated cuerdas can be predicted from 1927/73 with a reasonable degree of accuracy and precision; and the spatial SAR model furnishes a respectable description of the 1973/74 geographic distribution. Bothersome results include a discrepancy, again, between MC and GR values for the OLS temporal model residuals, considerably ill behaved OLS temporal residuals, and OLS temporal residuals that display nonconstant variance across the range of mean response levels (Figure 6.18a); the corresponding SAR residual plot is much better behaved (Figure 6.18b).

In summary, the density of cuerdas of sugar cane production in Puerto Rico benefits substantially from, approximately, an untranslated square-root power transformation for all time periods studied; the only glaring discrepancy is $\hat{\delta}$ for 1958/59. A weak, simple east-west map pattern, which increasingly becomes pronounced with the passing of time, characterizes the density of cuerdas of sugar cane production across the island. The spatial SAR model increasingly furnishes a better statistical explanation of geographic distributions with the passing of time. Strong inertia appears to be present in the agricultural production system, with the geographic distribution in time t being highly predictable from the corresponding geographic distribution in time t−1. But residuals from the OLS temporal specification become increasingly ill behaved with the passing of time.

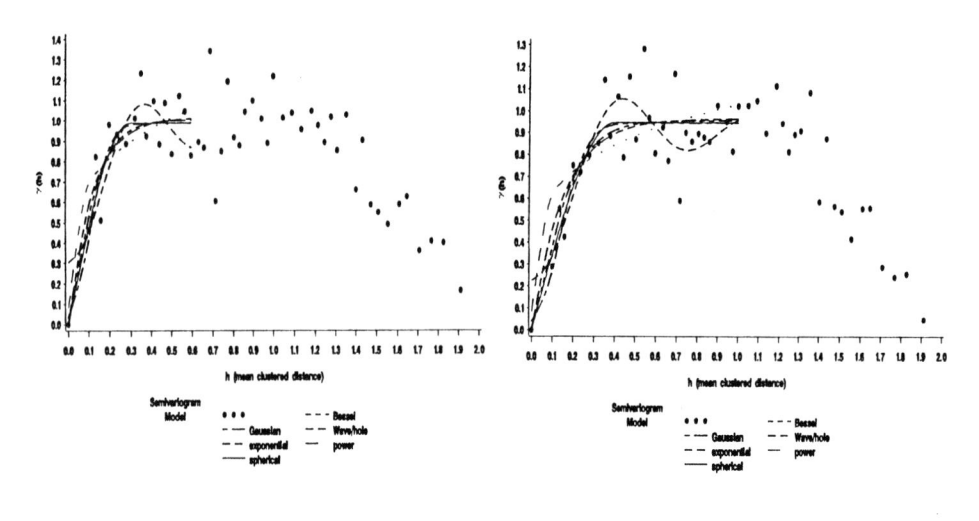

Figure 6.19. Top-left: (a) Semivariogram plot of OLS residuals for 1959/60 density of cultivated sugar cane CPR data, with selected semivariogram model curves superimposed. Top-right: (b) Semivariogram plot of OLS residuals for 1966/67 density of cultivated sugar cane CPR data, with selected semivariogram model curves superimposed. Bottom-right: (c) Semivariogram plot of OLS residuals for 1973/74 density of cultivated sugar cane CPR data, with selected semivariogram model curves superimposed.

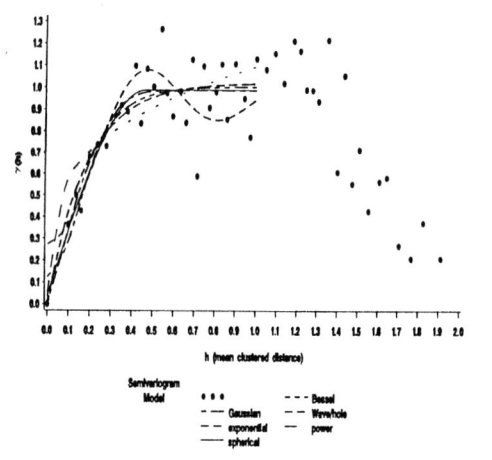

6.4.2. GEOSTATISTICAL ANALYSES OF PUERTO RICAN SUGAR CANE PRODUCTION

The geostatistical analyses for the three time periods consisted of analyzing the residuals from the respective OLS spatial regression equations described in the preceding output tables for the spatial autoregressive analyses. For each time period, 50 distance classes were specified to construct the respective semivariogram plots. These class have the following properties:

year	number of distance pairs	maximum distance	minimum number of distance pairs/distance class
1959/60	5329	0.6	64
1966/67	5329	1.0	62
1973/74	5329	1.0	64 .

Maximum distance was determined separately for each semivariogram plot modeled, and although the number of distance pairs per distance class was over 100, the minimum number reported here is above the minimum desired for each distance class (§1.6). Modeling of the semivariogram plots (s 6.19a-c) for each time period results in the following parameter estimates:

1959/60 parameter estimates

model	RSSE	C_0	C_1	r	
exponential	0.30506	0.00000	1.01688	0.11539	
Bessel	0.28534	0.00000	1.00467	0.07484	
Gaussian	0.27369	0.03272	0.95606	0.13751	
spherical	0.27279	0.00000	0.98858	0.29602	
wave/hole	0.34174	0.30446	0.64047	0.08114	
power	0.40035	0.00000	1.19087	*	$\lambda = 0.23321$

1966/67 parameter estimates

model	RSSE	C_0	C_1	r	
exponential	0.39627	0.00000	0.96632	0.15240	
Bessel	0.36631	0.00000	0.95751	0.09887	
Gaussian	0.34550	0.03926	0.90553	0.18222	
spherical	0.34269	0.00000	0.94617	0.39715	
wave/hole	0.34598	0.22558	0.68285	0.09718	
power	0.53593	0.00000	1.03238	*	$\lambda = 0.21899$

1973/74 parameter estimates

model	RSSE	C_0	C_1	r	
exponential	0.29864	0.00000	1.02201	0.18268	
Bessel	0.27854	0.00000	1.00415	0.11373	
Gaussian	0.27524	0.12451	0.86287	0.21674	
spherical	0.26261	0.03668	0.95257	0.46178	
wave/hole	0.33342	0.27282	0.66396	0.10491	
power	0.41128	0.00000	1.09550	*	$\lambda = 0.26994$.

Generally speaking, parameter estimates for models fit to each of the three semivariogram plots are similar in terms of their respective RSSE values. As indicated by the range parameter values, spatial autocorrelation tends to increase (i.e., range parameter values increase) over the time periods, paralleling findings reported for the preceding spatial autoregressive analyses. For each of the time periods the spherical semivariogram model achieves the best fit based on the RSSE values, although in most cases the RSSEs for the other transition models are not very much different from that for the spherical model.

6.5. MILK PRODUCTION IN PUERTO RICO

The U.S. Department of Agriculture (USDA), with a comprehensive quinquennium census that began in 1939, also collects agricultural production data for the island of Puerto Rico. As mentioned in §6.4, the island of Puerto Rico is partitioned into 72 areal units (municipios), which aggregate into five agricultural administrative regions (n_j = 19, 11, 16, 14, 13). From time to time, Puerto Rico suffers periods of drought that diminish its agriculture production; 1964 was a year of severe drought. In 1964, about 199,000 head of dairy cattle existed on the island; by 1987, this number had increased to 221,500. The growth in Puerto Rico's dairy industry has been a response to increasing demand through time for better nutrition. Dairy herds are mainly concentrated in the coastal lowland areas, having displaced sugar cane production and maintaining close proximity (i.e., milk sheds) to urbanized and urbanizing areas. One relevant question associated with the USDA data is whether or not the geographic distribution of milk production in time t can be used to predict its geographic distribution in time t+k (k = 4, 5; the next census year). This is a generalization of the questions addressed in §6.3, which deals with annual geographic distributions for small numbers of areal units, and §6.4, which deals with annual geographic distributions for large numbers of areal units.

The USDA milk production data essentially are replicates of the CPR data for the years 1964, 1969, 1974, 1978, 1982, and 1987, as well as other years; Griffith and Amrhein (1991) report tabulations and analyses for the 1974 CPR data. Because the CPR data are complete, they have been used in bivariate regression analyses to impute any missing (suppressed) USDA production values for a given year (see Chapter 9), where the predictor variable is the corresponding CPR production figures. Discrepancies do exist between these data sets, though, because each government agency has a slightly different definition of farm unit and production year, and the USDA suppresses selected data for reasons of confidentiality, whereas CPR does not. Nevertheless, these pairs of data should be extremely similar. The SAR spatial autoregressive model is employed for these data because autocorrelation should arise from missing variables, such as rainfall and quality of pastureland. All milk production has been converted to density of milk production to adjust for differences in municipio sizes. As mentioned in §6.4, because the AARs partition the island into five reasonably compact regions, they are used to evaluation homogeneity of geographic variance. And, because the AARs are mutually exclusive, Bonferroni adjustments are utilized when statistics are evaluated separately for each of these regions.

6.5.1. SPATIAL AUTOREGRESSIVE ANALYSES OF PUERTO RICAN MILK PRODUCTION

The 1964 USDA density of milk production data benefit considerably from the application of a power transformation:

raw	power transformation
	$\hat{\delta}$ = 0.08161
	$\hat{\gamma}$ = 0.21321
	RSSE = 7.2×10^{-3}
S-W = 0.66566[†††]	S-W = 0.94649[†††]
Bartlett = 67.079[†††]	Bartlett = 8.374[†]
Levene = 3.049[††]	Levene = 1.475

AGRICULTURAL YIELD 339

```
          MC = 0.20113                    MC = 0.31048
              (ô_MC ≈ 0.07625)            GR = 0.65663
          GR = 0.64964

AAR:          San Juan     Arecibo     Mayaguez     Ponce       Caguas
S-W(raw):     0.83639††    0.63565†††  0.76931†††   0.84506†    0.79277††
S-W(p-t):     0.91158      0.95879     0.93706      0.93591     0.93878
```

Considerable symmetry is gained from applying approximately a quarter-root power transformation, although the resulting empirical frequency distribution still deviates markedly from a normal distribution due to the persistence of skewness (Figure 6.20a-c). How closely these individual AAR empirical frequency distributions conform to a normal distribution plays an important role in geographic variance homogeneity assessment. Nevertheless, normality of the individual AAR frequency distributions is attained with this power transformation. In addition, geographic variance heterogeneity is all but eliminated by this power transformation. Marked positive spatial autocorrelation is latent in these data (Figure 6.21) and is somewhat obscured by the skewed frequency distribution for the raw data. Meanwhile, the components of spatial autocorrelation latent in these 1964 USDA densities are uncovered through the following analyses:

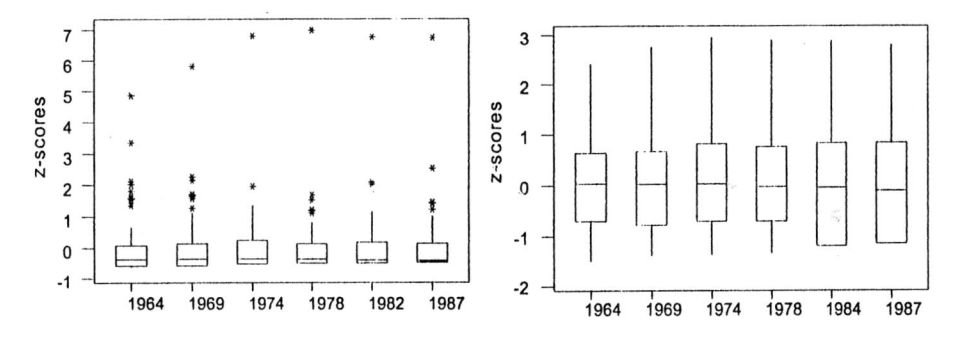

Figure 6.20. Top-left: (a) Boxplots of the raw DMLK for the agricultural census years 1964 to 1987 (USDA data). Top-right: (b): Boxplots of the power-transformed DMLK for the agricultural census years 1964 to 1987 (USDA data). Bottom-right: (c) Boxplots of the power-transformed 1964 DMLK for the AARs USDA data.

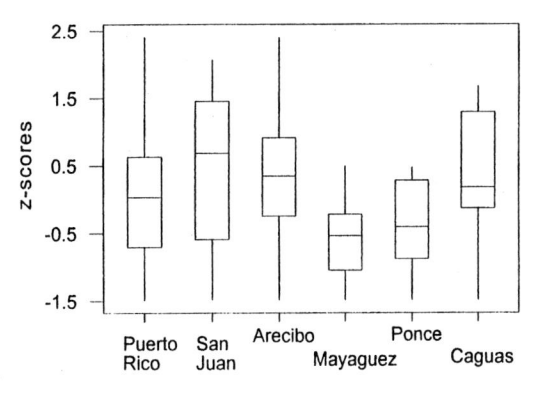

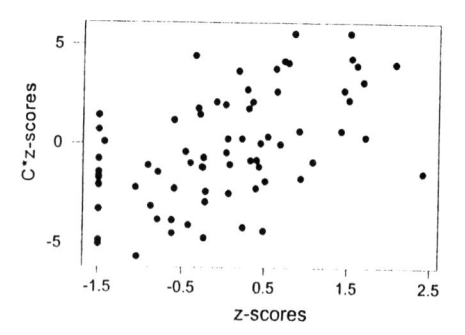

Figure 6.21. MC scatterplot of the power-transformed 1964 DMLK (USDA data).

	parameter estimates (from z-scores)	
	OLS	SAR(W)
ρ	*	0.41635
b_0	-0.41248	-0.48572
$b_{u-\bar{u}}$	0.11763	
$b_{v-\bar{v}}$	0.52283	0.36600
$b_{(u-\bar{u})(v-\bar{v})}$	0.08459	
$b_{(v-\bar{v})^2}$	0.48504	0.36119
(pseudo-) R^2	0.384	0.351
residuals		
S-W	0.98694	0.97407
A-D	0.296	0.282
R-J	0.9936	0.9935
K-S	0.061	0.061
Bartlett	5.351	3.954
Levene	0.889	0.508
MC	0.03535	0.00509
GR	0.95549	0.98937

These results reveal that most of the apparent positive spatial autocorrelation is attributable to simple map pattern, with the east-west trend and the north-south-east-west interaction trend being surrogates for other sources of spatial autocorrelation. Both the OLS and SAR residuals are well behaved, exhibiting conformity to a normal frequency distribution, constant geographic variance across the five AARs, and an absence of spatial autocorrelation.

The 1969 USDA density of milk production data also benefit considerably from the application of a power transformation:

raw	power transformation
	$\hat{\delta} = 0.01919$
	$\hat{\gamma} = 0.29294$
	RSSE = 7.8×10^{-3}
S-W = 0.64499†††	S-W = 0.93922†††
Bartlett = 87.525†††	Bartlett = 9.156†
Levene = 3.163††	Levene = 1.312
MC = 0.17010	MC = 0.32290
GR = 0.73745	GR = 0.65473

AAR: San Juan   Arecibo   Mayaguez   Ponce   Caguas
S-W(raw):   0.84056††   0.62614†††   0.73327†††   0.89462   0.88704
S-W(p-t):   0.91665   0.93683   0.91662   0.90371   0.94053 .

Considerable symmetry is gained from application of approximately a cube-root power transformation, although the resulting empirical frequency distribution still deviates markedly from a normal distribution, while each of the individual AAR frequency distributions does conform to a normal distribution after the power transformation has been applied. In addition, geographic variance heterogeneity is dramatically reduced and may well be eliminated by this power transformation. Weak to moderate positive spatial autocorrelation is latent, and is noticeably obscured by the skewed frequency distribution for the raw data. Meanwhile, the components of spatial autocorrelation latent in these 1969 USDA densities parallel those uncovered in the 1964 data and are uncovered through the following analyses:

	parameter estimates (from z-scores)		
	OLS spatial	SAR(W)	OLS temporal
ρ	*	0.45141	*
b_0	-0.24879	0.01733	-0.69739
b_{t-5}	*	*	1.71009
$bu-\bar{u}$	0.12883		
$bv-\bar{v}$	0.50320	0.31512	
$b(v-\bar{v})^2$		0.30442	
(pseudo-) R^2	0.330	0.328	0.880
residuals			
S-W	0.99412	0.99058	0.97913
A-D	0.282	0.217	0.696†
R-J	0.9928	0.9923	0.9794††
K-S	0.078	0.059	0.094
Bartlett	2.945	6.898	14.302†††
Levene	0.333	0.793	1.354
MC	0.06678	0.00780	0.09612
			E(MC) ≈ -0.01846
GR	0.94209	1.04657	0.80064 .

These results basically are a duplicate of those obtained with the 1964 data but present a less complex north-south map pattern. The t-5 temporal OLS model also has been estimated here. As with the sugar cane analyses reported in the preceding section, this temporal OLS model accounts for a much larger percentage of the 1969 geographic variability of these density measures than does the SAR model, although the longer time lag period may well have diminished this explanatory power. The temporal OLS residuals are somewhat less well behaved than are their SAR counterparts. In contrast with the preceding sugar cane analyses, because a Box-Cox transformation already has been applied to the 1964 density of milk production data, its linear predictive power is not optimized here with a Box-Tidwell transformation (Figure 6.22).

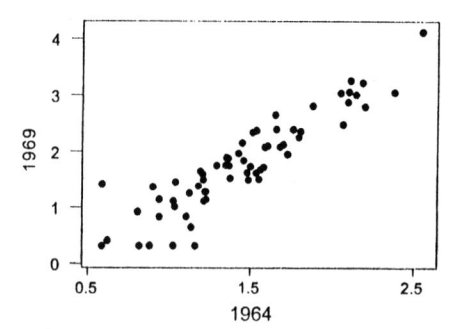

Figure 6.22. Scatterplot of the power-transformed 1969 versus 1964 DMLK by municipio (USDA data).

As is the case with the 1964 and 1969 USDA density of milk production data, the 1974 measures also benefit considerably from the application of a power transformation:

```
raw                              power transformation
                                 δ̂ = 0.00028
                                 γ̂ = 0.29837
                                 RSSE = 1.2×10⁻²
S-W = 0.56148†††                 S-W = 0.93475†††
Bartlett = 109.315†††            Bartlett = 6.696
Levene = 2.863††                 Levene = 0.746
MC = 0.15135                     MC = 0.35382
GR = 0.73564                     GR = 0.57889
```

AAR:	San Juan	Arecibo	Mayaguez	Ponce	Caguas
S-W(raw):	0.83384††	0.57993†††	0.56928†††	0.86289	0.88425
S-W(p-t):	0.94290	0.90669	0.85660†	0.86710	0.89927

As before, these results suggest that considerable symmetry is gained from application of approximately a cube-root power transformation, again with the empirical frequency distribution still deviating markedly from a normal distribution. Now, one of the individual AAR frequency distributions fails to conform adequately to a normal distribution after the power transformation has been applied; moreover, each of these five empirical frequency distributions fails overall to mimic a normal distribution, just as its 1969 counterpart. In addition, even a more pronounced geographic variance heterogeneity than in 1969 is removed in this case. Weak to moderate positive spatial autocorrelation is latent, and is noticeably obscured by the skewed frequency distribution for the raw data. Meanwhile, the components of spatial autocorrelation latent in these 1974 USDA densities parallel those found in the 1964 and 1969 data and are uncovered through the following analyses:

```
            parameter estimates (from z-scores)
            OLS spatial      SAR(W)       OLS temporal
ρ           *                0.51311      *
b₀          0.00000          0.01135      - 0.18185
```

b_{t-5}	*	*	1.07200
$bu-\bar{u}$	0.10843		
$bv-\bar{v}$	0.51465	0.34150	
(pseudo-) R^2	0.326	0.411	0.886
residuals			
S-W	0.98039	0.98132	0.91558†††
A-D	0.598	0.557	2.957†††
R-J	0.9849†	0.9840†	0.9497†††
K-S	0.084	0.080	0.157†††
Bartlett	3.663	6.313	22.815†††
Levene	0.628	0.565	2.231†
MC	0.16554	0.02001	-0.00548
GR	0.83463	1.01562	1.01204 .

These results are very similar to those reported for the 1969 data analysis, but the normality feature of the temporal OLS residuals essentially is completely lost. The statistical explanatory power of the SAR model appears to be slowly improving over time.

In keeping with the behavior of the preceding three annual geographic distributions, the 1978 measures also benefit considerably from the application of a power transformation:

```
raw                                power transformation
                                      δ̂ = 0.03925
                                      γ̂ = 0.23880
                                      RSSE = 1.3×10⁻²
  S-W = 0.53276†††                  S-W = 0.92877†††
  Bartlett = 126.715†††             Bartlett = 5.347
  Levene = 3.507††                  Levene = 0.686
  MC = 0.16946                      MC = 0.33585
  GR = 0.70328                      GR = 0.56425

AAR:        San Juan   Arecibo   Mayaguez   Ponce     Caguas
S-W(raw):   0.82084††† 0.63170††† 0.67633††† 0.83897†  0.88997
S-W(p-t):   0.92815    0.94205    0.88098    0.82520†† 0.87364 .
```

Once again, considerable symmetry is gained from application of approximately a quarter-root power transformation, but the empirical frequency distribution still deviates markedly from a normal distribution. As for the 1974 density measures, one of the individual AAR frequency distributions fails to conform adequately to a normal distribution after the power transformation has been applied, compromising the Bartlett's homogeneity of variance test interpretation. Although raw data geographic variance heterogeneity appears to be increasing through time, again, the power transformation successfully stabilizes this variance. Weak to moderate positive spatial autocorrelation is latent, once more being noticeably obscured by the skewed frequency distribution for the raw data. Meanwhile, the components of spatial autocorrelation latent in these 1978 USDA densities mirror those already exposed in the three preceding milk density data analyses and are uncovered through the following analyses:

```
              parameter estimates (from z-scores)
                    OLS spatial        SAR(W)       OLS temporal
    ρ                   *              0.49236          *
    b₀               0.00000          -0.01770       0.41758
    b_{t-4}             *                 *          0.67927
    bu-ū             0.07766
    bv-v̄             0.56251          0.37053
    (pseudo-) R²     0.323            0.418          0.921

    residuals
    S-W              0.99184          0.98740        0.90083†††
    A-D              0.243            0.340          3.127†††
    R-J              0.9935           0.9917         0.9410†††
    K-S              0.058            0.068          0.177†††
    Bartlett         4.030            3.670         18.620†††
    Levene           1.050            0.859          1.390
    MC               0.17671          0.01332        0.29787
    GR               0.80525          0.98893        0.83159 .
```

These results primarily replicate the preceding ones, but with prominent positive spatial autocorrelation appearing in the temporal OLS residuals. The persistence of residual variance heterogeneity and the potential of spatial autocorrelation being present suggest that a more serious analysis of these data would employ a space-time model specification. The statistical explanatory power of the temporal model specification has increased, as is expected because the time lag has decreased from five to four years.

As should be anticipated given the preceding power transformation results, the 1982 measures also benefit from the application of a power transformation, although not as much as previous years:

```
    raw                              power transformation
                                         δ̂ = 0
                                         γ̂ = 0.29688
                                         RSSE = 9.5×10⁻³
    S-W = 0.53761†††                 S-W = 0.90582†††
    Bartlett = 126.033†††            Bartlett = 8.173†
    Levene = 4.495†††                Levene = 2.335†
    MC = 0.20145                     MC = 0.32920
    GR = 0.66885                     GR = 0.56421


    AAR:         San Juan    Arecibo    Mayaguez    Ponce      Caguas
    S-W(raw):    0.78427†††  0.69004††† 0.63964†††  0.50538††† 0.88330
    S-W(p-t):    0.86479†    0.94603    0.89920     0.71227††† 0.92741 .
```

These results suggest, too, that considerable symmetry is gained from application of approximately a quarter-root power transformation, but the empirical frequency distribution still deviates markedly from a normal distribution. Conformity of the individual AAR frequency distributions appears to be deteriorating over time, now with two empirical distributions failing to conform adequately to a normal distribution after the power transformation. Since the previous point in time, namely 1978, raw data geographic variance heterogeneity has become more complex rather than having increased, with the power transformation being able to dramatically decrease it but unable to fully stabilize it. Weak to moderate positive spatial autocorrelation is latent in these data, once more being noticeably obscured by the skewed frequency distribution for the raw data. Meanwhile, the components of spatial autocorrelation latent in these 1982 USDA densities are very similar to those already reported for the four preceding milk density data analyses, but with the east-west map pattern component disappearing from the OLS spatial specification, and are uncovered through the following analyses:

	parameter estimates (from z-scores)		
	OLS spatial	SAR(W)	OLS temporal
ρ	*	0.53876	*
b_0	0.00000	-0.03208	- 0.77043
b_{t-4}	*	*	1.49354
$b_v - \bar{v}$	0.55855	0.32975	
(pseudo-) R^2	0.259	0.423	0.887
residuals			
S-W	0.98131	0.97900	0.77343[†††]
A-D	0.228	0.406	6.154[†††]
R-J	0.9950	0.9913	0.8581[†††]
K-S	0.055	0.080	0.233[†††]
Bartlett	5.673	4.765	29.574[†††]
Levene	2.012	1.599	1.108
MC	0.22946	0.02892	0.06893
GR	0.74200	0.95586	0.85787 .

These results are very similar to those for the 1974 data. In contrast with results for 1978, the presence of non-0 spatial autocorrelation is not detected here in the temporal OLS residuals.

Finally, the 1987 measures again benefit from application of a power transformation, although not as much as do any of the preceding five sets of density data (Figure 6.23):

raw	power transformation
	$\hat{\delta} = 0$
	$\hat{\gamma} = 0.28083$
S-W = 0.52419[†††]	RSSE = 8.2×10^{-3}
Bartlett = 151.231[†††]	S-W = 0.89469[†††]
Levene = 5.228[†††]	Bartlett = 11.213[††]
MC = 0.23049	Levene = 2.780[††]
GR = 0.63649	MC = 0.34012
	GR = 0.56061

AAR:	San Juan	Arecibo	Mayaguez	Ponce	Caguas
S-W(raw):	0.73633[†††]	0.72720[†††]	0.59531[†††]	0.63166[†††]	0.90667
S-W(p-t):	0.81825[†††]	0.96715	0.89215	0.69251[†††]	0.93288 .

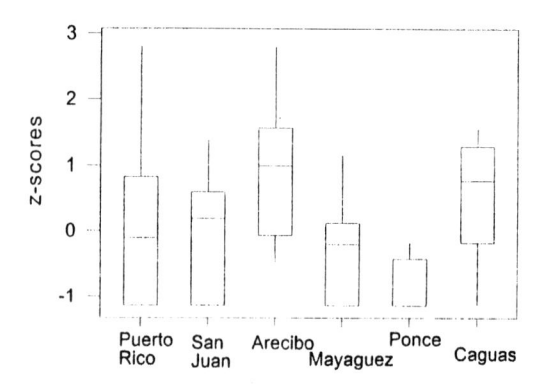

Figure 6.23. Boxplots of the power-transformed 1987 DMLK for the AARs (USDA data).

Considerable symmetry is gained from application of approximately a quarter-root power transformation, but with the resulting empirical frequency distribution still deviating markedly from a normal distribution. Conformity of the individual AAR frequency distributions appears to have continued deteriorating over time, now with one empirical distribution failing to even remotely conform to a normal distribution after the power transformation. Since 1982, raw data geographic variance heterogeneity has both increased and become more complex, with the power transformation being able to decrease it dramatically, but unable to fully stabilize it. Weak to moderate positive spatial autocorrelation is latent, again, being noticeably obscured by the skewed frequency distribution for the raw data. Meanwhile, the components of spatial autocorrelation latent in these 1987 USDA densities are the same as those of 1982 and are uncovered through the following analyses:

	parameter estimates (from z-scores)		
	OLS spatial	SAR(W)	OLS temporal
ρ	*	0.56651	*
b_0	0.00000	-0.02354	- 0.03102
b_t	*	*	0.93529
$b_v - \bar{v}$	0.55984	0.32020	
(pseudo-) R^2	0.260	0.446	0.966
residuals			
S-W	0.97546	0.98532	0.86984[†††]
A-D	0.314	0.400	3.460[†††]
R-J	0.9931	0.9928	0.9322[†††]
K-S	0.060	0.093	0.214[†††]
Bartlett	7.554	5.665	13.859[†††]
Levene	2.022	1.980	2.530[††]
MC	0.24019	0.02174	0.07798
GR	0.73744	0.98417	0.99819

These results are similar to those for the 1974 and 1982 data sets.

In summary, all six geographic distributions of the USDA milk production densities suggest the need for a power transformation to improve the symmetry of their respective frequency distributions and to stabilize their corresponding geographic variances; the implied power transformation across these six cases is essentially the same. With the passing of time, the USDA has suppressed an increasing number of production figures for reasons of confidentiality; these figures have been estimated here with an EM-type of algorithm (see Chapter 9). Replacing missing values with conditional expectations tends to suppress the accompanying variance estimate, which may well account for an increase here in the degree and complexity of geographic variance heterogeneity and an increased failure of individual AAR frequency distributions to conform to a normal distribution with the passing of time. Pooling these six results implies that part of the spatial autocorrelation detected in density of milk production is attributable to a simple east-west trend map pattern, but with other quite pronounced sources of spatial autocorrelation being concomitant ($\hat{\rho} \approx 0.5$). The SAR model specification is consistent over time, with its statistical explanatory power increasing. The temporal OLS specification repeatedly achieves an impressive level of statistical explanation, with t-4 geographic distributions being better predictors than are t-5 geographic distributions, while generating very ill behaved residuals. Coupling these spatial and temporal findings suggests that a space-time specification may well be in order here, a topic that is beyond the scope of this book. Comparing findings reported in this section with those for agricultural experimental results (§6.3) and for other observational agricultural studies (§6.4) helps confirm the important role played by inertia in a space-economy.

6.5.2. GEOSTATISTICAL ANALYSES OF PUERTO RICAN MILK PRODUCTION

Geostatistical analyses for density of milk production were performed separately for each of the six time periods, for which the residuals from the OLS spatial regression of the power-transformed measures have been employed. Similar to the specifications for Puerto Rico sugar cane production, 50 distance classes were specified for each of the six time periods. For all six data sets, the number of distance pairs is equal ($73^2 = 5329$), the maximum rescaled distance used for modeling purposes was kept at 1.0, and the minimum number of distance pairs per distance class for each of the six data sets is 62, slightly above the minimum target value of 50 distance pairs per class that we prefer to use. Selected models fitted to each of the six semivariogram plots (Figures 6.24a-f) render the following parameter estimates:

		1964 parameter estimates		
model	RSSE	C_0	C_1	r
exponential	0.58914	0.00000	0.61527	0.04044
Bessel	0.58933	0.00000	0.61626	0.03394
Gaussian	0.59391	0.00133	0.61263	0.06075
spherical	0.59467	0.00000	0.61317	0.11497

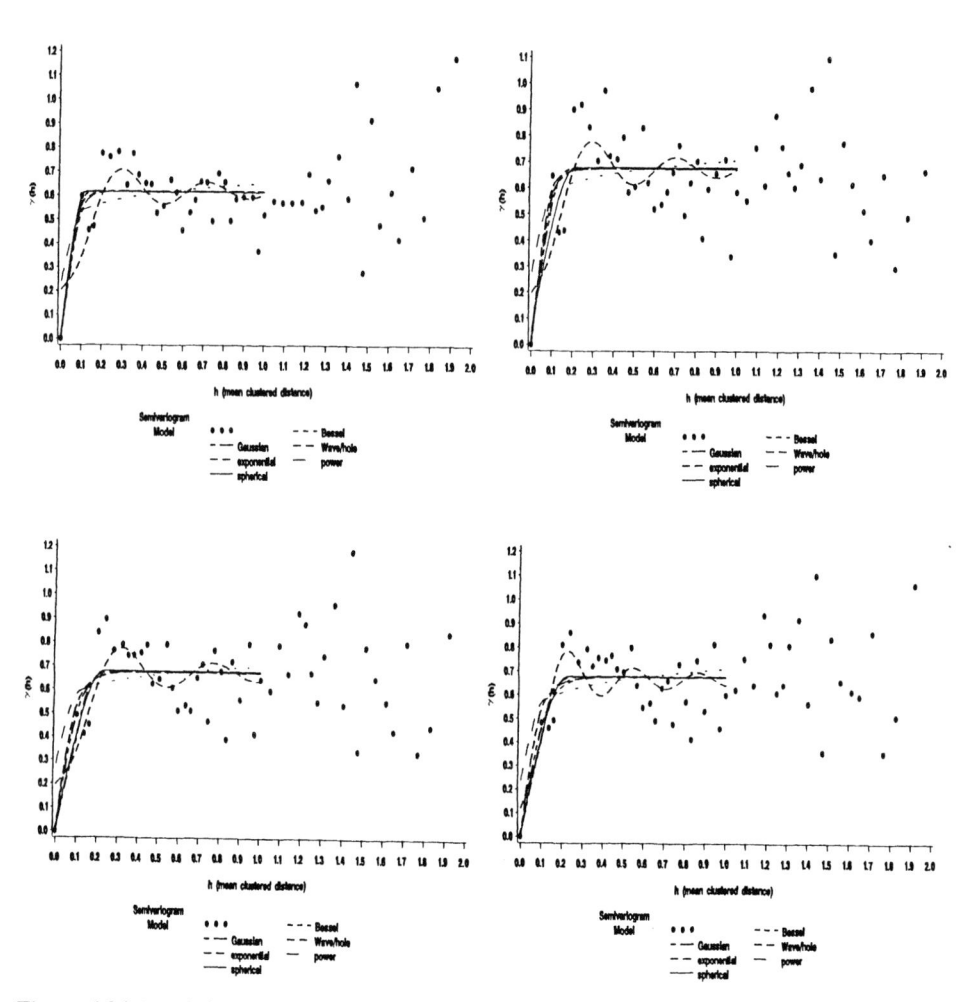

Figure 6.24. Top-left: (a) Semivariogram plot of 1964 DMLK USDA data, with selected semivariogram model curves superimposed. Bottom-left: (b) Semivariogram plot of 1969 DMLK USDA data, with selected semivariogram model curves superimposed. Top-right: (d) Semivariogram plot of 1978 DMLK USDA data, with selected semivariogram model curves superimposed. Bottom-right: (e) Semivariogram plot of 1982 DMLK USDA data, with selected semivariogram model curves superimposed.

```
wave/hole    0.54797    0.20374    0.41276    0.06544
power        0.82491    0.00000    0.64992       *         λ = 0.08750
```

1969 parameter estimates

model	RSSE	C_0	C_1	r
exponential	0.70602	0.00000	0.67337	0.04503
Bessel	0.70502	0.00000	0.67484	0.03677
Gaussian	0.71092	0.01621	0.65759	0.07835
spherical	0.70211	0.04839	0.62984	0.21029
wave/hole	0.65607	0.19785	0.47593	0.06398
power	0.88922	0.00000	0.70672	*

$\lambda = 0.08178$

AGRICULTURAL YIELD   349

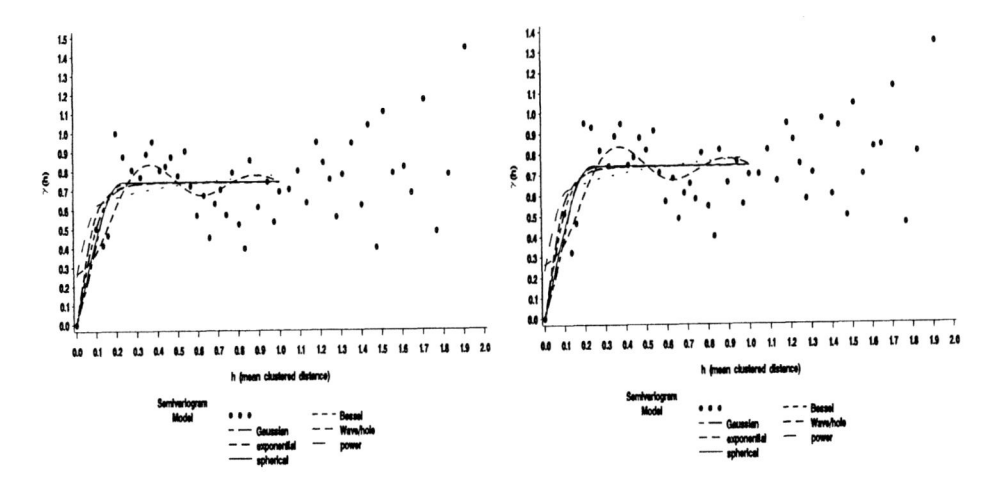

Figure 6.24: continued. Left: (c) Semivariogram model curves superimposed. Right: (f) Semivariogram plot of 1987 DMLK USDA data, with selected semivariogram model curves superimposed.

1974 parameter estimates

model	RSSE	C_0	C_1	r
exponential	0.64055	0.00000	0.67356	0.06119
Bessel	0.62938	0.00000	0.67485	0.04682
Gaussian	0.61124	0.01580	0.66155	0.10482
spherical	0.59560	0.00161	0.67581	0.22884
wave/hole	0.56596	0.19504	0.47525	0.07000
power	0.82047	0.00000	0.70113	*

$\lambda = 0.07968$

1978 parameter estimates

model	RSSE	C_0	C_1	r
exponential	0.60022	0.00000	0.67542	0.06105
Bessel	0.59633	0.00000	0.67844	0.05354
Gaussian	0.57191	0.01177	0.66596	0.10217
spherical	0.55886	0.00310	0.67535	0.22359
wave/hole	0.59078	0.11834	0.54648	0.05006
power	0.81343	0.00000	0.71199	*

$\lambda = 0.09676$

1982 parameter estimates

model	RSSE	C_0	C_1	r
exponential	0.67012	0.00000	0.74084	0.06542
Bessel	0.65925	0.00000	0.74402	0.05346
Gaussian	0.63512	0.00701	0.73706	0.10833
spherical	0.62172	0.00000	0.74379	0.22986
wave/hole‡	0.62131	0.26781	0.46725	0.08073
power	0.85957	0.00000	0.77883	*

$\lambda = 0.09721$

‡ Initial parameter estimate values for the wave/hole semivariogram model were specified explicitly as $C_0 = 0.0001$, $C_1 = 0.8$, and $r = 0.4$.

1987 parameter estimates

model	RSSE	C_0	C_1	r
exponential	0.65678	0.00000	0.74102	0.06841
Bessel	0.64260	0.00000	0.74270	0.05180

Gaussian	0.61736	0.01020	0.73430	0.11254
spherical	0.60214	0.00000	0.74413	0.23827
wave/hole†	0.60370	0.26535	0.46904	0.08183
power	0.84388	0.00000	0.78172	* $\lambda = 0.10397$.

†Initial parameter estimate values for the wave/hole semivariogram model were specified explicitly as $C_0 = 0.0001$, $C_1 = 0.8$, and $r = 0.4$.

All of the RSSE values for each time period are moderate to high, indicating substantial variation in the scatter of points constituting the respective semivariogram plots modeled. As seen in the semivariogram plots for each of the six time periods (Figures 6.24a-f), a spike in the scatter of points occurs between the 0 and the maximum rescaled distance specified (1.0). This pattern of points undoubtedly is the main contributing factor to the relatively high RSSE values, although evidence of a well behaved (§2.6.2) semivariogram scatter of points can be seen. Based strictly upon RSSE values, the wave/hole model achieves the best fit of the selected semivariogram models in all but the 1978 and 1987 time periods, where the spherical model achieves the best fit. The success of the wave/hole model in most of these data sets is more than likely due to the particular pattern of the scatter of points for each of the sets (i.e., the spike of points that generally lies halfway between the 0 and the 1.0 distance classes). The level of differentiation between the best-fitting semivariogram model for each of the six time periods is low and, in fact, tends to decrease with increasing time. Selection of an appropriate semivariogram model to describe each data set is not clear, although based on the spatial autoregressive results there is no reason to conclude that any of the data sets contain negative spatial autocorrelation. Values of the range parameter suggest relatively low amounts of spatial autocorrelation in each of the data sets and, whereas the amount of autocorrelation for the Puerto Rico sugar cane production increased over time, an increase in the range parameter values for the density of milk production over time is less conspicuous.

6.6. Concluding Comments: Spatial Autocorrelation and Agricultural Yield Variables

Because of the important role spatial autocorrelation plays in geographic distributions of agricultural production, such georeferenced data virtually always should benefit from being subjected to spatial statistical analyses. If in field experiments one goal is to neutralize effects of spatial autocorrelation, spatial statistical analyses will help document the achievement of this objective. Of historical note along these lines of discussion, findings reported here for the sizeable space-time Broadbalk data set do not seem to furnish convincing evidence that spatial autocorrelation had been neutralized by 1923 by Fisher's randomization practices. But contrasting these findings with those for a much shorter time and larger space space-time set of Puerto Rican agricultural production figures may be more illuminating. When inertia in farming practices prevails, rather than the designing of agricultural field experiments, both temporal and spatial correlation discernibly increase.

The eight sequential pairs of Broadbalk data render temporal correlations that describe weak to moderate positive tendencies; these data render spatial correlations that mostly are weak (with some being negligible). Wheat and straw temporal correlation results are consistent, whereas two pairings of spatial correlations for these phenomena are inconsistent (wheat suggests a weak relationship while straw suggests no relationship). Residuals from successive time period regressions fail to have spatial autocorrelation filter out of them, as well. In contrast, temporal correlation for Puerto Rican sugar cane production is very strong, whereas that for Puerto Rican milk production is strong or very strong. The corresponding spatial autocorrelation is either weak to moderate or moderate. In other words, experimental design may well reduce, but not really neutralize, the effects of spatial autocorrelation in agricultural field data.

Findings reported in this chapter also expand upon the set of empirical applications being compiled in this book by establishing empirical links between the semivariogram models occasionally used in agricultural field data analyses, and the set of popular spatial autoregression specifications (§3.3) that are available. Linkages reported in this chapter include

Table 6.1. Summary of linkage findings for Chapter 6

phenomenon	areal unit	autoregressive model	semivariogram model
Mercer-Hall wheat	field plot	CAR	anisotropic exponential
Mercer-Hall straw	field plot	CAR	anisotropic exponential
Wiebe wheat	field plot	SAR	power; exponential; Bessel
Wiebe straw	field plot	SAR	power; exponential; Bessel
Puerto Rican sugar cane	municipio (i.e., county)	SAR	spherical; Gaussian; Bessel; exponential
Puerto Rican milk	municipio (i.e., county)	SAR	spherical; Gaussian; Bessel

These articulations of relationships between geostatistical and spatial autoregressive models continue to reflect findings of §4.6, namely, a frequent inability to relate simply and easily a single semivariogram model to a particular spatial autoregressive specification.

Anisotropy is a focal point of discussion in the beginning of this chapter. Several features of this treatment merit attention in a concluding commentary. First, the estimation procedure becomes more complicated, because the georeferenced data must be kept at a disaggregated level to address issues of axis rotation. This requirement is not problematic when one is dealing with a standard field experiment but certainly could be for much of the other georeferenced data analyzed in this book. Second, the isotropic semivariogram model range parameter estimate roughly equals the weighted average of the anisotropic range parameter estimates; this relationship holds for the spatial autoregressive parameter estimates, too. Third, because estimating the axis rotation angle is not too cumbersome in geostatistics (although it is quite formidable in spatial autoregression), one effective strategy would be to estimate this angle using an appropriate geostatistical model specification, followed by rotating the axes and then undertaking a spatial autoregressive analysis; moreover, geostatistics would become a preprocessing step in the analysis. This strategy further emphasizes the need to articulate sound linkages between these two sets of models. And, fourth, analysis of the rook's versus the bishop's adjacency specification for the spatial autoregression undertaken with the Wiebe data reveals that the simpler rook's specification furnishes a description whose spatial autocorrelation is equivalent to that captured by the more complicated bishop's specification.

Finally, results reported in this chapter help demonstrate what statistics for spatial data can offer to the agricultural sciences. Besides contributing to a better understanding of whether or not experimental design actually neutralizes spatial autocorrelation effects, analysis of the Mercer-Hall and the Wiebe data disclose that straw yields sometimes may serve as acceptable replicate surrogates for grain harvests. Analysis of these two data sets also demonstrates that considerably different spatial dependencies can characterize experimental agricultural field data. On the one hand, prediction over time and/or over space is achievable with agricultural production data; on the other hand, taking into account geographic patterns underlying agricultural production at one point in time does not necessarily filter out latent spatial autocorrelation for a succeeding year. Perhaps, then, restricted randomization that acknowledges the presence of non-0 spatial autocorrelation warrants closer scrutiny; after all, blocking is a third important principle of statistical experimental design. Ideally, an optimal spatial experimental design should be the rule, rather than the exception, for conducting agricultural field trials; in practice, this infrequently has been the case. Regardless, as Grondona and Cressie (1991) so aptly argue, incorporating latent spatial dependencies will improve description and precision of the statistical analysis of treatments in uniformity trials.

7

ANALYSIS OF GEOREFERENCED POLLUTION VARIABLES

Geography often is defined in terms of human-land relationships and as such is concerned with the surroundings of human beings, namely, the natural environment. Part of this concern pertains to the manner in which humans alter their surroundings, either directly or indirectly adding substances or energy forms to the natural environment in such a way that higher concentration levels occur at specific locations than otherwise would be found there (background noise), resulting in environmental pollution. Data collected about these interactions and relationships often are spatial in nature—arguing for them to be georeferenced and making their analysis amenable to geostatistical and spatial autoregressive techniques. Cressie and ver Hoef (1993) provide an historical overview of the use of statistics for spatial data within the context of environmental and ecological modeling.

The phenomenon of pollution includes many naturally occurring substances, some of which are emitted with little or no apparent effect and others whose emission can drastically alter balances in the natural environment. We include in this chapter pollution examples of the second type, classifying them in terms of the resource categories of air, water, and land that they affect. Air pollution may be defined as the accumulation of any natural or artificially composed material—solid particles, liquid droplets, and/or gases—in the air that are either directly or indirectly harmful or dangerous to the natural environment.[1] Air pollution, especially tiny particles, kills about 50,000 people in the United

---

[1] Currently many nations are concerned about the geographic distribution of emissions of carbon dioxide and other greenhouse gases.

States each year and aggravates asthma in millions of others.[2] The benchmark data for this chapter comprise measures of particulate stack emissions from coal burning, presented in Cressie and Clark (1994). Water pollution may be defined as the accumulation in oceans, lakes, rivers, streams, and wetlands of substances or energy forms in sufficient concentration to produce a measurably negative effect on the associated natural environments. The 1997 *Reader's Digest Good Beach Guide* highlights one problem affiliated with this category of pollution by indicating that only 136 of 763 British beaches tested met tough Marine Conservation Society water quality standards; not one beach in all of northwest England is rated "good," reflecting the spatial nature of environmental pollution. Land pollution may be defined as the contamination of soil and near-surface ground principally by physical, chemical, or biological solid waste or radioactive waste. The most infamous case of chemical pollution was that of Love Canal, in Niagara Falls, New York.[3] A primary source for this category of pollution is agricultural fertilizers and pesticides. Interestingly, the early Broadbalk agricultural field plot trials discussed in Chapter 6 actually initiated data collection about this type of pollution. Although good natural drainage existed at Broadbalk, the drainage was improved with tile drains put under the center furrow separating the halves of each plot. Because there was one drain under each plot, discharge through these drains was used to measure losses of plant nutrients. Tolerance of any of these three categories of pollutants by human beings depends upon the nature and degree of environmental effects they generate and the extent of human activity at or near a site of pollution—yet another indication of the spatial nature of pollution.

Following the Love Canal calamity in the late 1970s, the plethora of toxic monuments prompted governments to draft and promulgate legislation aimed at eliminating the health and environmental threats posed by years of intensively and/or continuously dumping hazardous waste at warehouse, manufacturing facility, processing plant, and landfill sites. In 1980 the U.S. Congress enacted the Comprehensive Environmental Response, Compensation, and Liability Act, which was amended by the Superfund Amendments and Reauthorization Act in 1986. This legislation proposed to clean up the nation's worst toxic waste sites, which now number nearly 1,400. The argument is that the redevelopment of contaminated, or potentially contaminated, sites provides substantial benefits to the

---

[2] This situation has been documented by what might be called a natural experiment, supplementing correlations across North American communities between dirty air and health problems. A labor dispute causing a Provo, Utah, steel mill—the major source of air pollution in a mountain valley—to be closed for 13 months was reflected in counts of admissions for respiratory illness to the three local hospitals. All counts declined substantially during the plant's shutdown.

[3] In the late 1970s, chemicals disposed of by Hooker Electrochemical Co. in the 1940s and 1950s, including the carcinogens dioxin (see §7.5) and benzene, began seeping into adjoining backyards and basements of houses. Concomitantly, residents began displaying alarming incidences of miscarriages, birth defects, cancer, and respiratory ailments. As a result, the federal government declared Love Canal a disaster area, evacuated the neighborhood, and then created the Superfund program, which funded the clean up of Love Canal as one of its first projects.

protection of community health and safety, the environment, and local economies. Not surprisingly, the U.S. EPA (Environmental Protection Agency) has been a champion of these efforts. Its Office of Solid Waste and Emergency Response has developed a list of 22 chemicals—from arsenic, benzene, and cadmium, through lead and mercury, to zinc—of specific interest to Superfund cleanup efforts (also see the annual U.S. EPA *Toxic Release Inventory* publications). In part, the EPA EMAP (see Chapter 5) project arose from these concerns and efforts. In addition, numerous scientific expositions, such as Hird's *Superfund: The Political Economy of Environmental Risk* (1994) and Ott's *Environmental Statistics and Data Analysis* (1995), have appeared.

Spatial statistics play a central role in much of the proper scientific analysis of environmental pollution data. Cressie (1991, 4) notes that spatial prediction (the missing data problem of Chapter 1) is very important because people living in nonsampled locations also need to be informed about the quality of their environment. With regard to the Love Canal case, for example, the EPA took thousands of soil samples from the area, which it then needed to map in order to understand the geography of the problem there. In 1995, the National Center for Environmental Decision-making Research (NCEDR) was established to provide national leadership for strengthening environmental decision making (Dobson 1996). One of its goals is to improve understanding of the data and models information components of the decision-making process; NCEDR will be using GISs (geographic information systems)—a technology that facilitates adequate quantitative assessments of spatial variability and spatial heterogeneity—to help achieve this goal. Christakos (1992, 170) notes that environmental research is dominated by "traditional methods of classical statistics, which are incapable to capture important features of" the geography of pollution, hence ignoring "the relative distances of the sample locations/instances over space ... that enter the analysis of the correlation structure." Spatial statistical practices outlined in this chapter are responsive to the primary goal of NCEDR, especially by addressing current weaknesses in environmental modeling endeavors.

A number of diverse pollution data sets are explored in this chapter, ranging from dioxin contamination, near-coastal waters pollution, chemical elements in groundwater, coal-ash content, mountain water acidity, and the benchmark particulate stack emissions from coal-burning data. Lead is a toxic element, but is analyzed in the next chapter in terms of pediatric lead poisoning (in 1991 the U.S. EPA officially launched a concerted attack against lead poisoning). Christensen (1991) uses lead as one of his empirical illustrations, commenting that lead contamination in the soil around a smelter is likely to be correlated. He also notes that the linear semivariogram model (an additional one to those listed in §3.2.1) appears to be appropriate for such a situation, where the diffusion of the contaminate is away from a logical origin, and resembles a random walk in that its variability increases with increasing distance from the smelting facility. Other georeferenced pollution data sets also are readily available. For example, Bailey and Gatrell (1995, 145, 249) furnish such data for radon gas levels in Lancashire,

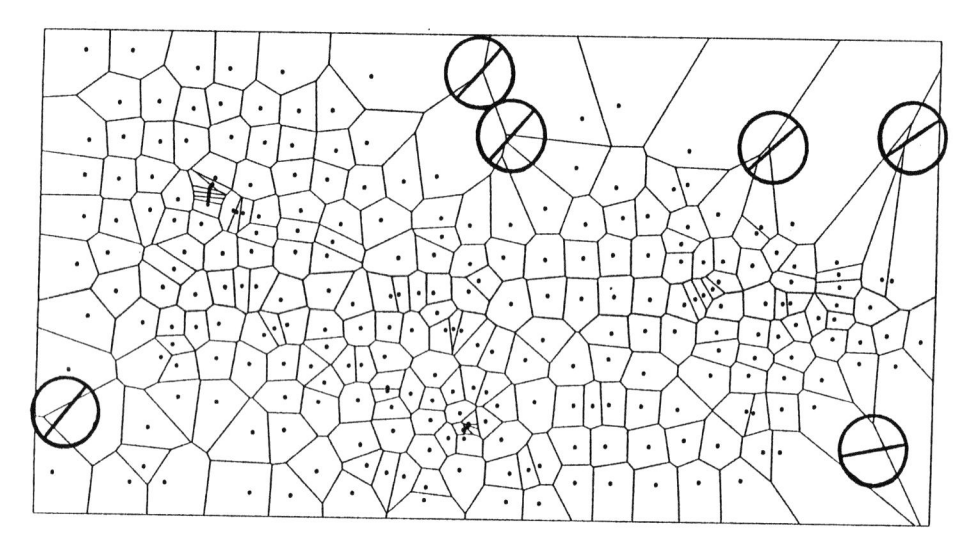

Map 7.1. Thiessen polygon surface partitioning for the smokestack locations (Cressie and Clark's data).

England, PCBs in south Wales, and emissions of nitrogen and ammonia in Europe. Both Hamilton (1992, 207) and Chatterjee et al. (1995, 164-76) reprint PCB concentrations measured in 1984 and 1985 for 37 U.S. bays and estuaries, data extracted from the Council on Environmental Quality's *Environmental Quality 1987-88* report. Walford (1995, 407-18) furnishes a data set that combines measures reported from a British national river quality survey and measures obtained from water samples from rivers in the North West Water Authority region—that were collected on randomly selected days from river segments and nearby factories. Hamilton (1992) presents tabulated georeferenced data on radon (p. 93) and on hydrocarbon pollution potential (pp. 186-87). Rasmussen (1992) reprints percentage of dwellings units with lead paint for Massachusetts communities (p. 70).

    Cressie and Clark (1994, 4.15-4.20) report georeferenced data for particulate stack emissions from the burning of coal at 272 locations in South Africa; similar to the single error of omission from the Wolfcamp aquifer data they reprint (mentioned in Chapter 5), 5 observations have been omitted from these tabulated data, reducing the sample size to 267. Environmental scientists are interested in this type of pollution phenomenon because each year thousands of lives are lost, tens of thousands of hospital visits are made, and billions of dollars in medical costs are incurred due to dangerous microscopic particles of soot and other air contaminants. The geographic distribution of stack locations, together with their associated Thiessen polygon surface partitioning, appears in Map 7.1 (each ignored polygon border is denoted by a slash crossing it). Analysis of these data serves as a benchmark in this chapter. The S-W (Shapiro-Wilk) statistic for this variable equals 0.97957, the MC (Moran Coefficient) equals 0.30206 [E(MC) ≈ -0.00369; $\sigma_{MC} \approx 0.03649$], the GR (Geary Ratio) equals 0.64719, Bartlett's

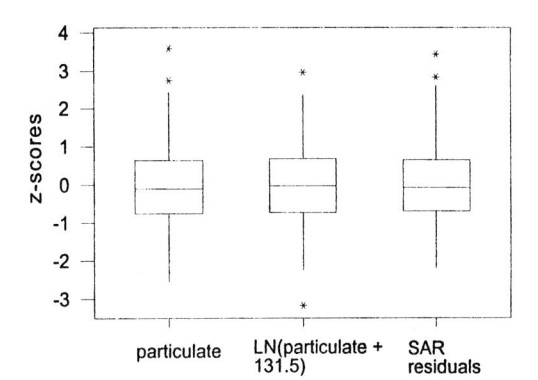

Figure 7.1. Boxplot metamorphosis of Cressie and Clark's particulate emissions pollution data: raw measures to SAR residuals.

homogeneity of variance test statistic[4] equals 8.619††, and Levene's equals 2.591†. These diagnostics imply that (1) the frequency distribution for the particulate emissions adequately conforms to a normal distribution, (2) the particulate emissions data exhibit marked positive spatial autocorrelation, and (3) geographic variation across the four planar quadrants of this South African landscape is quite heterogeneous.

The best power transformation uncovered for these data is the logarithm— $\hat{\delta} = 131.49891$, $\hat{\gamma} = 0$, RSSE = 3.6×10^{-3}—yielding S-W = 0.98913 (slightly improving conformity to a normal frequency distribution), Bartlett = 7.072† and Levene = 2.510† (implying persistence of nonconstant variance), and MC = 0.30689 (GR = 0.65198). The accompanying boxplots (Figure 7.1) reveal that this logarithmic transformation replaces two extreme high emission values with both an extreme high and an extreme low emission value. Consequently, a power transformation should not be used with these pollution data, as it fails to dramatically improve upon the detected nonstationarity in spread, while with either measurement scale the frequency distribution closely mimics a bell-shaped curve.

Presence of the extreme high emission value (with a z-score of roughly 3.6), located along the north-central periphery of the geographic arrangement of stacks, was adjusted for in the OLS (ordinary least squares) specification by creating a binary indicator variable for it. Results are based upon the standardized normal deviate for particulate emissions (z_y). On the Moran scatterplot (Figure 7.2), this outlier is the extreme positive horizontal axis point; notably its position within the scatter of points is not truly inconsistent with the remainder of the scatterplot cloud. The introduction of this indicator variable essentially stabilizes the variance across the four quadrants of the geographic landscape but at the cost of slightly deteriorating conformity with a normal frequency distribution. A very poor statistical description of these data is furnished by a standard OLS specification, with the outlier accounting for nearly 5% of the variation in particulate emissions, while the statistical description furnished by an SAR model specifica

---

[4] Statistical significance is denoted throughout by † for a 10%, †† for a 5%, and ††† for a 1% level.

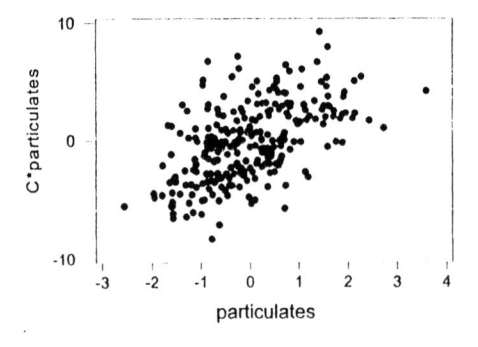

Figure 7.2 (left). MC scatterplot of OLS residuals (Cressie and Clark's data)..

Figure 7.3 (right). Traditional homogeneity of variance SAR residual plot for particulate emissions pollution (Cressie and Clark's data).

tion is modestly good:

	parameter estimates (for z_y-scores)		
	OLS	SAR(W)-1	SAR(W)-2
ρ	*	0.59035	0.60987
b_0	-0.01351	0.00813	0.02370
b_{1176}	3.60673	2.64139	*
(pseudo-)R^2	0.049	0.321	0.306
residuals			
S-W	0.97570[†]	0.97722	0.97982
A-D	0.850[††]	0.450	0.499
R-J	0.9947[††]	0.9969	0.9967
K-S	0.045	0.042	0.043
Bartlett	5.505	2.154	2.976
Levene	1.954	0.598	0.682
MC	0.29579	-0.00229	-0.00862
GR	0.66527	0.96648	0.96303

These results indicate that both the suspected outlier and the heterogeneity of geographic variance are attributable to spatial dependency effects. Once the presence of non-0 spatial autocorrelation is taken into account, the suspected outlier accounts for less than 2% of the variance in particulate emissions [SAR(W)-1]. The residuals of the pure spatial autoregressive model [SAR(W)-2] are well behaved, displaying a bell-shaped frequency distribution profile (their boxplot is nearly identical to the one for the raw data; Figure 7.1), constant geographic variance, essentially constant variance across the range of mean responses (Figure 7.3), and spatial independence. Because the outlier is less conspicuous in the context of a SAR specification, the pure SAR model [SAR(W)-2] is preferred here (invoking the scientific law of parsimony). Finally, these South African coal-burning particulate stack emissions are moderately positively spatially autocorrelated. A spatial SAR model specification has been employed here because from a conceptual point of view a researcher should expect this spatial auto

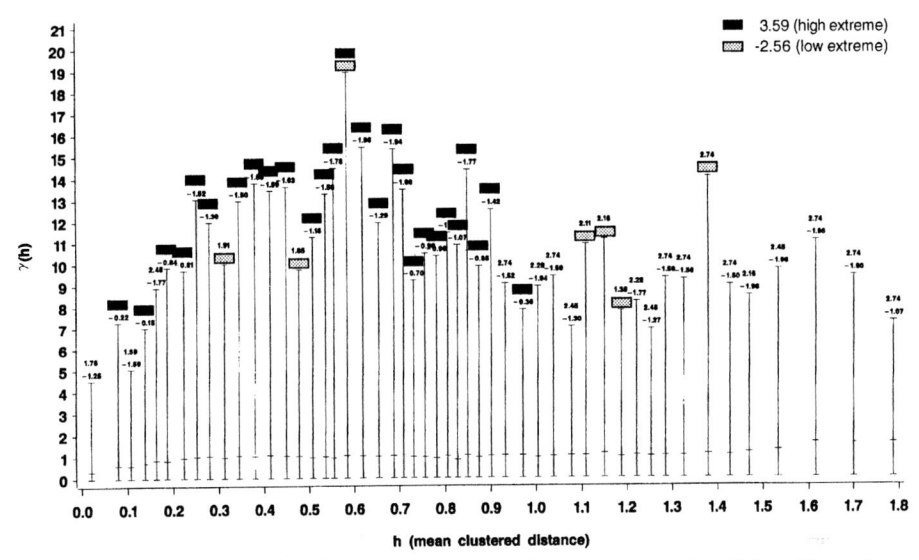

Figure 7.4. Semivariogram plot displaying the mean and extreme values for OLS residuals for particulate emissions pollution (Cressie and Clark's data).

correlation to arise from missing variables, such as quantity and grade of coal burned.

A geostatistical analysis of residuals from the OLS estimated regression equation also reveals the presence of moderate spatial autocorrelation latent in these data. The importance of including an outlier indicator variable again can be seen in Figure 7.4, where the gamma value of 3.59 is present in 27 of 33 distance classes, up to the rescaled[5] distance class of 0.971. That this value is dominant throughout this distance class and absent beyond it indicates that the outlier value lies toward the center of the map (§2.6.2). In fact, a map of the emissions variable shows that the position of this value is located almost exactly at the center of the region. The dominance of this outlier is removed when its presence is uniquely accounted for in the OLS regression equation.

Fifty distance classes were specified, which resulted in over 600 distance pairs per class up to the class containing 0.7 (Figure 7.5a). Results for the selected semivariogram model fits to this scatter of points are as follows:

model	RSSE	C_0	C_1	r
exponential	0.15398	0.11104	0.84363	0.08971
Bessel	0.13888	0.24557	0.70670	0.06678
Gaussian	0.12167	0.38003	0.56821	0.13951
spherical	0.10710	0.29328	0.65459	0.28734
wave/hole	0.15086	0.50476	0.42621	0.07270
power	0.43146	0.00000	1.05640	*	$\lambda = 0.15083$.

---

[5] Distances are rescaled by dividing all (U, V) coordinate values by the maximum absolute U or V coordinate value in the set of georeferencings attached to the sample of areal units.

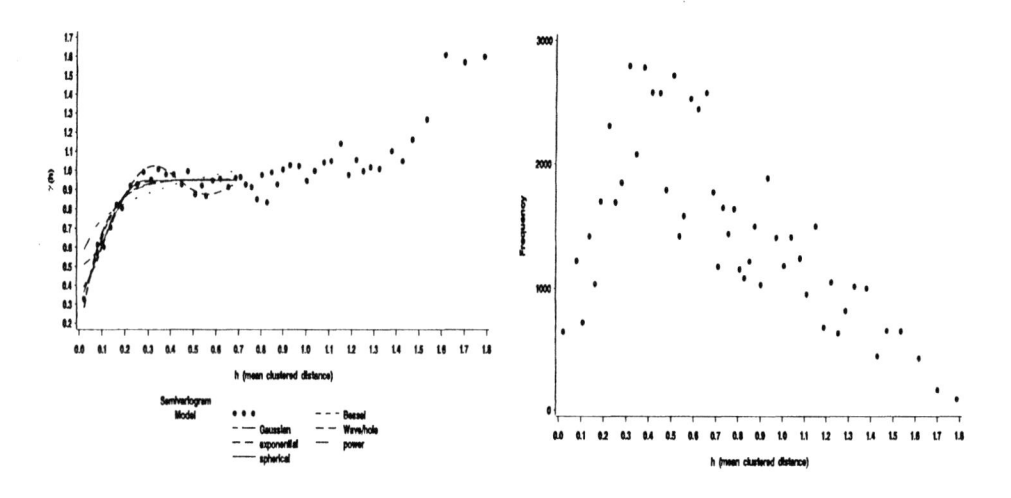

Figure 7.5. Left: (a) Semivariogram-type plot of the frequency of semivariance values in each distance class for Cressie and Clark's data. Right: (b) Semivariogram plot of OLS residuals for Cressie and Clark's data, with selected semivariogram model curves superimposed.

Based on the lowest RSSE, the spherical model achieves the best fit and also can be considered the model of best fit in terms of its range parameter value, given the scatter of points in the semivariogram plot (Figure 7.5b). Examination of the pattern of gamma values in Figure 7.5b reveals increasing gamma values as the distance from the origin increases, up to a rescaled distance of approximately 0.25. Such a pattern suggests that the preferred transition semivariogram model for describing these points also should have a relatively large range parameter value.

As an aside, air pollution measures are furnished by Sokal and Rohlf (1981) for 41 selected U.S. cities and are reprinted in *A Handbook of Small Data Sets* (Hand et al. 1994).

7.1. SOUTHWESTERN PENNSYLVANIA COAL ASH

Cressie (1991, 47) reprints 208 coal-ash core percentages reported by Gomez and Hazen (1970) for the Robena Mine Property in Greene County, Pennsylvania. These data come from the Pittsburgh coal seam that covers much of southwestern Pennsylvania, northeastern Ohio, and northern West Virginia, and derive from selected locations forming a regular square tessellation grid. A contour map of these data appears in Map 7.2. A typical coal core content analysis measures the weight percentage of moisture, ash, volatile matter, and fixed carbon (the material, other than ash, that does not vaporize when heated in the absence of air). Chemical analyses often are done, too, in order to determine the carbon, hydrogen, sulfur, nitrogen, oxygen, moisture content, and trace elements (e.g., mercury, chlorine) present, particularly when subsequent coal burning occurs where environmental pollution needs to be reduced. Ash is created by the burning of a variety of minerals contained in coal in varying proportions. This material fails to contribute

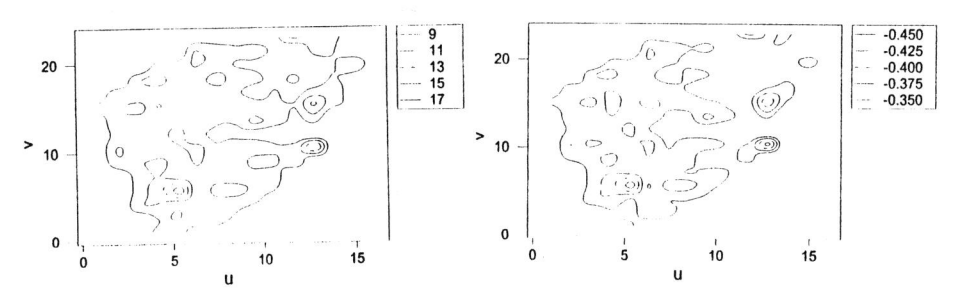

Map 7.2. Contour map of raw coal-ash values (Cressie's data).

Map 7.3. Contour map of arcsine-transformed coal-ash values (Cressie's data).

to the heating value of coal and in most cases becomes an undesirable residue (at high temperatures coal ash becomes sticky and eventually forms molten slag) and a source of pollution. These coal-ash data also could have served as the benchmark data set for this chapter had they not contained a missing value. Rather, first the known values are analyzed in this section, and then this unknown missing value is estimated in Chapter 9.

The coal-ash data are recorded as percentages and as such should not be expected to have a frequency distribution that conforms to a normal distribution. An optimal power transformation could be applied (although on empirical grounds, $\frac{y_{max}}{y_{min}} < 3$ argues against doing so), as could a logit transformation. Another popular transformation is $SIN^{-1}(\sqrt{proportion_i})$, which transforms a sample proportion into the angle whose sine is $\sqrt{proportion}$, expressed in either degrees or radians. Advantages of this arcsine transformation are that it induces a constant variance and it tends to improve conformity with a normal distribution; in contrast, the logit transformation renders a variance that is a function of the expected value. A shifting of the resulting angles, of the form $SIN^{-1}(\hat{\alpha}\sqrt{proportion} + \hat{\delta})$, is somewhat more effective in stabilizing variance (after Kotz et al. 1982, vol. 1, 88-89). The angle of rotation, represented by $\hat{\delta}$, is restricted to $\pm 45°$. These data appear to benefit from either a power or an arcsine transformation:

raw	power transformation	$SIN^{-1}(\hat{\alpha}\sqrt{proportion}+\hat{\delta})$
	$\hat{\delta}$ = -3.56597	$\hat{\delta} = -\frac{\sqrt{2}}{2}$
	$\hat{\gamma}$ = 0	$\hat{\alpha}$ = 0.88180
	RSSE = 2.0×10^{-2}	RSSE = 5.1×10^{-3}
S-W = 0.95263[†††]	S-W = 0.98529	S-W = 0.97499[†]
Bartlett = 12.480[†††]	Bartlett = 7.849[†††]	Bartlett = 9.218[††]
Levene = 1.072	Levene = 1.855	Levene = 1.644
MC = 0.31324	MC = 0.36047	MC = 0.33405
($\hat{\sigma}_{MC} \approx 0.05206$)		
GR = 0.70191	GR = 0.64390	GR = 0.67534 .

Boxplots for these three measurement scales (Figure 7.6) reveal the presence of at least one extremely high percentage outlier. Although, conceptually speaking, the arcsine transformation should be the preferred one here, nonlinear parameter estimation convergence was complicated by the presence of a marked outlier value, a feature of the data highlighted by Cressie (1991). To obtain sensible power transformation parameter estimates here, then, this outlier was treated temporarily as a missing value and estimated with an EM-type of procedure (see Chapter 1) after attaching a constraint that it had to be larger than the second largest value in the data set. The discarded estimate obtained was 0.13296 (versus an observed value of 0.1761), with an accompanying statistic of S-W = 0.98107. This transformation gave sensible subsequent results, and hence these coal-ash data are analyzed here in their arcsine measurement scale version. A contour map for the arcsine transformation values appears in Map 7.3 and displays a map pattern almost identical to that appearing in Map 7.2.

Only a modest statistical description of these data is furnished by standard OLS estimation, with the outlier accounting for nearly 12% of the variation in coal-ash percentages, while the statistical description furnished by an SAR model is just slightly better:

	parameter estimates OLS	s.e.	SAR(W)	s.e.
ρ	*		0.16940	0.11802
b_0	-0.44717	0.00109	-0.44706	0.00129
$bu-\bar{u}$	-0.00280	0.00033	-0.00278	0.00039
I_{max}	0.09628	0.01567	0.09255	0.01547
(pseudo-) R^2	0.363		0.380	
residuals				
S-W	0.97640		0.97595	
A-D	0.576		0.611	
R-J	0.9933††		0.9928††	
K-S	0.045		0.057†	
Bartlett	6.049		6.539†	
Levene	1.461		1.690	
MC	0.09989		-0.00617	
GR	0.89536		1.00329 .	

In a graphical interpretation of the OLS results (Figure 7.7), the upper regression line describes a mean response for the single outlier, whereas the lower regression line describes a mean response for the remaining 207 observed coal-ash values. These results reveal that the weak positive spatial autocorrelation detected in these coal-ash data largely arises from a map pattern aligned with a southwest-northeast axis (see Cressie 1991, 32), possibly concomitantly with some other minor sources of spatial autocorrelation. The OLS residuals are well behaved, exhibiting marginally significant positive spatial autocorrelation (Figure 7.8) and relatively constant variance across the range of mean responses (Figure 7.9); the spatial autoregressive parameter estimate for the SAR specification may not be statistically significant (recall that the standard error of $\hat{\rho}$ may be unduly inaccurate; §2.7). A

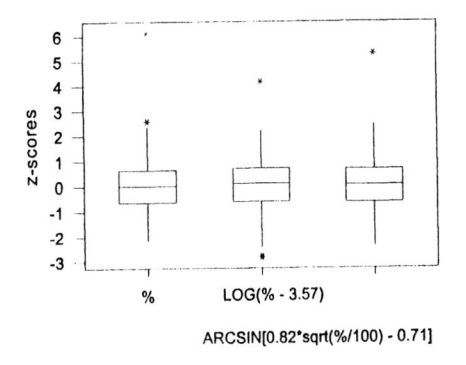

Figure 7.6 (left). Boxplots for the raw and two power-transformed coal-ash measures (Cressie's data)...

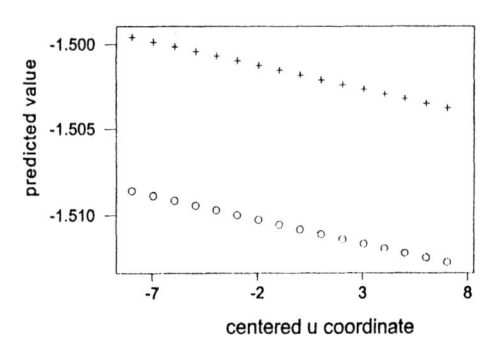

Figure 7.7 (right). Coal ash as a function of the georeferenced east-west coordinates regression lines for (top) the outlier population and (bottom) the common population (Cressie's data).

conspicuous outlier is present in these data, one that severely compromises the evaluation of statistical assumptions. The SAR specification is conceptually more appealing than is an AR specification because coal ash found at one location should not directly influence quantities of coal ash found at nearby locations. This SAR model yields reasonably well behaved residuals that are virtually void of spatial autocorrelation, but this is at the cost of slightly more asymmetry in the accompanying frequency distribution and slightly more variance heterogeneity. A CAR specification may be more suitable here, given that MC ≈ 0.1 is an approximation for its $\hat{\rho}$ estimate.

The outlier just described is also conspicuous when the arcsine transformed reexpression of the coal-ash variable is subjected to a geostatistical analysis. Examination of Figure 7.10, where the residuals were generated by removing the indicator variable representing the outlier from the OLS regression equation, reveals that a dominant value of 0.1 is present in all but the first and last few distance classes, indicating that the position of the outlier is somewhere near the edge of the map (§2.6.2). Specification of 50 distance groups provided a minimum of 208 distance pairs per class up to the maximum distance class

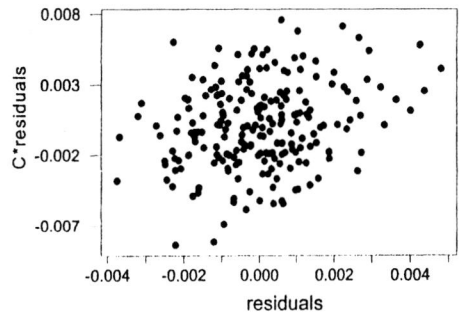

Figure 7.8 (left). MC scatterplot of outlier-adjusted OLS residuals for coal ash (Cressie's data).

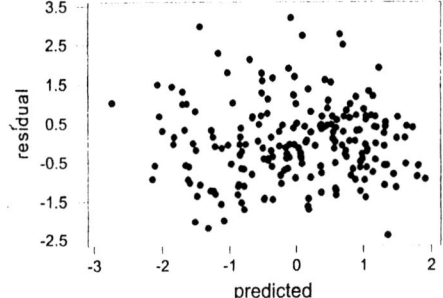

Figure 7.9 (right). Traditional homogeneity of variance outlier-adjusted OLS residual plot for coal ash (Cressie's data).

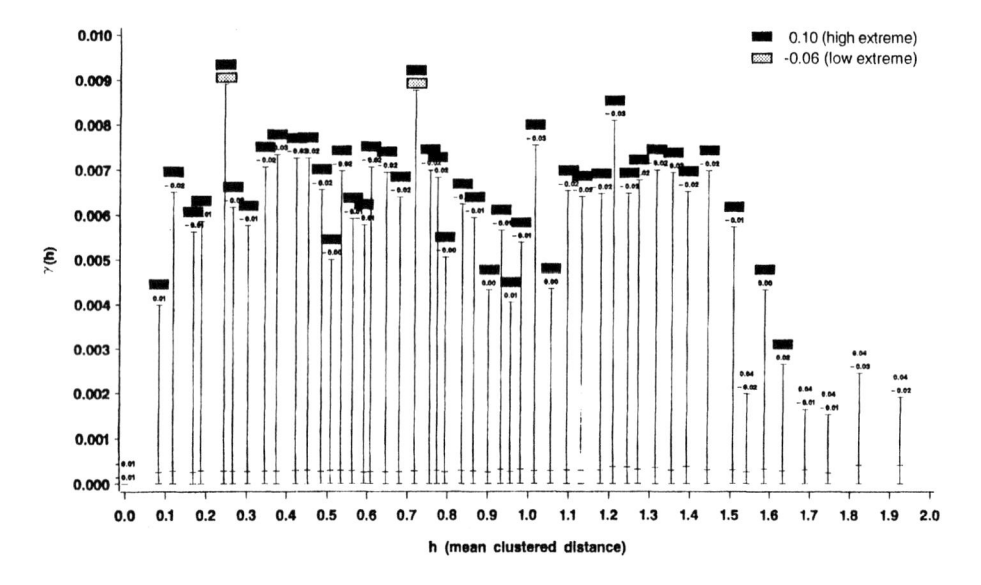

364   A CASEBOOK FOR SPATIAL STATISTICAL DATA ANALYSIS

Figure 7.10. Semivariogram plot displaying the mean and extreme values for arcsine-transformed coal-ash measures (Cressie's data).

including the rescaled distance value of 0.4 (Figure 7.11a). The parameter estimates for selected fits to the scatter of points in the semivariogram plot (Figure 7.11b) include

```
model        RSSE      C₀            C₁          r
exponential  0.03211   0.00000       2.303×10⁻⁴  0.02866
Bessel       0.03248   3.430×10⁻⁷    2.290×10⁻⁴  0.02028
Gaussian     0.03101   0.00000       2.300×10⁻⁴  0.04981
spherical    0.03076   0.00000       2.300×10⁻⁴  0.10581
wave/hole    0.25127   4.959×10⁻⁵    1.767×10⁻⁴  0.02863
power        0.20695   0.00000       2.779×10⁻⁴  *       λ = 0.13761 .
```

These estimation results indicate the presence of weak positive spatial autocorrelation, which mainly is attributable to a lack of any intermediate gamma values occurring between the 0 distance and its juxtaposed distance class. Low RSSE values in this case indicate a small amount of variation in gamma values across all but the 0 distance class included in the semivariogram model-fitting exercise. Based strictly on RSSE values, the spherical specification achieves the best fit, although there is very little difference between goodness-of-fit results for this and the exponential, Bessel, and Gaussian semivariogram models.

As an aside, although the 208 locations of these coal-ash data are a subset of a 16-by-23 regular square tessellation surface partitioning, their failure to coincide with a rectangular region introduces additional computational complexities. Furthermore, in analyses undertaken for this section, the missing value location, (7, 6), has been treated as a hole in the map. Data values analyzed here

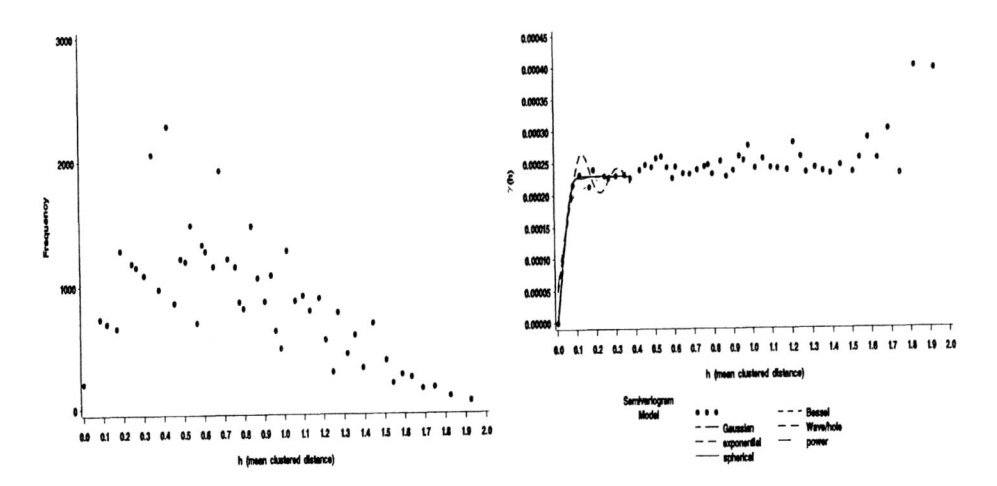

Figure 7.11. Left: (a) Semivariogram-type plot of the frequency of semivariance values in each distance class for Cressie's data. Right: (b) Semivariogram plot of OLS residuals for coal ash (Cressie's data), with selected semivariogram model curves superimposed.

deliberately were not converted to z-scores, because in Chapter 9 a raw data missing value estimate is computed for location (7, 6).

Another interesting study of particulate pollution appears in Kahn (1997), in which the relationship between manufacturing activity and particulate levels using U.S. county-level data is documented. Unfortunately, he overlooks the georeferenced nature of those data.

7.2. EMAP INDICATORS OF ECOLOGICAL CONDITION

Selected natural resources inventory EMAP data are analyzed in Chapter 5. The EMAP project also has collected georeferenced environmental quality indices in an attempt to describe and monitor the overall condition of the biota in ecosystems, including dissolved oxygen (DO) for water in the bays of the Virginian near-coastal biogeographical province. DO is used in the biochemical reactions by which the nutrients in wastes present in the water are broken down. Hence, as wastes are discharged into the bays of this near-coastal region, the amount of DO in their water declines. This variable has been subjected to ANCOVA (analysis of covariance). In this context, DO seems to benefit somewhat from being subjected to a power transformation:

raw	power transformation
	$\hat{\delta}$ = 2.48350
	$\hat{\gamma}$ = 2.65843
	RSSE = 2.8×10^{-2}
S-W = 0.86424†††	S-W = 0.96012†
Bartlett = 2.456	Bartlett = 2.234
diff. of means F = 1.00	diff. of means F = 1.48 .

Slightly better conformity with a normal distribution and slightly more stable variance are attained with the power-transformed measures. The ANOVA (analysis of variance) assumption of normal distributions within groups, not simply across all groups, requires inspection of the following S-W diagnostic statistics:

```
bay acronym                      D         LIGSN       N          C
S-W:  raw                     0.80270     0.84240   0.80170    0.90603
S-W:  power-transformed       0.93226     0.94844   0.92075    0.93469 .
```

Not only does the overall frequency distribution better mimic a normal distribution when subjected to a power transformation, but individual bay frequency distributions also dramatically improve in their respective conformity to normality.

ANCOVA results for these data are as follows:

```
                     ANCOVA
                  sum of squares        parameter estimates
b₀                     *                     - 200.51
u                   100181                      2.44
temp                549554                     15.30
bay                 350462
  I₁ - I₉                                     261.98
  I₂ - I₉                                    - 67.31
  I₅ - I₉                                    - 392.78
I_outliers          336576                     347.81
  (#19, #53, #55—after removal of Chincoteague Bay)
error               644288
TOTAL              1981060
R²                   0.675

residuals                                 MC          GR
A-D         0.374       Chesapeake     0.08130     0.99979
R-J         0.9912                    (0.15617)
K-S         0.056       Delaware       0.02436     0.94568
Bartlett    5.919                     (0.13608)
Levene      2.668†      Long Island-
                        Nantucket    -0.03757     0.78590
                                     (0.19245) .
```

These results insinuate that any latent spatial autocorrelation detected in these data is attributable to simple map pattern—more specifically, to a pronounced east-west trend. A second source of variation in DO is water temperature. And a third is the bays themselves, with Delaware Bay having substantially more DO than does Chesapeake Bay, with the Long Island near-coastal region having approximately the same level of DO as does Chesapeake Bay, and with Nantucket Bay having substantially less DO than does Chesapeake Bay. The ANCOVA results indicate that only the average level of DO varies across these bays, yielding parallel regression lines for them. In terms of residuals, heterogeneity of variance across the bays is tolerable, once conspicuous DO outliers have been taken into account. The residuals conform quite well to a normal frequency distribution, and different levels of trivial, trace spatial autocorrelation are present in this set of bays.

Geostatistical analyses for DO were performed separately for the Chesapeake, Delaware, and the Long Island Coast (Long Island, Great South, and Nantucket bays combined) bays, using the residuals from the preceding OLS regression analysis of the power-transformed measures. All three bays combined give a total of 69 observations; because of its isolation from the other locations, the observation for Chincoteague Bay was dropped. To take into account the relatively small sample sizes for each of the three data sets, 10, rather than the usual 50, distance classes were specified. Details for number of distance pairs, maximum rescaled distances used for modeling purposes, and minimum number of distance pairs per distance class for each of the three data sets are as follows:

bay	number of distance pairs	maximum distance	minimum number of distance pairs/distance class
Chesapeake	26^2	0.30	54
Delaware	25^2	0.08	25
Long Island Coast	18^2	0.60	20

Selected models fitted to each of the three semivariogram plots (Figures 7.12a-c) yield the following parameter estimates for these three data sets:

Chesapeake Bay

model	RSSE	C_0	C_1	r	
exponential	0.52476	0.00000	16450.4	0.02652	
Bessel	0.49299	0.00000	16478.7	0.01779	
Gaussian	0.47187	0.00000	16424.6	0.03225	
spherical‡	0.47161	0.00000	16424.3	0.06909	
wave/hole	0.27705	9731.60	6714.9	0.03062	
power	0.88036	0.00000	18278.6	*	$\lambda = 0.08153$

‡Initial parameter estimate values for the spherical and wave/hole semivariogram models were respectively specified explicitly as $C_0 = 0.0001$, $C_1 = 16000$, and $r = 0.1$; and, $C_0 = 9730$, $C_1 = 6720$, and $r = 0.03$.

Delaware Bay

model	RSSE	C_0	C_1	r	
exponential	0.00150	0.65600	7758.99	0.01195	
Bessel	0.00256	10.2183	7698.45	0.00821	
Gaussian	0.00459	1.0091	7648.86	0.01658	
spherical‡	0.00471	0.00000	7646.70	0.03436	
wave/hole	0.00007	11.7779	7495.37	0.00290	
power‡	0.04058	0.00000	14884.30	*	$\lambda = 0.22611$

‡Initial parameter estimate values for the spherical and power semivariogram models were specified explicitly as $C_0 = 0.0001$, $C_1 = 7000$, and, respectively $r = 0.04$ and $r = 0.5$.

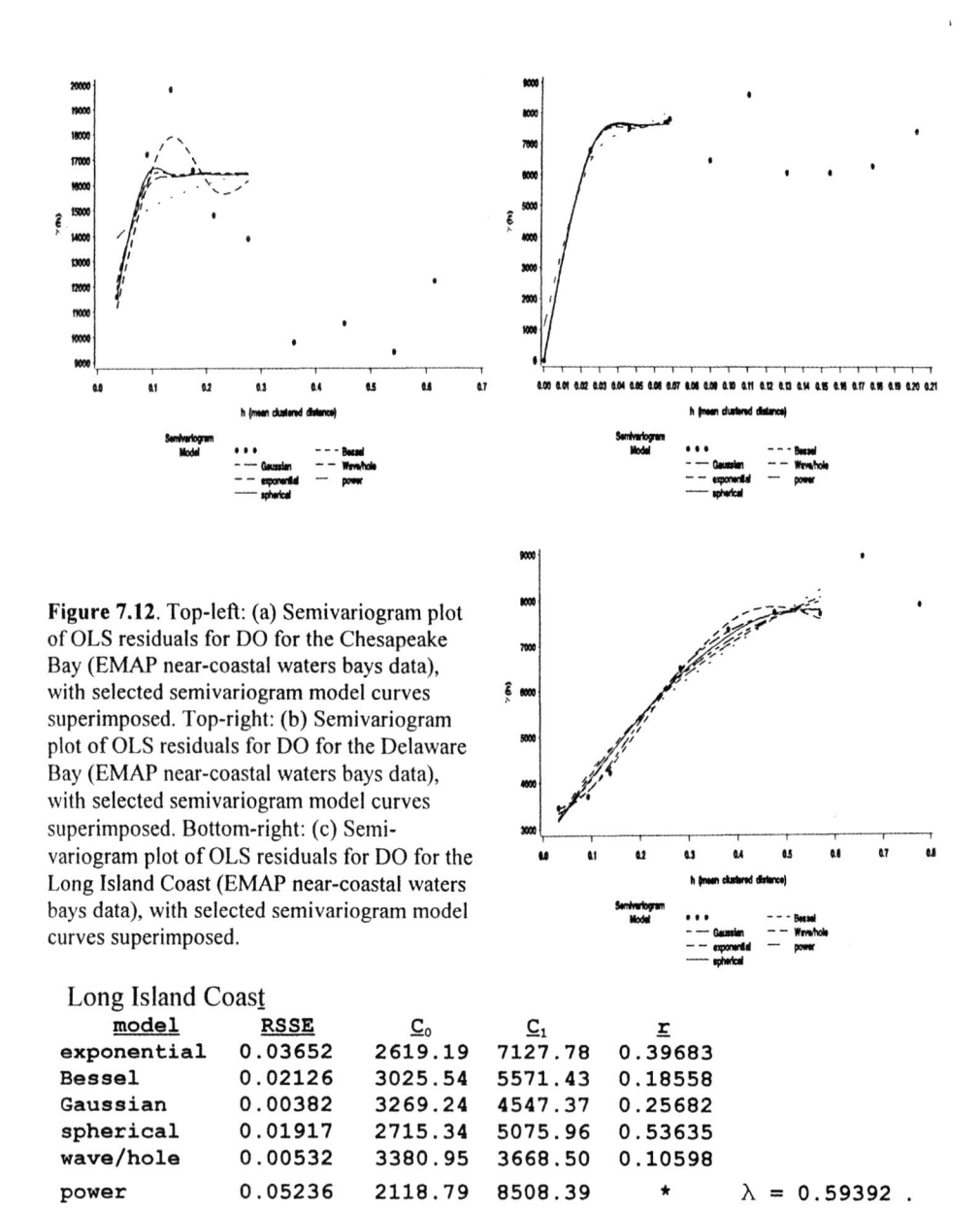

Figure 7.12. Top-left: (a) Semivariogram plot of OLS residuals for DO for the Chesapeake Bay (EMAP near-coastal waters bays data), with selected semivariogram model curves superimposed. Top-right: (b) Semivariogram plot of OLS residuals for DO for the Delaware Bay (EMAP near-coastal waters bays data), with selected semivariogram model curves superimposed. Bottom-right: (c) Semivariogram plot of OLS residuals for DO for the Long Island Coast (EMAP near-coastal waters bays data), with selected semivariogram model curves superimposed.

Long Island Coast

model	RSSE	C_0	C_1	r
exponential	0.03652	2619.19	7127.78	0.39683
Bessel	0.02126	3025.54	5571.43	0.18558
Gaussian	0.00382	3269.24	4547.37	0.25682
spherical	0.01917	2715.34	5075.96	0.53635
wave/hole	0.00532	3380.95	3668.50	0.10598
power	0.05236	2118.79	8508.39	*

$\lambda = 0.59392$.

Results from these geostatistical analyses do not parallel exactly those from the preceding ANCOVA analysis. The scatter of points in the semivariogram plot for the Chesapeake Bay (Figure 7.12a) indicates that model fits should be poor, an expectation confirmed by the relatively high RSSE values. Low levels of spatial autocorrelation are indicated by the small range parameter values. The MC for Chesapeake Bay is indicative of the presence of weak positive spatial autocorrelation; notably the wave/hole semivariogram model has the lowest RSSE. The

MC for Delaware Bay indicates the presence of even weaker positive spatial autocorrelation in its residuals; inspection of its scatter of points (Figure 7.12b) reveals a pattern typical of low to no spatial autocorrelation, which is characterized by the absence of intermediate gamma values between the 0 and the next largest distance classes. Also, when compared to the range parameter values for the Chesapeake Bay data set, lower range parameter values are obtained for Delaware Bay. Again, however, the wave/hole semivariogram model achieves the best correspondence (lowest RSSE) among the models fitted to this semivariogram plot, even though the results from the preceding ANCOVA fail to indicate that negative spatial autocorrelation is latent in these data. The MC for the Long Island Coast data set indicates the presence of a weak to moderate level of negative spatial autocorrelation. The presence of negative spatial autocorrelation, however, is not obvious from an inspection of Figure 7.12c, where the scatter of points displayed is more indicative of relatively strong positive spatial autocorrelation being present in this data set. In fact, the Gaussian semivariogram model has the lowest RSSE value here, indicating positive spatial autocorrelation in the residuals. The RSSE value for the wave/hole model, however, is very close to that of the Gaussian model (relative to the RSSE for the other semivariogram model fits), corroborating the results from the preceding ANCOVA analysis.

A second environmental quality indicator relates to sediment toxicity and is provided by average percentage of composite silt-clay content. Bay sediment comprises clay and silt, among other materials. The relative abundance of these two sediments in part depends upon the nature of the local drainage basin and the prevailing climate. Soil erosion increases the silt content of water, which upon deposit reduces surface areas and depths of bays. In addition, many materials that are detrimental to the ecology of these bays (e.g., excessive quantities of nutrients, heavy metals, pesticides, oil, certain bacteria, and toxic algae such as *Pfiesteria*) are deposited in these bays during the settling of silt and remain potential pollutants available for regeneration into a bay's water as long as they are present.

Mean percentage of silt also has been subjected to analysis of covariance and likewise seems to benefit from being subjected to a power transformation:

```
raw                          power transformation
                                  δ̂ = -0.84802
                                  γ̂ =  0.26197
                                  RSSE = 3.8×10⁻²
S-W = 0.80129†††             S-W = 0.94185††
Bartlett = 7.861††           Bartlett = 4.481
diff. of means F = 3.67††    diff. of means F = 6.11††† .
```

Noticeably better conformity with a normal distribution is attained with the power transformation, as is markedly more stable variance. Again, the ANOVA assumption of normal distributions within groups, not simply across all groups, requires inspection of the following S-W diagnostic statistics:

bay acronym		D	LIGSN	N	C
S-W:	raw	0.70140	0.91749	0.68363	0.78870
S-W:	power-transformed	0.88189††	0.84484†	0.95333	0.94698

These results indicate that the power transformation improves agreement between the overall empirical frequency distribution and a normal distribution, while dramatically improving conformity to normality of data from each of the four bay areas. Frequency distributions for the Delaware Bay and the near-coastal Long Island region still deviate significantly from normality, though. This power transformation helps stabilize the variance of mean percentage of silt and suggests that differences between the bays may be far more striking than is indicated by the raw data.

ANCOVA results for these data are as follows:

	ANCOVA sum of squares		parameter estimates
b_0	*		2.3720
u	0.5963		− 0.0061
Bay	10.4690		
$I_1 - I_9$			− 0.5397
$I_2 - I_9$			0.9581
$I_5 - I_9$			0.3621
error	29.3229		
TOTAL	40.3882		
R^2	0.274		

residuals				MC	GR
A-D	0.804††	Chesapeake		−0.11271	0.92721
R-J	0.9831†			(0.15617)	
K-S	0.099†	Delaware		0.08481	0.62957
Bartlett	5.573			(0.13608)	
Levene	2.163	Long Island−Nantucket		0.00770	0.84357
				(0.19245)	

These results imply that these data contain little spatial autocorrelation. The bays themselves are the principal source of variation in mean percentage of silt, with the starkest contrast being between the Long Island near-coastal region and Chesapeake Bay. These ANCOVA results indicate that only the average mean percentage of silt varies across these bays, yielding parallel regression lines for them. Residuals generated with this analysis are reasonably well behaved, with variance across the bays being tolerably stable, with a frequency distribution that conforms well to a normal frequency distribution, except possibly in the tails, and different levels of trivial, trace spatial autocorrelation being presents.

Residuals from the regression equation

$$\hat{Y} = 2.3720 - 0.0061 \times (u - \bar{u}) - 0.5397 \times (I_{Delaware} - I_{Chesapeake}) + 0.9581 \times (I_{Long Island} - I_{Chesapeake}) +$$

$$0.3621 \times (I_{Nantucket} - I_{Chesapeake})$$

were used to conduct geostatistical analyses. Like the previous dissolved oxygen geostatistical analyses, the total data set containing 69 observations was subdivided into three separate data sets, namely Chesapeake Bay, Delaware Bay, and the Long Island Coast. Once again, because of the relatively small sample sizes for each of the three data sets, 10, rather than the usual 50, distance classes were specified. Details for number of distance pairs, maximum rescaled distances used for modeling purposes, and minimum number of distance pairs per distance class for each of the three data sets are as follows:

bay	number of distance pairs	maximum distance	minimum number of distance pairs/distance class
Chesapeake	26^2	0.15	80
Delaware	25^2	0.10	25
Long Island Coast	18^2	0.60	20

Selected models fitted to each of the three semivariogram plots (Figures 7.13a-c) yielded the following parameter estimates for these three data sets:

Chesapeake

model	RSSE	C_0	C_1	r	
exponential	0.03692	0.00000	0.79646	0.05232	
Bessel	0.01192	0.00000	0.74120	0.02730	
Gaussian	0.00000	0.08594	0.62050	0.04516	
spherical	0.00000	0.03359	0.67276	0.10135	
wave/hole	0.00000	0.25563	0.39168	0.02517	
power	0.10379	0.00000	1.95559	*	$\lambda = 0.47408$

Delaware

model	RSSE	C_0	C_1	r	
exponential	0.01440	0.00586	0.25989	0.02937	
Bessel	0.03200	0.00701	0.24298	0.01610	
Gaussian	0.08565	0.00903	0.22671	0.02463	
spherical	0.06736	0.02316	0.22023	0.06401	
wave/hole‡	0.20848	0.08995	0.13642	0.01712	
power	0.00317	0.00000	0.64780	*	$\lambda = 0.37549$

‡Initial parameter estimate values for the wave/hole semivariogram model were specified explicitly as $C_0 = 0.0001$, $C_1 = 0.14$, and r = 0.02.

Long Island Coast

model	RSSE	C_0	C_1	r	
exponential	0.22451	0.00000	0.42085	0.06495	
Bessel	0.21803	0.07189	0.34678	0.04754	
Gaussian	0.19709	0.13114	0.28791	0.09967	
spherical	0.18318	0.08227	0.33770	0.20742	
wave/hole	0.10466	0.13202	0.27438	0.04500	
power	0.39375	0.00000	0.52651	*	$\lambda = 0.23666$.

Figure 7.13. Top-left: (a) Semivariogram plot of OLS residuals for the mean percentage of silt for the Chesapeake Bay (EMAP near-coastal waters bays data), with selected semivariogram model curves superimposed. Top-right: (b) Semivariogram plot of OLS residuals for the mean percentage of silt for the Delaware Bay (EMAP near-coastal waters bays data), with selected semivariogram model curves superimposed. Bottom-right: (c) Semivariogram plot of OLS residuals for the mean percentage of silt for the Long Island Coast (EMAP near-coastal waters bays data), with selected semivariogram model curves superimposed.

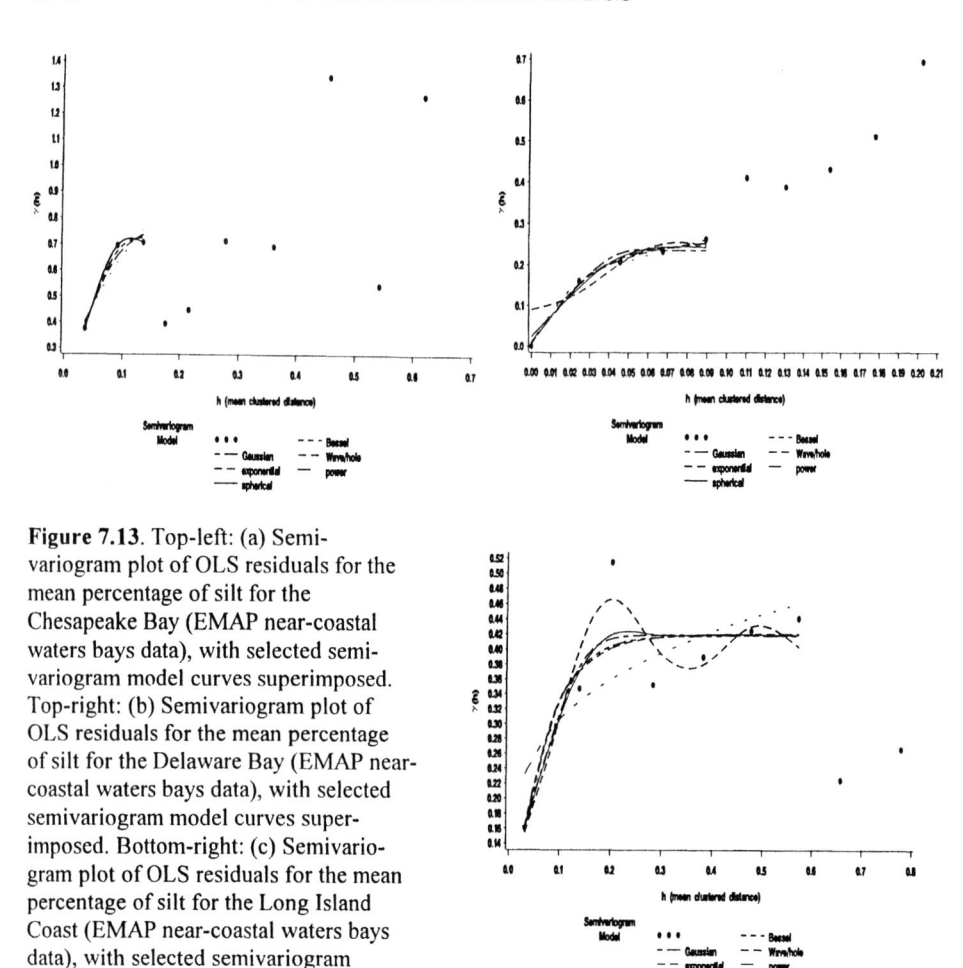

In this silt variable case, the MC for the Chesapeake Bay data set indicates observable negative spatial autocorrelation in the OLS regression residuals. As seen in the preceding semivariogram model fits for these data, the Gaussian, spherical, and wave/hole models all achieve perfect fits as well as RSSE values of 0. Such phenomenal RSSE values do not allow discrimination among these three models based on the RSSE values alone and actually result from using gamma values for only the first three distance classes to construct the semivariogram plot (Figure 7.13a)—three nonlinear equations in three unknowns. A choice of the spherical semivariogram model might be most appropriate here, considering the slope of the fitted curve between the first and second distance classes. The slope would belie a relatively large value of the range parameter, which is equivalent to the range for the spherical model, the single specification that most closely corresponds to this expectation. Examination of the range parameter for the wave/hole model does not support the level of negative spatial autocorrelation implied by the MC. The MC for the Delaware Bay data set indicates the presence of a weak level of positive spatial autocorrelation in the OLS regression residuals. Inspection of the scatter of points in the semivariogram plot (Figure 7.13b), however, could be interpreted in two different ways: high positive spatial autocorrelation, or omission of an explanatory variable in the OLS regression leading to a nonconstant mean across the surface— typical of the case when a power semivariogram model should be employed. In fact, the power semivariogram model has the lowest RSSE, although the exponential model seems to provide an acceptable alternative specification if a nontransition semivariogram model is desired, as might be implied if attention is restricted to only the first five distance classes when modeling the semivariogram plot. The scatter of points in the semivariogram plot for the Long Island Coast (Figure 7.13c) represents a transition model, excepting the gamma value for the fourth distance class. The MC from the spatial autoregressive analysis indicates that a low to negligible level of positive spatial autocorrelation is present in the residuals from the OLS regression. For this data set, semivariogram model selection based on the lowest RSSE suggests that the wave/hole model is the preferred specification; but the scatter of points displayed by the semivariogram plot does not indicate that spatial dependency is necessarily negative.

7.3. GREAT SMOKY MOUNTAINS WATER PH

As a quantitative index of the acidity of water, pH represents a real thermodynamic quantity, the Gibbs free energy of formation of positive hydrogen ions in a solution. It measures the negative logarithm of the concentration of this hydrogen ion in kilograms per cubic meter, converting this measure to a numeric scale ranging from 0 to about 14. The concentration of hydrogen ions in pure water, which is neutral in terms of its acidity, registers a pH value of 7. Water with a pH value less than 7 is considered acidic; water with a pH value greater than 7 is considered alkaline. The pH of naturally occurring water can range from very acidic conditions for which pH is roughly 3 (e.g., peat swamps) to very alkaline conditions for which pH is roughly 9; most lakes have a pH value lying between 6 and 8, while areas such as Whiteface Mountain, New York, have fog that often is 10 or more times as acidic as the local precipitation. Specific pH values for various lakes, streams, estuaries and bays is of interest to environmental scientists because most organisms are unable to live in conditions of extreme acidity or alkalinity. In addition, the pH levels of water bodies have been changing in recent decades due to the phenomenon known as acid rain, which generally begins with emissions into the atmosphere of sulfur dioxide and nitrogen oxide—pollution. These gases combine with water vapor in clouds to form sulfuric and nitric acids, which in turn become highly acidic precipitation when falling from clouds, often having a pH value not exceeding 5.

As an empirical example, Schmoyer (1994) prints a tabulation of georeferenced water pH and elevation data for 75 locations forming a network across the Great Smoky Mountains. These data are accompanied by a graphic displaying

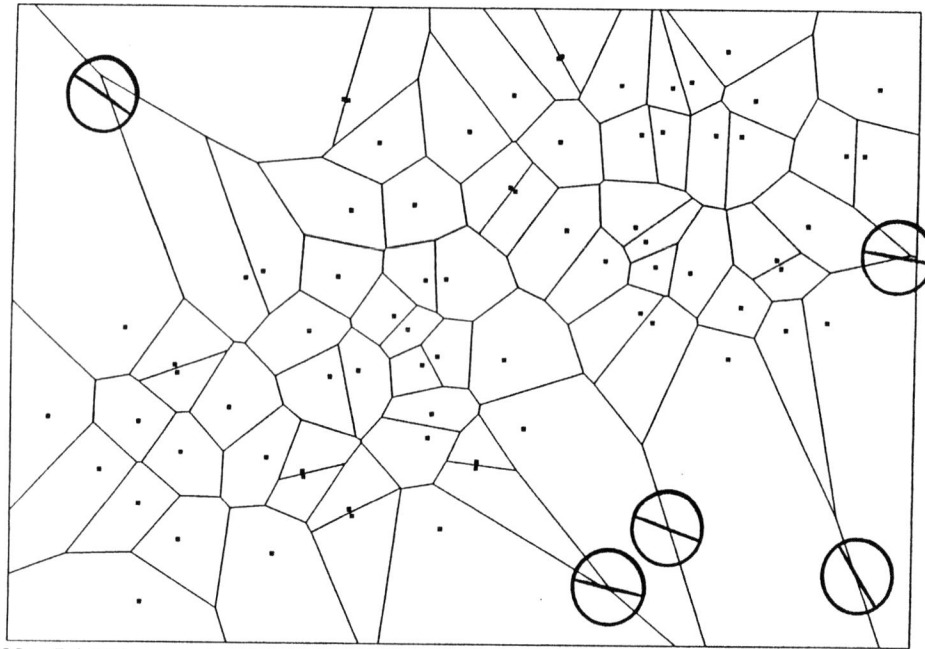

Map 7.4. Thiessen polygon surface partitioning for the Great Smoky Mountains pH monitoring locations (Schmoyer's data).

a Thiessen triangulation of the geographic coordinates, without any of the edge connections being deleted. Schmoyer (1994, 1513) notes that "[t]wo kinds of edges can be distinguished" and hence weights them differentially. The analysis reported in this section continues utilizing the modus operandi we promote in this book, namely to delete nonsensical linkages along the edge of a study region (see Chapter 5). The geographic distribution of pH monitoring locations, together with their associated Thiessen polygon surface partitioning, appears in Map 7.4 (each ignored polygon border is denoted by a slash crossing it). In this landscape pH seems to benefit from being converted to its original measurement scale (because $\frac{y_{max}}{y_{min}} < 3$, little should be gained from employing a Box-Cox power transformation):

```
raw                              exponentiation [1000×e^-pH]
S-W = 0.85966†††                 S-W = 0.96640
Bartlett = 9.025††                Bartlett = 6.790†
Levene = 1.369                   Levene = 1.572
MC = 0.21658  (ô_MC ≈ 0.07255)   MC = 0.18733
GR = 0.60739                     GR = 0.70196 .
```

The frequency distribution of the power-transformed version of these measures conforms to a normal distribution, the variance is more stable across the four planar quadrants of the region, and the degree of positive spatial autocorrelation detected in the raw data seems to be exaggerated. Boxplots for these data (Figure 7.14a)

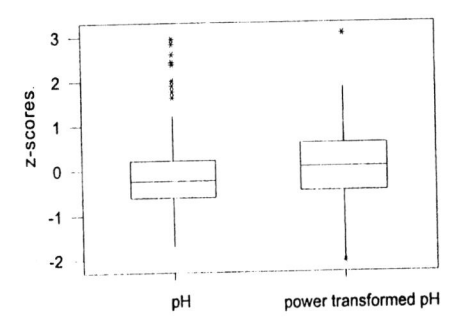

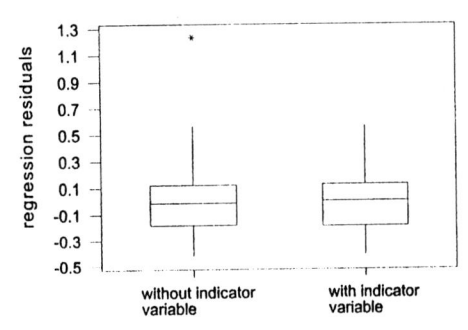

Figure 7.14. Left: (a) Boxplots for the raw and power-transformed pH measures (Schmoyer's data). Right: (b) Boxplots for the power-transformed pH measures (left) unadjusted for and (right) adjusted for the single outlier that is present, (Schmoyer's data).

disclose that a single outlier pH value persists after transformation; this extreme value accounts for a large percentage of the variability in the transformed version of pH but only a small portion of the remaining variance heterogeneity (Figure 7.14b).

A respectable statistical description of these data is furnished by standard OLS estimation, with the outlier accounting for nearly 19% of the variation in the power-transformed pH data, while the statistical description furnished by an SAR model is just slightly better:

```
           parameter estimates [1000×f(y)]
                    OLS        s.e.        SAR(W)       s.e.
ρ                    *                     0.36080     0.23802
b₀                  0.38904    0.07413     0.33105     0.09120
b_{v-v̄}            -0.00565    0.00095    -0.00607     0.00138
b_elevation         0.74617    0.11450     0.83867     0.13630
I_max               1.05791    0.20969     0.90643     0.20248
(pseudo-) R²        0.541                  0.575

residuals
S-W                 0.95786††              0.95408††
A-D                 0.564                  0.805††
R-J                 0.9871                 0.9829††
K-S                 0.095†                 0.093
Bartlett            6.948†                 4.923
Levene              2.105                  1.362
MC                  0.12415                0.01181
GR                  0.82590                0.93812 .
```

A graphical interpretation of the OLS results appears in Figure 7.15, supporting the notion that two regimes govern these data; the upper regression line describes a mean response for the single outlier, whereas the lower regression line describes a mean response for the remaining 74 transformed pH values. These results reveal that the positive spatial autocorrelation detected in these pH data partly is due to

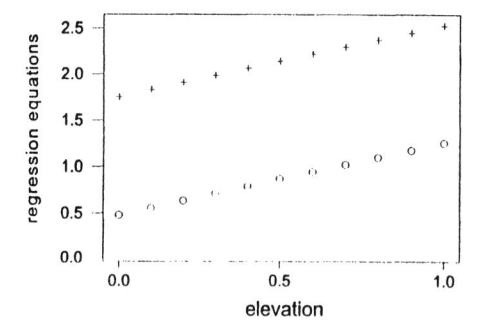

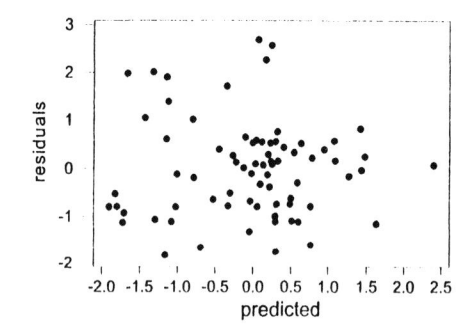

Figure 7.15 (left). pH as a function of elevation regression lines for (top) the outlier population and (bottom) the common population (Schmoyer's data).

Figure 7.16 (right). Traditional homogeneity of variance SAR residual plot for pH (Schmoyer's data).

a simple north-south map pattern gradient, possibly coinciding with some other sources of spatial autocorrelation. The OLS residuals are not too ill behaved, nearly conforming to a normal distribution, containing some positive spatial autocorrelation, and exhibiting somewhat disturbing variance heterogeneity across the four quadrants of the region. The spatial SAR specification produces a decrease in residual variance heterogeneity, at the expense of a slight increase in deviation from a normal distribution, but may not be statistically significant (recall that the standard error of ρ may be unduly inaccurate; §2.7). The SAR specification also generates residuals that are nearly void of spatial autocorrelation; this specification is conceptually more appealing within the context of acid rain, because a principal source of spatial autocorrelation should be the cloud patterns carrying acid rain to a region. Finally, the SAR residuals display a reasonable degree of variance homogeneity across the range of mean responses (Figure 7.16). The SAR residuals uncover a second possible outlier.

Prior to a geostatistical analysis of the OLS regression residuals, an outlier value was identified (see the preceding parameter estimation output), and then a decision was made to adjust for its presence by including an indicator variable for it in the OLS regression equation specification. As can be seen in Figure 7.17, which was constructed with the residuals from the OLS regression equation whose specification did not include an outlier indicator variable, the value of 1.02 dominates 20 of the first 24 distance classes. In fact, this outlier value is present only in the first two-thirds of the distance classes, indicating that its georeferenced position is located toward the middle of the map. Inclusion of an outlier indicator variable in the OLS regression equation resulted in removal of this dominant outlier pattern from the plot of high-low semivariogram values.

Retaining the standard number of 50 distance classes yielded a minimum of 26 distance pairs per class (Figure 7.18a) up to the maximum distance class including the rescaled distance value of 0.8, although the number of pairs per class is around 100 or more for most classes of this portion of the semivariogram plot. The scatter of points displayed by this plot (Figure 7.18b) indicates that a transition

POLLUTION 377

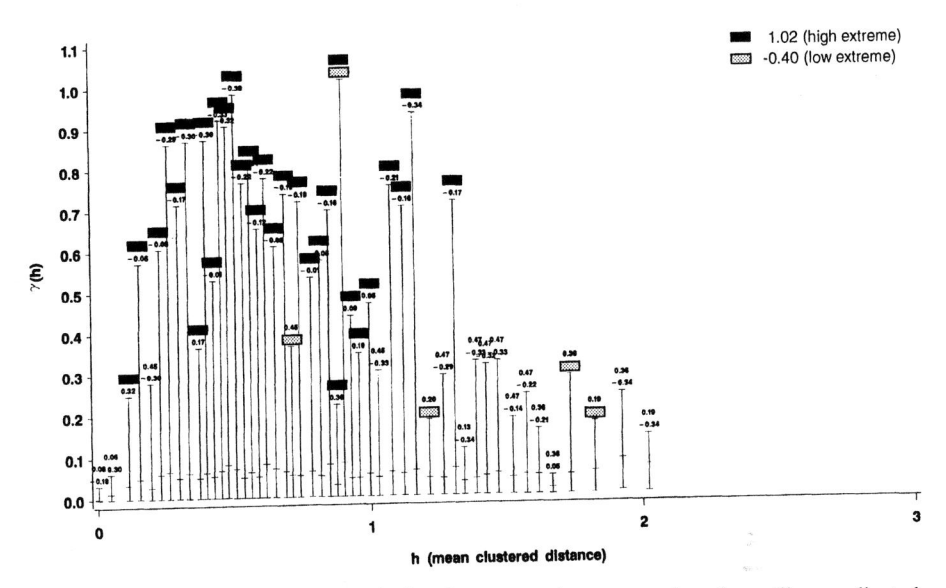

Figure 7.17. Semivariogram plot displaying the mean and extreme values for outlier-unadjusted OLS residuals for pH (Schmoyer's data).

model should fit these data well and that a moderate level of spatial autocorrelation should be present—an expectation corroborated by the MC and $\hat{\rho}$ values reported in the preceding spatial autoregressive analysis results. Estimation results for the selected semivariogram model fits to this scatter of points include

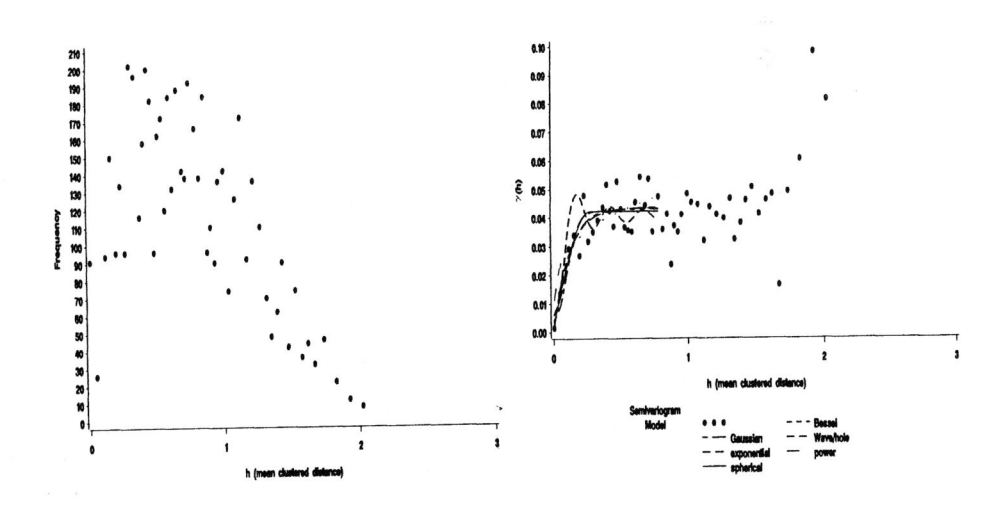

Figure 7.18. Left: (a) Semivariogram-type plot of the frequency of semivariance values in each distance class for Schmoyer's data. Right: (b) Semivariogram plot of outlier-adjusted OLS residuals for pH (Schmoyer's data), with selected semivariogram model curves superimposed.

model	RSSE	C_0	C_1	r
exponential	0.41946	0.00118	0.04247	0.12133
Bessel	0.42761	0.00359	0.03968	0.08137
Gaussian	0.45473	0.00436	0.03800	0.13226
spherical	0.45063	0.00238	0.03994	0.27730
wave/hole	0.55222	0.00591	0.03490	0.03914
power	0.46901	0.00000	0.05038	*

$\lambda = 0.25420$.

Here the exponential semivariogram model achieves the best fit, based on the RSSE values, although the difference in RSSE values between this and the Bessel semivariogram model is slight. Moderate RSSE values in this case result from the relatively wide scatter among the points in the semivariogram plot, particularly in distance classes whose average distance lies beyond the range parameters.

As an aside, Oliver and Webster (1986) find that the spatial dependency in subsoil pH may be described with a linear semivariogram model, whereas the spatial dependency in topsoil pH may be described with a power semivariogram model.

7.4. CHEMICAL ELEMENTS IN NORTHWEST TEXAS GROUNDWATER

The preceding section outlines some reasons why pH is of concern when studying environmental pollution. SO_4, the sulfate (also spelled sulphate) ion composing salts, is a naturally occurring chemical compound of sulfuric acid. The chemistry of groundwater divulges that the average concentration of sulfate, one of the ten most important dissolved matter constituents, is roughly 11 out of 120 parts per million. This amount is being modified by human activities, which have increased its quantity in this source of fresh water by as much as 30 percent in some instances. In the introduction to this chapter, arsenic is identified as an element of concern for Superfund remediation efforts. Also of interest is radioactivity, particularly within the context of hazardous waste. Andrews and Herzberg (1985) reprint georeferenced data from a northwest Texas pilot geochemical survey, consisting of 12 measurements taken on each of 127 groundwater samples (pp. 124-26); notably, they use the negative sign to denote censored concentration values. Measurements include pH, SO_4, arsenic, and uranium, furnishing yet another opportunity to study these elements, even if only in terms of achieving a better understanding their respective background rates of occurrence. The geographic distribution of these sample locations, together with their associated Thiessen polygon surface partitioning, appears in Map 7.5 (once again, each ignored polygon border is denoted by a slash crossing it).

As with the case of the Great Smoky Mountains data (§7.3), pH in this Texas landscape seems to benefit from being converted to its original measurement scale (because $\frac{y_{max}}{y_{min}} < 3$, again, little should be gained from employing a Box-Cox power transformation):

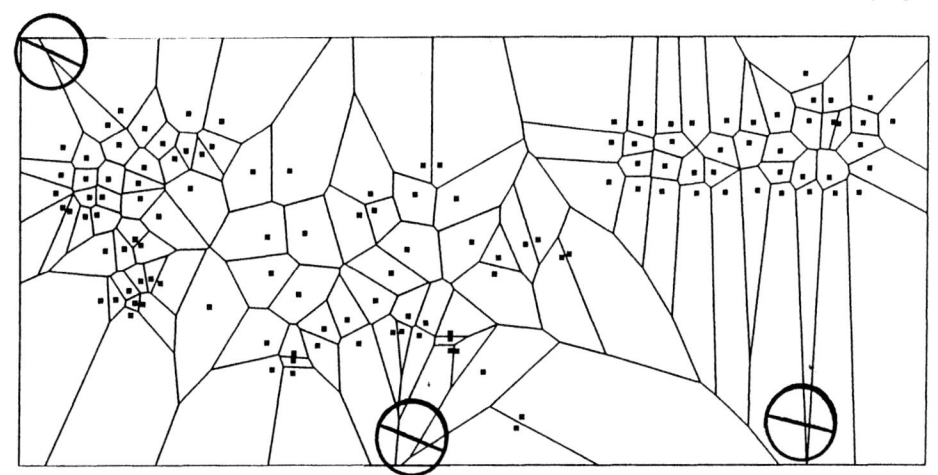

Map 7.5. Thiessen polygon surface partitioning for the northwest Texas groundwater pilot geochemical survey sampling locations.

```
raw                                    exponentiation [1000×e⁻ᵖᴴ]
S-W = 0.92326†††                       S-W = 0.97081†
Bartlett = 8.204†                      Bartlett = 5.986
Levene = 2.027                         Levene = 2.378†
MC = 0.09291 (σ̂_MC ≈ 0.05581)          MC = 0.05608
GR = 1.00435                           GR = 0.98727 .
```

Similar to that for the Great Smoky Mountains pH data, the frequency distribution of the power-transformed version of these measures better conforms to a normal distribution, the variance is stabilized across the four planar quadrants of the region, and the degree of positive spatial autocorrelation detected in the raw data seems to be both low (Figure 7.19) and exaggerated. Boxplots for these data (Figure 7.20) disclose that the power transformation removes all but one of the initially detected outlier pH values.

A very poor statistical description of these data is furnished by standard OLS estimation, with the outlier accounting for roughly 8% of the variation in the power-transformed pH data:

```
        OLS parameter estimates (based on z_y-scores)
                                    residuals
ρ            *                      S-W        0.94855†††
b₀          -0.02526                A-D        1.111†††
I_outlier    3.20841                R-J        0.9951
            (#119)                  K-S        0.054
R²           0.081                  Bartlett   4.367
                                    Levene     2.028
                                    MC         0.05481
                                    GR         1.01046 .
```

These results corroborate some of the findings of the Great Smoky Mountains pH analysis, namely that spatial autocorrelation is very low to nonexistent and that

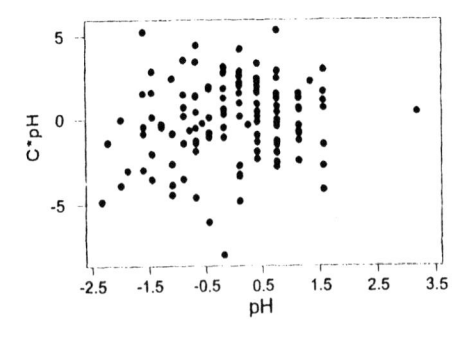

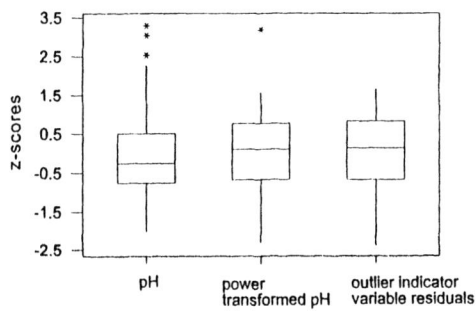

380   A CASEBOOK FOR SPATIAL STATISTICAL DATA ANALYSIS

Figure 7.19 (left). MC scatterplot of pH (northwest Texas groundwater data).

Figure 7.20 (right). Boxplot metamorphosis of the northwest Texas groundwater pH data: raw measures to outlier-adjusted power-transformed measures.

perhaps a CAR specification would be more suitable for characterizing georeferenced pH data. The OLS residuals obtained for this case are reasonably well behaved.

For a geostatistical analysis of the standardized-exponential transformed pH variable, exclusion of the indicator variable for the outlier results in the value of 3.18 being present in 9 of the first 18 distance classes used to construct the plot of high-low OLS regression values (Figure 7.21). In fact, this value appears in many of the larger distance classes, too, indicating that this outlier value is located toward the edge of the map. A plot of the analysis values verifies this expectation, as the outlier value is located in the lower left-hand region of Map 7.6. Inclusion of an indicator variable for this outlier removes its dominance; disturbingly,

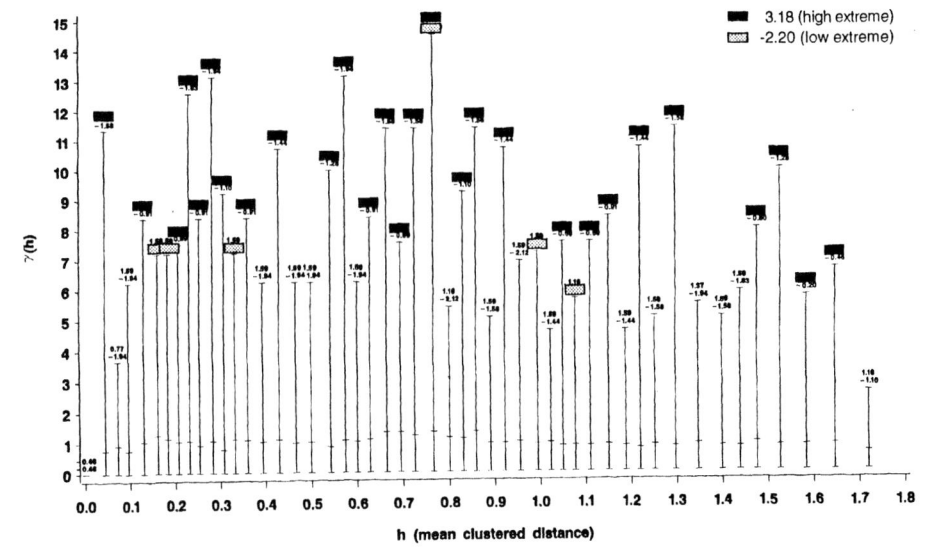

Figure 7.21. Semivariogram plot display-ing the mean and extreme values for outlier-unadjusted OLS residuals for pH (northwest Texas groundwater data).

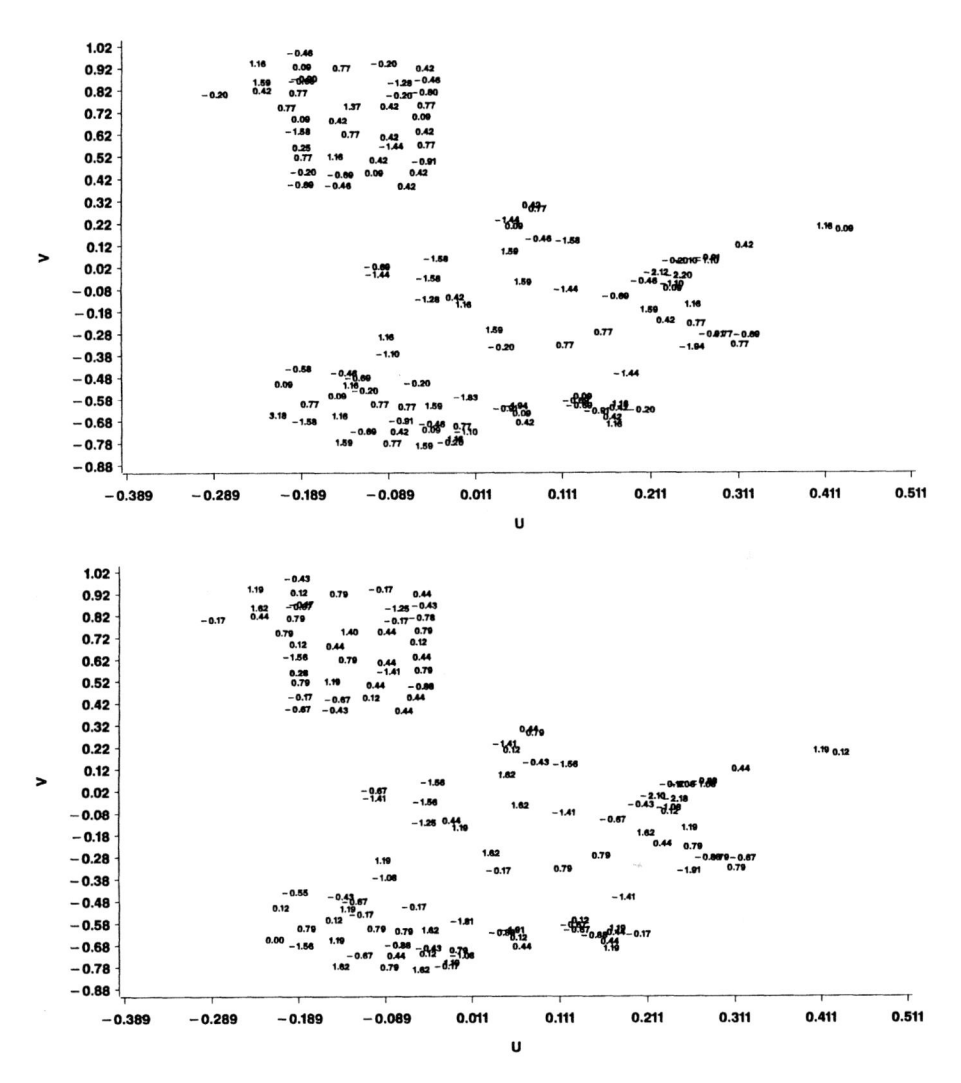

Map 7.6 (top). Geographic distribution of standardized exponential-transformed pH measures for the northwest Texas pilot geochemical survey.
Map 7.7 (bottom). Geographic distribution of standardized outlier-adjusted OLS residuals for Map 7.6.

accounting for 3.18 results in another value, 1.62, appearing in 13 of the first 18 distance classes, which suggests the presence of a second outlier value. Fortunately, Map 7.7 reveals that this value actually is georeferenced to four different locations on the map, three positions near the center and one position in the upper left-hand region; actually this value is not a univariate outlier. Particularly for the three sites for this value near the center of the map, inspection of the surrounding values suggests that these values are local spatial outliers, supporting the earlier claim about the utility of using the high-low plot to detect both aspatial and spatial outliers (§2.6.2). To be consistent with results for the spatial autoregressive

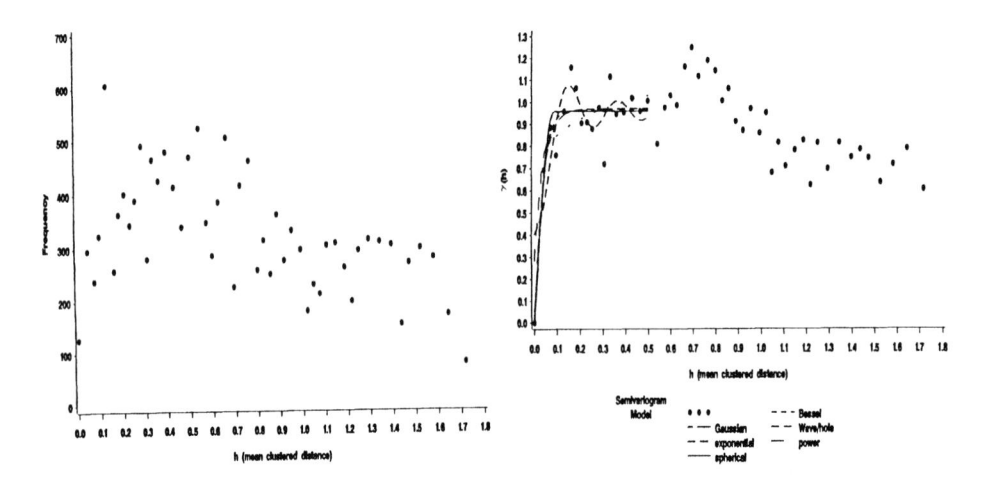

Figure 7.22. Left: (a) Semivariogram-type plot of the frequency of semivariance values in each distance class for the northwest Texas groundwater data. Right: (b) Semivariogram plot of outlier-adjusted OLS residuals for pH (northwest Texas groundwater data), with selected semivariogram model curves superimposed.

analysis, though, the model was not adjusted further in an attempt to account for these spatial outliers.

Specification of the usual 50 distance classes yielded a minimum of 26 distance pairs per class up to the maximum rescaled distance of 0.5 (Figure 7.22a), although the majority of classes contained approximately 100 pairs. The selected model specification fits to the semivariogram plot (Figure 7.22b) result in the following parameter estimates:

model	RSSE	C_0	C_1	r
exponential	0.29175	0.00516	0.96533	0.03629
Bessel	0.29783	0.00002	0.96335	0.02304
Gaussian	0.32000	0.01049	0.94913	0.03957
spherical	0.32212	0.01199	0.94736	0.08507
wave/hole	0.41510	0.40363	0.55323	0.03349
power	0.39735	0.00000	1.12023	*

$\lambda = 0.12029$.

Similar to results obtained for the Great Smoky Mountain pH variable, low levels of spatial autocorrelation are observed for the Texas groundwater pH data, characterized by very few intermediate gamma values between the 0 distance class and the succeeding first few distance classes (Figure 7.22b), coupled with low range parameter values. Based on RSSE values, the exponential semivariogram model achieves the best goodness-of-fit, with a trivial difference between this and the Bessel model. The earlier conclusion about these data that perhaps a CAR spatial statistical model provides the best description of them in terms of spatial autoregression is supported, at least in part, by the exponential model registering the lowest RSSE among the set of selected models fit to the semivariogram plot.

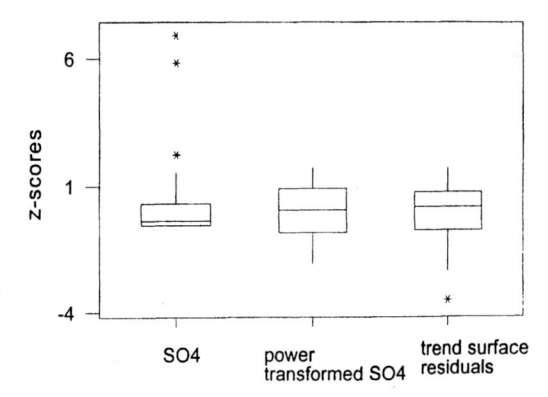

Figure 7.23. Boxplot metamorphosis of the northwest Texas groundwater SO₄ data: raw measures to linear trend surface residuals.

Sulfate measures (i.e., SO_4) in this Texas landscape also seem to benefit from being subjected to a power transformation:

```
         raw                                  power transformation
                                              δ̂ = -30
                                              γ̂ = 0.35422
                                              RSSE = 3.4×10⁻²
S-W = 0.86591†††                              S-W = 0.93824†††
Bartlett = 47.334†††                          Bartlett = 10.439††
Levene = 15.960†††                            Levene = 2.294†
MC = 0.52247 (σ̂_MC ≈ 0.05581)                 MC = 0.50025
GR = 0.43755                                  GR = 0.44381 .
```

The approximately cube-root transformation of these sulfate data renders a frequency distribution that more closely mimics a bell-shaped curve and dramatically reduces geographic variance heterogeneity. Unfortunately, both heavy tails and some nonconstant variance remain in the power-transformed measures (Figure 7.23). In either case, a significant, moderate level of positive spatial autocorrelation prevails.

A modest statistical description of these data is furnished by standard OLS estimation, which reveals a camouflaged outlier, located along the periphery of the landscape, that accounts for nearly 7% of the variation in the power-transformed SO_4 values. This aberrant SO_4 value also is a spatial outlier (Figure 7.24). The statistical explanation achieved here substantially improves when the OLS specification is replaced with a spatial autoregressive specification:

		parameter estimates				
	OLS-1	OLS-2	s.e.	AR(W)	s.e.
ρ	*			0.54323	0.11832
b₀	10.24700	10.36224	0.34268	10.48081	0.65136
bu-ū	− 8.58482	− 9.70190	2.20285	− 5.33572	2.12949
bv-v̄	− 5.19695	− 5.46075	0.64450	− 2.79636	0.80471
b_outlier (#4)	*	−14.63591	3.90662	−15.66683	3.38644
(pseudo-) R²	0.331	0.399		0.557	

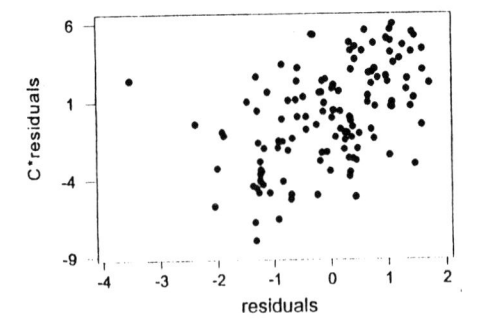

384   A CASEBOOK FOR SPATIAL STATISTICAL DATA ANALYSIS

Figure 7.24 (left). MC scatterplot of OLS residuals for SO$_4$ (northwest Texas groundwater data).

Figure 7.25 (right).Traditional homogeneity of variance AR residual plot for SO$_4$ (northwest Texas groundwater data).

```
residuals
S-W                0.96058†††      0.95897†††            0.97493


A-D                0.959††         0.784††               0.694†
R-J                0.9841†††       0.9898†               0.9928
K-S                0.083††         0.068                 0.102††
Bartlett           5.490           4.318                 1.528
Levene             0.897           1.002                 0.602
MC                 0.31707         0.38140               0.02966
GR                 0.65062         0.60747               0.94173 .
```

A sizeable portion of the latent spatial autocorrelation is attributable to simple map pattern and a considerable amount of it is attributable to other sources (nearly 16% of the variance in the power-transformed SO$_4$ values, at the margin); a large amount of the geographic variance heterogeneity is attributable to the presence of spatial autocorrelation; and AR residuals essentially are well behaved. Some deviation from normality, specifically in the tails, seems to persist in the AR residuals; another outlier emerges here, one that is located very close to the center of the landscape. And, some nonconstant variance is detectable across the range of mean responses (Figure 7.25). Conceptually speaking, an AR specification is preferable to an SAR specification in this case because groundwater migrates, resulting in direct impacts of SO$_4$ at one location upon values of SO$_4$ in nearby locations.

Inspection of Figure 7.26 reveals that exclusion of the outlier indicator variable (specifying the fourth observation, a value of −30) from the OLS regression analysis results in the presence of the value −14.2 in 17 of the first 21 distance classes, as well as the presence of it in most of the remaining distance classes (with the exception of the last 6), indicating that this outlier is located near the edge of the map. Inclusion of the outlier indicator variable in the OLS regression equation uncovers the presence of a second outlier, the low value of −9.63, which appears in 12 of the first 21 distance classes, revealing that this outlier is located more toward the center than toward the edge of the map.

POLLUTION 385

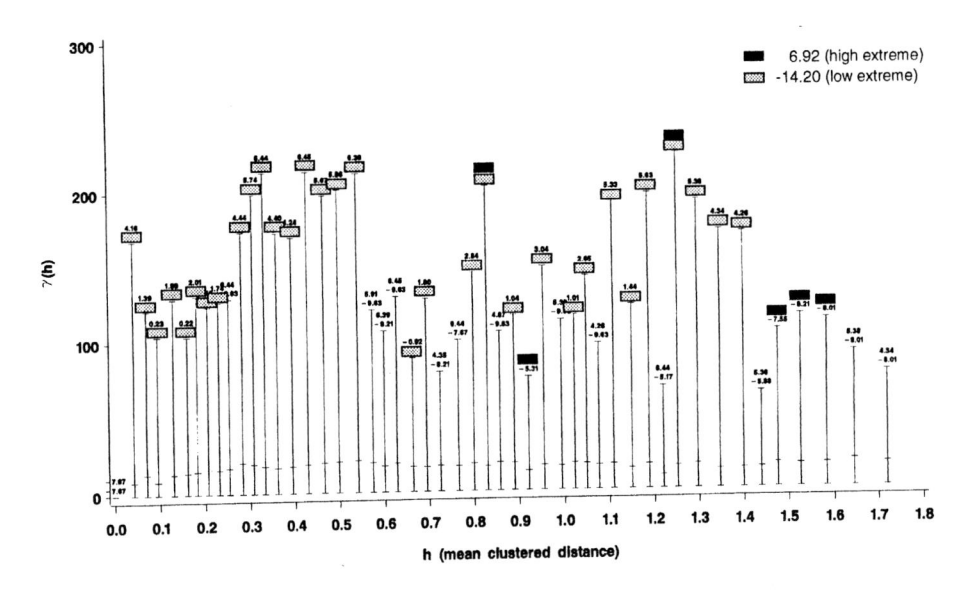

Figure 7.26. Semivariogram plot displaying the mean and extreme values for outlier-unadjusted OLS residuals for SO_4 (northwest Texas groundwater data).

Division of the 127^2 distance pairs into 50 distance classes yielded a minimum of 127 pairs per class (Figure 7.22a). Model fits to the semivariogram plot (Figure 7.27) using a maximum rescaled distance of 0.6 resulted in the following parameter estimates:

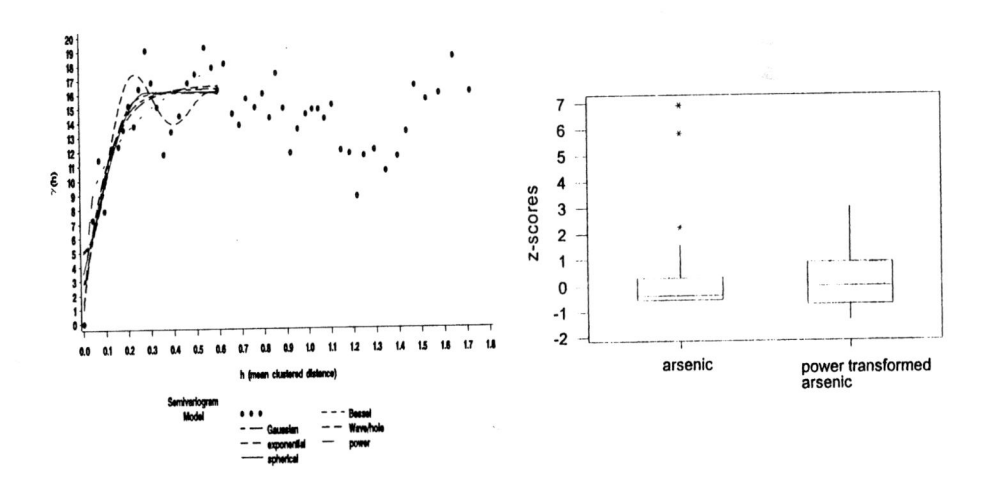

Figure 7.27 (left). Semivariogram plot of outlier-adjusted OLS residuals for SO_4 (northwest Texas groundwater data), with selected semivariogram model curves superimposed.

Figure 7.28 (right). Boxplots for the raw and power-transformed arsenic measures (northwest Texas groundwater data).

model	RSSE	C_0	C_1	r
exponential	0.29769	1.39635	15.2359	0.10722
Bessel	0.31354	2.83631	13.5579	0.07015
Gaussian	0.33705	5.10360	11.0161	0.13520
spherical	0.30895	3.56402	12.6017	0.28559
wave/hole	0.34026	4.97108	10.2340	0.05221
power	0.30867	0.00000	20.5262	* $\lambda = 0.26461$

The scatter of points in Figure 7.27 is typical of a semivariogram plot that may be described as a transition semivariogram model, which contains a moderate level of spatial autocorrelation (indicated by the presence of intermediate gamma values within the first few distance classes, up to the sill value), corroborating the values of $\hat{\rho}$ and MC reported in the preceding spatial autoregressive results. The exponential semivariogram model achieves the best fit to the semivariogram plot, based on RSSE results, even though the pattern of gamma values in Figure 7.27 up to the rescaled distance value of 1.5 suggests a decreasing wave/hole function. The presence of this pattern across distance classes up to the rescaled distance value of 0.6 results in a relatively poor fit of the different selected models to the semivariogram plot; even for these distance classes the wave/hole model has the poorest fit of all the transition models. Selection of the exponential semivariogram also is supported by its estimated range parameter value, as a higher range parameter value is more indicative of the spread of points for the first few distance classes; the spherical model also appears to provide a good description of the spatial autocorrelation latent in these OLS regression residuals, as does the power model.

Arsenic measures made on the Texas groundwater samples benefit tremendously from being subjected to a power transformation:

raw	power transformation
	$\hat{\delta} = -0.48832$
	$\hat{\gamma} = 0.28057$
	RSSE $= 1.2 \times 10^{-2}$
S-W $= 0.60174^{\dagger\dagger\dagger}$	S-W $= 0.92589^{\dagger\dagger\dagger}$
Bartlett $= 171.774^{\dagger\dagger\dagger}$	Bartlett $= 1.450$
Levene $= 5.466^{\dagger\dagger\dagger}$	Levene $= 1.004$
MC $= 0.40961$ ($\hat{\sigma}_{MC} \approx 0.05581$)	MC $= 0.57368$
GR $= 0.76924$	GR $= 0.42373$.

The approximately cube-root transformation of these arsenic data renders a frequency distribution that much more closely resembles a normal distribution and extraordinarily reduces geographic variance heterogeneity across the four quadrants of this Texas landscape. Unfortunately, although three marked outliers disappear on the transformation measurement scale, boxplots reveal that some positive skewness remains (Figure 7.28). In either case, a significant moderate level of positive spatial autocorrelation is detected, with the extent of this autocorrelation being obscured by the dirtiness of the raw arsenic data.

A reasonable description of these data is furnished by standard OLS estimation, which renders somewhat well behaved but spatially autocorrelated residuals. The statistical explanation achieved by replacing the OLS specification with a spatial AR specification improves slightly, with the autoregressive terms accounting for roughly 7% of the variance in the power-transformed arsenic measures:

	parameter estimates			
	OLS	s.e.	AR(W)	s.e.
ρ	*		0.42092	0.13831
b_0	1.23466	0.04512	1.22661	0.07256
$b_{u-\bar{u}}$	-0.79587	0.28852	-0.44995	0.29128
$b_{v-\bar{v}}$	0.74907	0.08469	0.42327	0.13288
(pseudo-) R^2	0.478		0.553	
residuals				
S-W	0.98349		0.98640	
A-D	0.435		0.563	
R-J	0.9954		0.9931	
K-S	0.059		0.075*	
Bartlett	0.653		2.120	
Levene	1.063		1.630	
MC	0.20630		-0.04007	
GR	0.81074		1.08806	

These results reveal that at least part, but not all, of the positive spatial autocorrelation latent in arsenic is attributable to simple map pattern and that adjusting for the remaining spatial autocorrelation introduces both some geographic variance heterogeneity and some tail deviation from normality into the residuals. Spatial autocorrelation present in the OLS residuals appears to cause an east-west gradient to be uncovered that seems to be nothing more than an artifact of the latent spatial autocorrelation. Nevertheless, the AR model residuals are well behaved too.

Specification of 50 distance classes for these OLS regression residuals resulted in a minimum of 127 pairs per class (Figure 7.22a), and, specifying a maximum rescaled distance of 0.4, yielded the following model parameter estimation results for the semivariogram plot in Figure 7.29:

model	RSSE	C_0	C_1	r
exponential	0.27065	0.00592	0.23502	0.03452
Bessel	0.28523	0.00000	0.23941	0.02135
Gaussian	0.31481	0.00238	0.23477	0.03481
spherical	0.32068	0.00000	0.23661	0.06839
wave/hole	0.42774	0.10626	0.12954	0.03314
power	0.27944	0.00000	0.29928	* $\lambda = 0.15582$

Low range parameter values—as is the case for each of the transition semivariogram models here— furnish evidence of low levels of spatial autocorrelation in these data. The semivariogram plot displayed in Figure 7.29 reveals a paucity of intermediate $\gamma(h)$ values across the first few distance classes, up to the sill value, which also is indicative of data with low levels of spatial autocorrelation. These

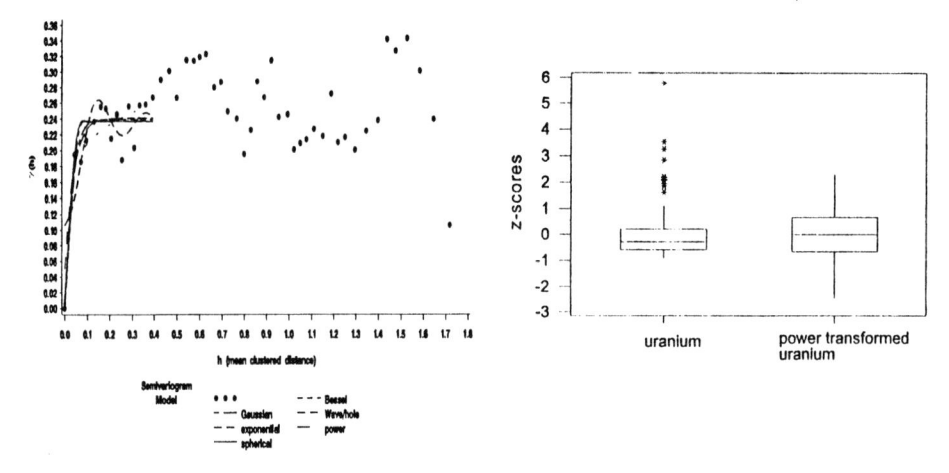

Figure 7.29 (left). Semivariogram plot of outlier-adjusted OLS residuals for arsenic (northwest Texas groundwater data), with selected semivariogram model curves superimposed.

Figure 7.30 (right). Boxplots for the raw and power-transformed uranium measures (northwest Texas groundwater data).

two results corroborate the preceding spatial autoregressive analysis in which most of the spatial autocorrelation detected in the raw data is accounted for by simple map pattern. The exponential semivariogram model achieves the lowest RSSE, with the Bessel model providing a comparably good description of the spatial autocorrelation present in the OLS residuals; the goodness-of-fit of the power model lies between those for Bessel and exponential.

The last variable selected from this data set for spatial analysis purposes is uranium, whose data originally were collected to explore the ability to separate background samples from samples anomalous in uranium content. Extending this viewpoint, analysis of these data should afford a more informative understanding of sites contaminated by radioactive materials. The uranium data benefit considerably from application of a negative power transformation:

```
raw                              power transformation
                                 δ̂ = 4.68759
                                 γ̂ = -0.27726
                                 RSSE = 3.6×10⁻³
S-W = 0.73511†††                 S-W = 0.98067
Bartlett = 45.697†††             Bartlett = 2.757
Levene = 3.401††                 Levene = 0.570
MC = 0.12569 (σ̂_MC ≈ 0.05581)    MC = 0.19210
GR = 1.02932                     GR = 0.88175 .
```

This power transformation dramatically reduces variance heterogeneity across the four quadrants of this geographic region, removing all apparent outliers identifiable in its raw data form (Figure 7.30), converts the empirical frequency distribution to

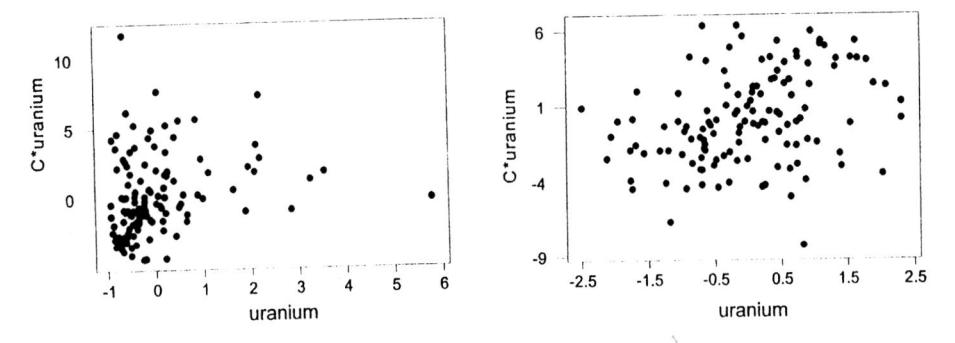

Figure 7.31. Left: (a) MC scatterplot of raw uranium measures (northwest Texas groundwater data). Right: (b) MC scatterplot of power-transformed uranium measures (northwest Texas groundwater data).

one that closely resembles a normal frequency distribution, and indicates that the poor statistical behavior of the raw data conceals the true degree of positive spatial autocorrelation latent in these data (Figures 7.31a-b).

These power-transformed uranium data are best described with a pure autoregressive model (which means the SAR and AR specifications are indistinguishable):

```
              parameter estimates
              OLS             SAR(W), based on z_y-scores
ρ             *               0.34614
b_0           0.45872         0.45722
(pseudo-) R²  *               0.110

residuals
S-W           0.98067         0.98439
A-D           0.169           0.170
R-J           0.9983          0.9980
K-S           0.039           0.043
Bartlett      2.757           2.731
Levene        0.570           0.402
MC            0.19210         -0.01942
GR            0.88175         1.10220 .
```

Spatial autocorrelation accounts for roughly 11% of the variance exhibited by the power-transformed uranium data, and the SAR residuals are well behaved, void of spatial autocorrelation (Figure 7.32). In other words, there is a weak to moderate tendency for uranium concentrations found in a given groundwater sample to be similar to those concentrations found in nearby groundwater.

Similar findings are obtained through geostatistical analysis. Specification of the 127^2 distance pairs into 50 distance classes gave a minimum of 127 pairs per class (Figure 7.22a) for the distance classes up to the rescaled distance value of 0.75, exactly that amount for the three previous attribute variables contained in this data set. The standardized, power-transformed version of uranium renders the semivariogram plot scatter of points displayed in Figure 7.33. Following are the parameter estimation results for the selected semivariogram models:

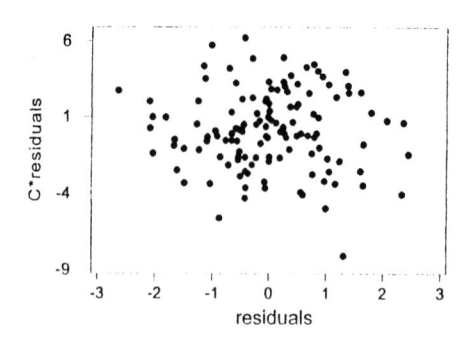

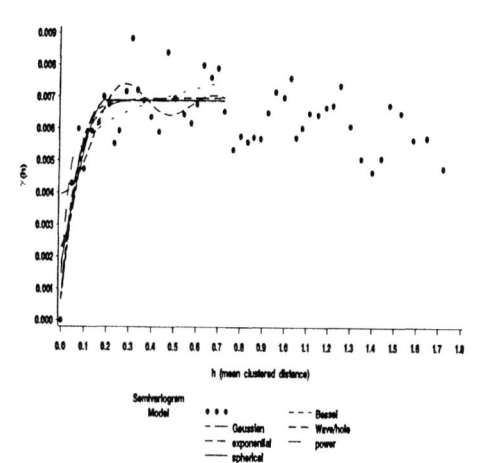

Figure 7.32 (left). MC scatterplot of SAR residuals for uranium (northwest Texas groundwater data).

Figure 7.33 (right). Semivariogram plot of uranium measures (northwest Texas groundwater data), with selected semivariogram model curves superimposed.

model	RSSE	C_0	C_1	r
exponential	0.37546	0.00065	0.00631	0.06421
Bessel	0.40459	0.00067	0.00621	0.03851
Gaussian	0.46762	0.00231	0.00457	0.08750
spherical	0.42989	0.00181	0.00510	0.19136
wave/hole‡	0.54102	0.00396	0.00283	0.06190
power	0.40679	0.00000	0.00784	* $\lambda = 0.14613$

‡Initial parameter estimate values for the wave/hole semivariogram model were specified explicitly as $C_0 = 0.0001$, $C_1 = 0.7$, and $r = 0.25$.

Although the exponential semivariogram model achieves the lowest RSSE, the poor fit of all selected models and a wide scattering of γ(h) values across all distance classes used to fit these models suggest the presence of a moderate level of spatial autocorrelation in the OLS regression residuals, corroborating the findings of the spatial autoregressive analysis. Although the relatively low range parameter values may suggest otherwise, sufficient numbers of intermediate γ(h) values within the first few distance classes support a contention that moderate levels of positive spatial autocorrelation are latent in these data.

7.5. Hazardous Waste Contamination of Soil: Dioxin

The final data set analyzed in this chapter is printed in Zirschky et al. (1986, 774-76). These data were derived from samples collected in November 1983 by the U.S. EPA along the two sides of a rural Missouri dirt farm road where, in 1971, a truck driver had dumped an undisclosed amount of dioxin (TCDD)-contaminated waste material in an attempt to avoid further tickets for being overweight. Not surprisingly, this toxin migrated from the original dump site, and, in addition, numerous chickens and other animals died from exposure to it. In this case, the

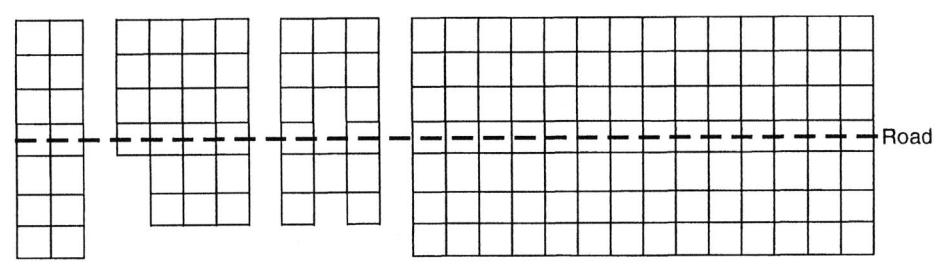

Map 7.8. Idealized sampling grid and actual dirt farm road map for the EPA Missouri roadside data.

horizontal axis, U, is defined as being parallel to the road, while the vertical axis, V, is defined as being perpendicular to the road. The authors furnish a map of the area under study; its sampling grid generalization appears in Map 7.8. Not all pixels of this grid are exactly the same size; the EPA analysis involved aggregating many 50-foot transects into 200-foot transects; and sample values were set to 0.05 when the TCDD concentrations were below the detection limit (a censored datum, of which there were many). The censored data are differentiated here with an indicator variable. The objective of the EPA study was to determine which sections of the road shoulders would require clean up.

When dealing with TCDD, EPA scientists posit from theoretical considerations that dioxin measures conform to a log-normal frequency distribution in the population. Hence, this particular power transformation is used here; but, of note, $\hat{\delta}$ = -0.5 with $\hat{\gamma}$ = 0.23110 yields better sample diagnostic statistics:

raw	power transformation
	$\hat{\delta}$ = -0.03778
	$\hat{\gamma}$ = 0
	RSSE = 2.7×10^{-2}
S-W = 0.47564[†††]	S-W = 0.88263[†††]
Bartlett = 379.840[†††]	Bartlett = 25.913[†††]
Levene = 14.551[†††]	Levene = 9.533[†††]
MC = 0.29740 ($\hat{\sigma}_{MC}$ ≈ 0.07352)	MC = 0.62750
GR = 0.41240	GR = 0.31932 .

The logarithmic transformation spectacularly reduces variance heterogeneity across the four planar quadrants of this landscape, it dramatically remolds the empirical frequency distribution of dioxin so that its profile looks more bell-shaped, and it unveils a markedly stronger degree of latent positive spatial autocorrelation. This measure of spatial autocorrelation remains complicated by the conspicuous presence of the censored data values (Figure 7.34).

A good description of these data is furnished by standard OLS estimation, which renders heterogeneously varying and spatially autocorrelated residuals. The statistical explanation achieved by replacing the OLS specification with a spatial

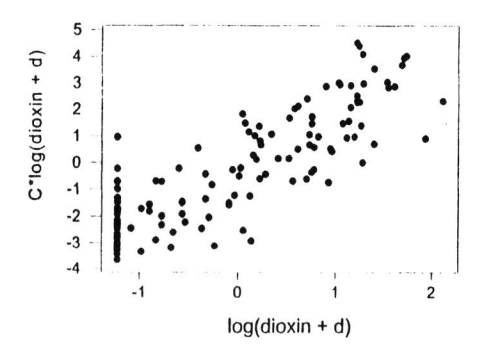

Figure 7.34. MC scatterplot of power-transformed dioxin measures (EPA Missouri roadside data).

AR specification improves slightly, with the autoregressive term accounting for roughly 5% of the variance in the power-transformed dioxin measures:

```
              parameter estimates (AR from z_y-scores)
                     OLS      s.e.        AR(W)     s.e.
ρ                     *                  0.39739   0.06474
b₀                 2.01103   0.24014     2.87728   0.40762
b_distance from road -0.10441 0.01332   -0.07681   0.01202
bu-ū              -0.00038   0.00010    -0.00012   0.00009
I_censored        -3.01648   0.24665    -2.32923   0.23489
(pseudo-) R²         0.827                 0.880

residuals
S-W                  0.97624              0.97418
A-D                  0.558                1.045†
R-J                  0.9950               0.9901†
K-S                  0.059                0.096†††
Bartlett             7.905††              1.255
Levene               2.633†               0.329
MC                   0.32539             -0.06956
GR                   0.65805              1.08019 .
```

These results reveal that the censored data come from a different statistical population (i.e., two parallel regression lines describe these data), that roughly half of the spatial autocorrelation latent in these data is attributable to simple map pattern, and that the AR residuals are reasonably well behaved forming a frequency distribution with perceptible deviations from a normal distribution in its tails.

The pattern of points displayed in Figure 7.35a reveals a near-linear trend across these data (especially when it is compared with similar plots for the Mercer-Hall data set; Chapter 6). This pattern also discloses an excess of well over 100 distance pairs per distance class. Inspection of the scatter of points constituting the semivariogram plot (Figure 7.35b) reveals an almost random scattering beyond the first 5 distance classes. The placement of these first few γ(h) values implies the presence of some spatial dependency in the OLS regression residuals; in contrast, the following model parameter estimates suggest a data set containing very low to no spatial autocorrelation:

POLLUTION 393

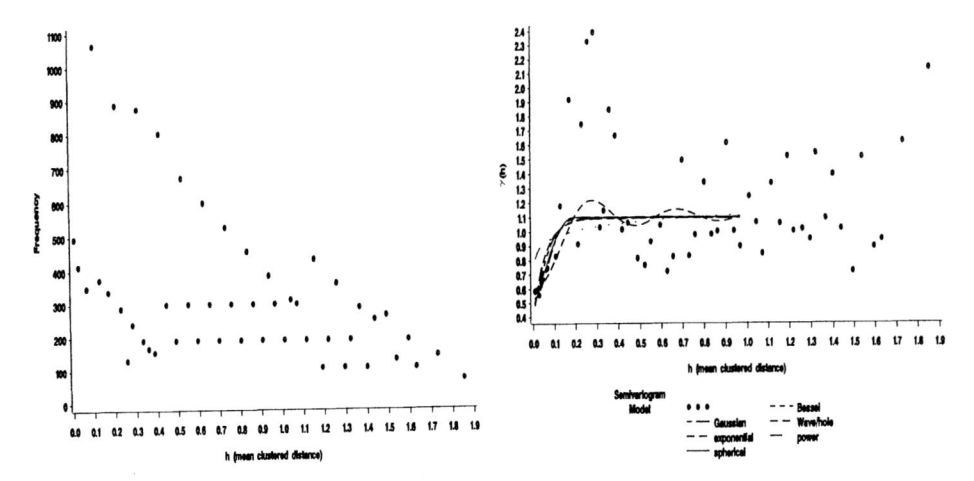

Figure 7.35. Left: (a) Semivariogram-type plot of the frequency of semivariance values in each distance class for the EPA Missouri roadside data. Right: (b) Semivariogram plot of OLS residuals for dioxin (EPA Missouri roadside data), with selected semivariogram model curves superimposed.

model	RSSE	C_0	C_1	r
exponential	0.86388	0.40161	0.69386	0.05561
Bessel	0.85635	0.48365	0.61485	0.04280
Gaussian	0.84160	0.53075	0.57266	0.09159
spherical	0.84007	0.46016	0.64164	0.18223
wave/hole	0.80560	0.60428	0.50498	0.06162
power	0.95839	0.00000	1.12217	*

$\lambda = 0.06567$.

As is indexed by the RSSE values, poor fits are the case for all of these selected semivariogram models, with the wave/hole model achieving the lowest RSSE. Furthermore, as can be seen in the variance-to-mean ratio plot appearing in Figure 7.36, low to no spatial autocorrelation is expected in the OLS regression residuals: the mean ratio is roughly 2.0 (§2.6.2), taking into account the wide fluctuations of ratio values up to the maximum rescaled distance value of 1.0. Although negative spatial autocorrelation is not detected with the spatial autoregressive analysis, the wave/hole model may well be the preferred semivariogram model here due to the sizeable fluctuations in the γ(h) values for the first few distance classes.

As an aside, a similar type of problem, namely, lead contamination adjacent to roadways, is studied by Newsome et al. (1997).

7.6. CONCLUDING COMMENTS: SPATIAL AUTOCORRELATION AND POLLUTION VARIABLES

Because of their contagious diffusion nature, pollutants almost always should be describable with spatial statistical model specifications, a contention attested to by the considerable effort invested in GEO-EAS software development and the EMAP project by the U.S. EPA. Findings reported in this chapter expand upon

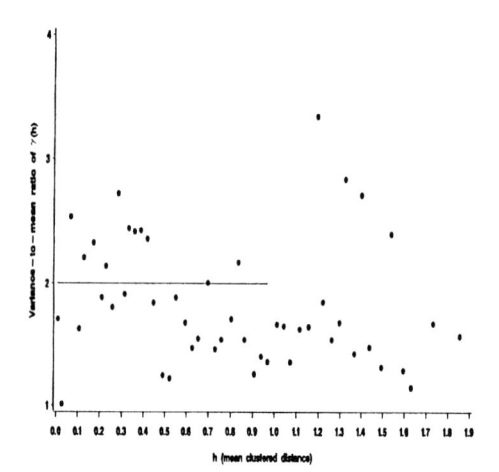

Figure 7.36. Semivariogram plot displaying the mean and extreme values for OLS residuals (EPA Missouri roadside data).

work of this type by furnishing data applications establishing empirical links between the semivariogram models commonly included in software such as GEO-EAS, and the popular spatial autoregression specifications (§3.3) that are available. Linkages reported in this chapter include

phenomenon	areal unit	autoregressive model	semivariogram model
particulate emissions	point	SAR	spherical
% coal ash	square lattice point	SAR	spherical; Gaussian; exponential; Bessel
dissolved oxygen (EMAP)	hexagonal llattice point	OLS	wave/hole
silt (EMAP)	hexagonal lattice point	OLS	wave/hole & power
pH (#1)	point	SAR	exponential; Bessel
pH (#2)	point	OLS	exponential; Bessel
sulfate	point	AR	exponential; power; spherical
arsenic	point	AR	exponential; Bessel; spherical
uranium	point	SAR	exponential; Bessel; power

Table 7.1. Summary of linkage findings for Chapter 7

| dioxin | point | AR | wave/hole |

These articulations of relationships between geostatistical and spatial autoregressive models continue to reflect findings of Chapters 4-6, namely, a frequent inability to relate simply and easily a single semivariogram model to a particular spatial autoregressive specification. In the cases where negligible spatial autocorrelation is detected, such as the EMAP pollution variables, the wave/hole often furnishes a best fit, even when trace positive spatial autocorrelation is detected by the corresponding autoregressive models. This inconsistency simply may be attributable to measurement error variability, because geostatistical models measure nearness in terms of Euclidean distance, whereas spatial autoregressive models measure nearness in terms of standardized topological distance. This inconsistency seems to emerge only in the trivial case of insignificant spatial autocorrelation.

Interestingly, but perhaps not surprisingly, outliers appear to play a more prominent role in pollution data (e.g., the particulate emissions), suggesting that these georeferenced data may well be dirtier than other types of spatial data. This class of georeferenced data may well be noisier than most other types, too, as is suggested by the relatively high RSSE values for the dioxin variable (although one needs to keep in mind that this variable does have a large percentage of its variance accounted for). Moreover, R^2 values for these data tend to be much lower than those for, say, socioeconomic (by a factor of roughly 1.7) or natural resources (by a factor of roughly 1.4) georeferenced data.

8

ANALYSIS OF GEOREFERENCED EPIDEMIOLOGICAL VARIABLES

For years the geographic dimension of community health has been considered important when studying the incidence, distribution, and control of diseases. Medical researchers and practitioners repeatedly have found the mapping of diseases to be illuminating. One of the most frequently cited examples of this cartographic benefit is the 1855 map of London cholera cases drafted by John Snow, a pioneer in the field of epidemiology, who then was able to establish the source of this water-born disease to be the Broad Street pump (Golden Square). Another celebrated geographic observation, made in 1909 by a community of dentists, was that naturally occurring fluoride in the water of Colorado Springs, Colorado, was fortifying the teeth of its residents against decay. The outcome of this geographic observation was the widespread fluoridation of municipal drinking water. More recently, spatial statistics—principally point pattern analysis—has become very topical in medical circles, with one of its goals being to discriminate between random geographic clusters and hot spots (i.e., locations of exceptionally high incidence due to the presence of a particular source) of disease incidences (see, *inter alia*, Diggle 1993). One of the more famous medical geography studies involving a spatial autoregressive perspective was completed by Cliff et al. (1981), which describes the geographic diffusion of measles in Iceland. More recently, a CAR specification has been used to explore lung cancer death rates in Ohio (Waller et al. 1997). Geostatistical approaches also can be found in the literature (e.g., Carroll et al. 1997).

    Georeferenced epidemiological data sets are somewhat accessible for plant diseases, such as Shaw's (1990) lesions in wheat, or for transmitters of diseases, such as Lloyd's (1967) wood lice (part of which was collected on a regular hexagonal lattice); both of these data sets are reprinted in *A Handbook of Small*

Data Sets (Hand et al. 1994). Equivalent data sets for human populations are more difficult to access, in part because of numerous fears surrounding disease and in part because of confidentiality issues. But many such data sets exist, as is attested to by the publication of sundry medical atlases, such as Pickle et al.'s *Atlas of United States Mortality* (1996) and the Dartmouth Medical School's *Dartmouth Atlas of Health Care 1998* (1998). And others, such as recorded georeferenced data from the 1991–92 dengue fever outbreak in the Puerto Rican municipio of Florida analyzed by Morrison et al. (1998), are carefully guarded by agencies such as the U.S. Centers for Disease Control. Nevertheless, some georeferenced health data sets are readily available, including the following provided in digital form by Bailey and Gatrell (1995): mortality rates in English Health Districts, cases of cancer of the larynx and lungs in Lancashire, England, cases of Burkitt's lymphoma in Uganda, and various classes of patients treated at a large London hospital. In addition, Hills et al. (1986) provide data on mortality and water hardness for 61 large towns in England and Wales, a data set that is reprinted in *A Handbook of Small Data Sets* (Hand et al. 1994). Hamilton (1992) reprints county lung cancer data (p. 93) adapted from Archer (1987) and Cohen (1988) and urban mortality rates (pp. 186–87), together with their concomitant hydrocarbon pollution potentials. Cressie (1991, pp. 537–38) prints Scotland lip cancer data. And Rasmussen (1992) reprints child mortality rates by country (p. 173). Meanwhile, Kennedy (1988) provides a spatial analysis of lung cancer data by county for those states that abut the Gulf of Mexico and the southeast Atlantic coast of the United States. Her study illustrates that with some effort a researcher can compile the necessary georeferenced data from available sources to conduct such a study. Health data sets explored in this chapter include pediatric blood lead levels in Syracuse, New York, Germany rabies cases, Haining's (1990) Glasgow standardized mortality rates, and, for the benchmark analysis of this chapter, Cressie's North Carolina sudden infant deaths (SIDs) syndrome cases.

SIDs is defined as the sudden, unexpected (given medical history), and inexplicable (after an autopsy has been performed) death of an infant no older than one year of age. Cressie (1991, pp. 386–89) reports georeferenced data for SIDs cases during 1974–78 and 1979–84 in North Carolina, accompanied by a choropleth map (p. 399) of these two geographic distributions. Portions of these data were first presented by Atkinson (1978) and subsequently augmented by Symons et al. (1983), Cressie and Read (1985, 1989), and Cressie and Chan (1989). Entries in Cressie's column labeled "County and Its Neighbors" are sorted in ascending order rather than beginning with the number of the corresponding county listed in the column labeled "Name" ; also #68 in the Yancey County listing should be #58. Epidemiologists are interested in SIDs because, at present, it is a leading category of post-neonatal death (accounting for roughly 2 deaths per 1,000 live births) and has causes that remain elusive. Geographical clustering of SIDs cases already has been documented for this North Carolina data set. Reasons for this clustering include the ethnic group to which a baby belongs, local health care practices (e.g., reporting bias, public education programs), and socioeconomic attributes of the local area. Cressie (1991, p. 429) comments that an auto-Poisson model would constitute a reasonable spatial autoregressive specification here and that a geographic trend in SIDs cases occurs across the set of 100 North Carolina counties. He also notes that a suitable power transformation—suggesting a preference for the logarithm and the square-root functions in this case—can be applied to the rates version of these data so that a Gaussian autoregressive specification can be estimated (p. 434). Cressie identifies an outlier in the 1974–78 data set, and adds unity to each value to better computationally handle 0 values (p. 391).

The question of whether or not these data benefit more from a square-root or a logarithmic power transformation—or, in fact, from any power transformation—can be ascertained by evaluating the following test statistic[1] results:

	1974–78	
	logarithm	exponent
raw	transformation	transformation
	$\hat{\delta} = 0.34235$	$\hat{\delta} = 0.08519$
	$\hat{\gamma} = 0$	$\hat{\alpha} = 0.54464$
	RSSE = 7.7×10^{-2}	RSSE = 2.6×10^{-2}
S-W = 0.89555[†††]	S-W = 0.98400	S-W = 0.96997
Bartlett = 16.804[†††]	Bartlett = 7.007[†]	Bartlett = 7.174[†]
Levene = 1.508	Levene = 1.533	Levene = 1.119
MC = 0.24465	MC = 0.27328	MC = 0.25921
($\hat{\delta}_{\partial *} \approx 0.06615$)		
GR = 0.67590	GR = 0.68871	GR = 0.65356

where S-W is the Shapiro-Wilk statistic, MC is the Moran Coefficient, and GR is the Geary Ratio.

	1979–84	
	logarithm	exponent
raw	transformation	transformation
	$\hat{\delta} = 0.39662$	$\hat{\delta} = 0.24964$
	$\hat{\gamma} = 0$	$\hat{\alpha} = 0.12160$
	RSSE = 1.5×10^{-1}	RSSE = 1.5×10^{-1}
S-W = 0.38207[†††]	S-W = 0.94179[†††]	S-W = 0.92233[†††]
Bartlett = 113.196[†††]	Bartlett = 2.933	Bartlett = 2.388
Levene = 0.883	Levene = 0.488	Levene = 0.352
MC = 0.01544	MC = 0.10231	MC = 0.11100
GR = 1.03675	GR = 0.84671	GR = 0.82367

Boxplots for these two sets of three different measurement scale values (Figures 8.1 and 8.2) confirm the presence of an abnormally high rate county in both time periods, reveal the presence of at least an uncommonly low rate county in 1974–78, and indicate that power-transformed versions of these data indeed have more symmetric frequency distributions. The 1974–78 graphics confirm that a square-

---

[1] Statistically significant values are denoted throughout by [†] for a 10%, [††] for a 5%, and [†††] for a 1% level.

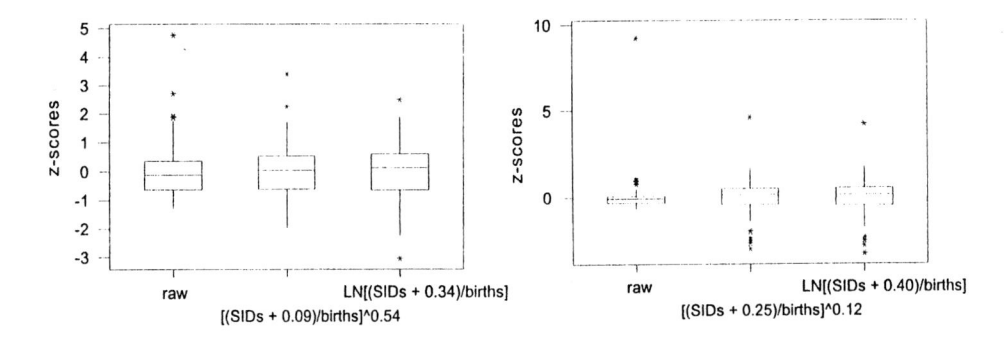

Figure 8.1 (left). Boxplots for the raw and two power-transformed 1974-78 SIDs measures (Cressie's data).

Figure 8.2 (right). Boxplots for the raw and two power-transformed 1979-84 SIDs measures (Cressie's data).

root transformation effectively competes with a logarithmic transformation; the 1979–84 graphics refute this finding, and offer evidence in support of the logarithmic transformation. Based upon these results, coupled with the conceptual expectation that the frequency distribution for SIDs conforms to a Poisson distribution, Gaussian analyses in this section employ the logarithmic transformation (log-Gaussian).

The foregoing diagnostics for the pair of logarithmic transformations imply that (1) Cressie's selection of unity to add to each county SIDs value may be excessive (i.e., both time periods render an optimal constant value of roughly $\frac{1}{3}$), (2) the frequency distribution for the logarithmic-transformed SIDs rates adequately conforms to a normal distribution in 1974–78 and dramatically moves toward a bell-shaped curve in 1979–84, (3) SIDs rates are weakly positively spatially autocorrelated, and (4) geographic variation across the four planar quadrants of this North Carolina landscape is far more homogeneous when SIDs rates are subjected to a logarithmic transformation.

Only a modest statistical description of these 1974–78 data is furnished by standard OLS (ordinary least squares) estimation, with the high rate outlier being reclassified as nonaberrant and with a camouflaged low rate outlier, accounting for roughly 7% of the variation in SIDs rates, appearing. Respecifying this OLS model to account for spatial autocorrelation by introducing eigenvectors of expression (1.10) as predictors noticeably improves the statistical description of these data, with the high SIDs rate outlier reemerging. But, the statistical description furnished by a Poisson specification incorporating the OLS-identified prominent eigenvectors, to adjust for spatial autocorrelation, is considerably better:

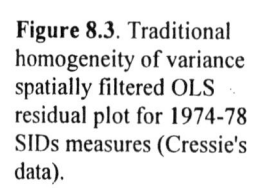

Figure 8.3. Traditional homogeneity of variance spatially filtered OLS residual plot for 1974-78 SIDs measures (Cressie's data).

	parameter estimates				
	OLS-1	OLS-2	s.e.	Poisson	s.e.
b_0	-6.25211	-6.26602	0.0497	-6.31137	0.0460
$b_{u-\bar{u}}$	0.00173	0.00180	0.0004	0.00248	0.0005
$b_{v-\bar{v}}$	-0.00382	-0.00599	0.0017	-0.00457	0.0017
I_{min}	-1.77965	-1.81613	0.5052	-13.38106	346.0466
I_{max}	*	1.42739	0.5043	1.51990	0.2803
E_4	*	1.45647	0.6323	1.28682	0.5797
E_{15}	*	-1.06673	0.4946	-0.86467	0.3708
E_{16}	*	-2.20995	0.4943	-2.43255	0.4652
E_{19}	*	-1.14049	0.5060	-1.12825	0.4585
(pseudo-) R^2	0.231	0.482		0.884	
residuals					
S-W	0.98407	0.98313		*	
A-D	0.219	0.347		*	
R-J	0.9972	0.9955		*	
K-S	0.047	0.065		*	
Bartlett	5.896	3.568		*	
Levene	1.504	0.948		0.962	
MC	0.10206	-0.07178		-0.01321	
		E(MC) = -0.05657			
GR	0.79618	1.01866		1.07165	

These results reveal that the excessively low rate outlier uncovered with the log-Gaussian specification analysis is inconsistent with a Poisson specification. The four eigenvectors of matrix **C** included as predictors account for virtually all of the spatial autocorrelation latent in these SIDs data, allowing the Poisson specification to be implemented in its standard form, without having to solve the nearly intractable normalizing constant problem. The OLS-2 residuals are well behaved; the residuals from this analysis also exhibit relatively constant variance across the range of mean responses (Figure 8.3). The frequency distribution profile for the Poisson model residuals is quite symmetric, allowing the Levene statistic to be interpreted; hence, constant geographic variance is detected across the four quadrants of the North Carolina landscape here, too.

Cressie concludes that the SIDs situation in North Carolina appears to have changed drastically between the two time periods. The following results for

1979–84, when compared with the preceding results for 1974–78, corroborate his findings:

	parameter estimates		
	OLS-1	OLS-2	Poisson
b_0	-6.17036	-6.17740	-6.22976
I_{max}	2.80230	2.68599	2.67216
I_{min}	-1.92248	-1.71743	*
E_3	*	-1.38958	-0.75240
E_{16}	*	1.50389	1.59628
E_{23}	*	-1.04944	*
E_{28}	*	-1.03097	*
E_{29}	*	0.85743	0.70757
E_{31}	*	-0.88194	-0.99303
(pseudo-) R^2	0.495	0.665	0.858
residuals			
S-W	0.97668	0.97990	*
A-D	0.606	0.376	*
R-J	0.9921	0.9943	*
K-S	0.094††	0.084†	*
Bartlett	0.904	1.828	*
Levene	0.159	0.730	0.708
MC	0.10169	-0.00096	0.03358
GR	0.78435	0.93566	1.05206

As with the 1974–78 data, the Poisson regression specification fails to deem the minimum extremes uncovered with the log-Gaussian specification to be statistically significant. In addition, two of the six eigenvectors found to be significant in the OLS-2 log-Gaussian specification drop out of the Poisson specification. Residuals for the OLS-2 log-Gaussian specification basically are well behaved, with some deviation from normality in the tails of the empirical frequency distribution; residuals for the Poisson specification are well behaved, too, but with conspicuous negative outliers. As a point of information, computing a back-transform for the log-Gaussian results yields a pseudo-R^2 value of 0.783, which is inferior to the Poisson result; this computation involved truncating negative back-transformed estimates at 0. Eigenvector #16 (MC = 0.55593) is common to both time period specifications. While regression coefficient estimates tend not to change between log-Gaussian and Poisson specifications, some of the OLS-2 standard errors appear to be substantially underestimated.

One difference between the estimation completed for this section and that reported in Cressie (1991) is that we incorporate outliers with indicator variables, whereas Cressie sets them aside (p. 558). Besides Anson County (#4) acting as an outlier in 1974–78, we also identify Jones County (#52) as an outlier in 1979–84. Our model specifications are much simpler than those posited in Cressie (p. 446). Cressie notes that the rates of SIDs are strongly correlated with people of color's live births. Specifically, for 1974–78, r = 0.579, whereas for 1979–84, r = 0.131. Interestingly, $\frac{\text{nonwhite births}}{\text{total live births}}$ is correlated with eigenvectors #4, #15, and #19 in

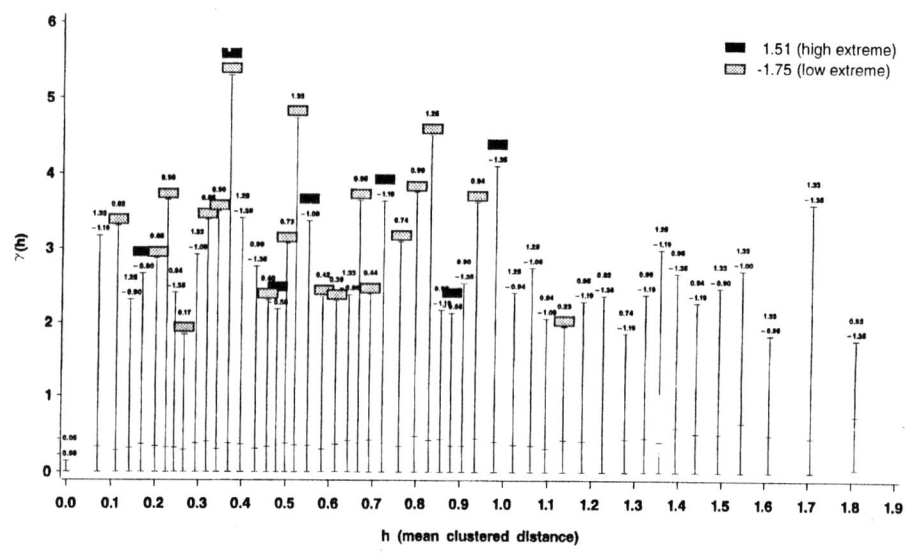

Figure 8.4. Semivariogram plot displaying the mean and extreme values for outlier-unadjusted OLS residuals for 1974-78 SIDs measures (Cressie's data).

1974–78 (r = 0.334), and with eigenvector #3 in 1979–84 (r = 0.246), which is consistent with Cressie's findings. Finally, Cressie comments that his auto-Poisson model is clumsy (p. 555); ours is not. He found evidence of possible positive spatial autocorrelation (which he declares unfortunate, since the auto-Poisson model is restricted to negative spatial autocorrelation situations); our identification of significant eigenvectors relating to map patterns with significant positive spatial autocorrelation is equivalent to this finding by Cressie.

Residuals generated by the OLS-1 regression equation specifications for the 1974–78 and 1979–84 data are used in the geostatistical analyses. In the 1974–78 data, a minimum value outlier is detected; the value of −1.75 is present in 11 of the first 21 distance classes (Figure 8.4). Inclusion of an outlier indicator variable in the OLS regression equation specification resulted in removal of this dominant outlier pattern from the plot of high-low semivariogram values.

Retaining the standard number of 50 distance classes when constructing the semivariogram plot yielded a minimum of 102 distance pairs per class (Figure 8.5a) up to the class including the specified maximum rescaled[2] distance value of 0.6. The scatter of points constituting the semivariogram plot (Figure 8.5b) indicates that a negligible amount of spatial autocorrelation is present; there is an absence of intermediate gamma values between the first and second distance classes. The selected semivariogram model fits to this scatter of points yield

---

[2] Distances are rescaled by dividing all (U, V) coordinate values by the maximum absolute U or V coordinate value in the set of georeferencings attached to the sample of areal units.

EPIDEMIOLOGICAL VARIABLES   403

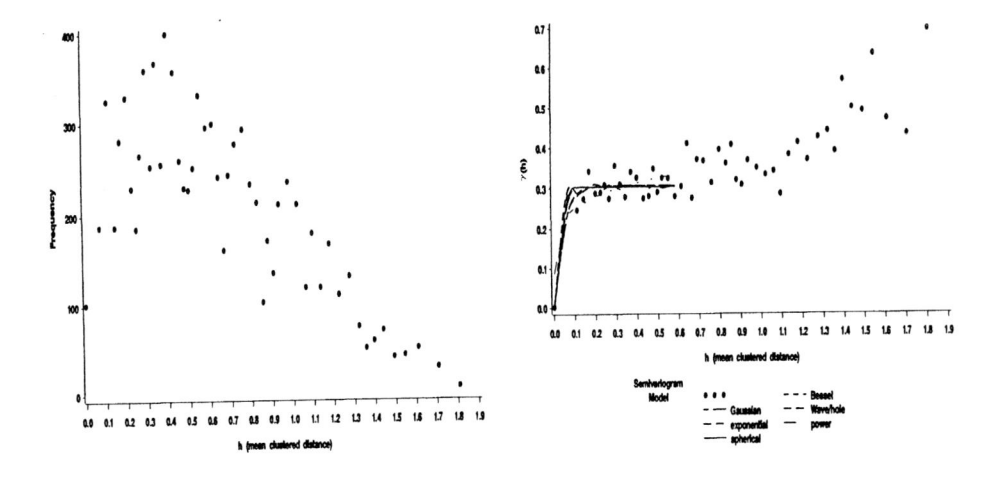

Figure 8.5. Left: (a) Semivariogram-type plot of the frequency of semivariance values in each distance class for Cressie's North Carolina data. Right: (b) Semivariogram plot of OLS residuals for 1974-78 SIDs measures (Cressie's data), with selected semivariogram model curves superimposed.

model	RSSE	C_0	C_1	r
exponential	0.32391	0.00658	0.30101	0.04530
Bessel	0.33396	0.01093	0.29540	0.02998
Gaussian‡	0.35688	0.00375	0.30021	0.04699
spherical	0.35868	0.00194	0.30171	0.09247
wave/hole‡	0.32358	0.00000	0.30335	0.00547
power	0.48184	0.00000	0.36580	*

$\lambda = 0.16666$

‡Initial parameter estimate values for the Gaussian and wave/hole semivariogram models respectively were specified explicitly as $C_0 = 0.0001$, $C_1 = 0.3$, and $r = 0.1$; and, $C_0 = 0.0001$, $C_1 = 0.3$, and $r = 0.005$.

These results, especially the relatively small range parameter values, corroborate the inference drawn from the semivariogram plot itself of negligible spatial autocorrelation being present. In addition, the average variance-to-mean ratio for the gamma values across the distance classes, through the class containing a maximum rescaled distance of 0.6, is 1.74 (s.e. = 0.10); this pattern is indicative of the presence of very weak spatial autocorrelation in these residuals (Figure 8.6). Despite the wave/hole semivariogram model having the best goodness-of-fit based on the RSSE values, the difference between its RSSE value and that for the exponential semivariogram model is trivial. According to the pattern of gamma values(Figure 8.5b), for those distance classes through the maximum rescaled distance of 0.6, there is no reason to expect negative spatial autocorrelation to be present. This expectation is corroborated by the MC value reported in the spatial autoregression analysis for these same data. Thus, the exponential semivariogram model may be considered the most appropriate specification here to describe the spatial autocorrelation present in the residuals for the SIDs 1974–78 data.

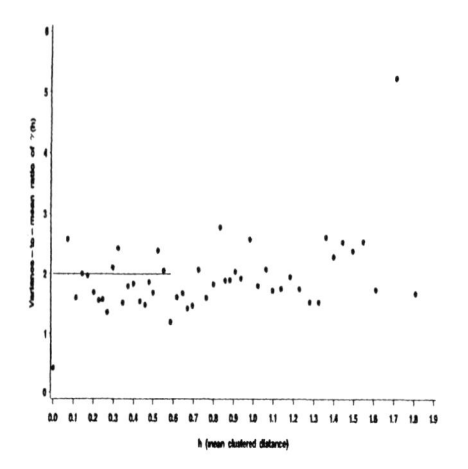

Figure 8.6 (left). Semivariogram-type plot of the distance class variance-to-mean ratio for Figure 8.5b.

Figure 8.7 (right). Semivariogram plot of OLS residuals for 1979-84 SIDs measures (Cressie's data), with selected semivariogram model curves superimposed.

For the 1979–84 SIDs data, both low and high outliers are conspicuous. Without an indicator variable for the high outlier value being included in the OLS regression equation specification, the value of 2.77 appears in 20 of the first 21 distance classes (the number of distance classes up to and including that class containing the specified maximum rescaled distance of 0.6). Likewise, when the indicator variable for the high outlier is introduced into the OLS regression equation specification, while the indicator variable for the low outlier values is not, the value of −2.21 appears in 14 of the first 21 distance classes. The presence of the high outlier value in almost all of the distance classes indicates a georeferenced position for this value near an edge of the map, whereas the presence of the low value outliers is more indicative of a georeferenced position near the center of the map. Exclusion of both the high and low outlier indicator variables from the OLS regression equation specification results in only a high value (and no low values) dominating the first 21 distance classes.

Again, retaining the standard number of 50 distance classes when constructing the semivariogram plot yields a minimum of 102 distance pairs per class, for the first 21 distance classes (Figure 8.5a). As indicated by the MC value computed with the OLS-1 residuals for this data set, little to no spatial autocorrelation is suggested by the scatter of points displayed by the semivariogram plot (Figure 8.7), especially given the absence of intermediate gamma values between the first and second distance classes (essentially between the 0 gamma value and the sill value). Results from fitting the selected semivariogram models include

model	RSSE	C_0	C_1	r
exponential	0.36170	0.00697	0.21016	0.00410
Bessel	0.36233	0.01502	0.20296	0.00393
Gaussian	0.36174	0.01568	0.20145	0.02326
spherical	0.36171	0.01494	0.20219	0.07010
wave/hole[‡]	0.28898	0.02333	0.19358	0.01489
power	0.64189	0.00000	0.25137	*

$\lambda = 0.13051$

[‡]Initial parameter estimate values for the wave/hole semivariogram model were specified explicitly as $C_0 = 0.0001$, $C_1 = 0.22$, and $r = 0.015$.

These results, particularly the relatively low range values, support the conclusion that negligible amounts of spatial autocorrelation are present. This finding also is indicated by the variance-to-mean ratio plot (Figure 8.8), as well as the mean (2.05) and standard error (0.13) of these ratios for the distance classes up to the maximum rescaled distance of 0.6. As is the case for the 1974–78 data set, the wave/hole model achieves the best goodness-of-fit of any of the transition models, although, again, the semivariogram plot for the pattern of gamma values modeled does not clearly indicate the presence of negative spatial autocorrelation in these residuals. The exponential semivariogram model achieves the best goodness-of-fit for positive spatial autocorrelation, based on the RSSE values, although the difference in RSSE values between it and the Bessel, Gaussian and spherical semivariogram models is trivial. Any one of these models furnishes an adequate description of the spatial autocorrelation in these residuals. Moderate RSSE values in this case result from the relatively wide scatter among the points in the semivariogram plot, particularly in distance classes whose average distance lies beyond the respective range parameters.

8.1. GLASGOW STANDARDIZED MORTALITY RATES

Haining (1990, p. 200) reports 1980–82 standardized mortality rates (adjusted for the age-sex composition of the population) for 87 community medicine areas

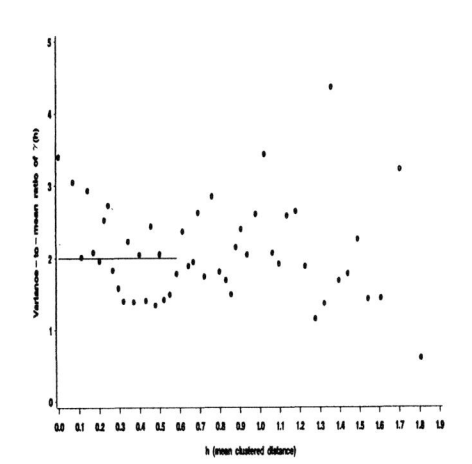

Figure 8.8. Semivariogram-type plot of the distance class variance-to-mean ratio for Figure 8.7.

(CMA) in Glasgow, which had been constructed by aggregating British enumeration districts in such a way to preserve socioeconomic homogeneity. Haining furnishes a map for the cancer data (p. 202), digitized coordinates for the CMAs (pp. 268–69), and a contiguous neighbors listing (pp. 266–67). Several discrepancies (some of which are errors) exist in this printed contiguity listing[3]: #16 is not adjacent to #12, #12 and #19 are not adjacent (we use the rook's linkage definition), #48 is not adjacent to #43, #60 and #62 are not adjacent (we use the rook's linkage definition), and #81 is not adjacent to #80; #35 is adjacent to #19, #41 is adjacent to #24, and #71 is adjacent to #72.

The standardized data are defined as $100 \times O_i/E_i$, where O_i is the observed number of deaths for CMA i, and E_i is the expected number for CMA i. As with the SIDs data analyzed in the previous section, the number of deaths due to cancer is assumed to be from a Poisson distribution with $\mu_i = n_i \times p_i$. This conceptual expectation indicates that the appropriate power transformation is the logarithm, with the log-Gaussian approximation having a nonconstant variance given by $\sigma^2 + \frac{1}{O_i} = \sigma^2(1 + \frac{1}{\sigma^2 \times O_i})$. The optimal power transformation appears to be a square-root one, though, but this implication is compromised by an extremes ratio that is less than 3:

raw	logarithm transformation	exponent transformation
	$\hat{\delta} = -22.78518$	$\hat{\delta} = -60.55074$
	$\hat{\gamma} = 0$	$\hat{\alpha} = 0.55460$
	SSE = 8.6×10^{-3}	RSSE = 7.8×10^{-3}
S-W = 0.96229††	S-W = 0.97868	S-W = 0.98227
Bartlett = 7.130†	Bartlett = 5.748	Bartlett = 6.032
Levene = 3.193††	Levene = 2.933††	Levene = 2.942†
MC = 0.40923	MC = 0.39782	MC = 0.39709
($\hat{\sigma}_{\partial*} \approx 0.06623$)		
GR = 0.56563	GR = 0.57118	GR = 0.59618

Boxplots (Figure 8.9) for these three measurement scales are quite symmetric, with no conspicuous outliers emerging. Behavior of values from the square-root transformation is not truly better than that for values rendered with the logarithmic transformation. Hence, the log-Gaussian specification is employed here. The frequency distribution for this power-transformed version adequately conforms to a normal distribution, geographic variance is somewhat heterogeneous across the four quadrants of the landscape, as expected, and moderate positive spatial autocorrelation is present.

Haining (1990, p. 366) reports that two useful descriptors of log-cancer rates are percentage of pensioners living alone and percentage of local population

---

[3] We wish to thank Dr. Robert Haining for providing us with a corrected surface partitioning map so that we could make the necessary corrections to our connectivity matrix.

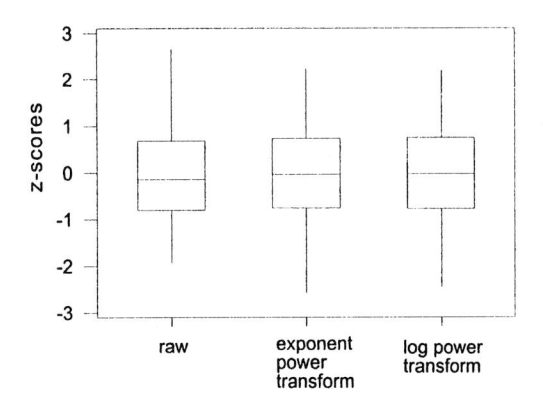

Figure 8.9. Boxplots for the raw and two power-transformed standardized cancer measures (Haining's data).

in social classes 4 and 5. The log-Gaussian estimated regression equation here is:

$$L\hat{N}(\text{cancerrate} - 22.78518) = 3.84532 + 0.01888 \times (\text{\% pensioners}) + 0.02308 \times (\text{\% in social classes 4 and 5}),$$

with $R^2 = 0.678$. This equation slightly differs from the one reported by Haining because of the constant, -22.78518, that is added here. The frequency distribution of the OLS residuals generated by this log-Gaussian specification appear to conform to a normal distribution (S-W = 0.98352, A-D = 0.382, R-J = 0.9911, K-S = 0.072), the variance of the residuals appears to be constant across the four quadrants of the region (Bartlett = 0.959, Levene = 0.069) as well as across the range of mean responses, and only trace positive spatial autocorrelation remains in these residuals (MC = 0.05692, GR = 0.90579). Haining also reports a lack of significant spatial autocorrelation in the residuals (p. 366) and uncovers a complicated geographic trend that is supposed to be latent in these cancer data. We find that this trend component is some sort of artifact, especially once two extreme low residual outliers (#12 and #16) are differentiated from the remaining 85 CMAs with an indicator variable. Our final estimation results include

	parameter estimates	
	OLS	Poisson
b_0	3.86207	4.24039
$b_{\text{\% pensioners}}$	0.01487	0.01164
$b_{\text{\% in social classes 4 and 5}}$	0.02411	0.01850
$I_{\min}$	-0.40044	-0.30443
(pseudo-) R^2	0.744	0.729
residuals		
S-W	0.98188	*
A-D	0.209	*
R-J	0.9949	*
K-S	0.048	*
Bartlett	3.219	*
Levene	0.613	1.053

MC	0.03596	0.06318
GR	0.93944	0.91516

The robustness of the log-Gaussian results is attested to by their near equivalency with the Poisson regression results. The log-Gaussian residuals are well behaved (including variance that essentially is constant across the range of mean responses), having a pseudo-R^2 for the back-transformed predicted values of 0.729—the same as the Poisson regression results—and rendering residuals that are slightly less spatially autocorrelated than those generated by Poisson regression. In contrast with the SIDs findings, the extreme, low rate outliers indicator variable does not become unimportant in the Poisson specification here.

Not introducing an indicator variable into the OLS regression equation specification to designate the low outlier values before undertaking a geostatistical analysis fails to reveal a clear dominating minimum gamma value for the initial distance classes up to the maximum rescaled distance of 0.6 (Figure 8.10). Specification of 50 clusters for the 87^2 distance pairs when constructing the semivariogram plot yields a minimum of 34 pairs per distance class, up to the specified maximum rescaled distance of 0.6, although the majority exceed 100 (Figure 8.11a). The scatter of points displayed by the semivariogram plot in Figure 8.11b indicates the presence of very weak positive spatial autocorrelation in these residuals, in part due to the lack of intermediate gamma values occurring between the initial distance class and the sill value. Goodness-of-fit measures and parameter estimates for the selected semivariogram models include

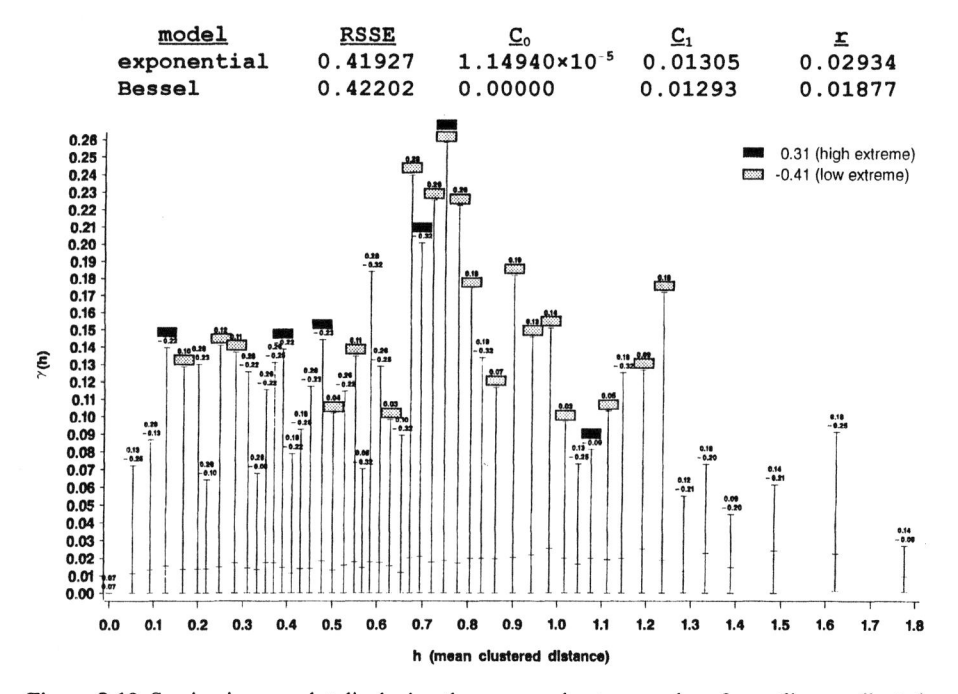

model	RSSE	C_0	C_1	r
exponential	0.41927	1.14940×10^{-5}	0.01305	0.02934
Bessel	0.42202	0.00000	0.01293	0.01877

Figure 8.10. Semivariogram plot displaying the mean and extreme values for outlier-unadjusted OLS residuals for standardized cancer measures (Haining's data).

EPIDEMIOLOGICAL VARIABLES 409

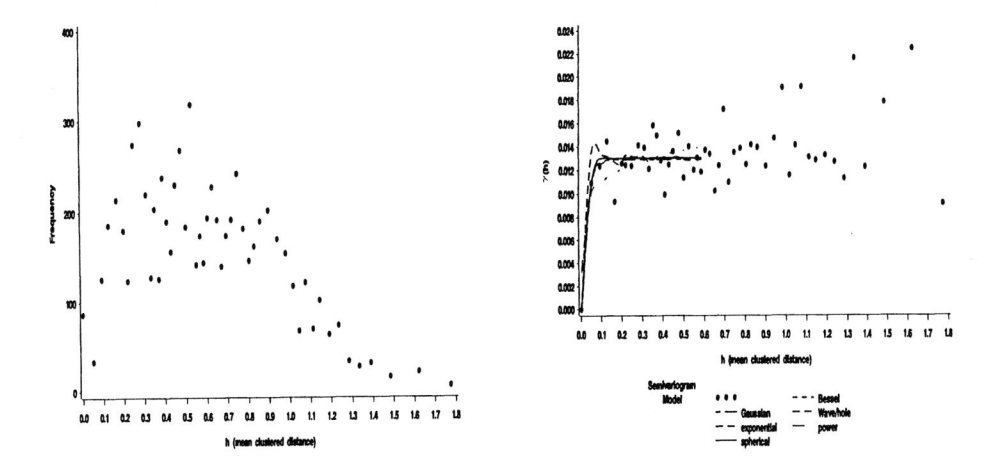

Figure 8.11. Left: (a) Semivariogram-type plot of the frequency of semivariance values in each distance class for Haining's Glasgow data. Right: (b) Semivariogram plot of OLS residuals for standardized cancer measures (Haining's data), with selected semivariogram model curves superimposed.

```
Gaussian      0.42045    6.27400×10⁻⁶   0.01303    0.03822
spherical‡    0.42069    0.00000        0.01304    0.07787
wave/hole‡    0.40573    0.00000        0.01297    0.00516
power         0.49842    0.00000        0.01503       *
                                              λ = 0.13408
```

‡Initial parameter estimate values for the spherical and wave/hole semivariogram models were specified explicitly as $C_0 = 0.0001$, $C_1 = 0.013$, and, respectively, $r = 0.1$ and $r = 0.005$.

The low range parameter values and the average (1.92, s.e. 0.14) of the variance-to-mean ratios for the gamma values, for distance classes up to the maximum rescaled distance of 0.6 (Figure 8.12), suggest that these residuals contain virtually no spatial autocorrelation. This finding is corroborated by the near-0 MC value

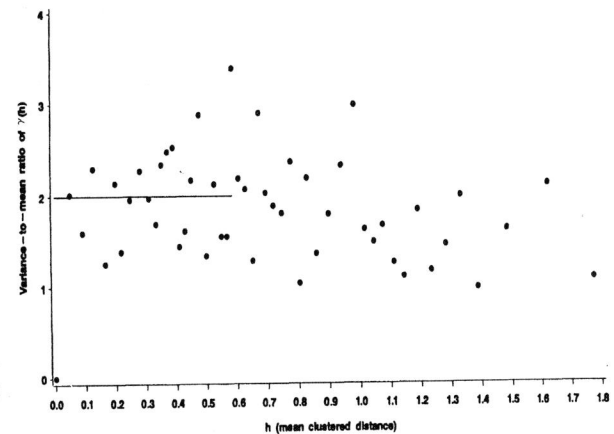

Figure 8.12. Semivariogram-type plot of the distance class variance-to-mean ratio for Figure 8.11b.

reported in the spatial autoregression analysis. As is indicated by the RSSE values, moderately poor goodness-of-fits characterize all of these semivariogram model estimations, with the wave/hole model achieving the lowest RSSE value. Although negative spatial autocorrelation is not detected in the spatial autoregressive analyses, the wave/hole model appears to furnish the best semivariogram description, presumably because of marked fluctuations displayed by the $\gamma(h)$ values for those distance classes up to the maximum rescaled distance of 0.6.

8.2. PEDIATRIC LEAD POISONING IN SYRACUSE, NEW YORK

One pollution theme addressed in Chapter 6 is the toxic heavy metals that have been geographically concentrated or whose elevated levels have been made ubiquitous in the environment because of human activities. This kind of pollution often is accompanied by community health consequences. Lead contamination of the environment falls under this heading, with the two principal sources of this pollution being the widespread use of lead-based paints and lead emissions in gasoline in earlier years. Remnants of lead-based paint still can be found in many older homes, resulting in geographic concentrations of this environmental pollutant. Gasoline emissions deposited high concentrations of lead into soils somewhat uniformly throughout especially urban areas. The community health issue here concerns childhood lead poisoning, which is completely preventable, and is the subject of research across the nation (e.g., Bailey et al. 1998).

The geography of pediatric blood lead levels[4] for Syracuse, New York, has been a topic of considerable interest in recent years; McDade (1994) provides a census tract (CT) tabulation of the first portion of these data made available in digital form. Griffith and Amrhein (1997, p. 313) furnish a map of the CT surface partitioning for this geographic landscape, together with logistic regression results for frequency of elevated blood lead levels. The local importance of this topic is attested to by the inclusion of dot maps in the City of Syracuse's 1997–98 *Consolidated Plan*—constructed by address matching blood lead levels to U.S. Census Bureau TIGER files[5]—depicting the locations of high numbers of elevated cases. Johnson et al. (1996) furnish background discussion of this community health problem, concentrating primarily on aspatial features of these data. The data analyzed in this section are based on 61 CT aggregates of 60,804 measures taken on 15,277 separate children under the age of five, for a two-year period (4/1/92–3/31/94), who had not changed residence during this period (Griffith et al.

---

[4] These data have been made available to us through Ms. Mary Burdick and Mr. Gary Urquhart, of the Onondaga County Health Department, which monitors childhood lead exposure through its Lead Poisoning Control Program. We are indebted to Mary and Gary, as well as to Dr. David Johnson, Department of Chemistry, SUNY College of Environmental Science and Forestry, and Dr. Andrew Hunt, SUNY Health Science Center/Syracuse, whom we have worked closely with in our scientific analyses of these data.

[5] These maps are one product of a project completed with Mr. Philip Doyle and Mr. David Wheeler during their undergraduate programs of study in the Department of Geography, Syracuse University.

1997, pp. 13–16) and whose addresses were successfully matched using Arc/Info on a RISC/6000 computer; the reduced number of distinct measures used here is 12,927. The two specific variables explored in this section are the (geometric) mean, because scientists have concluded that blood lead levels conform to a lognormal distribution in the population, and its variance (more specifically, the standard deviation) for the CT blood lead levels; spatial analysis results for these two variables have not been published before. While, conceptually speaking, the numbers of cases of elevated lead levels beyond selected threshold values should be described with an auto-binomial model, restricting attention here to means and standard deviations sanctions the choice of an auto-Gaussian model.

Lead (Pb; a heavy, soft, malleable, bluish-gray metal) is a ubiquitous element—found in rocks, soil, plants, animals, and human beings—usually naturally occurring in quite low levels. In an unpolluted environment, human blood contains approximately 0.5 micrograms of lead per deciliter (i.e., tenth of a liter) of blood (expressed in scientific shorthand as 0.5 µg dl^{-1}). Therefore, the "natural background level" of lead in human blood, 0.5 µg dl^{-1}, is a good benchmark against which to relate any lead screening and remediation efforts. In addition, current laboratory technology used to quantitatively measure the lead content of blood (i.e., laboratory tests done on blood samples acquired either from a finger prick or venipuncture) is unable to reliably detect concentrations between 0 and 5 µg dl^{-1}, frequently causing blood lead levels in this interval to be recorded as either 0 or 5. The raw data of 60,804 cases yields K-S = 0.16389[†††], indicating a marked deviation from a normal frequency distribution. Not surprisingly, this deviation occurs in both tails; the lower tail distribution is distorted by an inability to measure accurately quantities in the interval [0, 5], whereas the upper tail distribution is distorted by death-threatening levels that trigger intervention. The optimal logarithmic transformation of these data involves adding 0.6 to each blood lead level measure (which is very close to the theoretical value of 0.5 often suggested in the literature), which then yields K-S = 0.12005[†††]; tail deviations persist here, especially in the range of 0-5 µg dl^{-1}. Regardless, conformity with a bell-shaped curve is markedly improved by employing the logarithmic transformation.

Both the mean and the standard deviation have a $\frac{y_{max}}{y_{min}}$ ratio less than 3, implying that neither of these variables would benefit from being subjected to a power transformation. Summary statistics for these two raw variables include

```
S-W = 0.96567                          S-W = 0.98895
weighted K-S = 0.10004†††              weighted K-S = 0.11048
Bartlett = 12.035†††                   Bartlett = 3.900
Levene = 2.041                         Levene = 0.984
MC = 0.38469 (ô_MC = 0.07762)          MC = 0.17184
GR = 0.54683                           GR = 0.89523 .
```

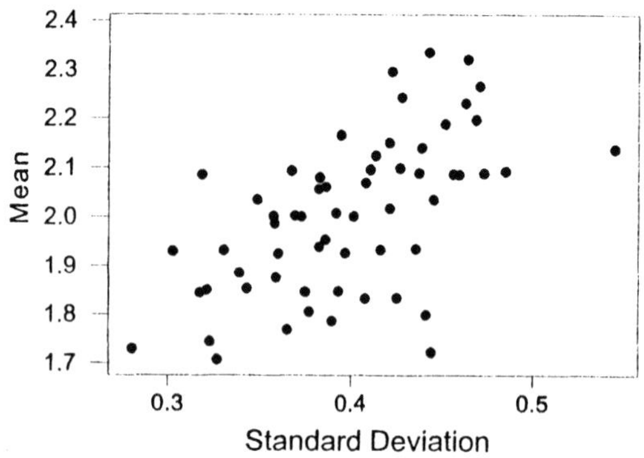

Figure 8.13. Scatterplot of the mean versus the standard deviation of pediatric blood lead levels by census tract (Onondaga County Health Department data).

Each of these two variables appears to be somewhat well behaved, from a statistical point of view. Disaggregate conformity to normality remains elusive, but adequate conformity is achieved in the aggregate. Some evidence exists indicating the presence of nonconstant geographic variance for the mean across the four quadrants of this Syracuse landscape. While this evidence is not very compelling, it is bolstered by the presence of a conspicuous relationship between the mean and the variance (Figure 8.13). And, each of these variables contains weak positive spatial autocorrelation.

The standard deviation of blood lead level by CT can be described with a modest degree of success by

$$\hat{\sigma} = 0.04154 + 0.17741 \times \bar{x} + 0.12396 \times I_{max} \;, R^2 = 0.427 \;,$$

where I_{max} is a binary indicator variable discriminating between CT #29 and the remaining 60 CTs, with this extreme high outlier accounting for roughly 7% of the variance in the standard deviation of aggregate blood lead levels across the CTs. This equation is accompanied by the following diagnostic statistics:

```
S-W = 0.99164        Bartlett = 1.233
A-D = 0.108          Levene   = 0.601
R-J = 0.9980         MC       = -0.02729
K-S = 0.049          GR       = 0.97399
```

These statistics are indicative of a set of well behaved OLS residuals. The moderate capability of predicting variability with this estimated regression equation is somewhat disappointing, but its interpretation is quite appealing. The standard deviation—which is the square root of a chi-square variate and hence theoretically should be normally distributed—has an empirical frequency distribution whose profile adequately conforms to a bell-shaped curve. As average geometric mean blood lead levels increases in a CT, there is a moderate tendency for variability within CTs to increase. In other words, with a background blood lead level of nearly 0, increased variability occurs only when relatively high blood lead levels coexist in geographic locations with low and intermediate blood lead levels. The weak positive spatial autocorrelation detected in the standard deviation of Syracuse

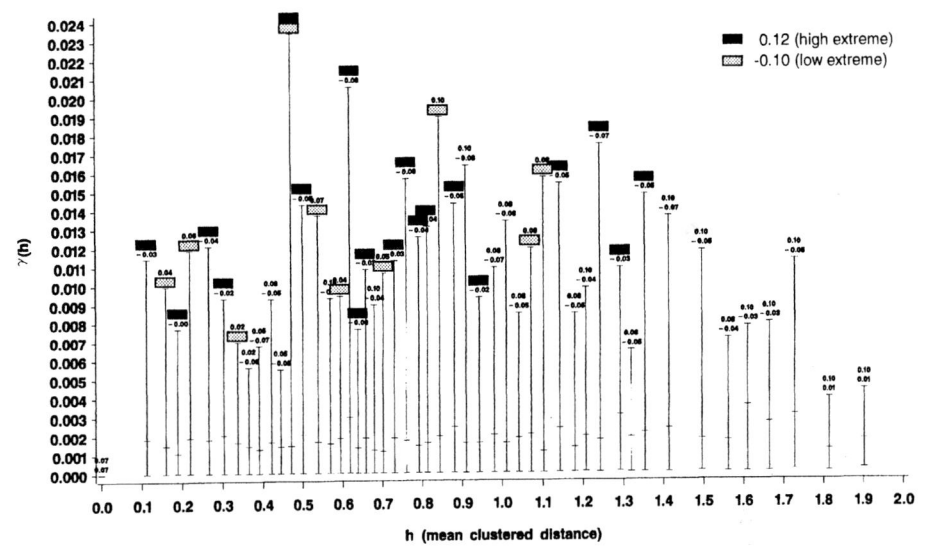

Figure 8.14. Semivariogram plot displaying the mean and extreme values for outlier-unadjusted OLS residuals for pediatric blood lead level standard deviations (Onondaga County Health Department data).

pediatric blood lead levels is virtually completely accounted for by the geographic distribution of the geometric mean levels; these geometric means are used to compute the corresponding standard deviations, which in turn may well be introducing spatial autocorrelation into the georeferenced standard deviation values. Therefore, an epidemiologist who measures a relatively high average blood lead level in a sample of children in a given area should expect the variability in that area to be relatively high, too.

When residuals generated by the OLS regression equation

$$\hat{Y} = 0.04154 + 0.17741 \times \bar{x}_{pb} + 0.12396 \times I_{max},$$

where Y denotes the standard deviation of individual log-blood lead levels aggregated by CT, are subjected to a geostatistical analysis, negligible spatial autocorrelation is detected. Removal of the indicator variable, designating the outlier value associated with CT #29 (I_{max}), from the OLS regression equation specification fails to reveal a clear pattern of a dominant high outlier value, especially for those distance classes up to the specified maximum rescaled distance of 1.0 (Figure 8.14); it is included in the specification because it is a significant predictor variable. Once again retaining the standard number of 50 distance classes when constructing the semivariogram plot yields a minimum of 36 distance pairs per class up to the maximum rescaled distance value of 1.0 (Figure 8.15a). The parameter estimates for selected semivariogram model fits to the scatter of points in the semivariogram plot (Figure 8.15b) include

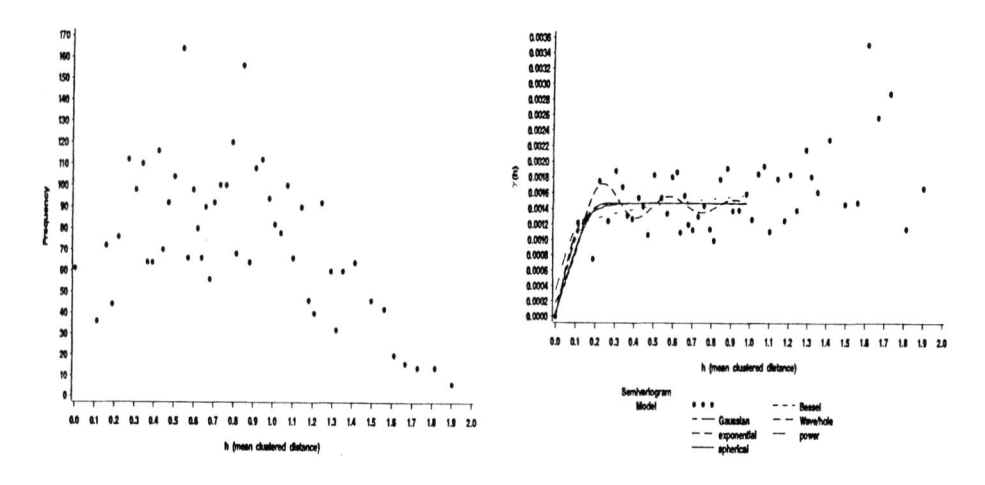

Figure 8.15. Left: (a) Semivariogram-type plot of the frequency of semivariance values in each distance class for the Onondaga County pediatric blood lead levels data. Right: (b) Semivariogram plot of OLS residuals for pediatric blood lead level standard deviations (Onondaga County Health Department data), with selected semivariogram model curves superimposed.

model	RSSE	C_0	C_1	r
exponential	0.59671	0.00000	0.00148	0.07341
Bessel	0.59579	0.00000	0.00147	0.05324
Gaussian	0.60093	3.191×10^{-5}	0.00144	0.10967
spherical	0.59654	3.749×10^{-5}	0.00144	0.24569
wave/hole	0.59598	1.828×10^{-4}	0.00126	0.05312
power	0.67999	0.00000	0.00158	*

$\lambda = 0.13399$

These estimation results indicate the virtual absence of spatial autocorrelation; intermediate gamma values are not present between the 0 distance class and the sill value, and range parameter values are relatively low. Unfortunately, conflicting evidence is provided by the average variance-to-mean ratio for all of the distance classes up to the maximum rescaled distance of 1.0: the mean is 1.66 (s.e. 0.12; Figure 8.16), which is significantly less than the expected value of 2.0 for an unautocorrelated data set. Moderately high RSSE values indicate a fairly large amount of variation in gamma values across all the distance classes included in the semivariogram model-fitting exercise. Based strictly on RSSE values, the Bessel specification achieves the best goodness-of-fit, although there is very little difference between its result and those for the remaining five transition semivariogram model specifications.

Because a sample mean is calculated before a sample standard deviation is calculated, and the sample mean is of interest, using the geocoded version of this statistic to predict georeferenced standard deviation values makes sense. Understanding the geographic distribution of the mean is more complicated, though. The geometric mean is explored in this section, because, as already mentioned, theo-

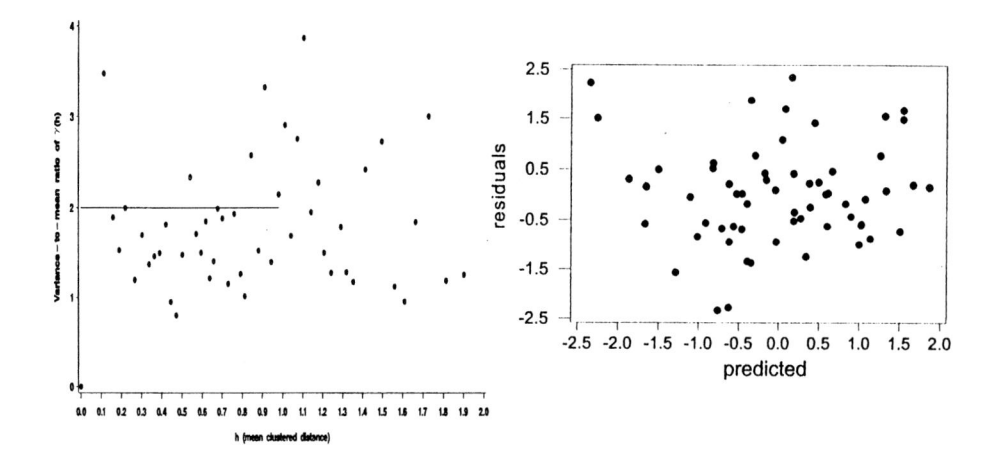

Figure 8.16 (left). Semivariogram-type plot of the distance class variance-to-mean ratio for Figure 8.15b.

Figure 8.17 (right). Traditional homogeneity of variance SAR residual plot for pediatric blood lead level geometric means (Onondaga County data).

retically speaking, blood lead level values come from a log-normal distribution. The above aggregate raw data S-W statistic (i.e., 0.96567) is consistent with this contention. Estimation results here include

parameter estimates

	OLS	s.e.	SAR(W)	s.e.
$\hat{\rho}$	*		0.35649	0.28531
b_0	2.02008	0.04921	2.03271	0.04947
$b_{u-\bar{u}}$	-1.12589	0.47322	-0.97320	0.66139
$b_{pop\ density}$	0.30649	0.07584	0.28717	0.07718
$b_{house\ value}$	-0.00371	0.00050	-0.00371	0.00049
$b_{\%>18}$	0.53238	0.11852	0.49039	0.12073
I_{min}	-0.27922	0.08414	-0.27474	0.08181
(pseudo-) R^2	0.758		0.777	

residuals

S-W	0.97299	0.96993
A-D	0.485	0.585
R-J	0.9894	0.9883
K-S	0.092	0.091
Bartlett	0.937	0.410
Levene	0.460	0.269
MC	0.12813	0.00988
GR	0.85983	0.98304

These results reveal that positive spatial autocorrelation latent in the geographic distribution of geometric mean pediatric blood lead levels in Syracuse is not attributable to simple map pattern (the significant east-west trend identified with the OLS specification becomes nonsignificant with the SAR specification). This autocorrelation is weak to moderate in degree, basically accounting for the same percentage of variance in mean blood lead levels as does the east-west Cartesian coordinate in the OLS specification. These results also indicate that the SAR residuals are well behaved, exhibiting constant variance across the four quadrants of the Syracuse landscape and essentially no spatial correlation. The traditional ê-versus-ŷ scatterplot (Figure 8.17) suggests reasonably constant variance for the residuals across the range of mean responses. A failure to have complete coverage of the rectangular region in this graphic does raise some worrisome concerns; a more comprehensive study of these particular data should incorporate the nonconstant variance, noted in the preceding paragraphs.

The SAR results imply the presence of a moderate tendency for CTs with similar mean blood lead levels to be contiguous and for mean blood lead levels to increase as (1) population density (a surrogate for poverty) increases, (2) mean housing value (a surrogate for housing age and maintenance quality) decreases, and (3) percentage of children in the 018 age cohort (surrogate for the size of the group potentially at risk) increases. CT #42 merits follow-up investigation, given that its mean level is abnormally low, once adjustments have been made for the aforementioned set of socioeconomic factors.

Introducing an indicator variable for the abnormally low mean blood lead level associated with CT #42 into the OLS regression equation specification eliminates the prevalence of a single, low value (i.e., −0.26) in 17 of the 31 distance classes, up to the specified maximum rescaled distance of 1.0, used to construct the semivariogram plot (Figure 8.18). Specification of 50 distance classes for the 61^2 distance pairs used to construct the semivariogram plot yields a

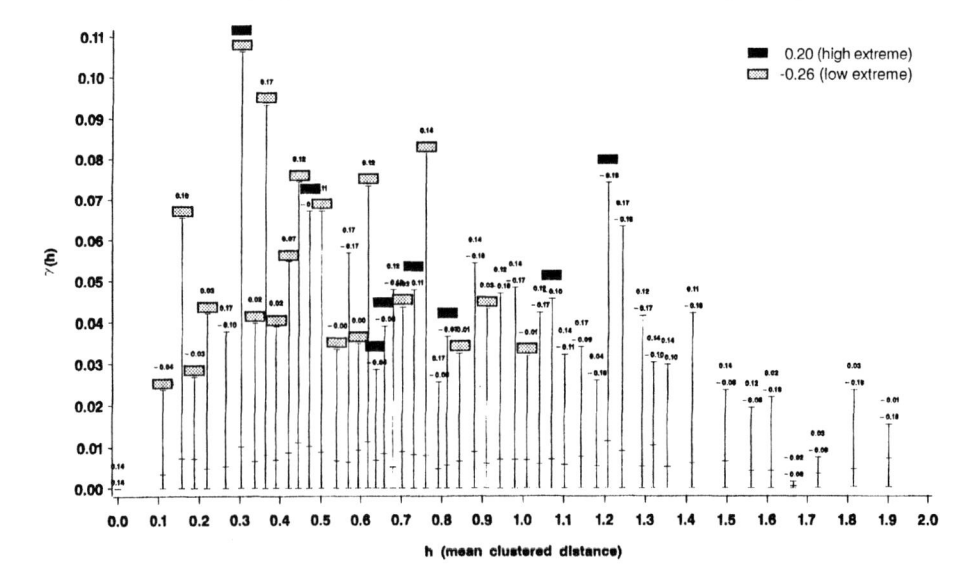

Figure 8.18. Semivariogram plot displaying the mean and extreme values for outlier-unadjusted OLS residuals for pediatric blood lead level geometric means (Onondaga County data).

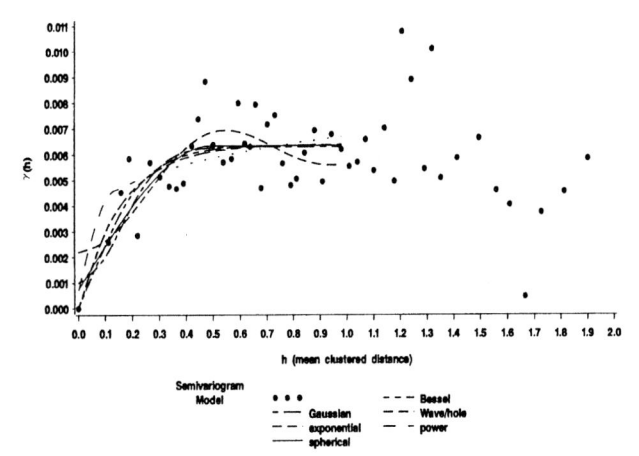

Figure 8.19. Semivariogram plot of OLS residuals for pediatric blood lead level geometric means (Onondaga County data), with selected semivariogram model curves superimposed.

minimum of 36 pairs per class (Figure 8.15a). As evidenced by the scatter of points displayed by the semivariogram plot (Figure 8.19), positive spatial autocorrelation appears to be latent in these residuals. Estimation of selected semivariogram model parameters and goodness-of-fit measures renders the following:

model	RSSE	C_0	C_1	r
exponential	0.48874	0.00000	0.00640	0.15837
Bessel	0.48442	6.050×10^{-5}	0.00630	0.10467
Gaussian	0.50296	1.005×10^{-3}	0.00530	0.21785
spherical	0.47541	7.478×10^{-4}	0.00560	0.49095
wave/hole	0.49368	2.216×10^{-3}	0.00388	0.12172
power	0.56339	0.00000	0.00672	*

$\lambda = 0.19365$

These estimation results indicate the presence of weak positive spatial autocorrelation, as seen by the moderate range parameter values and the number of intermediate $\gamma(h)$ values between the 0 distance class and the first class composing the sill value (Figure 8.19). The spherical semivariogram model achieves the best goodness-of-fit, based on the RSSE results, but all the transition models seem to describe the scatter of points in the semivariogram plot equally well.

8.3. FOX RABIES IN GERMANY

Andrews and Herzberg (1985, pp. 296–98) reprint recorded 1963–70 fox rabies cases in a selected 133 km² region in southern Germany, georeferenced by a 32-by-32 grid superimposed on the study area. The geographic distribution of these cases appears on Map 8.1. This wildlife epizootic spread in a southeastern direction through the study area during the eight-year time period; the location of the initial case appears at location (2, 5) on Map 8.1. A more comprehensive study of this phenomenon would investigate the disaggregated version of these data and specify a space-time model to describe them.

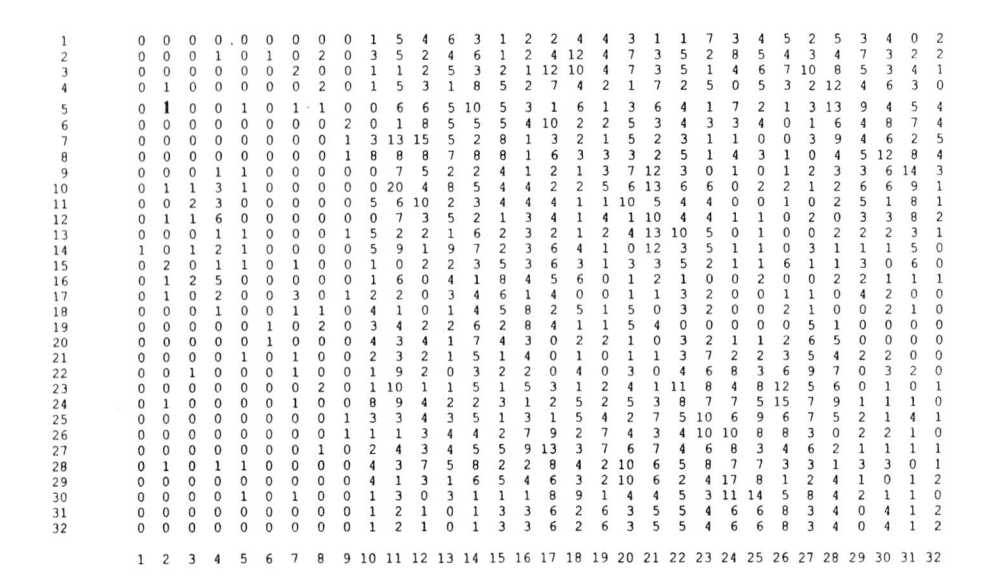

418   A CASEBOOK FOR SPATIAL STATISTICAL DATA ANALYSIS

	1	2	3	4	5	6	7	8	9	10	11	12	13	14	15	16	17	18	19	20	21	22	23	24	25	26	27	28	29	30	31	32
1	0	0	0	0	.0	0	0	0	0	1	5	4	6	3	1	2	2	4	4	3	1	1	7	3	4	5	2	5	3	4	0	2
2	0	0	0	1	0	1	0	2	0	3	5	2	4	6	1	2	4	12	4	7	3	5	2	8	5	4	3	4	7	3	2	2
3	0	0	0	0	0	0	2	0	0	1	1	2	5	3	2	1	12	10	4	7	3	5	1	4	6	7	10	8	5	3	4	1
4	0	1	0	0	0	0	2	0	1	5	3	1	8	5	2	7	4	2	1	7	2	5	0	5	3	2	12	4	6	3	0
5	0	1	0	0	1	0	1	1	0	0	6	6	5	10	5	3	1	6	1	3	6	4	1	7	2	1	3	13	9	4	5	4
6	0	0	0	0	0	0	0	0	2	0	1	8	5	5	5	4	10	2	2	5	3	4	3	3	4	0	1	6	4	8	7	4
7	0	0	0	0	0	0	0	0	1	3	13	15	5	2	8	1	3	2	1	5	2	3	1	1	0	0	3	9	4	6	2	5
8	0	0	0	0	0	0	0	0	1	8	8	8	7	8	8	1	6	3	3	3	2	5	1	4	3	1	0	4	5	12	8	4
9	0	0	0	1	1	0	0	0	0	7	5	2	2	4	1	2	1	3	7	12	3	0	1	0	1	2	3	3	6	14	3
10	0	1	1	3	1	0	0	0	0	0	20	4	8	5	4	4	2	2	5	6	13	6	6	0	2	2	1	2	6	6	9	1
11	0	0	2	3	0	0	0	0	0	5	6	10	2	3	4	4	4	1	1	10	5	4	4	0	0	1	0	2	5	1	8	1
12	0	1	1	6	0	0	0	0	0	0	7	3	5	2	1	3	4	1	4	1	10	4	4	1	1	0	2	0	3	3	8	2
13	0	0	0	1	1	0	0	0	1	5	2	2	1	6	2	3	2	1	2	4	13	10	5	0	1	0	0	2	2	3	1
14	1	0	1	2	1	0	0	0	5	9	1	9	7	2	3	6	4	1	0	12	3	5	1	1	0	3	1	1	1	5	0
15	0	2	0	1	1	0	1	0	0	1	0	2	2	3	5	3	6	3	1	3	3	5	2	1	1	6	1	1	3	0	6	0
16	0	1	2	5	0	0	0	0	1	6	0	4	1	8	4	5	6	0	1	2	1	0	0	2	0	0	2	2	1	1	1
17	0	1	0	2	0	0	3	0	1	2	2	0	3	4	6	1	4	0	0	1	1	3	2	0	0	1	1	0	4	2	0	0
18	0	0	0	1	0	0	1	1	0	4	1	0	1	4	5	8	2	5	1	5	0	3	2	0	0	2	1	0	0	2	1	0
19	0	0	0	0	0	1	0	2	0	3	4	2	2	6	2	8	4	1	1	5	4	0	0	0	0	5	1	0	0	0	0
20	0	0	0	0	0	1	0	0	0	4	3	4	1	7	4	3	0	2	2	1	0	3	2	1	1	2	6	5	0	0	0	0
21	0	0	0	0	1	0	1	0	0	2	3	2	1	5	1	4	0	1	0	1	1	3	7	2	2	3	5	4	2	2	0	0
22	0	0	1	0	0	0	1	0	0	1	9	2	0	3	2	2	0	4	0	3	0	4	6	8	3	6	9	7	0	3	2	0
23	0	0	0	0	0	0	2	0	1	10	1	1	5	1	5	3	1	2	4	1	11	8	4	8	12	5	6	0	1	0	1
24	0	1	0	0	0	0	1	0	0	8	9	4	2	2	3	1	2	5	2	5	3	8	7	7	5	15	7	9	1	1	1	0
25	0	0	0	0	0	0	0	0	1	3	3	4	3	5	1	3	1	5	4	2	7	5	10	6	9	6	7	5	2	1	4	1
26	0	0	0	0	0	0	0	0	1	1	1	3	4	4	2	7	9	2	7	4	3	4	10	10	8	8	3	0	2	2	1	0
27	0	0	0	0	0	0	0	1	0	2	4	3	4	5	5	9	13	3	7	6	7	4	6	8	3	4	6	2	1	1	1	1
28	0	1	0	1	1	0	0	0	0	4	3	7	5	8	2	2	8	4	2	10	6	5	8	7	7	3	3	1	3	3	0	1
29	0	0	0	0	0	0	0	0	0	4	1	3	1	6	5	4	6	3	2	10	6	2	4	17	8	1	2	4	1	0	1	2
30	0	0	0	0	1	0	1	0	0	1	3	0	3	1	1	1	8	9	1	4	4	5	3	11	14	5	8	4	2	1	1	0
31	0	0	0	0	0	0	0	0	0	1	2	1	0	1	3	3	6	2	6	3	5	5	4	6	6	8	3	4	0	4	1	2
32	0	0	0	0	0	0	0	0	1	2	1	0	1	3	3	6	2	6	3	5	5	4	6	6	8	3	4	0	4	1	2

Map 8.1. Geographic distribution of counts of fox rabies cases over a 32-by-32 lattice superimposed on a 133-by-133 km² region of southern Germany.

The geographic landscape is partitioned into 1,024 cells. The accompanying density of rabies cases is 2.63477. The frequency of cells by number of rabies cases occurring in each resembles a Poisson distribution (Figure 8.20). These data also appear to be moderately positively spatial autocorrelated (MC = 0.53399, GR = 0.47897), with a single conspicuous spatial outlier (Figure 8.21), which also is an aspatial outlier. These counts data benefit from a logarithmic power transformation, but perhaps not enough:

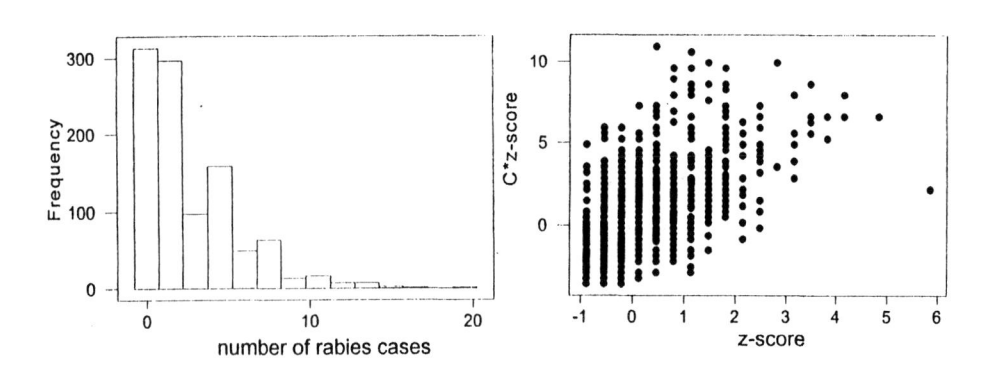

Figure 8.20 (left.) Histogram of counts of fox rabies cases (southern Germany data).

Figure 8.21 (right). MC scatterplot of raw counts of fox rabies cases (southern Germany data).

	raw	logarithm transformation

$\hat{\delta} = 2.22571$
$\hat{\gamma} = 0$
ESSE $= 6.4 \times 10^{-3}$

raw	logarithm transformation
S-W = 0.82196††	S-W = 0.88465
Bartlett = 47.332†††	Bartlett = 9.501††
Levene = 12.376†††	Levene = 4.273††
MC = 0.53399	MC = 0.60396
($\hat{\sigma}_{MC} \approx 0.02245$).	

A log-Gaussian specification may well suffer from the presence of considerable specification error, because the power-transformed frequency distribution remains quite skewed (the extreme high outlier remains conspicuous) and because considerable variance heterogeneity remains. In part, this positive spatial autocorrelation is attributable to an east-west trend across this particular German landscape, which is consistent with the observed space-time diffusion manifestations.

Following the approach outlined in §5.7, analytical eigenvectors have been generated for this 32-by-32 regular square tessellation, and those having significant positive spatial autocorrelation (i.e., those having associated eigenvalues greater than 0.17051) are used to predict the power-transformed rabies count measures. As before, for theoretical reasons, the first analytical eigenvector is not included in this set. Hence, the predictor set includes 465 (out of 1,024) eigenvectors, in addition to the east-west (U) coordinate.

Estimation results include the following:

parameter estimates

	OLS	s.e.	Poisson	s.e.	MC
b_0	1.47627	0.01255	0.70210	0.02684	
$b_{u-\bar{u}}$	0.02236	0.00132	0.05935	0.00387	
$bE_{3,1}$	-0.36827	0.02283	-1.02282	0.04980	1.00827
$bE_{7,2}$	0.20785	0.02181	0.50026	0.04893	0.91251
$bE_{1,3}$	0.19799	0.02283	0.40664	0.03858	1.00827
$bE_{2,3}$	-0.19491	0.02258	-0.30222	0.05277	1.00203
$bE_{11,1}$	-0.17195	0.02188	-0.48903	0.04494	0.77294
$bE_{5,2}$	0.14753	0.02181	0.33286	0.04820	0.96556

(Note: all eigenvectors to this point account for ≥ 2% of the variance.)

	OLS	s.e.	Poisson	s.e.
$bE_{6,1}$	0.13284	0.02255	0.28947	0.04770
$bE_{6,2}$	-0.12915	0.02181	-0.16796	0.05344
$bE_{20,1}$	0.12394	0.02184	0.34776	0.04075
$bE_{5,1}$	-0.12187	0.02218	-0.36440	0.04387
$bE_{4,4}$	-0.11503	0.02181	-0.24095	0.03779
$bE_{7,5}$	-0.11466	0.02182	-0.28167	0.03799
$bE_{3,5}$	-0.11264	0.02185	-0.31804	0.03982
$bE_{9,1}$	-0.11264	0.02191	-0.40307	0.04552
$bE_{17,1}$	-0.10384	0.02183	-0.17033	0.04119

(Note: all eigenvectors to this point account for ≥ 1% of the variance.)

$b_{E_{2,2}}$	0.09804	0.02181	0.20300	0.03963
$b_{E_{23,1}}$	-0.08639	0.02182	-0.16594	0.04062
$b_{E_{4,3}}$	-0.08770	0.02200	-0.20296	0.04378
$b_{E_{8,2}}$	-0.08514	0.02181	-0.08516	0.04322
$b_{E_{2,7}}$	-0.08142	0.02193	-0.21077	0.03998
$b_{E_{4,2}}$	-0.08001	0.02181	-0.09621	0.04195

(Note: all eigenvectors to this point account for ≥ 0.5% of the variance.)

These results are accompanied by S-W = 0.98427, R^2 = 0.578, Bartlett = 27.575[†††], Levene = 8.682[†††], and MC = 0.13857. Accordingly, they reveal that a substantial part of the originally detected spatial autocorrelation is attributable to simple map pattern and that the log-Gaussian residuals are not well behaved, having a frequency distribution that adequately conforms to a normal distribution but exhibiting marked heterogeneous variance as well as being spatially autocorrelated. The most prominent eigenvectors are associated with map patterns containing very strong positive spatial autocorrelation. The Poisson specification renders both coefficients and standard errors that differ quite noticeably from their log-Gaussian counterparts; no coefficient signs differ, and all coefficients are significantly different from zero in both specifications. This Poisson model generates expected values that are moderately correlated with the observed values (R^2 = 0.513), that have heterogeneous geographic variance (Levene = 17.876[†††]), and that contain weak positive spatial autocorrelation (MC = 0.12125; Figure 8.22). Therefore, a more comprehensive analysis of these georeferenced data would include a space-time specification based upon a Poisson model and would incorporate additional predictor variables describing the diffusion pattern.

The large number of eigenvectors registering as statistically significant here is consistent with the notion put forth (§1.2) that spatially autocorrelated data have fewer than (n - k) degrees of freedom; that is, the information content is substantially less than what it would be if georeferenced data were uncorrelated. These eigenvectors of matrix **C** are nearly uncorrelated; most have pairwise correlations of 0, with the maximum $|r| \approx 0.08$. Expression (1.10) yields eigenvectors with mean 0; 13 of the 21 analytical eigenvectors have a mean of 0, with the maximum $|\bar{x}| \approx 0.085$. Multicollinear diagnostics for the log-Gaussian OLS

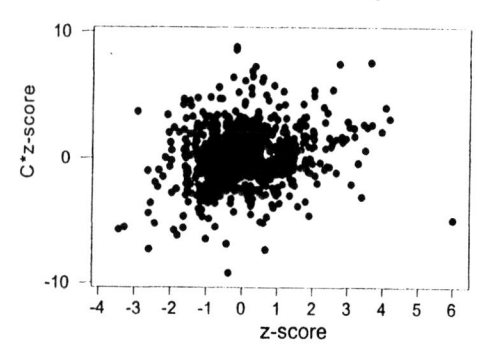

Figure 8.22. MC scatterplot of spatially filtered Poisson regression residuals for raw counts of fox rabies cases (southern Germany data).

EPIDEMIOLOGICAL VARIABLES 421

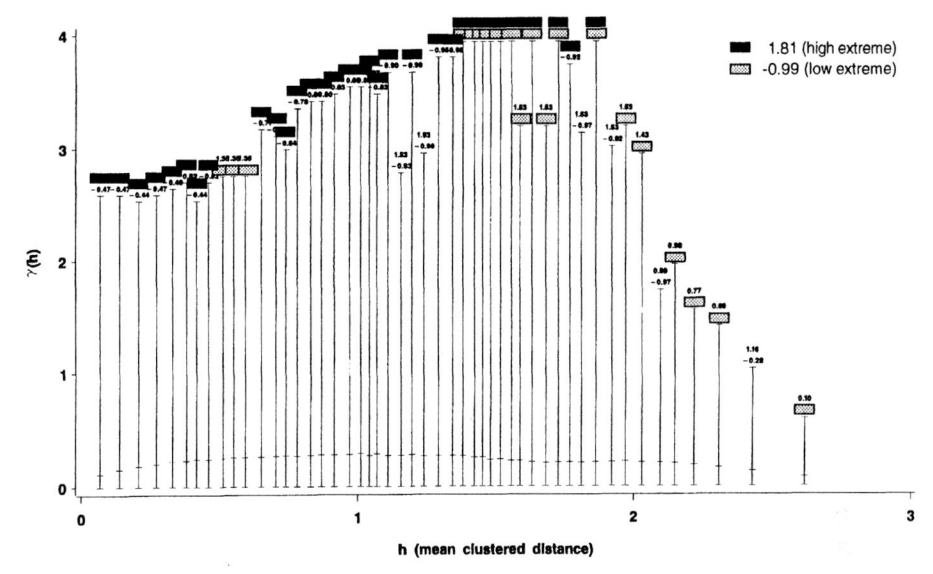

Figure 8.23. Semivariogram plot displaying the mean and extreme values for outlier-unadjusted OLS residuals for counts of fox rabies cases (southern Germany data).

results include a maximum VIF (variance inflation factor) of 1.17 and extreme eigenvalues of matrix X^TX of 1.44 and 0.56 (with 21 eigenvectors of unity), all of which suggest the absence of any serious multicollinearity. Moreover, since the term $\frac{1}{n}$ appearing in matrix expression (1.10) goes to 0 as n increases, the eigenvectors of matrix **C** and expression (1.10) should be very similar in this case, given that n = 1024 is reasonably large. As a comparison, 25 of those eigenvectors numerically extracted from expression (1.10) are statistically significant predictors of rabies; the Poisson residuals from this latter model estimation have MC = 0.11286.

Although an indicator variable for a high outlier value is not included in the spatial autoregressive analysis, examination of the high-low plot (Figure 8.23) of the residuals from the OLS regression equation

$$\hat{Y} = 1.4231 + 0.0240 \times (u - \bar{u})$$

reveals a dominant high value outlier (i.e., 1.81) in 18 of the first 22 distance classes—the number of distance classes included up to the specified maximum rescaled distance of 1.1. In fact, this value is the high value found in 31 of the 50 distance classes used to construct the semivariogram plot yields, and its absence in the larger distance classes indicates that its georeferenced position is located near the center of the map of residual values (Map 8.2). The outlier is associated with observation #299; introducing an indicator variable (I_{max}) into the OLS regression equation specification, yielding

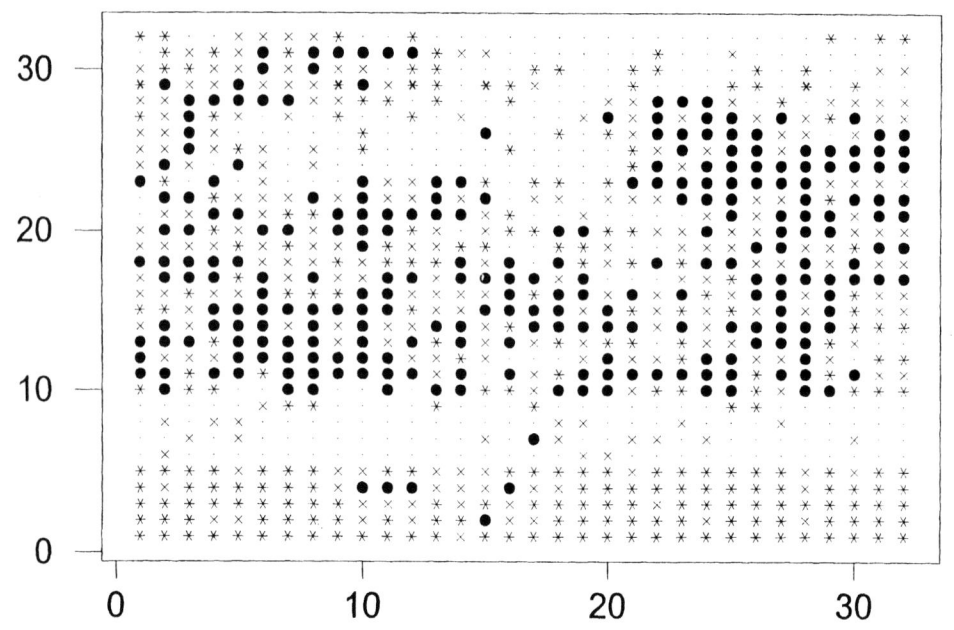

422   A CASEBOOK FOR SPATIAL STATISTICAL DATA ANALYSIS

Map 8.2. The geographic distribution of OLS residuals for counts of fox rabies cases (southern Germany data). Key: solid circle = large positive; solid triangle = moderate positive; asterisk = moderate negative; and, dot = large negative.

$$\hat{Y} = 1.4213 + 0.0241 \times (u - \bar{u}) + 1.8123 \times I_{max} ,$$

increases R^2 from 0.163 to 0.174, a very modest increase of approximately 1.1%. But the regression coefficient for variable I_{max} is significant (P=0.0003), prompting use of the revised OLS regression equation to generate residuals for the geostatistical analysis. Even after inclusion of the indicator variable for the high outlier in the OLS regression equation, however, the high-low values in Figure 8.24 reveal that other high value outliers potentially exist (e.g., 1.35 and 1.53); inspection of a frequency plot for these residuals does not indicate the presence of additional statistical outliers. Such dirty data complications are not unusual when Poisson random variables are being studied.

Although $1,024^2$ distance pairs exist for this data set, the standard 50 distance classes were used to group the residuals, yielding a minimum of approximately 9,000 pairs per class (Figure 8.25a). The selected semivariogram model fits to this scatter of points are as follows:

model	RSSE	C_0	C_1	r
exponential	0.01222	0.07103	0.21604	0.28831
Bessel	0.01715	0.10183	0.17967	0.19352
Gaussian	0.04774	0.13862	0.13738	0.37545
spherical	0.04395	0.11960	0.15702	0.79628
wave/hole	0.10226	0.16650	0.09425	0.19139
power	0.05541	0.00000	0.28818	*

$\lambda = 0.25940$.

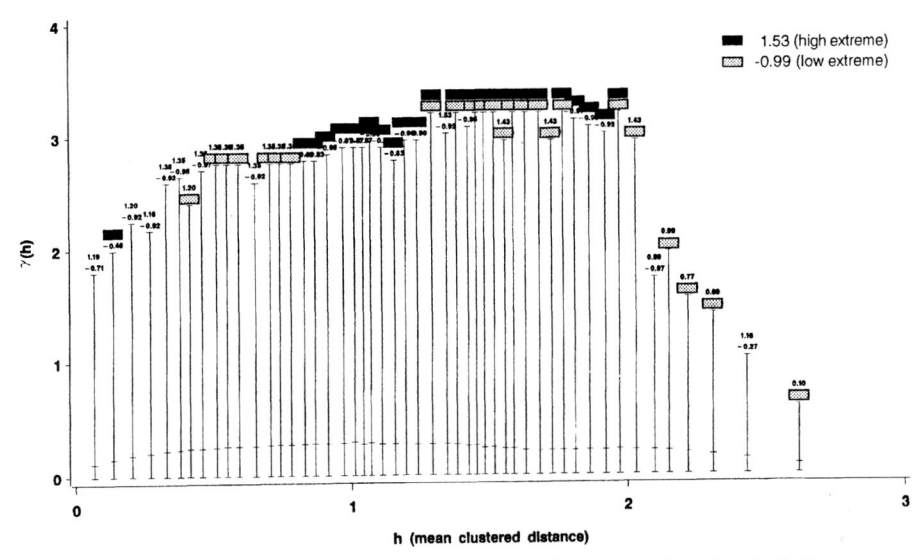

Figure 8.24. Semivariogram plot displaying the mean and extreme values for single (most conspicuous) outlier-adjusted OLS residuals for counts of fox rabies cases (southern Germany data).

The pattern of points in the semivariogram plot (Figure 8.25b) is typical of that for strong positive spatial autocorrelation: numerous intermediate gamma values are present between the first distance class and the sill value. This conclusion is corroborated by the nature and degree of spatial autocorrelation detected in the log-transformed data first used in the spatial autoregressive analyses (MC = 0.88465), coupled with the low amount of variance explained by the OLS regression equation used to generate the residuals that are subjected to a geostatistical analysis. Based on the lowest RSSE, the exponential model achieves the best description of spatial autocorrelation latent in these data, although the difference in RSSE values between it and the Bessel model is slight.

8.4. CONCLUDING COMMENTS: SPATIAL AUTOCORRELATION AND EPIDEMIOLOGICAL VARIABLES

Because they almost always are based on counts and because of their rare event nature—meaning they often are governed by laws of small numbers and include marked outliers—epidemiological variables almost always should be analyzed with Poisson regression techniques. But auto-Poisson models are plagued by two serious drawbacks, namely an intractable normalizing constant (Cressie 1991, pp. 428, 462) and restriction of cases to negative spatial dependence, with an inability to characterize positive spatial dependency (which also arises from requirements of the normalizing constant; Kaiser and Cressie 1997). One way to circumvent these two drawbacks is to estimate a log-Gaussian approximation model specification; another is to substitute auto-binomial approximation specifications. Classical

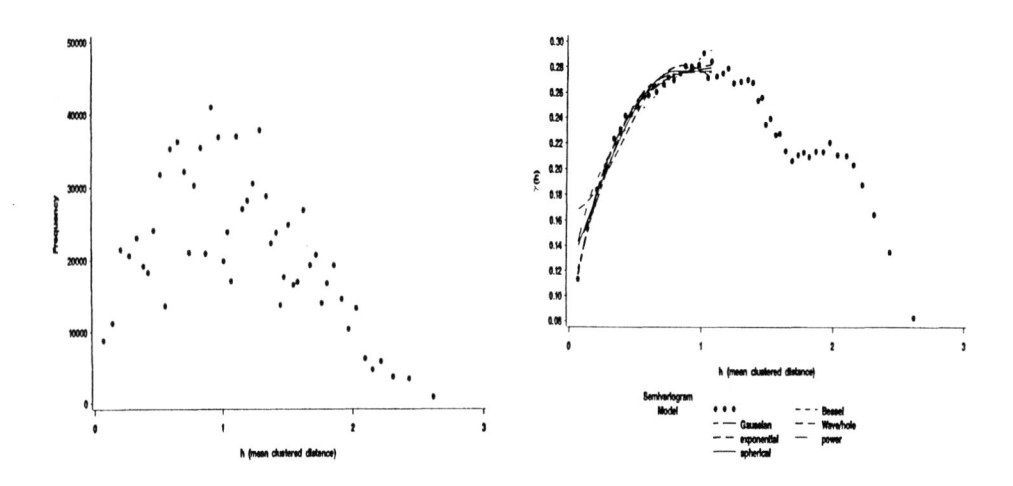

Figure 8.25. Right: (a) Semivariogram-type plot of the frequency of semivariance values in each distance class for the southern Germany data. Left: (b) Semivariogram plot of OLS residuals for counts of fox rabies cases (southern Germany data), with selected semivariogram model curves superimposed.

statistical theory indicates, however, that these approximations are close only when very large numbers are involved, currently an exceptional situation for many, if not most, georeferenced disease data sets. Therefore, a new mean response specification approach is explored in this chapter, one exploiting the eigenvector map pattern summarizations extractable from expression (1.10) (§1.5). This approach also is used in Chapter 5 to analyze numbers of plant species with Poisson regression techniques. Results reported in this present chapter indicate that this pseudo-auto-Poisson model specification furnishes a substantially better description for North Carolina SIDs georeferenced data and a roughly equivalent description for the Glasgow mortality and the German fox rabies georeferenced data. A noteworthy critical difference between the conventional log-Gaussian approximation and the pseudo-auto-Poisson specification promoted in this chapter is that the former incorporates spatial autocorrelation through a correlated errors specification whereas the latter incorporates spatial autocorrelation through the mean response function specification: "one [researcher's] deterministic mean structure may be another [researcher's] correlated error structure" (Cressie 1991, p. 114). These two specifications also are based on different assumptions and error structures.

Findings reported in this chapter continue to supply evidence concerning empirical links between the set of popular semivariogram models commonly employed in geostatistical analyses, and the set of popular spatial autoregression specifications (§3.3) that are available. Linkages reported in this chapter include

Table 8.1. Summary of linkage findings for Chapter 8

phenomenon	areal unit	autoregressive model	semivariogram model
SIDs (#1)	county	OLS-filtered; Poisson-filtered	exponential; Bessel; Gaussian; spherical
SIDs (#2)	county	OLS-filtered; Poisson-filtered	exponential; Bessel; Gaussian; spherical
standardized mortality rates	community medical district	OLS	wave/hole; Gaussian; Bessel; spherical; exponential
s^2 of blood lead level	census tract	OLS	wave/hole; Gaussian; Bessel; spherical; exponential
$\bar{x}$ of blood lead level	census tract	SAR	exponential; Bessel; spherical; Gaussian
fox rabies	pixel	OLS-filtered; Poisson-filtered	exponential; Bessel

These articulations of relationships between geostatistical and spatial autoregressive models continue to reflect findings of Chapters 4–7, namely, a frequent inability to relate simply and easily a single semivariogram model to a particular spatial autoregressive specification. In the cases where negligible spatial autocorrelation is detected, the wave/hole specification often furnishes the best goodness-of-fit, even when trace positive spatial autocorrelation is detected by the corresponding autoregressive models.

Considerable interest in the spatial analysis of health data persists, as is attested to by concerns about the diffusion of such diseases as AIDS (e.g., Getis and Ord 1998). Unfortunately, such data are less readily available, evident by the relatively few georeferenced data sets analyzed in this chapter. Nevertheless, concerns about complications introduced by spatial structure into epidemiological analyses, as well as the presence of spatial autocorrelation in georeferenced medical and health data, are being eloquently expressed (e.g., Walter 1992). It is hoped that the contents of this chapter, especially those pertaining to the pseudo-auto-Poisson model, will further promote the indispensable contributions to society that can be made by spatial analyses of epidemiological variables.

PART III

VISUALIZING WHAT IS NOT OBSERVED

9

EXPLODING GEOREFERENCED DATA WHEN MAPS HAVE HOLES OR GAPS: ESTIMATING MISSING DATA VALUES AND KRIGING

One of the pivotal links between geostatistical and spatial autoregressive analyses is the missing data problem (see §1.1). This is the prediction focus of geostatistics, falling under the heading of kriging, that deals with the interpolation problem of estimating attribute values at unsampled locations by exploiting the spatial autocorrelation detected in attribute values at sampled locations. By doing so, generalized isopleth maps can be constructed. It also is the parameter estimation focus of spatial autoregression when georeferenced data are lost in space, creating maps with holes in them, that addresses the interpolation problem of estimating the conditional mean for attribute values at data-absent locations. By doing so, spatial analysis that transcends spatial statistics, such as computing spatial operations research solutions (e.g., Griffith 1997), is unhampered by incomplete georeferenced data. In either case, the question being addressed is a forecasting type one:

> Can we see what we do not directly or indirectly observe in places located within some geographic landscape?

As in time series analyses, once a spatial model has been identified, estimated, and successfully subjected to a battery of diagnostics, and hence has been judged as adequately representing the geographic process generating a given georeferenced data series, this model can be used to supply spatial forecasts for unobserved locations. This sequence of steps furnishes an affirmative answer to the foregoing question, with the quality of this answer improving as spatial autocorrelation increases. Most traditional statistics summarize and hence are data reducing in nature; this approach is data explosive in nature and is analogous to applying a spatula to dollops of icing on a freshly baked cake (i.e., the observed data and locations) to spread (i.e., interpolate) it across the entire surface of the cake. This context directly relates spatial forecasting to the more conventional missing data literature and hence to the EM (estimation-maximization) algorithm.

Simply put, spatial forecasting says that data observed at one location are pretty much like those at nearby locations. But any forecast is simply an educated guess, and multiple, simultaneous educated guesses can be put forth (e.g., by changing the model specification). Without measuring and recording data at a given location, actual data values remain unknown, for georeferenced data are too complicated to be derived with certainty and without error through a simple spatial forecast. Not only do unobserved data remain unknown precisely in the presence of spatial forecasts, but observed georeferenced data also are not known precisely, because they can never be measured without error. Consequently, as is sound conventional statistical practice, kriging and spatial autoregression missing data estimates should be accompanied by measures of uncertainty. Although a researcher may well be familiar with traditional confidence intervals, which accompany parameter estimates, in this context, because unknown data values are being predicted, comparable uncertainty intervals are called prediction intervals. Paralleling the meaning of a confidence interval, a prediction interval contains the true data value, under the assumption of Gaussianity, in a prespecified percentage of all possible samples. For $(n-1)$ observed data values and a single conventional regression forecast, say $\hat{y}_m$ (where subscript m denotes "missing"), traditionally it is the interval defined as

$$\pm t_{\alpha/2, n-(P+1)-1} \sqrt{\hat{\sigma}^2 + \hat{\sigma}^2 \mathbf{x}_m (\mathbf{X}_{n-1}^T \mathbf{X}_{n-1})^{-1} \mathbf{x}_m^T},$$

where P is the number of predictor variables, $\mathbf{X}_{n-1}$ is the $(n-1)$-by-$(P+1)$ matrix of predictor values for the observed data (including a vector of ones for the intercept term), and $\mathbf{x}_m$ is the 1-by-$(P+1)$ vector of predictor values for the unobserved data. Basically, increased uncertainty is represented here by the left-hand term, $\hat{\sigma}^2$, under the radical sign. A confidence interval for the predicted mean response of a single observed value, say $\hat{y}_o$, is defined for $(n-1)$ observed values as the right-hand term under the radical sign, namely,

$$\pm t_{\alpha/2, n-(P+1)-1} \sqrt{\hat{\sigma}^2 \mathbf{x}_o (\mathbf{X}_{n-1}^T \mathbf{X}_{n-1})^{-1} \mathbf{x}_o^T},$$

where $\mathbf{x}_o$ is the 1-by-$(P+1)$ vector of predictor values for the observed data denoted by subscript o. Of course, a second, almost always minor component of uncertainty here is that $\mathbf{x}_m$ is not a row vector of matrix $\mathbf{X}_{n-1}$, whereas $\mathbf{x}_o$ is.

But georeferenced data contain autocorrelation and entail prediction intervals that are more similar to ones for time series forecasts. Several properties of these prediction intervals are worth highlighting here. First, one can never know more than the information distilled from observed data as one looks into the future; concatenating forecasted values with a data set does not increase its information content. Second, the prediction intervals are monotonically nondecreasing in width as forecasts are made further and further into the future; in most cases, plots of these prediction intervals look like the profile of a trumpet. Spatial forecasting possesses similar properties. First, one can never supplement the information contained in a georeferenced data set without collecting additional data values—the aforementioned dollops of icing are a fixed quantity that is being smeared across a cake's surface. Second, because many locations in a spatial series have numerous neighboring data values surrounding them, as the spatial forecasting prediction interval widens with increasing distance from a given location, beyond some distance it will begin to shrink as the location of another observed data value is approached. This dampening effect results in closely spaced observed georeferenced data values having prediction intervals with more platykurtic profiles, which become more leptokurtic with increasing distance between sample points or with decreasing spatial dependency. This tendency is further dampened by dissipation of spatial dependency effects with increasing distance, resulting in the accompanying predicted values being governed by the mean response term. The trumpet profile associated with time series data will be more characteristic of spatial predictions made along the edges of a region when forecasting into unobserved territory beyond the study region (i.e., spatial extrapolation).

A desire for optimal precision, and hence the smallest possible prediction intervals, is one motivation behind the use of the cross-validation technique for model assessment (see §1.6). Setting aside one data value at a time, the geostatistical or spatial autoregressive model is fitted to the remaining (n − 1) data values without altering its specification. Every data value is left out in turn. Each newly fitted model then is used to forecast the data value that has been set aside. The average squared difference between the n pairs of observed and forecasted values here provides an error measure—in an ideal situation, this average error equals the mean squared error for the model—which reflects upon the ability of a given model to forecast adequately a new georeferenced data value. This error quantity also can be used to help choose between competing model specifications. It is relevant to both geo- and spatial statistics, since both deal with the best linear unbiased predictor of an unknown data value. In this context, the role played by the normal distribution becomes clearer. Both kriging and spatial autoregressive equations can be derived without assuming a normal distribution, although this assumption often is used to obtained selected analytical results—a source of confusion. The normal distribution assumption simplifies the calculation of prediction intervals. A valid multivariate normal assumption means that the simple kriging estimator is the same as the conditional mean, as is the spatial autoregressive missing data estimator, and the kriging variance is the same as the variance of the conditional distribution.

9.1. An Introduction to EM Estimation

This chapter is motivated, in part, by the extensive literature that exists concerning the EM algorithm and statistical estimation of missing data (Little and Rubin 1987;

McLachlan 1997). The first problem in this book that benefits from an EM type of approach is estimating an optimal power transformation in the presence of dominating outliers. Often outliers are encountered in raw georeferenced data, either disappearing when the variable under study is subjected to a power transformation or being counterbalanced by a power transformation creating additional outliers in the opposite tail of the empirical frequency distribution. But sometimes these outliers are excessive in magnitude, sabotaging the power transformation estimation procedure outlined in §2.2. This was the case with the arcsine transformation estimation undertaken with the coal-ash data analyzed in §7.1, where the excessively high percentage of 17.6 prevented nonlinear regression parameter estimation results from converging. Setting this particular outlier aside and then treating it as a missing value, subject to the constraint that it must remain the largest value in the data set, enables parameter estimation convergence to occur.[1] The estimated value of this severe outlier is 13.30%; with the same goal in mind, Cressie (1991, 160) proposes an edited value of 12.84%, which, unfortunately, is less than the second largest value of 13.07%. The SAS code[2] used to implement this estimation includes

```
PROC NLIN MAXITER=500 METHOD=MARQUARDT;
    PARM A= 24.02 B= 53.73 C=0.88 YMISS=0.133; D=-SQRT(2)/2;
    BOUNDS 0<C<1, 0.1307<YMISS;
    MODEL NY = A + B*((1-IY)*ARSIN(C*SQRT(Y) + D)
             + IY*ARSIN(C*SQRT(YMISS) + D));
RUN;
```

This SAS code predicts the normal scores (NY), uses an indicator variable (IY) to differentiate between the known and missing values, and replaces the culprit outlier with a missing value to be estimated (YMISS). The difference here between NY and NŶ for the missing value observation indicated by IY is 0, which eliminates its contribution to the error sum of squares.

The large sample variance of this estimator is the square root of

$$\hat{\sigma}^2 [I_m + x_m(X_n^T X_n - x_m^T x_m)^{-1} x_m^T]_{diag} = \hat{\sigma}^2 [I_m + x_m(X_{n-1}^T X_{n-1})^{-1} x_m^T]_{diag} \;, \quad (9.1)$$

where the subscript diag refers to the diagonal of this matrix. This result is equivalent to that for the aforementioned conventional prediction interval; with regard to the left-hand expression, x_m is a row vector that is contained in matrix X_n. Furthermore, when more than one missing value is estimated, a Bonferroni adjustment should be made when selecting the appropriate t-value for computing the margin of error. For a single estimated missing data value this term becomes

---

[1] This approach is similar in spirit to that of Winsorizing, the practice of replacing each of the k most extreme values in each tail of a frequency distribution with its corresponding nearest retained value.

[2] In SAS 6.12, one can elect to have analytical derivatives calculated by the software.

$$\hat{\sigma}^2[1 + \mathbf{x}_m(\mathbf{X}_n^T\mathbf{X}_n - \mathbf{x}_m^T\mathbf{x}_m)^{-1}\mathbf{x}_m^T] = \hat{\sigma}^2[1 + \mathbf{x}_m(\mathbf{X}_{n-1}^T\mathbf{X}_{n-1})^{-1}\mathbf{x}_m^T],$$

perhaps more clearly disclosing the relationship between this and the aforestated forecasting error term. Additional variability is introduced here by substituting $\mathbf{x}_m\beta$ for y_m, which in turn almost always slightly alters $\hat{\sigma}$. For the preceding coal-ash data example, $\hat{\sigma} = 0.00137$, $[1 + \mathbf{x}_m(\mathbf{X}_{208}^T\mathbf{X}_{208} - \mathbf{x}_m^T\mathbf{x}_m)^{-1}\mathbf{x}_m^T] = 1.04354$, and hence the margin of error roughly equals 0.07411. The observed value of 17.6% falls within the accompanying prediction interval; after performing the necessary back-transformations, the 95% prediction interval here is (8.32%, 19.66%).

This strategy yields slightly, but not dramatically, better results for the logarithmic transformation used with Cressie's 1979-84 SIDs data (Chapter 8): the outlier, 0.03385, is replaced with a missing value estimate of 0.01017; and $\hat{\delta}$ increases from 0.39662 to 0.46821, accompanied by S-W (Shapiro-Wilk) increasing from 0.94179 to 0.96148. Similarly, for the coal-burning particulate emissions data (Chapter 7): the outlier, 227.70, is replaced with a missing value estimate of 220.97; and $\hat{\delta}$ increases from 131.49891 to 136.95561, accompanied by S-W increasing from 0.98913 to 0.98799. For the Isaaks and Srivastava's DEM variable (§5.6): the outlier, 55, is replaced with a missing value estimate of 46; and $\hat{\delta}$ increases from 43.07364 to 68.65248, accompanied by S-W changing from 0.97880 to 0.97386. Little, if anything, is gained in these three specific data analyses, suggesting that this more complicated approach is not warranted. In other cases, such as Cressie's 1974-78 SIDs data and the German fox rabies data, the missing value estimate actually is greater than the extreme observed value. This counterintuitive result may well arise from peculiarities of these particular data: the right tail outlier in Cressie's 1974-78 SIDs data becomes offset by a left tail outlier through the power transformation, the German fox rabies data have far too many 0s, and the DEM variable has two 0s that become left tail outliers after it is subjected to a power transformation. Still other data sets, such as that for the Wolfcamp aquifer (Chapter 5), the Galapagos Islands (§5.3), and the Texas groundwater SO_4 (§7.4), contain outliers that are not initially conspicuous that become apparent only after statistical analysis is underway.

These findings suggest that the EM algorithm approach to selecting an optimal power transformation provides a viable method for handling domineering outliers. But it should be used sparingly, with caution, and not indiscriminately because its statistical properties remain to be established. Certainly, it should be employed when parameter estimation convergence globally fails.

9.2. ESTIMATING MISSING VALUES: TWO SIMPLIFIED GEOREFERENCED DATA ILLUSTRATIONS

Noteworthy here is that the *iid* (independent and identically distributed) notion is critical to the missing data link connecting geostatistics and spatial autoregression. Consider equation (1.1) together with the missing variables interpretation of spatial autocorrelation. The Puerto Rican agricultural productivity data offer a rich database for illustrating this linkage. The CPR (Commonwealth of Puerto Rico) annually collects agricultural productivity data. In addition, the USDA (U.S. Department of Agriculture) collects agricultural productivity data on a periodic basis, having census results that coincide with some of those for the CPR.[3] Consider Table 13 in the 1996 *Facts and Figures of Agriculture in Puerto Rico* (CPR 1997), in which fiscal year 1991/92 delivered milk figures are missing for the Ponce region. First, given the time series context of these figures and the absence of dairy farms in Juana Diaz during 1991/92, this municipio can reasonably be allocated a production figure of 0. Second, subtracting the remaining four AAR (agricultural administrative region) totals from the island total leaves a production figure for the Ponce region of 1,918, which is consistent with this region's totals for other years.

The first step in this analysis is to identify a power transformation for the 1990/91 milk delivery variable that optimizes its conformity with a normal frequency distribution (see §2.3); this exercise yielded

Step 1: $(DMLK91/AREA + 0.03889)^{0.25723}$, $RSSE = 4.4 \times 10^{-3}$.

S-W increases from 0.50088 to 0.89309 with the use of this power transformation. The next step, in keeping with equation (1.1), is to regress the transformed predictor variable defined in Step 1 on the power-transformed Y specification (see van Gastel and Paelinck 1995) using only the observed data, which in this exercise yielded

Step 2: $(DMLK92/AREA + 0.07881)^{0.24245}$, $RSSE = 1.6 \times 10^{-3}$,
$b_1 = 1.13580$, $b_0 = 0.03889^{0.25723} - 1.13580 \times 0.07881^{0.24245}$.

The intercept term, b_0, is specified in a constraining way, ensuring that when there is no production in either year, the predicted value will be 0. This power transformation approach better ensures that the missing value estimates are not determined largely by only a few data points (Figure 9.1a-b). The third step is based upon equation (1.1) and iteratively estimates the three missing productivity values, substituting predicted values from the regression for the missing values in each iteration, while constraining these values to be equal to the residual from the island total. SAS code to implement this step includes

```
PROC NLIN DATA=STEP1 MAXITER=500 METHOD=MARQUARDT;
  PARMS B1=0.89 M20=8.3 M73=1.6;
  BOUNDS 0<M20, 0<M73;
  DX=0.03889; GX=0.25723; DY=0.07881; GY=0.24245;
  Z = IOBS*YTR +
         (1-IOBS)*((DY**GY - B1*(DX**GX)) + B1*XTR);
```

---

[3]Differences in results are due to both sampling error and measurement error (e.g., discrepancies in definitions of farm and production year between the two government agencies).

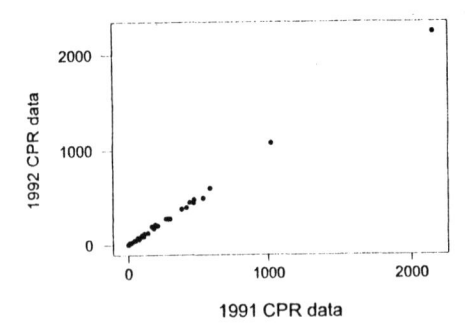

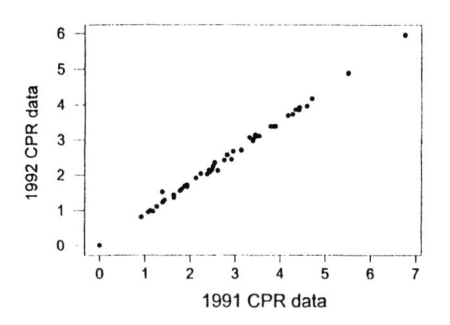

Figure 9.1. Right: (a) Scatterplot of raw 1992 DMLK versus raw 1991 DMLK measures (CPR data). Left: (b) Scatterplot of power-transformed 1992 DMLK versus power-transformed 1991 DMLK measures (CPR data).

```
MODEL Z = IOBS*((DY**GY - B1*(DX**GX)) + B1*XTR) +
                       (1-IOBS)*(IM1+IM20+IM73)*
      (1918*(1*IM1+M20*IM20+M73*IM73)/(1+M20+M73)/AREA
                                          + DY)**GY;
OUTPUT OUT=TEMP1 PRED=YHAT R=YRESID;
RUN;
```

The missing data values are differentiated from the observed data values with binary indicator variables (IOBS, IM1, IM20, and IM73 in the SAS code); both the right- and left-hand sides of the regression equation are the same for the missing data entries; and because the three entries must be non-negative values that sum to 1, only two missing values (M20 and M73) need to be explicitly estimated. This step produced

Step 3: $(DMLK92/AREA + 0.07881)^{0.24245}$, $RSSE = 1.5 \times 10^{-3}$,
$b_1 = 0.87984$, $b_0 = 0.07881^{0.242453} - 0.87984 \times 0.03889^{0.2572}$,
$DMLK92_{Adjuntas} = 186$, $DMLK92_{Coamo} = 1442$,
$DMLK92_{Yauco} = 290$.

Annual figures reported for the previous and subsequent years for these three estimates are, respectively, 177 and 197 for Adjuntas, 1,462 and 1,426 for Coamo, and 282 and 294 for Yauco[4]; hence, these three estimates are consistent with their respective time series contexts. The next step involves diagnostics, some of which suggest whether or not equation (1.2) needs to be employed. The residuals from Step 3 yield the following spatial autocorrelation diagnostic results:

---

[4]The actual mistake in the publication, identified through personal correspondence with Mr. Roberto de Jesus Martinez, was a reporting of the fiscal year 1991/92 data values as the 1992/93 data values (some annual figures for the Ponce region had been shifted one column to the right). Hence, the correct values are, respectively, 197, 1,426, 0, and 294.

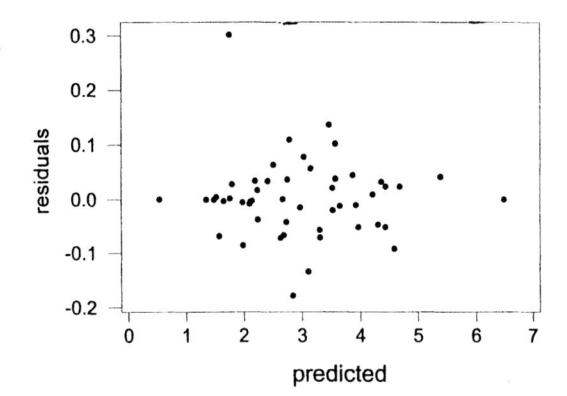

Figure 9.2. Traditional homogeneity of variance OLS residual plot for DMLK (CPR data).

Step 4: MC = −0.00230 , E(MC) = −0.00500 (see §1.4.2) ,
GR = 0.86366 ,

where MC denotes the Moran Coefficient and GR denotes the Geary Ratio. These statistics imply that there is virtually no spatial autocorrelation in the residuals from this regression. This finding is not surprising, because dairy production for two consecutive years should be governed largely by inertia in the system, and, as such, the two geographic distributions should be very similar. For DMLK91, MC = 0.35967, which indicates the presence of moderate, positive spatial autocorrelation. Because the two map patterns are almost identical, DMLK91 filters out spatial autocorrelation from the residuals in the regression (the missing variables interpretation of spatial autocorrelation). Therefore, equation (1.2) does not need to be used to obtain a spatial statistics solution (i.e., the problem is one that falls under the heading of conventional statistics). The completion of Step 4 diagnostics was

Step 4 (continued): Bartlett = 5.813, Leven = 0.538 .

These homogeneity of variance test statistics were computed for all but the Ponce region, because the latter's residuals all are nearly 0 by construction (the accompanying RSSE decreased from 1.6×10^{-3} to 1.5×10^{-3}, as it should, because using the expected values suppresses variability). Neither of these two statistics suggests nonconstant variance. The residual-versus-predicted scatter plot is displayed in Figure 9.2. This graphic reveals the presence of a possible outlier and variability that may well be less for the expected value extremes. Meanwhile, all three of the normality test statistics are highly significant, suggesting that the residuals do not adequately conform to a normal frequency distribution: A-D = 4.527, R-J = 0.8972, and K-S = 0.207. As an aside, most all of the diagnostic statistics improve slightly when the regional total constraint is removed from the missing values estimation procedure.

In this first example, missing data have been estimated in a standard EM fashion, and these estimates seem quite reasonable. The model they are based upon exhibits good, although not completely satisfactory, statistical behavior. The second example explored in this section uses the 1974 CPR sugar cane production data to estimate municipio values of sugar cane production suppressed (for reasons of confidentiality) in the 1974 USDA census publications. Once more ancillary information is used to help estimate missing data values; since San Lorenzo is the only municipio with a missing value in the Caguas region, its value is set equal to the difference between the sum of all other values and the reported regional total, namely 15,480.

The first step in this analysis is to identify a power transformation for the 1974 sugar cane variable that optimizes its conformity with a normal frequency distribution:

Step 1: $(CPRSC74/AREA)^{0.46382}$, $RSSE = 1.9 \times 10^{-2}$.

S-W increases from 0.79020 to 0.90422 with the use of this power transformation. The next step is to estimate equation (1.1), which, using only the observed data, yielded

Step 2: $(USDASC74/AREA + 0.69507)^{0.46059}$, $RSSE = 8.0 \times 10^{-2}$, $b_1 = 0.95450$.

The intercept term, b_0, is set equal to $0.69507^{0.46059}$, constraining the analysis and ensuring that when there is no production detected by either agency, the predicted value will be 0. The third step iteratively estimates the three missing productivity values, substituting predicted values from the regression for the missing values in each iteration. Following is part of the SAS code for implementing this step:

```
PROC NLIN DATA=STEP1 MAXITER=500 METHOD=DUD;
  PARMS B1=0.95 M41=2.0 M57=0.7 M59=0.6;
  BOUNDS 0<M41, 0<M57, 0<M59;
  DX=0.0; GX=0.46382; DY=0.69507; GY=0.46059;
  Z = IOBS*((Y + DY)**GY) +
                          (1-IOBS)*((DY**GY - B1*0) + B1*XTR);
  MODEL Z = IOBS*((DY**GY - B1*0) + B1*XTR) +
  (IM23+IM41+IM57+IM26+IM59)*
              (( 90660*(1*IM23+M41*IM41+M57*IM57)/(1+M41+M57) +
              336853*(1*IM26+M59*IM59)/(1+M59))/AREA + DY)**GY;
  OUTPUT OUT=STEP0 R=RSC74 PRED=YHATSC74;
RUN;
```

As with the milk production data set, missing values are constrained here to sum to the residual regional totals (i.e., 90660 and 336853), eliminating the necessity to explicitly estimate two of them (M23 and M26). This step produced

Step 3: $(USDASC74/AREA + 0.69507)^{0.46059}$, $RSSE = 8.4 \times 10^{-2}$, $b_1 = 0.95644$, $USDASC74_{Dorado} = 27586$,

Figure 9.3. MC scatterplot for NLS residuals for the power-transformed CPR sugar cane measures regressed on the power-transformed USDA DSGR measures.

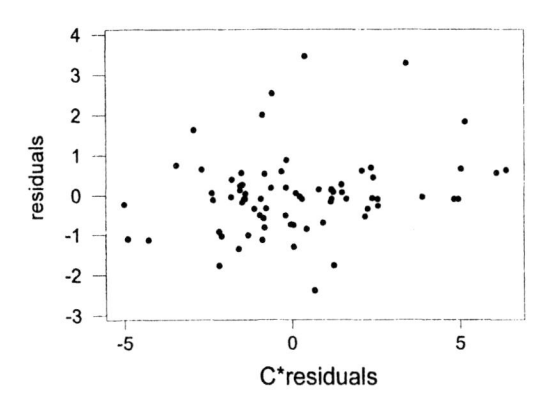

$USDASC74_{Loiza-Canovanas} = 46659$, $USDASC74_{Rio\ Grande} = 16415$, $USDASC74_{Guayama} = 203079$, $USDASC74_{Salinas} = 133774$.

These estimates seem quite reasonable; unfortunately, there is no time series context within which to assess them. The next step involves diagnostics, some of which suggest whether or not equation (1.2) needs to be employed. The residuals from Step 3 yielded the following spatial autocorrelation diagnostic results:

Step 4: MC = 0.10459, E(MC) = −0.00672, GR = 0.83976.

These statistics imply that there is no significant spatial autocorrelation in the residuals from this regression. Inspection of the MC scatterplot (Figure 9.3) supports this implication, too. This finding is not surprising either because both the CPR and the USDA were engaged in measuring exactly the same sugar cane production in 1974; this is a repeated measures type of situation, and, as such, the two geographic distributions should be practically the same. The completion of Step 4 diagnostics included

Step 4 (continued): Bartlett = 5.460, Levene = 0.608.

Because the missing values do not constitute all of the values in a single region, like in the case of 1992 milk production, they were not removed from the data set when computing these two statistics. As mentioned earlier, the use of expected values will suppress variability measures. Hence, if the residuals for the five induced values are removed, then

Step 4 (revised results): Bartlett = 5.949, Levene = 0.720,

which register only modest increases without moving them into their corresponding critical regions. Meanwhile, all three of the normality test statistics are significant, suggesting that the residuals do not adequately conform to a normal frequency distribution: A-D = 2.047, R-J = 0.9513, and K-S = 0.140. As with the density of milk production example, all of the diagnostic statistics improve when the regional total constraint is removed from the missing values estimation procedure.

Interestingly, a small part of this detected spatial autocorrelation is introduced by the intercept constraint; b_0 has been set to 0.84575 rather than to its OLS (ordinary least squares) estimate of 1.19815 (whose corresponding OLS b_1 value is 0.92077 rather than 0.95692). Removing this constraint reduces MC to 0.09608, without completely giving up variance homogeneity (Bartlett = 7.222, Levene = 0.975) but moving the residual frequency distribution further away from a normal one (A-D = 2.181, R-J = 0.9487, and K-S = 0.143). This trade-off of obtaining modestly improved diagnostic statistics at the expense of allowing negative agricultural production to be predicted seems ill advised.

In conclusion, missing data have been estimated in a standard EM fashion here. These estimates seem reasonable and are markedly improved by being constrained to equal their corresponding residual regional totals. Diagnostics indicate that the models they are based upon exhibit good, although not completely satisfactory, statistical behavior. They also suggest that the power transformation analyses are accompanied by deviations that appear to behave in an *iid* fashion, although their frequency distribution does not appear to conform adequately to a Gaussian one. These two exercises provide an introduction to the missing values estimation problem, fortunately in a conventional statistics setting, and establish a reference point for the ensuing spatial autoregressive and geostatistical (i.e., kriging) analyses in this chapter.

9.3. ESTIMATING A CONSPICUOUS MISSING DATA VALUE FOR THE COAL-ASH DATA SET

Cressie (1991, 160) notes that the complete coal-ash data set includes a missing value at coordinate location (7,6), which he forecasts to be 10.27%. Employing the spatial autoregressive model specification identified for these data in §7.1 renders a forecast of 10.17%, which compares very favorably with Cressie's forecast. SAS code for calculating this SAR forecast includes

```
PROC NLIN DATA=COMBINE MAXITER=250 METHOD=MARQUARDT;
  PARMS RHO=0.17 B0=-0.45 BU=-0.00028 BIMAX=0.093 M76=9.78;
  BOUNDS -1 < RHO < 1, 7.00< M76 <17.61;
  ARRAY LAMBDA{209}L1-L209;
  JAC = 0.0;
  DJAC = 0.0;
  DO I=1 TO 209;
    JAC = JAC + LOG(1 - RHO*LAMBDA{I});
    DJAC = DJAC - LAMBDA{I}/(1 - RHO*LAMBDA{I});
  END;
  DJAC = -DJAC/209;
  JAC = EXP(-JAC/209);
  X76 = ARSIN(0.88180*SQRT(M76/100) - SQRT(2)/2);
  DX76=0.88180/(2*SQRT(M76*100*(1 -
```

```
            (0.88180*SQRT(M76/100) - SQRT(2)/2)**2)));
     ZX = X*JAC;
     MODEL ZX = (RHO*(CX + X76*(I76N + I76S + I76E +
                I76W))/NEIGH - X76*I76 + B0 + BU*U +
                                         BIMAX*IMAX)*JAC;
     DER.B0=JAC;
     DER.BU=U*JAC;
     DER.BIMAX=IMAX*JAC;
     DER.M76=(DX76*RHO*(I76N + I76S + I76E + I76W)/NEIGH -
                                          DX76*I76)*JAC;
     DER.RHO=((RHO*(CX + X76*(I76N + I76S + I76E +
              I76W))/NEIGH - X76*I76 + B0 + BU*U +
              BIMAX*IMAX - X)*DJAC + (CX + X76*(I76N + I76S +
                                   I76E + I76W))/NEIGH)*JAC;
RUN;
```

The initial parameter estimates used here come from the estimation results reported in §7.1, except for M76, which is set equal to the mean coal-ash percentage. By inserting X76, which expresses M76 in its power-transformed form, a back-transformation calculation does not need to be performed. Not only must the missing value estimate be substituted for the missing value in the coal-ash data vector, but it also must appear in the autoregressive term, denoted here with

$$(CX + X76*(I76N + I76S + I76E + I76W))/NEIGH .$$

Because matrix **W** is used, the autoregressive parameter is restricted to the range ±1 (the geographic connectivity structure is based upon a regular square tessellation); for practical reasons, the missing value estimate is restricted to the range defined by the minimum and maximum observed coal-ash values. The derivative of the arcsine function was computed with MathCad 5.0—which is based upon the well-known Maple symbolic engine.

Parameter estimates in the presence of this estimated missing value include

	parameter estimates	
	SAR(W) without $\hat{y}_m$	SAR(W) with $\hat{y}_m$
ρ	0.16940	0.18968
b_0	-0.44706	$-0.44983 = \dfrac{-0.36451}{1-0.18968}$
$b_{u-\bar{u}}$	-0.00278	-0.00226
I_{max}	0.09255	0.09430
$\hat{y}_m$	*	10.16666 .

As can be seen from this example, parameter estimates do change slightly as the regression line rotates to ensure that $\hat{y}_m$ exactly equals its regression predicted value (Navidi 1997). This is the aforementioned source of increase in $\hat{\sigma}^2$.

In contrast with equation (9.1), the prediction interval equation for a CAR or an SAR model now is based upon

$$\hat{\sigma}^2 \{V_{mm}^{-1} + V_{mm}^{-1}[-(V_{mo}x_o + V_{mm}x_m)] \times$$
$$[X_n^T V X_n - (V_{mo}x_o + V_{mm}x_m)^T V_{mm}^{-1}(V_{mo}x_o + V_{mm}x_m)]^{-1}[-(V_{mo}x_o + V_{mm}x_m)]^T V_{mm}^{-1}\}_{diag} , \quad (9.2)$$

where matrix $\mathbf{V}/\sigma^2$ is the inverse covariance matrix—matrix $(\mathbf{I} - \rho\mathbf{C})$ for the CAR specification and matrix $(\mathbf{I} - \rho\mathbf{W})^T(\mathbf{I} - \rho\mathbf{W})$ for the SAR specification—with the matrix portion being partitioned as follows (§1.1):

$$\begin{bmatrix} \mathbf{V}_{oo} & \mathbf{V}_{om} \\ \mathbf{V}_{mo} & \mathbf{V}_{mm} \end{bmatrix}.$$

In practice these inverse covariance matrices are computed using $\hat{\rho}$. If the observations are independent (the traditional EM algorithm case), then $\mathbf{V} = \mathbf{I}$, meaning $\mathbf{V}_{kk} = \mathbf{I}_{kk}$ and $\mathbf{V}_{kl} = \mathbf{I}_{kl}$ ($k \ne l$); hence this expression reduces to the conventional one given by the right-hand expression of equation (9.1). If a single georeferenced missing value is present, then $\mathbf{V}_{mm} = 1$, while the rest of the expression remains the same. And, if a set of missing values is dispersed over a map, then $\mathbf{V}_{mm} = \mathbf{I}_{mm}$, while the rest of the expression remains the same. When computed by SAS, this asymptotic standard error for the coal-ash example is 1.00830, rendering (8.18%, 12.15%) as its 95% prediction interval.

Kriging computations for this single missing coal-ash data value were performed assuming an isotropic surface. In contrast with Cressie's estimate of 10.27% using a spherical semivariogram model, here we obtain, for the three models whose corresponding RSSE values are the lowest (§7.1),

model	residual estimate	back-transformed missing value estimate
spherical	$\dfrac{0.39281}{\sqrt{1000}}$	10.62%
Gaussian	$\dfrac{0.18662}{\sqrt{1000}}$	10.18%
exponential	$\dfrac{0.15803}{\sqrt{1000}}$	10.12% .

These results are very similar to each other, to the result reported by Cressie, and to the preceding spatial autoregressive missing value estimate result. One minor source of additional uncertainty here is the residual variance estimate of 0.00024 deviating from the C_0 parameter estimates of roughly 0.00023 (§7.1). A second additional source of uncertainty is the sequential rather than nonsimultaneous estimation of regression parameters with semivariogram model parameters; as the preceding spatial autoregression results suggest, the impact of this source should be trivial. Another reason for deviation between these results and Cressie's is our use of power transformations rather than a spatial median polish technique.

A second coal-ash missing value estimation problem concerns the extreme value of 17.61% present in these data, which has been handled in our model specifications by use of a binary indicator variable. Cressie (1991, 160) notes that he edited this outlier for location (5,6), replacing it with 12.84%. The power transformation analysis results summarized in §7.1 reveal that this value would need to be reduced to approximately 13.3% for it to be the largest value of the set of 208 attribute values if they are to conform adequately a normal distribution. Corresponding kriging estimates for this value are

model	residual estimate	back-transformed missing value estimate
spherical	$\dfrac{0.37515}{\sqrt{1000}}$	10.58%
Gaussian	$\dfrac{0.12526}{\sqrt{1000}}$	10.05%
exponential	$\dfrac{0.13182}{\sqrt{1000}}$	10.07% .

The spatial SAR estimate, using the preceding code, is 10.41%, with a 95% prediction interval—using the standard errors computed by SAS—of (8.38%, 12.42%); this value is closest to that rendered by the spherical semivariogram model specification. Notably, all of these missing value estimates fall within this prediction interval! Nevertheless, these values are markedly lower than the one computed by Cressie; his value falls just beyond the upper limit of this prediction interval, placing his estimate at roughly the same magnitude, and making it significantly greater than our predicted values.

9.4. ESTIMATING CONSPICUOUS MISSING DATA VALUES FOR AN AGRICULTURAL EXPERIMENT

Rayner (1969) reports results for a vandalized agricultural experiment, converting it from a latin square to an unbalanced design. Turnip production data for this field trial are reprinted in Hand et al. (1994, 61). In this experiment, six varieties of turnips (denoted by A, B, C, D, E, and F) were grown in field plots laid out as a 6-by-6 regular square lattice. Yield was measured as the fresh weight (roots plus tops) of turnips in pounds per field plot. A cluster of three plots in one corner of the experimental field—plots (5, 6), (6, 5) and (6, 6)—had been attacked by vandals, causing their yield data to be unusable. Because these plots are located along a corner edge, a spatial forecast of their respective missing values actually is spatial extrapolation, which is accompanied by the greatest degree of uncertainty. Because the ratio of extreme yield values for these plots exceeds 3 (i.e., 5.23), these data may well benefit from being subjected to a power transformation. One complication that cannot be overlooked here, though, is that these unobserved data come from six different populations, a situation that can confuse expectations about their aggregate frequency distribution; the underlying ANOVA assumption is that each of the six individual samples comes from a normal distribution.

Initial analysis of these yield data (n = 33) renders the following:

	raw data	unbalanced ANOVA
S-W	0.93596[†]	0.95473

Bartlett$_{variety}$	4.341	4.341
Levene$_{variety}$	0.732	0.732
Bartlett$_{geographic}$	2.374	3.040
Levene$_{geographic}$	1.516	0.988
MC ($\hat{\sigma}_{MC} \approx 0.13608$)	0.03275	0.23899
GR	0.71315	0.79527
R^2	*	0.623 .

The latin square design only appears to neutralize latent spatial autocorrelation in these yield data; the variance basically is homogeneous across varieties as well as across the four quadrants of the plane; a sizeable amount of the variance in yield is accounted for by turnip variety; an, indeed, differences in mean yields are a source of the detected deviation from a normal frequency distribution. When coupled with the ensuing S-W statistics, these judgments imply that these data should not be subjected to a variable transformation.

Conventional maximum likelihood estimates of these missing data would be the respective observed variety means, namely 18.8 for plot (5, 6), 28.9 for plot (6, 5), and 27.8 for plot (6, 6). Analysis of these yield data (n = 36) renders

	raw data	unbalanced ANOVA
S-W	0.93708[†]	0.96259
Bartlett$_{variety}$	4.845	4.845
Levene$_{variety}$	0.898	0.898
Bartlett$_{geographic}$	2.366	5.145
Levene$_{geographic}$	1.273	1.708
MC ($\hat{\sigma}_{MC} \approx 0.12910$)	-0.01480	0.21509
GR	0.98393	0.72217
R^2	*	0.643 .

A more detailed inspection, in which these data are disaggregated by variety, yields the following set of S-W statistics (Bonferroni critical regions need to be used here):

variety	A	B	C	D	E	F
raw data	0.98058	0.85862	0.88972	0.94419	0.83766	0.90440
ANOVA residuals	0.96217	0.85862	0.88972	0.95266	0.83766	0.92006 .

These disaggregated results corroborate that the mixing of values drawn from populations having different mean yields corrupts the aggregate S-W statistic; one should keep in mind that if missing values were not imputed for this (and the next) ANOVA, these results would be pairwise identical. Not all varieties exhibit significantly different yields (e.g., varieties B and F are very similar). Conclusions based upon these results essentially are the same as those for the observed data; only slight changes have occurred, with geographic variability being somewhat dampened for the raw data and noticeably amplified for the residual data, and with the prediction capability being somewhat improved. As before, not all varieties exhibit significantly different yields (e.g., varieties B and F continue to be very similar). Because detected spatial autocorrelation should be attributable to variables missing from the regression equation specification, and hence the error terms should be spatially autocorrelated, a spatial statistical SAR model specification is selected to describe these data; this specification results in $\hat{\rho} = 0.44071$. An SAR EM type estimation of these missing data values renders 17.66 for plot (5, 6), 29.98 for plot (6, 5), and 28.25 for plot (6, 6); each of these estimates is constrained to lie between the observed extremes of its respective turnip variety. Analysis of these yield data (n = 36) renders the following:

	completed raw data	ANOVA
S-W	0.93368†	0.91986††
Bartlett$_{variety}$	4.823	8.503
Levene$_{variety}$	0.826	1.161
Bartlett$_{geographic}$	1.949	6.047
Levene$_{geographic}$	1.083	1.458
MC ($\hat{\partial}_{MC} \approx 0.12910$)	-0.01378	-0.01260
GR	0.98071	0.97830
(pseudo-) R^2	*	0.790 .

These results reveal that little has changed. As can be expected, estimated missing values with and without taking spatial autocorrelation into account are roughly the same. At first glance, a failure to satisfy the normality assumption suggests that perhaps a variable transformation should be used here, especially since n is relatively small. But a more detailed inspection again counters this suspicion (Bonferroni critical regions need to be used here):

variety	A	B	C	D	E	F
raw completed data	0.96217	0.85862	0.88972	0.95266	0.83766	0.92006
spatially adjusted ANOVA residuals	0.97407	0.85642	0.97287	0.94867	0.80579	0.91130 .

As before, not all varieties exhibit significantly different yields (e.g., varieties B and F continue to be very similar). Of particular interest here are the ANOVA results, which may be tabulated as follows:

raw incomplete data
ANOVA

source	DF	Seq SS	Adj SS	Adj MS	F	P
variety	5	1289.00	1289.00	257.80	8.92	0.000
error	27	779.91	779.91	28.89		
total	32	2068.91				

spatially adjusted completed data (Griffith 1978, 1992)

ANOVA

source	DF	Seq SS	Adj SS	Adj MS	F	P
variety	5	1524.93	1524.93	304.99	12.22	0.000
error	30	748.83	748.83	24.96		
total	35	2273.76				

The number of error degrees of freedom should be adjusted in the second case, because four additional parameters have been estimated; regardless, with an F value of 12.22, the change in the accompanying probability is negligible. Nevertheless, this is the type of analysis for which spatial autoregression has a distinct advantage over geostatistics.

Kriging computations for the three missing turnip yield data values were performed assuming an isotropic surface. A geostatistical analysis of these turnip yield data renders the following goodness-of-fit measures, using disaggregated distances up to the maximum rescaled distance value of 0.6:

model	RSSE: OLS residuals	model	RSSE: OLS residuals
exponential	0.082	spherical	0.129
Bessel	0.094	wave/hole	0.313
Gaussian	0.131	power	0.069

The specification of choice here appears to be the power model, although the exponential and Bessel furnish good descriptions of the spatial autocorrelation latent in these data. Of these latter two specifications, the preceding SAR result of $\hat{\rho} = 0.44071$ favors selecting the Bessel specification, whereas the slightly lower RSSE favors selecting the exponential specification.

Parameter estimates for the two specifications achieving the lowest RSSE are

power: $C_0 = 0$, $C_1 = 27.78826$, $r = 0.26932$, and
exponential: $C_0 = 0.34579$, $C_1 = 22.89793$, $r = 0.13570$.

Using these two estimated models yields the following kriged values:

location	treatment	power residual	$\hat{y}_{power}$	exponential residual	$\hat{y}_{exponential}$
(5,6)	F	0.06697	18.87	-0.05877	18.74
(6,5)	A	1.59101	30.49	-0.09092	28.81
(6,6)	D	1.51646	29.32	-0.07812	27.72

The exponential kriging estimates compare more favorably with the aforementioned SAR missing value estimates of, respectively, 17.66, 29.98, and 28.25. This conclusion attests to the usefulness of cross-validation as an additional decision-making criterion, especially when kriging is involved.

A more visual analysis of these data using an exponential semivariogram model specification can be undertaken with GEO-EAS. Trial-and-error parameter estimation with this software yields a nugget effect of 0.35, a sill value of 30, and a range value of 4.5 (recall that this is not truly a range, as the exponential model has no range). Kriging results obtained with it render the following kriging estimates:

location	treat-ment	exponential residual	$\hat{y}_{exponential}$	standard deviations kriging	prediction OLS	SAR
(5,6)	F	-0.04186	18.76	4.45988	5.8880	5.79096
(6,5)	A	-0.43784	28.46	4.42900	5.8880	5.78921
(6,6)	D	0.15796	27.96	5.21220	5.8880	6.18859

The prediction interval is calculated with a modified version of equation (9.1). Since the residuals and their corresponding predicted values theoretically are pairwise independent, by construction, the resulting prediction intervals can be computed by replacing the identity matrix term in equation (9.1) with the kriging variance estimates; these are, for this turnips yield example,

location	treatment	$\hat{y}_{exponential}$	lower bound	upper bound
(5,6)	F	18.76	8.36	29.16
(6,5)	A	28.46	18.12	38.80
(6,6)	D	27.96	16.18	39.74

These prediction intervals are very similar to, but somewhat narrower than, their SAR counterparts, which respectively are (5.76, 29.57), (18.08, 41.88), and (15.53, 40.97). These slight differences are attributable to SAS using a t-distribution value based upon fewer degrees of freedom (additional degrees of freedom are lost for the three missing value estimates themselves as well as for the estimate of the spatial autocorrelation parameter ρ) and different standard errors (the "kriging" error term and the mean response are estimated simultaneously, rather than sequentially, with spatial autocorrelation being taken into account in the process). Regardless, all of these missing value estimates are very similar.

9.5. ESTIMATING MISSING MEDIAN FAMILY INCOME DATA FOR OTTAWA-HULL

Griffith (1992a) outlines how to use an AR model to forecast missing census data, specifically estimating suppressed 1986 median family income for 11 census tracts in Ottawa-Hull. These data also appear in Griffith (1993b). Interesting findings of this particular missing data analysis include (1) using the EM point estimates is better than using the mean of the observed data, (2) the SAR EM type point estimates are better than the OLS-based EM point estimates, and (3) the prediction intervals for the SAR EM type point estimates are the narrowest of the three. These findings are consistent with expectations promoted in this book: spatial autoregressive models tend to have an improved prediction capability and smaller parameter estimate standard errors. Those computed with the AR EM type

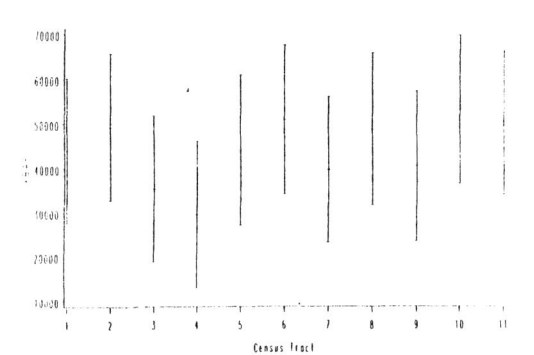

446 A CASEBOOK FOR SPATIAL STATISTICAL DATA ANALYSIS

Figure 9.4. Prediction intervals for missing income figures based upon the AR EM-type algorithm (Griffith's Ottawa-Hull data).

algorithm are graphically portrayed in Figure 9.4; comparative intervals for $\bar{x}_0$ and a standard EM solution appear in Griffith (1992a).

Parameter estimates for this analysis, accompanied by their SAS-computed standard errors, include

	OLS-based EM	s.e.	AR EM type	s.e.
$\hat{\rho}$	*		0.59327	0.10558
b_0	43513.34644	1133.54143	44459.00229 = $\dfrac{18082.81094}{1-0.59327}$	5061.56311
$b_{density}$	- 1.01781	0.29074	- 0.60622	0.26425
b_{m_1}	42071.83865	10057.11461	43979.35681	8237.33633
b_{m_2}	41746.56224	10053.45240	49231.92596	8312.90317
b_{m_3}	40203.56872	10047.77419	35354.25629	8341.06812
b_{m_4}	43447.91635	10082.07110	29565.79568	8380.09194
b_{m_5}	42324.13992	10060.54528	44058.98993	8547.19210
b_{m_6}	42437.02260	10062.24693	50804.02116	8442.16250
b_{m_7}	39643.74540	10050.49306	39610.16692	8318.72925
b_{m_8}	43143.04896	10075.22505	48647.21917	8597.13540
b_{m_9}	41754.69944	10053.53358	40354.04422	8554.67184
$b_{m_{10}}$	43512.56229	10083.61876	53076.00400	8354.01187
$b_{m_{11}}$	43509.66951	10083.54880	50130.47828	8194.68407

These results reveal that significant positive spatial autocorrelation is present in these data, that the intercept estimate is considerably more variable than is indicated by a conventional analysis, that the OLS-based EM solution confuses the population density pattern (which may be described with a negative exponential equation) with latent spatial autocorrelation, that one of the OLS-based EM point estimates verges on being markedly different from its AR EM type estimated counterpart ($z = -1.657$ for m_3), and that, on average, the OLS-based EM standard errors of the estimated missing values are inflated by roughly 20%. Including a compellingly correlated auxiliary variable (e.g., population density) tightens the individual prediction intervals and helps avoid nonsensical negative lower bounds like those requiring truncation in Griffith et al. (1989).

A geostatistical analysis of the OLS residuals (divided by 1,000) generated by regressing income on population density renders the following goodness-of-fit measures, using disaggregated distances up to a maximum rescaled distance value of 0.006125:

model	RSSE: OLS residuals	model	RSSE: OLS residuals
exponential	0.367	spherical	0.349
Bessel	0.359	wave/hole	0.272
Gaussian	0.352	power	0.544

The specification of choice here appears to be the spherical model; although the wave/hole specification achieved a lower RSSE, the preceding spatial autoregressive analysis indicates that the nature of the latent spatial autocorrelation is positive rather than negative. Parameter estimates for the spherical model using the sample of n = 181 unsuppressed georeferenced data points is

$$C_0 = 13.86547, C_1 = 96.62593, r = 0.00082 .$$

Based upon this model specification, the following kriging estimates of the missing income data were computed:

CT# 2.01	46237.6	CT#125.01	45157.1
CT# 6.00	38502.9	CT#130.01	58284.4
CT# 34.00	32817.1	CT#137.03	37248.6
CT# 47.00	34452.8	CT#140.01	51266.8
CT#110.00	45291.6	CT#160.03	58966.8
CT#120.01	43628.3		

These missing value estimates are much better than their aforementioned conventional EM counterparts and are markedly more similar to their aforementioned spatial autoregressive AR EM counterparts.

9.6. GENERALIZING A MAP SURFACE WITH KRIGING

Earickson and Harlin (1994, 116-17) furnish a map of Wayne County, Michigan, showing Cartesian coordinates and locations of the 14 automatic telemetered air sampling stations, and a table reporting annual geometric means of total suspended particulate (TSP) and sulfur dioxide (SO_2) levels recorded in 1977. This map was digitized (Map 9.1) to be used here for illustrative purposes. This readily available data set is employed in this section because its small sample size is convenient for reporting detailed computational results. Of interest here is the creation of a generalized map surface approximating the continuous geographic distribution of these two pollution variables. While the monitoring network covers Wayne County reasonably well, it is shifted slightly too far north as well as too far east (perhaps

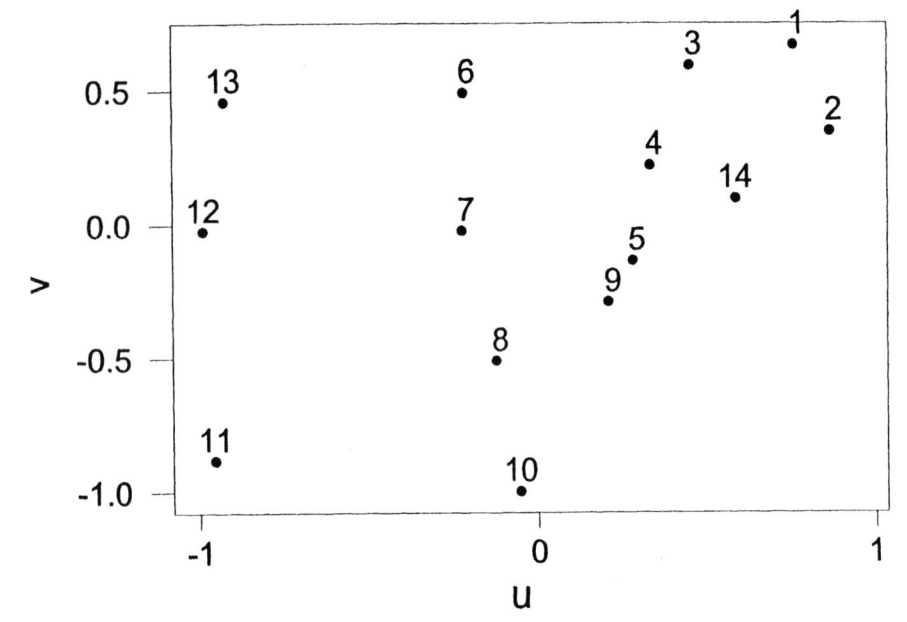

Map 9.1. Geographic distribution of telemeter air monitoring locations in Wayne County, Michigan (Earickson and Harlin's data).

better reflecting the boundaries of the city of Detroit); and monitors #5 and #9 may well be positioned too close together. In contrast with the inferential data analysis undertaken in §9.4, this sort of map generalization is the type of exercise for which geostatistics has a distinct advantage over spatial autoregression.

Each of the respective frequency distributions for both TSP and SO_2 adequately conforms to a bell-shaped curve; their corresponding S-W statistics are 0.93635 and 0.92035. The ratio of extreme values for TSP is less than 3, suggesting that it more than likely would not benefit from being subjected to a power transformation. The ratio of extreme values for SO_2 is greater than 3, suggesting that it might benefit from being subjected to a power transformation. Both pollution variables display a prominent east-west gradient; their respective affiliated scatterplots (Figure 9.5a-b), however, do not support the use of regression residuals. Considering the possibility that an international border effect exists, the five monitors along the Detroit River are grouped together for a homogeneity of variance analysis; with regard to both pollution variables, neither the Bartlett nor the Levene statistic for this grouping is significant. In addition, values for each variable are grouped as small or large, based upon the sample median. Using this grouping, some evidence of nonconstant variance is detected for TSP, while conflicting evidence of nonconstant variance is detected for SO_2. Consequently, because compelling evidence supporting the use of power transformations is elusive here, neither of these pollution variables is subjected to a variance-stabilizing transformation before undertaking kriging. Z-scores for the respective MCs are 0.92 for TSP and 1.39 for SO_2; each of these georeferenced variables

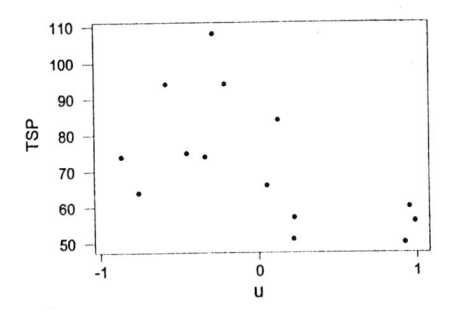

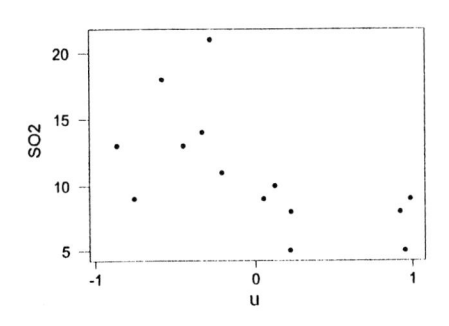

Figure 9.5. Right: (a) Scatterplot of TSP air pollution measures versus the east-west georeferenced coordinates axis (Earickson and Harlin's data). Left: (b) Scatterplot of SO_2 air pollution measures versus the east-west georeferenced coordinates axis (Earickson and Harlin's data).

contains very weak, statistically nonsignificant positive spatial autocorrelation. SO_2 was used to generate the kriged surface as well as its accompanying prediction error surface discussed in this section; TSP was used to generate the illustrative cross-validation results reported in the next section. One should keep in mind that the sample size[5] of n = 14 for this example is far to small for the reported statistical results to be truly meaningful.

A geostatistical analysis of these SO_2 data renders the following goodness-of-fit measures, using disaggregated distances up to a maximum rescaled distance value of 2.2:

model	RSSE: OLS residuals	model	RSSE: OLS residuals
exponential	0.000	spherical	0.016
Bessel	0.002	wave/hole	0.129
Gaussian	0.025	power	0.011

The specification of choice here appears to be the exponential model. Its parameter estimates for the complete sample of n = 14 georeferenced data points is

$$C_0 = 5.54339, \; C_1 = 19.38269, \; r = 0.60082 \;.$$

Using this model specification the kriging equations [forming matrix equation (1.5) in §1.1] were used to construct a generalized surface of SO_2 levels (Map 9.2). This generalized map depicts two valleys in the western part of the county, where SO_2 levels tend to be lower, and two peaks in the eastern part of the county, where SO_2

---

[5] A useful rule of thumb is to have a sample size that provides at least 20 observations per parameter estimated. In geostatistics, a sample size also should be sufficiently large to ensure at least 30 distance pairs are contained in each semivariogram interval; recall that throughout this book we attempt to have at least 50 pairs.

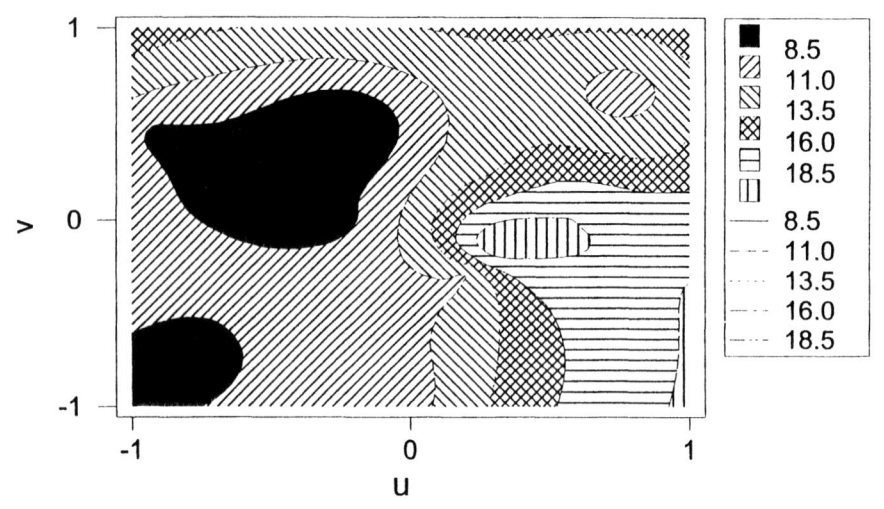

Map 9.2. Contour map of kriged SO₂ values for the Wayne County telemeter air monitoring measures (Earickson and Harlin's data).

levels tend to be very high. This generalized map was constructed by using a 21-by-21 grid of missing values, and for convenience SO_2 levels were multiplied by 1,000.

GEO-EAS also was used to construct a generalized map of SO_2 levels (Map 9.3). A visual fitting of the exponential semivariogram model with this software package produces the following estimates:

$$C_0 = 0, C_1 = 24, r = 0.8 .$$

Using this model specification for kriging purposes results in the contour map displayed in Map 9.3, which is very similar to the preceding one created with the analytical parameter estimates. One difference between the two analyses is that GEO-EAS does not use the 0 distance values; another is our use of clustering rather than uniform size distance intervals when constructing a semivariogram plot.

The second contour map constructed with GEO-EAS is for the kriging standard deviations. This map (Map 9.4) is a two-dimensional generalization of the previously discussed and calculated prediction intervals. For the Wayne County SO_2 levels, prediction uncertainty goes to 0 at the 14 monitoring locations and increases with increasing distance from each one of them. The greatest degree of uncertainty occurs along the edges of the map; most conspicuously, along the northern border of the county and in the southeast corner of the map. Especially these results allude to the problem of boundary effects in spatial statistical analysis (see Griffith 1983, 1985). In addition, uncertainty decreases when monitoring locations cluster, as in the eastern part of the county, and increases when monitoring locations become sparse, as in the western part of the county. The kriging standard errors are quite large, implying that the kriged surface (Map 9.3) basically

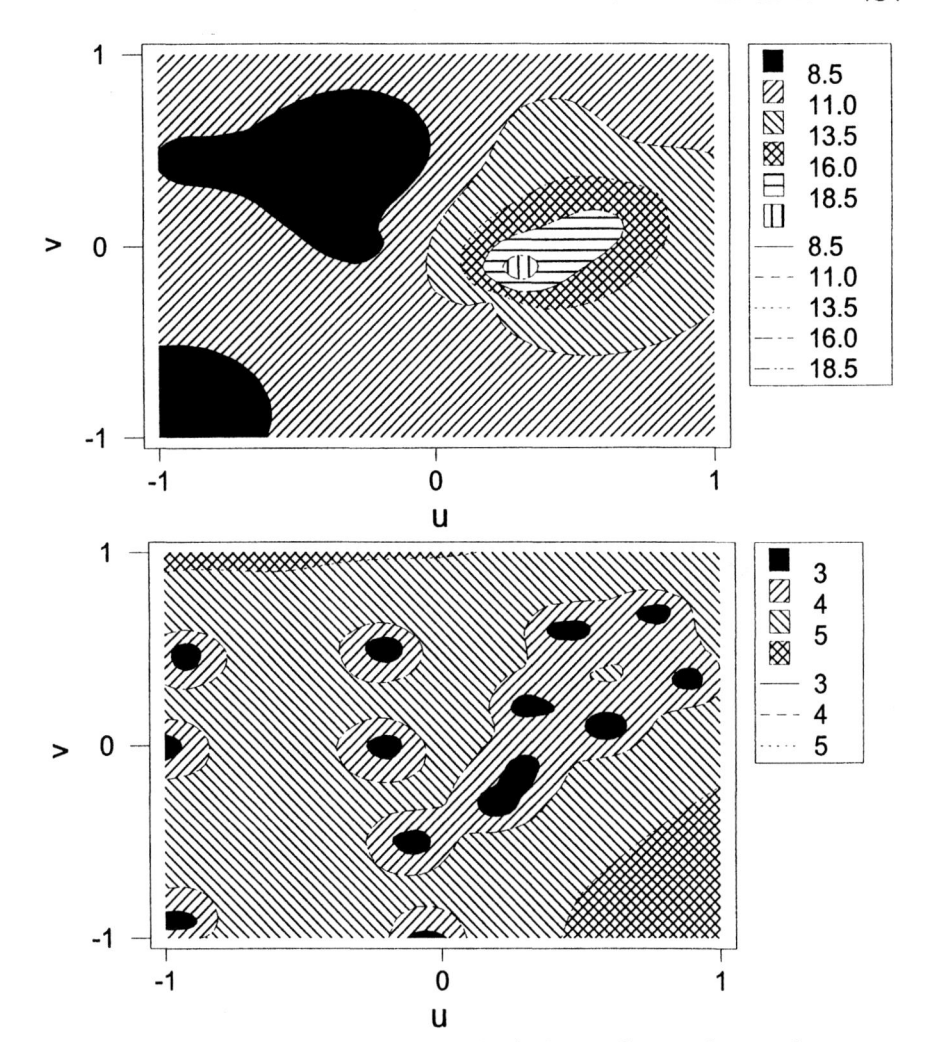

Map 9.3 (top). Contour map of kriged SO$_2$ values for the Wayne County telemeter air monitoring measures (Earickson and Harlin's data), produced by GEO-EAS.
Map 9.4 (bottom). Contour map of kriged SO$_2$ value standard errors for the Wayne County telemeter air monitoring measures (Earickson and Harlin's data), produced by GEO-EAS.

is unreliable; this finding primarily is attributable to the very small sample size of 14 that was used to construct the generalized map.

9.7. A CROSS-VALIDATION EXAMPLE

Cross-validation (§1.6), a popular form of statistical experimentation for more than 20 years, is concerned with assessing the fit of a model to the collected georeferenced data. As is described in the introduction to this chapter, in its most common form it involves fitting a semivariogram model n times (Christensen 1991; Tietjen 1986), temporarily deleting each of the observed georeferenced data points in turn; the removed data point then is treated as missing and hence predicted each time using geostatistical kriging. Next the sum of squares of deviations (observed minus kriged values) is calculated. This sum can be compared with the RSSE values reported throughout this book for goodness-of-fit evaluation purposes. As such, cross-validation is of the same flavor as the PRESS routine associated with traditional OLS regression and is similar to the resampling technique of jackknifing. Similarly, it furnishes another means for diagnosing problems with a given model fit. As Isaaks and Srivastava (1989, 351) note, "[cross-]validated residuals have important spatial information, and a careful study of the spatial distribution of cross-validated residuals, with a specific focus on the final goals of the estimation exercise, can provide insights into where an estimation procedure may run into trouble." Cross-validation supplies a check to help avoid modeling blunders; it helps establish whether or not a fitted model is grossly incorrect (but cannot establish whether or not the fitted model actually is correct!). Cross-validation also enables the variability of prediction error to be assessed. One should keep in mind, however, that estimation in this context for sampled georeferenced data is not identical to estimation for the voluminous amount of unsampled georeferenced data.

Consider the total suspended particles (TSP) data for Wayne County mentioned in the preceding section. Relevant semivariogram tabulations for these data include, for $k = 7$ distance classes,

$\gamma(\bar{d}_k)$	$\bar{d}_k$	frequency	distance minimum	maximum
0.000	0.00000	14	0.00000	0.00000
98.000	0.17147	2	0.17147	0.17147
103.500	0.31111	6	0.28023	0.33755
185.300	0.38160	10	0.36132	0.40007
266.227	0.51229	22	0.48013	0.54573
239.400	0.59860	10	0.55336	0.61752
277.667	0.71114	6	0.68050	0.74786

A geostatistical analysis of these TSP data renders the following goodness-of-fit measures, using aggregated distances up to the maximum rescaled distance value of 0.75:

model	RSSE: OLS residuals	model	RSSE: OLS residuals
exponential	0.062	spherical	0.055
Bessel	0.053	wave/hole	0.037
Gaussian	0.041	power	0.067

The specification of choice here appears to be that for the Gaussian model. Its parameter estimates for the complete sample of $n = 14$ georeferenced data points is

SPATIAL INTERPOLATION   453

$$C_0 = 0, C_1 = 289.26666, r = 0.37391 .$$

Using this model specification in a cross-validation exercise renders

observation moved	C_0	C_1	r	RSSE	kriged estimate	observed value
1	3.27879	269.22170	0.31593	0.034	72.42	64
2	0	300.26887	0.33369	0.107	71.96	74
3	0	262.28894	0.27062	0.240	67.01	75
4	0	375.90679	0.45905	0.146	98.24	74
5	0	217.10231	0.43499	0.106	90.74	108
6	11.87386	684.34607	0.75331	0.120	31.79	51
7	27.45509	981.91430	1.32592	0.365	95.48	57
8	0	275.37315	0.33755	0.060	66.70	84
9	0	276.48216	0.34919	0.043	102.60	94
10	0	292.49172	0.37266	0.054	72.84	66
11	0	289.26728	0.37391	0.043	70.31	60
12	0	298.10549	0.36988	0.072	67.32	56
13	4.04228	407.80740	0.50901	0.091	57.03	50
14	0	281.95773	0.34079	0.191	80.45	94
none	0	289.26666	0.37391	0.000	*	*

These tabulated cross-validation results furnish some resampling type of information, which could be used to explore the sampling distributions of the respective parameter estimates (similar to what is reported in §5.7). These results also reveal that the Gaussian model specification yields a cross-validation error of 268.66, which is roughly 20% less than the associated cross-validation error of 337.57 calculated by simply estimating each suppressed value, in turn, with the mean of the 13 undeleted values. The most conspicuous contributor to this cross-validation error is monitor #7; the temporary removal of this monitor produces a marked increase in the Gaussian semivariogram model goodness-of-fit RSSE value.

An alternative, simpler implementation of cross-validation uses the single semivariogram model fitted for the total set of n georeferenced locations (Ripley 1981; Cressie 1991; Carr 1995), avoiding reestimation of model parameters each time a data point is suppressed in the analysis. In other words, each of the n georeferenced attribute values is predicted, in turn, without including that data location only in the kriging equations. This version acknowledges that kriging is an exact interpolator; applying the kriging equations forming matrix equation (1.5) to the known georeferenced values returns these values exactly, except perhaps for rounding error. The goal here is to determine how closely kriging can estimate a known georeferenced data value using latent locational information. Ripley (1981, 57) cautions that forming the error sum of squares in this version may not be suitable. Cressie (1991, 498) echoes this sentiment, raising a caution about introducing additional interdependencies into a problem where data already are spatially correlated through repeated sample reuse. This rendition of cross-validation yields the following kriged estimates for the TSP data:

#1:   75.42   #5:   90.17   #9:   104.44   #12: 66.72
#2:   71.10   #6:   67.39   #10: 72.80   #13: 69.12
#3:   60.60   #7:   73.06   #11: 70.31   #14: 83.67
#4:   92.77   #8:   65.21 .

The cross-validation error associated with these estimates is 194.61, considerably less than the previous result of 268.66. This reduction in mean squared error is attributable to removal of additional variability introduced by reestimating model parameters for each of the n kriging exercises. Now the most conspicuous deviates occur for measures taken at monitor locations #4 and #5, which are neighbors (Map 9.1).

One should keep in mind that this Wayne County example has far too small a sample size, suggesting that results of the spatial statistical analyses reported here are only useful for illustrative purposes. However, in such practical situations (after all, this is a real-world monitoring network!) sample sizes may well be this small.

9.8. Concluding Comments: Exploding Georeferenced Data

One principle objective of conventional statistics is to summarize relevant information contained in large quantities of data, condensing voluminous amounts of numbers to a few meaningful statistics/parameters descriptors. Criteria for assessing the quality of parameter estimates reflect this goal. The property of sufficiency, for example, pertains to summarizing all of the information captured in a sample that is relevant to the corresponding population parameter. Especially with the advent of time series forecasting, statistics expanded its scope to embrace the unraveling, or exploding, of information. The term explosion is used here because concentrated pieces of information are detonated into particular patterns, much like fireworks. No new information is created; rather, old information is repackaged. For our purposes, such patterns are governed by the nature and degree of latent spatial correlation coupled with the underlying geographic configuration of areal units or georeferenced locations. In more recent years, data explosion has become very popular in statistics, its promulgation in part being due to a wide dissemination of the EM algorithm, its theory, and its practice. The usefulness of missing data imputations is widely accepted today. In this chapter, both spatial forecasting—particularly kriging—and spatial autoregressive versions of the EM algorithm are discussed within this data analytic context. Of special note is the illustration in §9.5 of how spatial statistical solutions improve prediction intervals: mean responses tend to improve slightly, whereas prediction errors decrease and hence prediction intervals shrink.

Although the principle link between spatial autoregression and geostatistics established in §1.1 is the kriging/missing values imputation equations, one noticeable difference between these two estimation methods merits emphasis. Kriging strictly is a spatial forecasting procedure. Once a semivariogram model is fitted to sampled data, values can be predicted for locations at which data are not sampled. This procedure is equivalent to forecasting in time series analysis or even to prediction of new observation values in OLS regression. The missing data imputation solution for spatial autoregression is iterative, essentially beginning with the calculation of this forecast as its first step. Then, in standard EM algorithm fashion, model parameters are reestimated, updated imputations computed, followed by another round of model parameter reestimation, and so on, until convergence of parameter estimates is obtained. In this setting, missing values actually are treated like parameters that are to be estimated. In practice, their estimates differ little from their forecast counterparts. This feature is seen repeatedly in examples presented in this chapter.

In terms of evidence associated with linkages between spatial autoregressive and geostatistical model specifications, the coal-ash missing data estimation exercise supports a connection between the SAR and Gaussian semivariogram models. Cressie (1991, 157) estimates the spherical semivariogram model for these data, a decision that is consistent with results reported in §7.1. His spatial prediction of the missing value for location (7,6) is 10.27%. Our SAR imputation for this missing value is 10.17%, whereas our Gaussian spatial prediction is 10.18%. This spatial prediction is closer to both our SAR one and Cressie's reported one than it is to our spatial forecast generated by the spherical model specification, which is 10.62%. The mean of the 208 known coal-ash percentages is 9.78%, whereas the mean of the four neighboring coal-ash percentages is 10.88%. All of the estimates fall into the interval (9.78%, 10.88%), which is defined by these two values. Cressie's spatial prediction is closer to the midpoint of this interval; our SAR imputation and Gaussian spatial prediction are closer to the mean percentage for the 208 sampled locations. In contrast, our spatial forecast derived from the spherical semivariogram model specification, namely 10.62%, is very close to the upper limit of this interval. Cressie subjects these data to a spatial median polish procedure and edits the location (5,6) outlier value; we subject these data to an arcsine transformation and include an indicator variable to account for the conspicuous outlier value.

Turning attention to the location (5,6) outlier value now, Cressie's data editing procedure renders a replacement value of 12.84%; this substitute is markedly greater than the mean of the four neighboring coal-ash percentages (10.86%). Here, when we treat this outlier as a missing value our SAR imputation is 10.41%, whereas our spherical spatial prediction is 10.58%. This agreement supports the initial decision reported in §7.1 of a linkage between the SAR and spherical semivariogram models. Again, the spherical model generates a spatial forecast that is closer to the mean of the neighboring values than it is to the mean of the 207 remaining percentages. These results, which are of the same flavor as ones rendered by cross-validation, suggest once more that cross-validation be used as one of the decision-making criteria for model selection.

The turnip field plot data, which display a somewhat weak level of spatial autocorrelation prior to imputation but after turnips variety means have been taken into account imply a linkage between the SAR and exponential semivariogram models. The Bessel function specification does not furnish a markedly poorer description of spatial dependencies latent in these georeferenced data. Furthermore, inspection of the raw data suggests that the latin square experimental design appears to neutralize spatial dependency effects; inspection of the residuals generated by an ANOVA of these data, however, suggests otherwise. This specific finding further supports Grondona and Cressie's (1991) argument that it is desirable to take spatial considerations into account in the analysis of agricultural field experiments.

The median family income missing data analysis advocates a linkage between the spatial autoregressive AR model and the spherical semivariogram model specifications.

Another important facet of the findings reported in this chapter concerns variability. Just like reporting polling results without their corresponding margins of error, neither spatial prediction nor spatial imputation is suitably informative without being accompanied by some sensible measure of its uncertainty. We provide such measures in this chapter, together with both conventional and spatial statistical theory background. For a given georeferenced data analysis, prediction intervals constructed for either spatial autoregressive or geostatistical estimation appear to be comparable and respectable. Of particular importance for the map generalization problem supported by kriging is a map of the uncertainty; we supply an isopleth example of this type of map in §9.6. It portrays two salient characteristics it portrays that spatial scientists need to keep in mind: (1) spatial prediction uncertainty increases as one moves from interpolation to extrapolation, and (2) spatial prediction uncertainty increases for interpolation as the number of georeferenced data points supporting it decreases and becomes sparsely spaced.

Finally, two versions of cross-validation are demonstrated. The first, which is a more conservative implementation, is roughly equivalent to the jackknife resampling procedure. It may well offer a means of better understanding sampling distribution issues affiliated with semivariogram model parameter estimates. Unfortunately, it involves reestimating the parameters of a selected semivariogram model specification n times. The second, which is far easier to implement, furnishes a measure that is analogous to the standard PRESS statistic of OLS regression. It renders another sensible criterion for model selection (§3.2.3), which should be used together with the RSSE value. An additional criterion such as the cross-validation error is needed to better argue quantitatively for (1) choosing between two RSSE values that essentially are the same, and (2) determining why the wave/hole model specification, which often achieves the lowest RSSE value (e.g., TSP in §9.7) when the accompanying MC value suggests that positive spatial autocorrelation is present, should not be selected.

10

CONCLUDING COMMENTS

Much of the emphasis in this book is on commonalities shared by geostatistical and spatial autoregressive methodologies: the pivotal concept of spatial autocorrelation, a principal linkage provided by the missing data problem, the dual relationship between spatial correlograms and partial correlograms, and the MC (Moran Coefficient) scatterplot and the semivariogram plot visualization tools. This is a theme that other spatial scientists are beginning to address, too (see Cressie, et al., 1999). One salient feature differentiating geostatistics from spatial autoregression is that the former deals with a punctiform sampling usually of more or less continuously distributed geographic phenomena, whereas the latter deals with a discretization of some geographic surface and hence an aggregation of geographic phenomena. Consequently, mutually exclusive and collectively exhaustive surface partitionings tend to render semivariogram plots that are more similar to those that would result from the superimposition of an essentially uniform (systematic) sampling network (demonstrated repeatedly with the various data sets analyzed in this book). Although a representative spatial sample requires a markedly uniform network, measures of spatial autocorrelation require at least a reasonable number of clustered sample georeferenced coordinates to cover adequately the entire range of interpoint distances. Example implementations appearing in GEO-EAS suggest use of a mixed sampling design to achieve this end, part of which involves systematic sampling and part of which involves simple random sampling. As is done repeatedly for the analyses summarized in the chapters of this book, such a sample of points can be converted to a tessellated surface for spatial autoregression by constructing Thiessen polygons. This perspective is consistent with Watson (1992), as well as with the common practice of using these polygons to describe a given sampling network.

On a fine enough scale of resolution, though, all geographic phenomena are discontinuous. Even "continuous" georeferenced attributes, such as soil type, are piecewise discontinuous at the mesoscale, tending to be distributed like chocolate swirls in a marble cake. Similarly, on a coarse enough scale of resolution, any discrete geographic phenomenon can appear to be continuous. Heuvelink (1993) furnishes an enlightening comparative discussion of these two viewpoints on spatial variation. This paradox is reminiscent of that currently unreconciled in physics between quantum mechanics and relativity theory: quantum mechanics deals with discreteness by treating it as continuous, while general relativity is a continuum theory. The implications here are that there are discoveries about geographic reality still to be made and that there often is more than one acceptable way of describing any given georeferenced set of attribute values. Geostatistics and spatial autoregression are lessons in how some superficially obvious notions, such as spatial autocorrelation, indeed can make reality less mysterious. Both of these spatial data analysis conceptualizations render geographic reality more explicable and predictable than it was before their developments. Unfortunately, the surface partitionings employed in most spatial autoregressive analyses often produce centroid spacings whose intervals are greater than the range for an associated semivariogram model, erasing that portion of the semivariogram plot that portrays an increasing trend over short distances and showing only a scatter of points about the variance across all distances.

Regardless of which of the two approaches to spatial data analysis is preferred by a scientist, the popular question addressed in traditional circles—So what?—is replaced by the question "So where?" In a quest to answer this second question, we promote in this book the art of reaching conclusions at the interface of scientific theory and georeferenced observation. Spatial data are complicated, virtually always being noisy, dirty, and messy, and with nontrivial complications arising from the presence of non-0 spatial autocorrelation lurking about. Not only are observed data that are tagged to specific locations in and of themselves often unintelligible, but they also often do not easily interpolate to unobserved locations; fine-resolution geographic landscapes are too big and too complex. Given this context, we outline some of the pearls and perils of first impressions gleaned from judiciously selected and readily available georeferenced data sets, emphasizing the frequent need to obtain second, third, or even further impressions of such data. Otherwise, one is faced with twisted interpretations supported by misused statistics; spatial scientists are held hostage by inappropriate traditional statistical formulae. But what difference do such bad statistical computations make? If spatial autocorrelation is near 0, not much. However, if spatial autocorrelation is moderate or strong—the rule rather than the exception for georeferenced data—parameter estimates may be biased, most surely inefficient, insufficient, and occasionally inconsistent: all the right ingredients to compromise an inferential basis. The data sets investigated in this book have been drawn from a wide range of sources and application domains, and every effort has been made to discover, correct, and report data-recording errors; accordingly, the focus is on sampling, specification, and stochastic error issues rather than on calculation or measurement error issues.

10.1. MORE ABOUT THE NATURE OF GEOREFERENCED DATA

Phenomena distributed across a geographic landscape are governed by some random function urn model, with their observed realizations being selected from this urn filled with "magnetized" (spatial autocorrelation) balls that, when drawn by the invisible hand of nature, produce samples similar to clusters of grapes; this same concept typifies the popular "magic box" magnetic sculptures. Pure empirical geographic facts, however, are corrupted prior to, during, and after they are collected, and this corruption potentially is propagated through arithmetic operations performed on the numerical data (see Arbia et al. 1998b). Types of corruption that erode the quality of information contained in georeferenced data include low-level noise (calculation, measurement, sampling, specification, and stochastic), messiness (data analytic technique complications arising from missing values), and dirtiness (nonlinear relationships and disagreements between replicate measurements). The goal of a spatial statistical analysis is to describe and predict georeferenced phenomena, distilling the level of uncertainty that can be attached to such descriptions and predictions. Noise is a hallmark of each data set analyzed in this book, which both requires and complicates such an assessment of uncertainty. Data sets such as the vandalized turnip experiment, the coal-ash geographic distribution, and the suppressed Ottawa-Hull median family income or Puerto Rican milk and sugar cane production figures furnish examples of messy georeferenced data. The rampant presence of non-0 spatial autocorrelation, as well as outliers, and discrepancies between comparable CPR and USDA replicate measurements, supply instances of dirty georeferenced data; perhaps the most glaring example is provided by the Cliff-Ord Eire data, in which latent non-linearities result in a curvilinear relationship misinterpreted as positive spatial autocorrelation.

The role of noise here is curious. Map patterns are systematically organized sets of numbers containing information that is trying to tell something about reality, and noise most often tends to interfere with this signal. Yet, sampling and stochastic noise, especially, can play a useful role in map pattern analysis. If the underlying map pattern is faint (basically appearing as a constant across some geographic landscape mixed with a small amount of white noise), because of a paucity of attribute information, which causes it to be contrastless, sampling and structured stochastic error can help map pattern detail emerge. Of course, the risk in such a situation is that only a sample-specific or a false pattern is discovered. In contrast, if the noise level is moderately high, geographic patchiness may well materialize, accompanied by a map pattern that begins to resemble an abstract art mosaic-type painting or perhaps a pointillist painting. If the noise level is too high, it becomes overwhelming, masking any pattern. In other words, the goal of a spatial scientist should be to effectively and efficiently manage noise contained in georeferenced data, not simply to eliminate it. This objective is achieved when spatial predictions are more accurate than is allowed by the rules of chance dictated by theoretical levels of stochastic and sampling error.

The presence of noise means georeferenced data exhibit variability. A semivariogram plot displays the similarity of values geographically distributed across locations, given typical variation in a data set, as a function of distance. Similarly, the MC and GR (Geary Ratio) spatial autocorrelation indices reference neighboring values to this typical variation, visualizing it with a MC scatterplot. If neighboring values tend to behave just as chaotically as the totality of georeferenced phenomena, which renders covariation levels comparable to the overall level of variability exhibited by the phenomena in question, spatial dependency is absent. However, if neighboring values tend to be similar relative to a fairly high level of variability displayed by the totality of some georeferenced phenomenon, spatial dependency is widespread. The modus operandi is that nearby values are more (dis)similar than are more distant values. The indices of MC and GR are limited to assessing merely the closest neighboring values regardless of separation distance. A Thiessen polygon partitioning of a surface rarely conforms to a regular tessellation. Of course, the semivariogram plot overcomes this weakness by inspecting similarity across the entire range of separation distances.

A valid first impression can be gained from classical statistics by calculating the central tendency exhibited by georeferenced data: the correct summary values are deemed typical. Although this central tendency is devised to be characteristic of some abstract mathematical space, it also is characteristic of geographic space. If georeferenced data are dirty, then this characterization may need to involve the application of a variable transformation and to accommodate spatial and/or aspatial outliers. Such variable transformations are selected in this book by maximizing the corresponding univariate S-W (Shapiro-Wilk) normality statistic. Meanwhile, such accommodations are accomplished through the inclusion of indicator variables in equation specifications. Virtually all georeferenced data are sufficiently dirty to benefit from being described with an autoregressive term; incorporating the presence of spatial autocorrelation often results in a 5-10% improvement in statistical description of a given georeferenced data set. In contrast, determining how typical the typical value is —uncertainty—does not fair as well when conventional statistics are computed with georeferenced data. Spatial statistics rectifies this situation and thus provides proper guidance about where typical georeferenced values are too low and where such values are too high; the emphasis here is on *where*, which is the jurisdiction of spatial statistics (Berry 1994, 22). Unfortunately, implementing the necessary geostatistics or spatial autoregression is computationally intensive; fortunately, the awkward and cumbersome arithmetic mechanisms to do so are quite simple for a modern computer to handle.

10.2. REFLECTIONS ON SPATIAL DATA MODEL SPECIFICATIONS

Certainly the issue of variable transformations merits careful and extensive thought when undertaking a georeferenced data analysis. Cressie (1991) basically avoids this issue by employing a spatial median polish procedure. Believing that the geographic dimension of spatial data is not necessarily compatible with this technique, we instead advocate the use of Box-Cox power transformations, especially in keeping with Griffith et al. (1998). The goal is to achieve a more bell-shaped frequency distribution of attribute values; accordingly, the power transformation is selected that maximizes the S-W normality diagnostic statistic. By employing such power transformations, a researcher is able to work with a more manageable range of data values and to fit semivariogram models to more stable semivariogram plots. The use of power transformations, however, is accompanied by serious drawbacks. Foremost is that estimators based upon the transformed values, rather than upon the original raw data, are the ones that have the desired statistical properties induced by a transformation, removing some precision from back-transformed confidence or prediction intervals. In addition, back-transformation often causes kriging spatial forecasts to become extremely sensitive to changes in sill values. But the skewness of a data set does tend to influence its variogram plot, frequently causing this plot to depict mostly a pure nugget effect. One interesting comparison of a pair of variogram plots, one for the raw and the other for the log-transformed version of these data, is given by Journel and Froidevaux (1982), who use the coefficient of variation as a selection criterion for choosing between these two possibilities. The potential for numerous specific comparisons is provided by Dutter and Reimann (1998), who have estimated semivariogram models for both untransformed and log-transformed versions of dozens of georeferenced variables, including terrestrial moss, humus, topsoil, reindeer lichen, podzol profiles (5 horizons), and lakewater. They used data collected for nearly 700 locations distributed across a 188,000 km^2 region comprising the entire area between 24° and 35.5° east longitude, and between the Barents Sea to the north and the Polar Circle to the south.

The issue of anisotropy also merits further contemplation. Findings reported in §6.1 suggest that anisotropy is easier to handle in geostatistical analyses. Accordingly, the angle of axis rotation in an anisotropic situation could be established with semivariogram model estimation, the axes rotated, and then a spatial autoregressive analysis undertaken using the rotated axes. For this procedure to be sound, however, clear linkages between spatial autoregressive and semivariogram model specifications are essential. This deduction, however, is not supported by White et al. (1997), who reveal that even the presence of pronounced anisotropy across the continental United States had little effect upon kriged soil zinc spatial predictions computed with omnidirectional variogram models. These authors contend that because kriging is a smoothing interpolator (and in most cases only a relatively few neighboring points determine a kriged estimate) similar variograms will generate very similar kriging outcomes. Another way to view this situation can be gleaned from Chapter 1: while the semivariogram model characterizes the spatial covariation matrix, its inversion essentially yields a quite sparse spatial autoregressive inverse-covariation matrix. The sparseness of this inverted matrix means that only a few neighboring data points will be involved in the kriging estimation. Furthermore, one should keep in mind that these weights do not directly depend upon either the georeferenced data or any distributional assumptions, such as normality.

A third salient issue worthy of pondering has to do with semivariogram model specifications. Many exist! A generalized form bridging the exponential and Gaussian specifications is the following specification, known as the stable model:

$$C_0 + C_1\{1 - EXP[-(\frac{d}{r})^\lambda]\}, 0 < \lambda \leq 2.$$

When $\lambda = 1$, this expression reduces to that for the exponential model; when $\lambda = 2$, it reduces to that for the Gaussian model. Oliver and Webster (1986) include a circular specification, whose functional form is

$$C_0 + C_1\{\frac{2}{\pi}\frac{d}{r}[\sqrt{1-(\frac{d}{r})^2} + SIN^{-1}(\frac{d}{r})]\}.$$

Cressie (1991, 61) lists a rational quadratic, whose functional form may be written as

$$C_0 + C_1 \frac{(\frac{d}{r})^2}{1 + (\frac{d}{r})^2},$$

or its related modified version, the Cauchy specification (Wackernagel 1995, 219), whose functional form is

$$C_0 + C_1\{1 - \frac{1}{[1 + (\frac{d}{r})^2]^\lambda}\}.$$

Olea (1991, 15, 57, 70) lists a penta-spherical and a cubic, whose respective functional forms are

PENTA-SPHERICAL

$$C_0 + C_1[\frac{15}{8}(\frac{d}{r}) - \frac{5}{4}(\frac{d}{r})^3 + \frac{3}{8}(\frac{d}{r})^5], 0 \leq d < r$$
$$C_0 + C_1, d > r.$$

CUBIC

$$C_0 + C_1[7(\frac{d}{r}) - \frac{35}{4}(\frac{d}{r})^3 + \frac{7}{2}(\frac{d}{r})^5 - \frac{3}{4}(\frac{d}{r})^7], 0 \leq d < r$$
$$C_0 + C_1, d > r.$$

These two specifications are very similar in functional form to that for the spherical model. Haining (1990, 97) lists a De Wijsian, whose functional form is

$$C_0 + C_1 \times LN(d) ,$$

and whose more generalized functional form is

$$C_0 + C_1 \times LN(d^2 + \delta^2) , \quad \delta \geq 0$$

(Wackernagel 1995, 220). And, of course, numerous authors list the linear specification, whose functional form is

$$C_0 + C_1 \times d .$$

The De Wijsian specification is the special case of the power specification for which $\lambda = 0$, whereas this linear specification is the special case of the power specification for which $\lambda = 1$. In addition, Deutsch and Journel (1992, 23), as well as Cressie and Clark (1994) and others, mention positive linear combinations of such specifications; such a hybrid specification essentially is what Olea(1991) labels nested. In other words, the number of candidate specifications is nearly unlimited; we have inspected only the standard ones, promoting the Bessel function[1] as an appealing member of this set (also see Wackernagel 1995, 219). Some authors question the utility of certain of these specifications. Zimmerman and Zimmerman (1991), for example, dismiss the spherical and Gaussian models. Tabulated linkage findings reported in each of the concluding sections of Chapters 4-9 in this book suggest otherwise; these two specifications repeatedly furnish satisfactory spatial dependency descriptions. Regardless, the spherical model specification is widely used in geology and soil science, with numerous empirical applications in geostatistics demonstrating its usefulness. But the spherical covariance function, which is the only one of the popular models with a range parameter literally being a range, has a tendency to produce multimodal likelihood profiles, making estimation of covariance parameters difficult, because it is not twice differentiable with respect to the range parameter (Mardia and Watkins 1989). Meanwhile, the Gaussian model specification tends to suffer from prediction intervals that are unrealistically narrow, and it often needs at least a small nugget value for the kriging equations forming matrix equation (1.5) to involve a numerically stable matrix inversion. The unconstrained linear semi-variogram model verges on being nonsense in all but perhaps just a selected set of applications. In addition, models without sills are associated with random functions with an unlimited capacity for dispersion.

The generalized functional form of the De Wijsian specification is of the same flavor as the power transformation specification used throughout this book, namely,

---

[1] Bessel functions can be either of the first-kind or of the third-kind (i.e., the Basset specification; Wackernagel 1995, 219). These latter functions also are known as modified Bessel functions of the second-kind.

$$LN(y + \delta) \text{ and } (y + \delta)^{\gamma},$$

where a < δ (a being the minimum observed attribute value). Moreover, any of the variogram models listed in this book may be modified by substituting ($\sqrt{d^2 + \delta^2} - \delta$) for distance d; when d = 0, $\sqrt{0^2 + \delta^2} - \delta = 0$. In this context, the parameter δ quantifies the curvature of an attribute map surface at the sampled locations (Kitanidis 1997, 64). This curvature can be described qualitatively for selected situations where δ = 0 as parabolic for the exponential model (Figure 10.1a); logistic for the Gaussian, Bessel, rational quadratic, Cauchy, and wave/hole models (Figure 10.1a-c); linear for the spherical, circular, penta-spherical, and cubic models (Figure 10.1a-c); and variable (depending upon λ) for the power, stable, and Cauchy models (Figure 10.1d-f). For a given C_0 and C_1, the steepest curve is generated by the cubic model but only to d = $\frac{r}{2}$, whereas the shallowest curve is generated by the wave/hole model but only to d=1.43042πr; the cubic specification also appears to capture negative spatial autocorrelation. The other three models having true ranges generate the next steepest curves; the Gaussian, Bessel, Cauchy, and power (λ < 1) models tend to generate the next shallowest curves, with some modified behavior preceding and following d = r (Figure 10.1a-c).

Parameter estimates for these six additional specifications, using the High Peak remotely sensed georeferenced data presented in §5.7, are

model	RSSE	C_0	C_1	r	λ
circular	0.00852	0.03699	0.30859	0.89716	*
rational quadratic	0.00915	0.04978	0.35244	0.42677	*
penta-spherical	0.00492	0.01763	0.33370	1.20306	*
cubic	0.00515	0.14765	0.12625	1.75674	*
stable	0.00477	0.00000	0.38850	0.48137	1.14130
Cauchy	0.00682	0.02044	0.64130	0.18492	0.21242

All RSSE values are relatively small, and both the penta-spherical and the cubic specifications achieve better fits than does the spherical specification (RSSE = 0.00569), which is declared the model of choice in Chapter 5. These results further emphasize the need for additional selection criteria for discriminating between model descriptions. Even invoking theoretical properties, such as the conditionally negative definiteness condition, which relates a semivariogram model to an intrinsically stationary stochastic process, reveals only that, for the power specification, 0 ≤ λ < 2. For this domain, then, both the De Wijsian and the linear models are valid, even though in their unrestricted forms they are associated with infinite variance situations. The power model result of $\hat{\lambda}$ = 0.57917 (§5.7), however, suggests that a De Wijsian model would be inappropriate here; parameter estimates of a De Wijsian specification render $\hat{\delta}$ = 0 and RSSE = 0.404, which are in agreement with this expectation. As an aside, the stable model achieves a lower RSSE (= 0.00447) value than do its exponential and Gaussian siblings, yields 0 < $\hat{\lambda}$

CONCLUSION 465

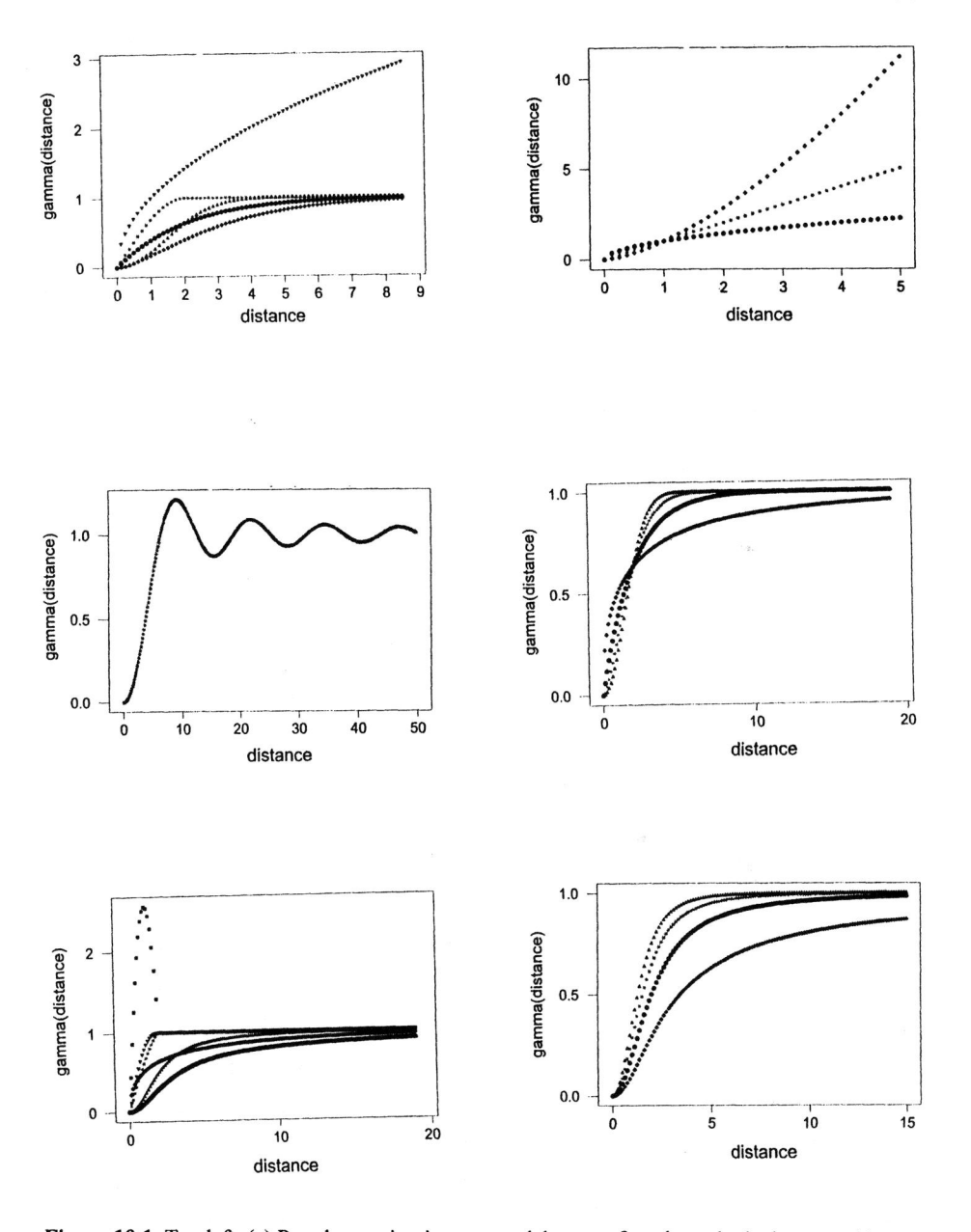

Figure 10.1. Top left: (a) Popular semivariogram model curves for a hypothetical geographic landscape: power (▼), spherical (✶), Gaussian (▲) exponential (●), and Bessel (◆) function. Middle left: (b) Popular semivariogram model curves for a hypothetical geographic landscape: wave/hole. Bottom left: (c) Other semivariogram model curves for a hypothetical geographic landscape: penta-spherical (▼), cubic (■), circular (✶), rational quadratic (▲), stable (◆), and Cauchy (●). Top right: (d) Power semivariogram model curves for a hypothetical geographic landscape: $\lambda < 1$ (●), $\lambda = 1$ (✶), and $\lambda > 1$ (◆). Middle right: (e) Stable semivariogram model curves for a hypothetical geographic landscape: $\lambda = 0.5$ (◆), $\lambda = 1$ (●), $\lambda = 1.5$ (✶), and $\lambda = 2$ (▲). Bottom right: (f) Cauchy semivariogram model curves for a hypothetical geographic landscape: $\lambda = 0.5$ (◆), $\lambda = 1$ (●), $\lambda = 1.5$ (✶), and $\lambda = 2$ (▲).

< 2 ($\hat{\lambda}$ = 1.14130), and produces a C_0 (= 0) value identical to that for the exponential model and a C_1 (= 0.38850) value that lies between those for the exponential and Gaussian models; its range parameter value (= 0.48137) is closest to that for the Gaussian model.

The cross-validation model selection procedure is demonstrated here now for the 1986 Ottawa-Hull geographic population density model (mentioned in §4.2 and studied by Griffith and Can 1996) using an AR autoregressive model specification ($\hat{\rho}$ = 0.21428). Applying a logarithmic transformation to these data yields LN(population density + 1757.20296) as the dependent variable. A regression equation for this phenomenon relates it to distance from the CBD (central business district), the (U, V) Cartesian coordinates, and two indicator variables, one identifying the three anomalously low densities and the other differentiating between Ottawa and Hull; the resulting equation produces R^2 = 0.426, Bartlett = 10.537, and Levene = 5.429—the latter two both are significant, implying variance heterogeneity—and S-W = 0.97933 for its residuals. Residuals from this regression model are the data explored. Initial parameter estimates for 11 of the semivariogram models depicting positive spatial autocorrelation are as follows, where the 36,158 distance pairs not exceeding a rescaled distance of 1.0 were grouped into 50 clusters and a maximum rescaled distance of 0.40 was used:

model	RSSE	C_0	C_1	r	λ
spherical	0.381	0.08116	0.08615	0.02670	*
Gaussian	0.381	0.09616	0.07117	0.01340	*
exponential	0.379	0.04409	0.12336	0.00748	*
power	0.780	0.12803	0.05333	*	0.17057
circular	0.381	0.08332	0.08399	0.02429	*
rational quadratic	0.373	0.05947	0.10839	0.00604	*
penta-spherical	0.381	0.07875	0.08856	0.03118	*
cubic	0.433	0	0.16665	0.01990	*
Bessel	0.380	0.07390	0.09350	0.00597	*
stable	0.378	0	0.16754	0.00517	0.78715
Cauchy	0.383	0.09572	0.72035	0.01658	1.98224

The original set of 36,864 distance pairs was reduced to 36,158 pairs by including only those distances roughly equal to half of the maximum rescaled distance of 1.94099; truncating the average distance at 0.4 further reduced this set to the 27,998 pairs forming 28 of the 50 initial clusters. The minimum number of distance pairs per cluster was 432. To gain more detail from clustering, these data were reanalyzed be excluding all distances beyond 0.4. This analysis reduced the number of distance pairs to 28,236, removed considerable smoothing from the semivariogram plot, created groups with a minimum number of distance pairs per cluster of 222, and yielded the following parameter estimates:

model	RSSE	C_0	C_1	r	λ
spherical	0.621	0.06057	0.10653	0.01774	*
Gaussian	0.619	0.07723	0.08992	0.00920	*
exponential	0.617	0.03826	0.12907	0.00595	*

power	0.864	0.12724	0.05511	*	0.17521
circular	0.621	0.06199	0.10512	0.01585	*
rational quadratic	0.612	0.06423	0.10355	0.00538	*
penta-spherical	0.620	0.06035	0.10678	0.02167	*
cubic	0.640	0	0.16682	0.01138	*
Bessel	0.617	0.00624	0.105	0.00454	*
stable	0.616	0	0.16747	0.00414	0.75496
Cauchy	0.609	0.03013	0.13843	0.00241	0.64083

As expected, these latter model fittings have markedly larger RSSE values associated with them than do their preceding counterparts. Not surprisingly, model parameters change with the use of less aggregated distance parings; C_0 tends to decrease, C_1 tends to increase, and r tends to decrease. The specification with the closest agreement between parameter estimates is for the power model; the specification with the largest deviation between parameter estimates is for the stable model, followed by the Bessel function model. The rank order of the model fits places the rational quadratic specification as most preferred and the power specification as least preferred; seven of these specifications essentially render the same RSSE value. The cross-validation mean squared error rates for these models are as follows:

model	cross-validation MSE	model	cross-validation MSE
spherical	0.16429	penta-spherical	0.16453
Gaussian	0.42460	cubic	8.88971
exponential	0.16714	Bessel	0.18321
power	0.15777	stable	0.16011
circular	0.16428	Cauchy	0.16421
rational quadratic	0.18622	wave/hole	3.49606×10^8

The wave/hole specification achieved RSSE values of 0.401 and 0.653 for the semivariogram model fits but a very large cross-validation MSE. In other words, for this georeferenced data set, cross-validation clearly eliminates the wave/hole specification from consideration. Cross-validation also clearly eliminates consideration of the cubic specification. The cross-validation exercise ranks the power specification first, which is absolutely opposite to its ranking according to the RSSE values, and the Gaussian specification last. As with the RSSE values, most of these MSE values are very similar. Basically combining RSSE and MSE results removes the Gaussian, power, cubic, and wave/hole models from the candidate pool, leaving nine specifications from which to select. The poor discriminatory power of the cross-validation procedure is disappointing! As an aside, this exercise exemplifies the often time-consuming and tedious nature of the semivariogram modeling task.

A fourth matter deserving reflection pertains to estimation. Zimmerman and Zimmerman (1991) compare seven different semivariogram estimators and find that no single estimator is uniformly superior for parameter estimation, regardless of whether georeferenced data are regularly or irregularly spaced. We have elected here to employ Cressie's (1985) weighted least squares nonlinear regression, in part mimicking Heuvelink's WLSFIT software. But fitting variogram models is partly a science and partly an art. Our most successful implementation strategy to date has been first to fit a model using unweighted NLS (nonlinear least squares), and then to use the estimates from this step to initiate fitting a model using weighted NLS. Problems arise because the nonlinear nature of the problem introduces complexities that lie dormant in linear estimation. Foremost is convergence upon local rather than global optima. We checked all of our SAS PROC NLIN results with WLSFIT; in a few cases, SAS finds a better solution (i.e., one with a lower RSSE) than does WLSFIT. Furthermore, in all but the spherical, circular, penta-spherical, and cubic specifications, a range parameter rather than the range itself is estimated. To interpret empirical results meaningfully, effective or practical ranges, say r^*, need to be established. Estimates suggested in the literature include, when $C_0 = 0$,

exponential	$\gamma(r^*) = 0.95(C_0 + C_1)$	$\rightarrow r^* = 3r$
Gaussian	$\gamma(r^*) = 0.95(C_0 + C_1)$	$\rightarrow r^* = (\sqrt{3})r$
Bessel function	$\gamma(r^*) = 0.95(C_0 + C_1)$	$\rightarrow r^* = 4r$

The following practical ranges can be established by extending this class of calculations:

rational quadratic	$\gamma(r^*) = 0.95(C_0 + C_1)$	$\rightarrow r^* = (\sqrt{19})r$
stable	$\gamma(r^*) = 0.95(C_0 + C_1)$	$\rightarrow r^* = [(3)^{\frac{1}{\lambda}}]r$
Cauchy	$\gamma(r^*) = 0.95(C_0 + C_1)$	$\rightarrow r^* = [\sqrt{(20)^{\frac{1}{\lambda}} - 1}]r$
wave/hole		when the range is interpreted as: the distance at which this function is maximized, $r^* = 1.43042\pi r$; in the cubic specification, $r^* = 2\pi r$; and, the maximum distance a given sine wave deviates from C_0 when it first falls within a $(1 \pm 0.05)C_0$ interval, $r^* = 6.48537\pi r$.

For any given empirical application, then, a researcher simply could determine the value of r^* for which $\gamma(r^*) = 0.95(C_0 + C_1)$ and use that as the effective/practical range for most any semivariogram model specification.

A final issue warranting comment in this section deals with the three popular spatial autoregressive model specifications. The CAR specification is a first-order model because its covariance matrix is a function of the geographic connectivity matrix alone (i.e., raised to the power 1). The SAR and AR models are second-order models because their respective covariance matrices are a function of some product of the geographic connectivity matrix. Biparametric versions of these matrices are found in the rare instances where anisotropy is incorporated into a specification. Higher-order models essentially are not found in practice. A principle difference between the SAR and AR specifications is the presence of spatially lagged predictor variables in the former and their absence in the latter. Haining (1990) provides some guidance for using the graphical device of added variable plots to decide which predictor variables should have their accompanying spatially lagged terms introduced as new explanatory variables into a spatial autoregressive specification. A liability of this approach is that the researcher might be confronted with estimation of a mixed SAR-AR model, further obscuring a linkage with some semivariogram model specification.

10.3. IMPLICATIONS REGARDING RELATIONS BETWEEN SPATIAL AUTOREGRESSIVE AND GEOSTATISTICAL MODELS

Arguments put forth in this book demonstrate that geostatistics and spatial autoregression are two subtly different faces of the same spatial statistics coin. In selected situations, they both seek to impute missing georeferenced values and then quantify the uncertainty associated with such estimated missing values. The principal goal of geostatistics, however, is to produce a generalized map surface, whereas the principal goal of spatial autoregression is to support spatially adjusted data analytic statistical techniques. Regardless, with spatial autocorrelation serving as the progenitor of both, a need to articulate relationships between their respective models is paramount to achieving an integration of these two subdisciplines that have for too long developed nearly autonomously but in parallel.

Results tabulated in the concluding sections of Chapters 4-9 furnish an empirical basis for articulating at least some linkages between geostatistical and spatial autoregressive models.

autoregressive model	exponential	Bessel	Gaussian	spherical	wave/hole	power
single linkage						
OLS					2	
CAR	2					
SAR	1			2		
AR			1	2	3	1
multiple linkages						
OLS	9	8	4	5	3	2
CAR						
SAR	11	11	7	9		4
AR	3	1	2	4		2

where OLS denotes ordinary least squares. Seven additional empirical cases, including one SAR case, fail to reveal any linkages whatsoever; one single linkage case involves a pure autoregressive model, which can and has been recorded here as both an SAR and an AR specification. The preponderance of empirical cases suggests that autoregressive models numerically link, based solely on an RSSE criterion, to multiple semivariogram models. These results are slightly biased by a rejection of wave/hole linkages when negative spatial autocorrelation is not revealed by a spatial autoregressive specification. Two trends in the preceding table meriting additional comment here: the scarcity of linkages to the power semivariogram model specification, and the roughly uniform frequency of linkages between each autoregressive model specification and the exponential, Bessel, Gaussian, and spherical semivariogram model specifications. In this book we promote the rule of thumb that the power semivariogram model specification should be used in cases of an explicitly unmodeled nonconstant mean, including the AR specification at times. Because virtually all nonconstant mean responses are modeled in this book, a failure to find many connections between spatial autoregressive and semivariogram model specifications is not surprising. The second pattern disclosed by the above tabulation suggests that restricting attention to the exponential and Bessel specifications may be warranted, or that ignoring concerns about use of the Gaussian and the spherical specifications may be warranted.

The preceding linkage evidence is inconclusive and hence disappointing. Rather than reveal clear and distinct pairings of models, this evidence alludes to the need for theoretical guidance. The work reported in Griffith and Csillag (1993) establishes links between the exponential and CAR models, and between the Gaussian and SAR models. Refinements reported in Griffith et al. (1996) revise this second connection to be between the Bessel and SAR models. Finally, Griffith and Layne (1997), exploring the back-transformation, find that both the exponential and CAR, and the Bessel and SAR pairings do not constitute perfect matches. The nature of the matrix inversion results, however, reflects the well-known fact that the kriging equations for a regular square tessellation produce the same weights for all locations in the same relative position within the lattice configuration; if we ignore edge effects, these may well be the binary entries of the CAR specification. One complicating factor is that the spatial correlation structure for an irregular lattice is quite complex. Another arises from spatial autoregression exploiting Σ^{-1}, which can result in relatively large order of magnitude autocorrelation index values and parameter estimates, whereas geostatistics exploits Σ itself, which as Bartlett (1975, 82-83) first showed can render corresponding semivariogram plots that suggest the presence of rather weak spatial autocorrelation. Consequently, given the preceding linkage confusion, the theoretical approach to be pursued apparently should concentrate on regular square tessellation benchmark findings. In addition, results summarized in this section need to be refined through more detailed evaluations of the foregoing model links using, for instance, cross-validation (§9.7) and the Akaike information criterion.

10.3.1. MODEL SELECTION BASED UPON THE AKAIKE INFORMATION CRITERION: THE WOLFCAMP AQUIFER GEOREFERENCED DATA REVISITED

First mentioned in §3.2.3, the Akaike information criterion (AIC) is an option available in the SAS MIXED procedure, a PROC (Version 6.12 and higher) that

CONCLUSION 471

can be used to estimate selected spatial autoregressive models. This statistic is useful for selecting from competing model specifications estimated with the same set of georeferenced data. It indexes the relative difference in a log-likelihood function—say log-L($\hat{\theta}$)—due to fitting two specific semivariogram models to the covariance function defined by latent spatial autocorrelation. Depending upon the isotropic semivariogram model specification employed, the AIC is computed in SAS for this particular PROC as

$$[\text{log-L}(C_0, C_1, r) - 3] \text{ or } [\text{log-L}(C_0, C_1, \hat{\lambda}) - 3] \text{ or } [\text{log-L}(C_0, C_1, r, \hat{\lambda}) - 4],$$

where the parameter estimates C_0, C_1, r, and $\hat{\lambda}$ are maximum likelihood estimates, and the subtracted number (i.e., 3 or 4) is the number of covariance function parameters being estimated. The model having the largest AIC is considered the best. The number of covariance function parameters often doubles with an anisotropic specification.

The menu of spatial covariance functions from which a researcher can select when executing PROC MIXED includes the exponential, Gaussian, spherical, power, linear, log-linear, anisotropic exponential, and anisotropic power; because of serious theoretical weaknesses, neither the linear nor the log-linear specification has been treated in this book. For illustrative purposes, again consider Cressie's Wolfcamp aquifer data (see the introduction to Chapter 5). The rank order of preference suggested by the AIC for selected semivariogram model specifications estimated with these data is as follows:

spherical	-10.8334	anisotropic	
Gaussian	-11.2781	power	-12.7808
exponential	-11.8406	exponential	-14.2873
power	-11.8406		

These semivariogram models were estimated with the same regression model specification appearing in Chapter 5: a north-south gradient variable, an east-west gradient variable, and an outlier indicator variable. Based upon RSSE results reported in Chapter 5, the spherical specification is followed in rank order by the exponential, Gaussian, and power. Agreement in identification of at least the best model is both encouraging and appealing.

Unfortunately such comparisons are somewhat tenuous. Semivariogram model parameter estimates rendered by PROC MIXED are as follows:

model	C_0	C_1	r
spherical	0.03882	0.06135	0.50049
Gaussian	0.04952	0.05462	0.28962
exponential	0.02513	0.07541	0.14128
power	0.02514	0.07540	*

$\lambda = 0.95727$.

These parameter estimates are very similar to those reported in Chapter 5, except for $\hat{\lambda}$; but, they do differ noticeably from their Chapter 5 counterparts. In part these discrepancies are attributable to the different way distances are clustered.

Accompanying regression model parameter estimates fair better, and are as follows:

model	parameter estimates			
	b_0	$b_{u-\bar{u}}$	$b_{v-\bar{v}}$	$b_{outlier}$
OLS	0.55665	-0.01060	-0.00986	0.57007
SAR(W)	0.59247	-0.01053	-0.01025	0.60987
spherical	0.61316	-0.01042	-0.01055	0.57625
Gaussian	0.61168	-0.01048	-0.01030	0.58048
exponential	0.60795	-0.01039	-0.01067	0.56806
power	0.60795	-0.01039	-0.01067	0.56806
anisotropic				
power	0.61955	-0.01035	-0.01067	0.58462
exponential	0.62039	-0.01028	-0.01066	0.57665

These results demonstrate, once more, that the way in which spatial autocorrelation is modeled has little effect upon regression parameter estimates. In contrast, comparisons of standard error estimates yield mixed implications:

model	standard error estimates			
	b_0	$b_{u-\bar{u}}$	$b_{v-\bar{v}}$	$b_{outlier}$
OLS	0.16090	0.00054	0.00062	0.16111
SAR(W)	0.15540	0.00069	0.00084	0.15067
spherical	0.16590	0.00104	0.00132	0.15326
Gaussian	0.16809	0.00114	0.00142	0.15029
exponential	0.17017	0.00100	0.00126	0.15977
power	0.17017	0.00100	0.00126	0.15976
anisotropic				
power	0.16894	0.00102	0.00132	0.15691
exponential	0.16133	0.00102	0.00127	0.14867

When examining these results, one needs to keep in mind the difference between employing an inverse covariance matrix for autoregressive specifications, and the covariance matrix itself for geostatistical specifications; the matrix inversion can play havoc with estimation results. The autoregressive result for s_{b_0} is about 3% smaller than its OLS counterpart, whereas all of the geostatistical results for this standard error are at least 3% larger. The autoregressive results for $s_{b_{u-\bar{u}}}$ and $s_{b_{v-\bar{v}}}$ are closer to the OLS results than are the geostatistical ones; but both the autoregressive and the geostatistical results are greater than their OLS counterparts. All of the spatial statistical results for $s_{b_{outlier}}$ are consistent, and are less than the OLS counterpart.

Therefore, PROC MIXED can be used in a limited way to obtain selected autoregressive results, including the AIC, a statistic that often will prove to be quite useful to a researcher. At present the MIXED procedure does not offer the options of a Bessel function or a wave/whole model, is not capable of being expanded to embrace additional valid semivariogram model specifications like those outlined in §10.2, and removes some of the specification, nonlinear estimation, and measurement control (e.g., defining the number and range of distance classes) from a spatial scientist. All of these drawbacks compromise the utility of the AIC statistic it calculates. Of note is that the SAS and SPSS code presented in this book does support these types of customized analyses, and hence one simply could code the AIC formula in these programs. PROC MIXED highlights, yet again, similarities between repeated measures analysis and spatial statistics. For the most part, though, it emphasizes linear model parameter estimation, even failing to furnish an overall summary estimate of the nature and degree of spatial autocorrelation, namely $\hat{\rho}$, latent in a georeferenced data set. Furthermore, it becomes extremely cumbersome-to-impossible to implement as n increases, being unable to exploit the Jacobian (i.e., Gaussian normalizing constant) approximation implementation appearing in §3.1.2, and thus being required to work with an n-by-n covariance matrix. The Wiebe data set, for example, whose analysis appears in Chapter 6 and for which n = 1,500, taxes PROC MIXED. Consequently the final conclusion for now is that a spatial scientist desiring the AIC statistic more than likely should obtain calculations for it elsewhere.

10.4. REFLECTIONS ON KRIGING

As Berry (1997e, 30) notes "spatial autocorrelation [is] the backbone of [many] interpolation techniques used to generate maps from point-sampled data." Although the degree of spatial dependence latent in a georeferenced data set to some degree determines how successful spatial interpolation will be, a spatial scientist cannot overlook the reliability of typical values generated by classical statistical techniques. This latter point is highlighted in §9.5, in which results are reported for interpolations based upon exploitation of pairwise attribute correlation combined with exploitation of spatial autocorrelation. In many instances, an existing attribute correlation will be substantially more pronounced than the routinely uncovered weak to moderate level of spatial autocorrelation. Exploiting attribute correlation structure yields good typical value imputations, on average; further exploiting spatial autocorrelation fine tunes these typical value imputations, in situ, reducing their variability across a map while preserving their good averageness. Berry (1996, 28-29) pens an incisive encapsulation of the issue by stating that "[a] pretty map can be generated regardless of the degree of dependency. But the map is just colorful gibberish if dependency is minimal ..." In other words, virtually any model will produce results; the useful solution is one that produces a map more affect by the data than tailored by the choice of model, method, or implementation algorithm. Dependency here must refer to both spatial and aspatial covariations. Furthermore, in the absence of spatial dependence, the best interpolator for constructing a generalized map is the OLS conditional mean, which may or may not be a function of positional coordinates. Accordingly, missing data estimates can be obtained with a standard EM algorithm, and spatial forecasts can be obtained with a standard regression prediction.

Both sampling design (including the geometric arrangement of georeferenced data locations[2]) and spatial statistical model specification can affect the resulting interpolated, generalized map surface. As mentioned in the previous section, the best sampling network at least partially conforms to a uniform distribution of sample points over a geographic landscape; interpolation can go berserk in regions of a map containing few or no sampling points as prediction intervals become infinitely wide. Interpolation essentially is smoothing; observed attribute values within a specified vicinity of a georeferenced point for which the attribute is not observed are used to compute a weighted average that in turn is assigned to this unobserved point. Although no single procedure is uniformly superior across all geographic cases involving the computing of this weighted average, some procedures consistently perform better than others. Goodchild et al. (1993) evaluate a number of more conventional interpolators for tessellated data; these candidate interpolators need to be compared with the one presented by Griffith et al. (1989), which algebraically is equivalent to the kriging interpolator. Other ad hoc procedures are surveyed in Bennett et al. (1984). Berry (1997c, 28) evaluates four popular interpolators within the context of geostatistical analysis, one of which is used in kriging; his numerical example renders the following ranking:

1. kriging (best)
2. inverse distance
3. minimum curvature
4. mapwide arithmetic mean .

This same ranking pertains to the subset of values that actually is extrapolation. The two extremes of these results are not surprising; the kriging estimator is the best linear unbiased estimator, whereas the mapwide average is the classical *iid* (independent and identically distributed) maximum likelihood estimator of a missing value. Griffith (1992a) reports similar results for the Ottawa-Hull income missing value estimates (§9.5). This differentiation is particularly noteworthy given the proliferation of software like SOLAS 1.0, which offers standard statistical solutions for the universal problem of missing values in data sets that researchers worldwide regularly confront.

Even if semivariogram model specifications for the kriging exercise most frequently render the best interpolation, their generalized map surfaces need to be subjected to diagnostic analysis. Otherwise, obfuscation may rein; the derived synthetic map may well be accepted as a perfect rendering of reality. Three

---

[2]The geometry of data locations and the pattern of interpolation locations can override impacts of the choice of model, methods or algorithms employed.

methods of verification are cross-validation (§9.7), mapping of prediction interval standard errors (§9.6), and ground-truthing fieldwork.

10.5. SPATIAL STATISTICS AND GIS

When the Clinton administration released its 1997 budget proposal, it proposed to bolster GIS work within the federal government by fiscally endorsing several GIS-related programs and agency initiatives (GIS World Inc. 1996, 26). Activities like this have dramatically raised the profile of GIS in the scientific and professional communities in recent years. As GIS becomes more of an everyday term, expectations of it are moving from fast map production to more general numerical treatments of georeferenced data, including spatial statistical analysis. Emphasizing this contention, Berry (1997f, 28) notes that spatial statistics can be viewed as part of GIS modeling. In turn, Nyman (1997, 86) notes that developments in automated spatial statistical analysis techniques are advancing the use of GIS in such fields as public health (Chapter 8). But GIS toolboxes are only beginning to assimilate and implement expressed expert opinion on the issue of which statistical routines to include to support spatial statistical analysis and interpretation of georeferenced data. Levine (1996) argues that, to date, the lack of such spatial statistical tools has hindered spatial scientists who extensively use GIS software. Recent attempts to remedy this situation include the development of macros, such as SAGE (Haining et al. 1996) or the one created by Zhang and Griffith (1997), to implement basic spatial statistical techniques within a GIS environment. Other remedies include specific couplings between reputable standard software packages and GIS software, such as S+ with Arx/Info and SPSS with MapInfo. A third solution has been the development of GIS modules within commercial statistical software packages, such as PROC GIS in SAS. And yet a fourth solution has been to design an array of interfaces between specialized spatial analysis software and specific GIS packages, a strategy adopted for SpaceStat (Anselin 1992). There is a proliferation of specialized packages for implementing geostatistics; meanwhile, Griffith (1989a, 1993b), for example, has written computer code for implementing spatial autoregressive models in MINITAB and SAS, code that is extended in §1.7 and in the appendices of Chapter 3.

Cressie et al. (1996, 118) confirm that only recently GISs have begun to incorporate spatial statistical procedures into their information-processing subsystems. Arguing for this type of development, Griffith (1993c, 112; also see Kemp and Wright 1997, Goodchild et al. 1992, Griffith 1991) contends that a GIS toolbox should include at least an elementary multiple linear regression algorithm, selected spatial statistical approximations, and a spatial weights generator. The spatial statistical work completed for this book indicates that this toolbox also should include a nonlinear regression algorithm and the ability to compute eigenfunctions. A regression procedure would support the calculation of MC and GR values (§1.3), and the approximations would simplify both calculation of the standard error of MC (§1.3) and calculation of the auto-Gaussian normalizing constant (§3.1.2). A nonlinear regression procedure would support the determination of variable transformation parameters (§2.2) and the estimation of both autoregressive and semivariogram model parameters (§3.1.1, §3.2.2). A numerical eigenfunction procedure would support computation of the auto-Gaussian normalizing constant approximation, the assessment of all possible orthogonal levels of spatial autocorrelation (§1.5), and the implementation of auto-logistic and auto-Poisson regression (§2.8, §3.1.3).

Engaging in spatial statistical analysis within a GIS environment more often than not will involve massive georeferenced data sets. Many of the data sets studied in this book would be considered by many spatial scientists as small in size; undoubtedly, one exception is the Adirondack Park (New York) remotely sensed image analyzed in §5.7. Cressie (1996, 116) asserts that there is less temptation to "chase noise" (see §10.1) when turning such massive amounts of georeferenced data into information; but this end may be achieved only at a cost of enormous amounts of computing time and memory. Cressie (1996, 117) further notes that "[s]ampling offers a way to analyze massive data sets with some statistical tools we currently have. Data that exhibit statistical dependence do not need to be looked at in their entirety for many purposes because there is much redundancy." Certainly the MC scatterplots displayed in §5.7 that were constructed for remotely sensed images confirm the presence of a high degree of redundant information—one interpretation of spatial autocorrelation (§1.2). Pursuing Cressie's suggested approach, however, would be like expunging geography from georeferenced data. If a remotely sensed image is being analyzed, theoretical and approximation results reported in §5.7 will help alleviate some of these computationally intensive requirements, for both the eigenfunction and the spatial statistical portions of an analysis, without removing the underlying geography. Furthermore, introducing eigenvectors into a mean response specification would allow Cressie's suggested approach to be implemented in a way that retains the geographic context of and spatial dependency effects latent in georeferenced data. Other solutions to handling computational intensities associated with spatial statistics are available. Anselin and Smirnov (1996) discuss efficient algorithms for constructing proper higher order spatial lag operators. Pace and Barry (1997) explore fast ways of computing spatial autoregressive parameter estimates. And Barry and Pace (1997) discuss fast ways of solving the kriging equations of geostatistics. As GIS packages embrace the data analytical capabilities furnished by spatial statistics, such efficient procedures for handling massive georeferenced data sets must be forthcoming and then implemented if practitioners are to utilize spatial statistics.

10.6. Now Is the Time for All Good Spatial Scientists to ...

A principal objective of this book is to illustrate, using a wide range of thematic georeferenced data sets, how to conduct a proper spatial statistical analysis, promoting sound spatial statistical practice and guidance on how to avoid spatial statistical malpractice. The steps of such an analysis for either exact or approximate auto-Gaussian specifications may be summarized as follows:

Step 1: construct a boxplot (identify aspatial outliers, check for symmetry), a histogram (inspect for bell-shapedness), and a quantile plot (check for normality);

Step 2: compute the S-W normality statistic, the Bartlett and Levene homogeneity of variance statistics, and the ratio of the extreme data values, $\frac{y_{max}}{y_{min}}$;

Step 3: if $\frac{y_{max}}{y_{min}} \geq 3$, then determine whether or not a power or logarithmic transformation markedly improves the S-W statistic, as well as the Bartlett and Levene homogeneity of variance statistics[3]; if one does, then employ the identified transformation,[4] and if neither does, then employ the raw data;

Step 4: for geostatistics, construct a semivariogram plot and a Thiessen polygon surface partitioning; for spatial autoregression, construct an MC scatterplot and compute the MC and GR values; search for evidence of the presence of spatial and/or aspatial outliers;

Step 5: compute OLS results, incorporating indicator variables to differentiate classes of outliers;

Step 6: using the OLS residuals, for geostatistics, construct a semivariogram plot, and for spatial autoregression, construct an MC scatterplot and compute the MC and GR values; again search for evidence of the presence of spatial and/or aspatial outliers;

Step 7: if significant spatial autocorrelation is detected in the residuals analyzed in Step #6, then select an autoregressive and a semivariogram model specification (see the pairings discussed in §10.3) and estimate their respective parameters; and

Step 8: for geostatistics, construct a generalized map surface using kriging; for spatial autoregression, interpret the pseudo-R^2 value

---

[3] If the principal concern is variance heterogeneity, the criterion to use becomes the ratio of an attribute's mean and its standard deviation. If $\frac{\bar{x}}{s} < 4$, determine whether a power or logarithmic transformation markedly improves the Bartlett and Levene homogeneity of variance statistics.

[4] If convergence problems are encountered, monitor the impact of outliers in an attempt to determine whether or not they are the culprit.

as well as the spatially adjusted statistical analysis results, and any data imputations that are necessary.

When the raw data are percentages or counts, the auto-Gaussian approximation results should be compared with those obtained from, respectively, logistic and Poisson regression procedures. To directly avoid auto-logistic and auto-Poisson model complications, while indirectly implementing either of these specifications, the analysis steps become

Step 1: complete the above auto-Gaussian approximation, when possible, replacing the Step 3 transformation search with an appropriately weighted log-odds ratio for logistic regression and the logarithm for Poisson regression;

Step 2: compute the eigenvectors of matrix expression (1.10) or, if n is sufficiently large, simply of matrix **C** itself;

Step 3: revise the auto-Gaussian approximation (an OLS solution) to include salient variables identified in Step #1 as well as those orthogonal and uncorrelated eigenvectors associated with prominent spatial autocorrelation extracted in Step #2 to filter spatial autocorrelation out of the regression residuals; and,

Step 4: use the results from Step #3 as input into a logistic regression or a Poisson regression routine.

Because the error structure of the model specification is altered in moving from an auto-Gaussian approximation to either a logistic or a Poisson regression, some differences between the two results can occur. A sizeable portion of the auto-Gaussian approximation results, however, often will transfer to these other specifications. Other solutions to this problem have been proposed (e.g., Kaiser and Cressie 1996). Therefore, *now is the time for all good spatial scientists to begin implementing appropriate spatial statistical model specifications.*

Both SAS and SPSS computer codes have been listed in this book to implement these foregoing steps (e.g., estimate model parameters and construct useful plots). The semivariogram construction technique advanced here (§1.7.1) differs from that used in common practice; it is based upon cluster analysis and hence views distances as containing two components, one an underlying distance value and the other a measurement error-type term. Thiessen polygon generation was executed with Arc/Info, with neighbors files being extracted from a resulting tessellation (see Zhang and Griffith 1997; Can 1996). Nonlinear regression is established as the workhorse of spatial statistics, superseding linear regression as the workhorse of classical statistics. Therefore, *now is the time for all good spatial scientists to begin producing appropriate spatial statistical visualizations, generalizing map surfaces, and imputing missing georeferenced data.*

In conclusion, a next developmental stage in spatial statistics will be to finalize articulations between spatial autoregressive and semivariogram model specifications, building upon results reported in this book, and the dissemination and implementation of spatial statistical procedures within standard statistical and GIS software packages. Therefore, *now is the time for at least some spatial scientists to begin exploring readily available georeferenced data sets*, such as those contained in the JASA Data Archives (contributed data sets from articles published in the *Journal of the American Statistical Association*), in support of these endeavors.

10.7. SOME QUESTIONS YET UNANSWERED: FUTURE RESEARCH

Besides the pressing task of incorporating spatial statistical techniques into GIS software packages, many other items can be compiled into a future research agenda list. We propose a number of novel statistical tools in this book that merit extensive subsequent evaluation. Our semivariogram plot of extreme gamma values per distance interval, to uncover spatial as well as aspatial outliers, needs to be more fully assessed. We have experienced this tool as more informative than the h-scattergram, which plots pairs of extreme attribute values for each distance group against the corresponding mean distance values. Our ratio of variance-to-mean gamma values for each distance class in a semivariogram plot, to see how they relate to the expected chi-square value of 2 for an independent georeferenced variable, needs to be developed into a formal diagnostic tool. Our approach of replacing uniform distance intervals by automated clustering results based upon Ward's algorithm needs to be subjected to a thorough comparison with already-published georeferenced data analysis results. Defining the number of clusters in this context is analogous to defining a lag tolerance. Our use of eigenvectors extracted from expression (1.10) (§1.5) as surrogate map pattern variates in OLS regression, to account for latent spatial autocorrelation without abandoning traditional regression theory (as is done repeatedly for comparative purposes in this book), deserves extensive investigation, especially in terms of statistical distributional theory. Moreover, by positing equation $\mathbf{Y} = \mathbf{E}_k \boldsymbol{\beta} + \boldsymbol{\epsilon}$, $\mathbf{b} = \mathbf{E}_k^T \mathbf{Y}$, for some subset of $k < n$ eigenvectors, the residual MC value is given by

$$\frac{n \quad \sum_{j=1}^{n} b_j^2 \lambda_j - \sum_{j=1}^{k} b_j^2 \lambda_j}{\sum_{i=1}^{n}\sum_{i=1}^{n} c_{ij} \quad \sum_{j=1}^{n} b_j^2 - \sum_{j=1}^{k} b_j^2},$$

which should tend toward its expected value because all the b_k statistics are the ones that are significantly different from 0, and should be affiliated only with eigenvectors associated with markedly positive spatial autocorrelation.[5] Adoption of this eigenvector approach for auto-logistic and auto-Poisson model situations, which is done a number of times in this book, merits a meticulous examination, particularly because it circumvents the theoretical constraint that it can accommodate only negative spatial autocorrelation, which is the exception rather than the rule for georeferenced data.

Our findings suggest that distributional theory for semivariogram model parameter estimates needs to be developed. Although Zimmerman and Zimmerman (1991) report that no estimator of these parameters is uniformly superior for the purpose of parameter estimation, we argue for use of Cressie's weighted least squares estimator. Intuitively, we believe that it is more appealing. We also do not find, as Zimmerman and Zimmerman (1991) do, that the estimators of semivariogram parameters perform best when the spatial dependence is weak. Rather, our experience is exactly the opposite, especially with remotely sensed data (§5.7),. We find that frequently most any semivariogram model specification will achieve a similar fit to a given set of georeferenced data; at times, the wave-hole model even achieves the best fit although the latent spatial autocorrelation is positive in nature.

As mentioned in §10.3, a clear and crisp articulation of relationships between spatial autoregressive and semivariogram models continues to be elusive. We conclude that model selection must depend upon parsimonious description of georeferenced data, and theoretical or conceptual considerations associated with the phenomenon under study. We believe that cross-validation can serve as a helpful tool here. Nevertheless, we echo Cressie's statement that "[a] fertile area for future research is in developing sensible model-selection criteria when the data are spatial" (1991). In keeping with this sentiment, an analysis of spatial autocorrelation covariations among index values, parameter estimates for a given semivariogram model specification, and the autocorrelation parameter estimate, $\hat{\rho}$, for a linked spatial autoregressive model specification of interest. Consider the SAR spatial autoregressive specification, which is employed in the spatial statistical analysis of 27 georeferenced data sets presented in this book. A principal components analysis (PCA) of $\hat{\rho}$, MC, and GR for these 27 empirical cases produces a single component that accounts for 95.3% of the variation. The following three semivariogram models achieve the best fit for geostatistical analyses of these data sets: power (n = 4), exponential (n = 6), and spherical (n = 17). The PCA scores conform to a normal distribution for the power (S-W = 0.89520) and spherical (S-W = 0.91365) classes, but not for the exponential class (S-W = 0.62494). A suitable power transformation is unable to correct this situation (the estimated exponent is approximately 1). The Bartlett (0.982) and Levene (0.023) statistics do not indicate heterogeneous variances. The resulting

---

[5]The sum of the eigenvalues, λ_j, of matrix C is exactly 0—the trace of the matrix is 0. Because the set $\{b_k\}$ will comprise positive values, the expected value will be negative. Because the remaining b_j values will be approximately 0, all of the remaining (n − k) λ_js will tend to be weighted equally in the sum.

ANOVA F-ratio is 0.31. This numerical result is consistent with the three individual variable ANOVA F-ratios (respectively, 0.50, 0.18, and 0.27), as well as with the Wilk's Λ MANOVA F-ratio (0.91), which also has a single dimension that accounts for 95.3% of variability. In other words, there is a very high degree of multicollinearity amongst $\hat{\rho}$, MC, and GR, allowing $\hat{\rho}$ to be predicted from MC or GR ($r_{MC,GR} = -0.988$) when georeferenced data essentially are detrended and conform to a normal distribution:

$$\hat{\rho} = 0.29201 + 0.85608 MC, \quad R^2 = 0.823,$$

with the residuals from this estimated regression equation appearing to conform to a normal distribution and exhibiting only a hint of nonconstant variance across the magnitudes of $\hat{\rho}$. In addition, the ANOVA and MANOVA results fail to expose any discriminatory power by levels of spatial autocorrelation latent in georeferenced data with regard to the corresponding semivariogram model that achieves the lowest RSSE. Thus, there is no evidence to indicate that particular semivariogram models are more suitable for different magnitudes of spatial autocorrelation.

We have succeeded in furnishing a wealth of detailed empirical applications, with illuminating contrasts between geostatistical and spatial autoregressive analyses. Such results should prove helpful in productively dealing with implementing solutions to problems such as the unconventional way authorities in Williamsport, Pennsylvania, are cleaning up a toxic waste site. This geographic landscape is dotted with devices, reminiscent of gigantic hypodermic needles, filled with diluted molasses. The molasses is being injected into groundwater that had been contaminated by an airplane engine factory to eradicate chemical and heavy metal pollutants. Microbes that naturally occur in soil consume the molasses, depleting the oxygen in the groundwater, which in turn causes a chemical change in chromium so that it becomes benign and binds to the soil. Although this toxic cleanup method is less expensive and less time consuming than more traditional approaches, it depends upon proper placement of the syringe-type devices; establishment of a locational grid should be followed by a proper spatial statistical analysis of the EPA-monitored wells. This is exactly the approach advocated by Leonte and Schofield (1996).

We also believe that better descriptions of spatial statistical techniques and spatial dimensions of georeferenced data are warranted. Many features of spatial statistics mystify researchers and practitioners alike. To this end, we have contributed considerable detailed discussion in Chapters 1-3. Griffith (1992b) presents a extensive discussion of the meaning of spatial autocorrelation, reviewed in Chapter 1. Kriging is the process of predicting geographically distributed attribute values by exploiting spatial autocorrelation; this prediction uses geographically adjacent values, and spatial autocorrelation is portrayed with a semivariogram. In explaining what the kriging equations forming matrix equation (1.5) are exactly, Haining et al. (1984) describe the multiplier effect affiliated with spatial dependency: georeferenced data point i influences its neighbor point j, which in turn influences i directly and returns part that the impact of the initial influence point i had exerted upon point j. This loop of mutual influence continues with diminishing amounts of the initial influences being circulated during each influence round until these amounts vanish. This depiction is reminiscent of the picture within a picture within a picture, and so on, curiously immortalized by Jonathan Swift:

> So, naturalists observe, a flea
> Hath smaller fleas that on him prey;
> And these have smaller fleas to bite 'em,
> And so proceed *ad infinitum*.

These multiplier effects are what make kriging possible. The kriging equations actually are weighted moving average equations. They dampen (by inverting matrix Σ) the relatively high variance of sample attribute values (the diagonal entries in matrix Σ) by utilizing data redundancies through including knowledge of the spatial covariance (the off-diagonal entries in matrix Σ). This dampening effect is extended to predicted values by postmultiplying matrix Σ_{mm} (§1.1), the statistical distance between unsampled locations and all of their geographic neighboring data, by matrix Σ^{-1}. Whereas the nugget has been interpreted as pure random noise or as the outcome of sampling error or measurement error, Griffith et al. (1996) demonstrate that it also can be interpreted as the outcome of edge effects or model misspecification errors that are amplified by this matrix inversion.

Another illuminating mnemonic is suggested by David (1977, 106; also see Wackernagel 1995, 43-45) to explain what is meant by a spherical semi-variogram model. Consider half-spheres superimposed over points on a planar surface. Each half-sphere, similar to an upside-down cup, centered on a given point, portrays the spatial dependency field surrounding a given location, with this dependency dropping off with increasing distance from the given central point in a manner that follows the outside surface of the half-sphere (or cup). Isotropy dictates the sphere shape; stationarity dictates that all spheres are the same. If points are sufficiently dense, these spheres will intersect. The volume of intersection between two spheres is the classical calculus problem of obtaining a solid by revolving a graph around the x-axis; for range r and an intersection of two spheres occurring at distance δ from a point:

$$\pi \int_0^r [\sqrt{r^2 - x^2}]^2 dx = \pi \int_0^r [r^2 - x^2] dx =$$

$$\pi (r^2 x - \frac{x^3}{3})\Big|_0^r = \pi (r^3 - \frac{r^3}{3} - r^2\delta + \frac{\delta^3}{3}) \quad ,$$

which is equivalent to a covariation term; this is a common mode of interpretation presented in traditional multivariate analysis. Dividing this result by the total volume of a half-sphere, namely $\frac{2\pi r^3}{3}$, converts it to a quantity that is analogous to a correlation and yields

$$1 - (\frac{3}{2}\frac{\delta}{r} - \frac{1}{2}\frac{\delta^3}{r^3}) \rightarrow \overline{\gamma(d)} = C_0 + C_1\{1 - [1 - [(\frac{3}{2}\frac{\delta}{r} - \frac{1}{2}\frac{\delta^3}{r^3})]\} ,$$

where subtraction from 1 [the correlation coefficient for γ(0)] converts this into its semivariogram model form. Not only does this derivation disclose the source of the coefficients and the functional form of the spherical semivariogram model, but it also furnishes a template for deriving other semivariogram model functional forms (e.g., penta-spherical, cubic). In addition, this derivation reveals the meaning of the model name, as well as the visual form of the spatial dependency field.

Finally, many more routine problems remain to be studied. Certainly, application of standard multicollinearity diagnostics (e.g., condition numbers; see Ababou et al. 1994) to the spatial connectivity matrix may well be fruitful. Because the eigenvectors of a covariance matrix and its inverse are the same, this contention is supported by recent advances in our understanding of what these eigenvectors mean for geographic connectivity matrices (e.g., Griffith 1996b). Peculiar problems, sometimes labeled paradoxes, such as obtaining ludicrously large kriging weights for the two values at the ends of a linear arrangement of georeferenced data (see Ali and Lall 1996), merit investigation. A similar type of complication arises in Chapter 6 for the Puerto Rico agricultural production data in part because the island is long and narrow and in part because its central mountain range creates a somewhat non-Euclidean landscape. Undoubtedly many other customary problems can be added to this list.

EPILOGUE:

A CARTOON GUIDE TO SPATIAL STATISTICS

The events of our lives happen in a sequence over time, but derive their significance to ourselves by finding an order across space
(adapted from Welty 1984)

Grégoire Dubois invented the useful geostatistical listserv called AI-GEOSTATS. Anyone interested in following illuminating debates, being informed about useful geostatistical reference resources and software, and/or learning some of the basics of spatial statistics should subscribe. The detailed contents of this book have benefited greatly from AI-GEOSTATS interactions. But although AI-GEOSTATS is for serious discussion, spatial autoregression issues rarely are dealt with.

Before reading this book, if you had been aimlessly wondering

> *who* are the mystics of spatial statistics?
> *what* is intriguing about kriging?
> *when* should you do a georeferenced data imputation, too?
> *where* are there lessons on spatial autoregressions?
> *why* deal with space in your data analysis case?
> *how* does map interpretation relate to spatial autocorrelation?

then you should have found this book informative, and it should put you on the road to spatial statistical literacy. Adding some humor to the clear, simple, and concise spatial statistical analyses you now will be presenting will help make them engaging. Whereas a public speaker is advised to introduce her/his talk with a joke, we suggest that you introduce your spatial statistical analysis presentation with a funny illustration; such an approach is motivated by *The Cartoon Guide to Statistics* (Gonick and Smith 1993). *Webster's New World Dictionary* (1986) defines a cartoon as "a humorous drawing, often with a caption." Cartoons are everywhere—in newspapers, magazines, books, and on the Internet. Some cartoons can be used directly, some can have a few words altered, and some can have their ideas adapted. Regardless of the source, an effective cartoon can help the audience better relate to your presentation.

One cartoon used in GEO 686: Advanced Spatial Statistics relates to the presentation of the concept of anisotropy (Chapter 5). A 1996 *Ralph* sketch depicts a job candidate being interviewed by a person in personnel for Map Makers, Inc. Hanging on the wall behind the interviewer is a company logo displaying the company's name and a standard map orientation icon showing the N, S, E, and W directions. The interviewee is asking, "Before you start the job interview, what does the N, W, S, and E stand for?" This cartoon is effective in initiating discussion about how the assumption of isotropy is ignorant of direction and is an example of using a cartoon in its original form. A 1996 *Non Sequitur* comic strip is about astronomy and astrology and was modified to be about spatial statistics and statistics. The first panel displays a professor at the front of the class saying, "Welcome to [GEO 686]. Before we start are there any questions?" The second panel shows five students, all raising their hands, with one asking, "Yea, like, what makes [spatial statistics] different from [statistics]?" The third panel shows the professor replying "Lots and lots of [computer programming]." The fourth and final box depicts the professor sitting in front of an empty classroom, with GEO 686 on the door to the room. This cartoon is useful in beginning a discussion about the numerical intensity of spatial statistics (Chapter 3). Figure E.1 presents a cartoon we devised that was inspired by a 1996 *American Airlines* magazine cartoon. The critical features of this graphic are the map of Puerto Rico that forms the table top, the histograms for each of the five agricultural administrative regions that sit on top of this table, the normal curve picture hanging on the wall behind the

Figure E-1. Cartoon: georeferenced data failing to conform to specific spatial statistical assumptions.

researcher, and the spatial analysis clinic sign on the wall. This cartoon is effective in initiating discussion about modeling assumptions (Chapter 2).

Other cartoons can be devised with some imagination; an ability to do so requires a relatively deep understanding of spatial statistics subject matter. One possibility for Chapter 1 is a modified version of Grant Wood's classic American Gothic picture, with a caption stating, perhaps, "You have been apart too long; it's time you were married!" Figure E.2 presents a second cartoon we created that was inspired by a cartoon reprinted in Moore's *Statistics: Concepts and Controversies* (3rd ed.; 1991, 172). The critical features of this graphic are the circumscription of numbers by areal unit outline maps—states in this version of the cartoon—and the altered maxim of "let the data speak for themselves." This cartoon is useful in initiating discussion about socioeconomic georeferenced data analysis (Chapter 4). A 1997 *B.C.* comic strip shows a genie being released from an oil drum retrieved from the ocean. In the fifth panel, the genie states, "Thank you for releasing me, master...." In the sixth panel he says, "You have your choice of three drilling sites." This cartoon can initiate discussion about natural resources (Chapter 6). A 1997 *Non Sequitur* cartoon is labeled "How air quality regulations steered American history: 1692-Salem, Massachusetts"; a pilgrim who is an EPA inspector is talking to other pilgrims raking leaves in preparation for a witch burning. This cartoon can be modified in a way that pertains to the geographic distribution of the leaf raking, and as such is effective in initiating discussion about pollution (Chapter 7). A 1998 issue of the *Institute of British Geographers* newsletter included an Austin cartoon depicting a doctor talking to a patient while pointing to a map of health areas,

"Today, we're going to let the geo-referenced data speak for themselves."
Figure E-2. Cartoon: letting the georeferenced data speak for themselves.

saying to the patient, "You have the disease in a very awkward place." This cartoon can be used for introducing discussion about epidemiology (Chapter 8). A 1998 *Pickles* comic strip shows a person in a construction hat digging holes who is being followed by a second person in a construction hat filling the holes. The fifth through eighth panels show the old man whose always appears in this cartoon talking with the second man, with appropriate adaptions introduced here:

old man:	"Uh ... excuse me. I couldn't help noticing that you're filling in all the holes that that other fellow just [made]."
second man:	"Yeah. [The government's working' on providing georeferenced data for public consumption]."
old man:	"[But why do you fill in a bunch of holes on the landscape?"
second man:	"Because I'm the kriger!"

This cartoon is effective in initiating discussion about missing data prediction and imputation (Chapter 9). Finally, a quote from a 1992 issue of *The Economist* states,

> Many people worry that the atmosphere is getting too warm. Perhaps it is. But finding out is not easy. Taking the world's temperature requires measurements in many different places, followed by some awkward averaging.

This quotation is ripe for the drafting of a cartoon for discussion about proper georeferenced data analysis and interpretation (Chapter 10).

"Spatial statistics is one of the most rapidly growing areas of statistics, rife with fascinating research opportunities" (National Research Council 1991, vii). We hope this book increases the awareness of many of these opportunities. It spans the wide scope of spatial statistics by integrating both spatial autoregressive and geostatistical analyses, and by interjecting humor into the subject through the use of cartoons. We trust that the readers of this book will go on to explore the many faces of spatial autocorrelation, promote sound spatial statistics data analysis practice pursue some of the likely fruitful directions for future research outlined in Chapter 10, and even devise a spatial statistics cartoon or two.

REFERENCES

Ababou, R., A. Bagtzoglou, and E. Wood. 1994. On the condition number of covariance matrices in kriging, estimation and simulation of random fields *Mathematical Geology* 26: 99-133.
Ali, A., and U. Lall. 1996. A kernel estimator for stochastic subsurface characterization from drill log data *Ground Water* 34: 647-58.
Allen, D., and F. Cady. 1982. *Analyzing Experimental Data by Regression*. Belmont, Calif.: Wadsworth.
Amrhein, C. 1993. Searching for the elusive aggregation effect: Evidence from statistical simulations *Environment and Planning A* 27: 105-19.
Amrhein, C., and G. Reynolds. 1996. Using spatial statistics to assess aggregation effects *Geographical Systems* 3: 143-58.
———. 1997. Using the Getis statistic to explore aggregation effects in Metropolitan Toronto Census data *Canadian Geographer* 41: 137-49.
Andrews, D., and H. Herzberg. 1985. *Data: A Collection of Problems from Many Fields for the Student and Research Worker*. Berlin: Springer-Verlag.
Anselin, L. 1988. *Spatial Econometrics*. Dordrecht: Kluwer.
———. 1992. *SPACESTAT TUTORIAL: A Workbook for Using Spacestat in the Analysis of Spatial Data* Technical Software Series S-92- 1. Santa Barbara, Calif: NCGIA.
———. 1995. Local indicators of spatial autocorrelation *Geographical Analysis* 27: 93-115.
Anselin, L., and A. Can. 1986. Model comparison and model validation issues in empirical work on urban density functions *Geographical Analysis* 18: 179-97.
Anselin, L., and D. Griffith. 1988. Do spatial effects really matter in regression analysis? *Papers of the Regional Science Association* 65: 11-34.
Anselin, L., and S. Hudak. 1992. Spatial econometrics in practice: A review of software options. *Regional Science and Urban Economic* 22: 509-36.
Anselin, L., and O. Smirnov. 1996. Efficient algorithms for constructing proper higher order spatial lag operators *J. of Regional Science* 36: 67-89.
Arbia, G. 1989. *Spatial Data Configuration in Statistical Analysis of Regional Economics and Related Problems*. Dordrecht: Kluwer.
Arbia, G., D. Griffith, and R. Haining. 1998a. Error propagation modelling in raster GIS: Overlay operations *International J. of Geographical Information Systems* 12:

145-67.
———. 1998b. Error propagation modelling in raster GIS: Addition and ratioing operations *Cartography and Geographic Information Systems*, forthcoming.
Archer, V. 1987. Association of lung cancer mortality with Precambrian granite *Archives of Environmental Health* 42: 87-91.
Arlinghaus, S., ed. 1996. *Practical Handbook of Spatial Statistics*. Boca Raton, Fla: CRC Press.
Atkinson, D. 1978. *Epidemiology of Sudden Infant Death in North Carolina: Do Cases Tend to Cluster?* PHSB Studies, no. 16. Raleigh: Division of Health Services, Public Health Statistics Branch, North Carolina Department of Human Resources.
Bailey, A., J. Sargent, and M. Blake. 1998. A tale of two counties: Childhood lead poisoning, industrialization, and abatement in New England *Economic Geography* Extra issue (AAG special issue): 96-111.
Bailey, T., and A. Gatrell. 1995. *Interactive Spatial Data Analysis*. Harlow: Longman Scientific.
Bamston, A. 1993. *Atlas of Frequency Distribution, Auto-Correlation and Cross-Correlation of Daily Temperature and Precipitation at Stations in the U.S., 1948-1991* Technical Report NOAA-ATLAS-11E. Washington, D.C.: Climate Analysis Center, National Weather Service.
Barry, R., and R. Pace. 1997. Kriging with large data sets using sparse matrix techniques *Communications in Statistics: Simulation and Computation* 26: 619-29.
Bartlett, M. 1975. *The Statistical Analysis of Spatial Pattern*. London: Chapman and Hall.
Bellhouse, D. 1977. Some optimal designs for sampling in two dimensions *Biometrika* 64: 605-11.
Bennett, R., R. Haining, and D. Griffith. 1984. The problem of missing data on spatial surfaces *Annals of the Association of American Geographers* 74: 138-56.
Bera, A., and P. Ng. 1995. Tests for normality using estimated score function *J. of Statistical Computation and Simulation* 52: 273-87.
Berman, A., and R. Plemmons. 1979. *Non-negative Matrices in the Mathematical Sciences*. New York: Academic Press.
Berry, J. 1994. Surf's up *GIS World* 7 (no. 2): 22.
———. 1996. Analyzing spatial dependency within a map *GIS World* 9 (no. 5): 28-29.
———. 1997a. Justifiable interpolation *GIS World* 10 (no. 2): 34.
———. 1997b. Compare map errors *GIS World* 10 (no. 3): 26.
———. 1997c. Move beyond a map full of error *GIS World* 10 (no. 4): 28.
———. 1997d. Spatial dependencies define the data *GIS World* 10 (no. 5): 26-27.
———. 1997e. Uncover the mysteries of spatial autocorrelation *GIS World* 10 (no. 6): 34.
———. 1997f. Varied applications drive GIS perspectives *GIS World* 10 (no. 8): 28.
Besag, J. 1974. Spatial interaction and the statistical analysis of lattice systems *J. of the Royal Statistical Society* 36B: 192-225.
———. 1981. On a system of two-dimensional recurrence equations *J. of the Royal Statistical Society* 43B: 302-09.
Bhatti, A., D. Mulla, F. Koehler, and A. Gurmani. 1991. Identifying and removing spatial correlation from yield experiments *Soil Science Society of America J.* 55: 1523-28.
Birkes, D., and Y. Dodge. 1993. *Alternative Methods of Regression*. New York: Wiley.
Box, G., and D. Cox. 1964. An analysis of transformations. *J. of the Royal Statistical*

Society 26B: 211-43.
Box, G., and P. Tidwell. 1962. Transformations of the independent variables *Technometrics* 4: 531-50.
Box, G., W. Hunter, and J. Hunter. 1978. *Statistics for Experiments*. New York: Wiley.
Buse, A. 1973. Goodness of fit in generalized least squares estimation *American Statistician* 27: 106-08.
Campbell, J. 1981. Spatial correlation effects upon accuracy of supervised classification of land cover *Photogrammetric Engineering and Remote Sensing* 47: 355-63.
Campbell, N., and H. Kiiveri. 1988. Spectral-temporal indices for discrimination *Applied Statistics* 37: 51-62.
Can, A. 1996. Weight matrices and spatial autocorrelation statistics using a topological vector data model *International J. of Geographical Information Systems* 10: 1009-17.
Can, A., and I. Megbolugbe. 1997. Spatial dependence and house price index construction *J. of Real Estate Finance and Economics* 14: 203-22.
Carr, J. 1995. *Numerical Analysis for the Geological Sciences*. Englewood Cliffs, N.J.: Prentice Hall.
Carroll, R, and D. Rupert. 1982. A comparison between maximum likelihood and generalized least squares in a heteroscedastic linear model *J. of the American Statistical Association* 77: 878-82.
Carroll, R., R. Chen, E. George, T. Li, H. Newton, H. Schmiediche, and N. Wang. 1997. Ozone exposure and population density in Harris County, Texas *J. of the American Statistical Association* 92: 392-404.
Casetti, E., and J. Jones. 1992. An introduction to the expansion method and to its applications In *Applications of the Expansion Method*, edited by J. Jones and E. Casetti, 1-41. New York: Routledge.
Chatterjee, S., M. Handcock, and J. Simonoff. 1995. *A Casebook for a First Course in Statistics and Data Analysis*. New York: Wiley.
Chorley, R. [chairman]. 1987. *Handling Geographic Information*. Report to the Select Committee on GIS. London: Her Majesty's Stationary Office.
Christakos, G. 1992. *Random Field Models in Earth Sciences*. New York: Academic Press.
Christensen, R. 1991. *Linear Models for Multivariate, Time Series, and Spatial Data*. Berlin: Springer-Verlag.
Cliff, A, and J. Ord. 1975. Model building and the analysis of spatial pattern in human geography *J. of the Royal Statistical Society* 37B: 297-328.
―――. 1981a. *Spatial Processes: Models and Applications*. London: Pion.
―――. 1981b. Spatial and temporal analysis: Autocorrelation in space and time In *Quantitative Geography: A British View*, edited by N. Wrigley and R. Bennett, 104-10. London: Routledge and Kegan Paul.
Cliff, A., P. Haggett, J. Ord, and G. Versey. 1981. *Spatial Diffusion: An Historical Geography of Epidemics in an Island Community*. Cambridge, U.K.: Cambridge University Press.
Cohen, B. 1988. Letter to editor *Archives of Environmental Health* 43: 313-14.
Commonwealth of Puerto Rico 1950-96 (selected years). *Facts and Figures of Agriculture in Puerto Rico*. Santurce, P.R.: Office of Agricultural Statistics, Department of Agriculture.
Congalton, R., ed. 1994. *Proceedings: International Symposium on the Spatial Accuracy*

of Natural Resource Data Bases. Bethesda, Md: American Society for Photogrammetry and Remote Sensing.
Copeland, J., C. Smith, S. Hale, P. August, and R. Latimer. 1994. EPA program monitors U.S. coastal environments *GIS World* 7 (no. 8): 44-47.
Cordy, C., and D. Griffith. 1993. Efficiency of least squares estimators in the presence of spatial autocorrelation *Communications in Statistics: Simulation and Computation* 22: 1161-79.
Cressie, N. 1985. Fitting variogram models by weighted least squares *J. of the International Association for Mathematical Geology* 17: 563-86.
———. 1986. Kriging nonstationary data *J. of the American Statistical Association* 81: 625-34.
———. 1989. Geostatistics *American Statistician* 43: 197-202.
———. 1991. *Statistics for Spatial Data*. New York: Wiley.
Cressie, N., and N. Chan. 1989. Spatial modeling of regional variables *J. of the American Statistical Association* 84: 393-401.
Cressie, N., and I. Clark. 1994. *Spatial Statistics for Environmental and Earth Sciences Data* CE9405 Workshop Manual of the 1994 Joint Statistical Meetings (Toronto), American Statistical Association.
Cressie, N., and T. Read. 1985. Do sudden infant deaths come in clusters? *Statistics and Decisions* 3 (supplement issue no. 2): 333-49.
———. 1989. Spatial data analysis of regional counts *Biometrical J.* 31: 699-719.
Cressie, N., and J. ver Hoef. 1993. Spatial statistical analysis of environmental and ecological data In *Environmental Modeling with GIS*, edited by M. Goodchild, B. Parks, and L. Steyaert, 404-13. Oxford, U.K.: Oxford University Press.
Cressie, N., A. Olsen, and D. Cook. 1996. Massive data sets: Problems and possibilities, with application to environmental monitoring In *Massive Data Sets: Proceedings of a Workshop*, edited by the National Research Council, Committee on Applied and Theoretical Statistics, 115-19. Washington, D.C.: National Academy Press.
Cressie, N., M. Kaiser, M. Daniels, J. Aldworth, J. Lee, S. Lahiri, and L. Cox. 1999. Spatial analysis of particulate matter in an urban environment, *GeoEnv98 Proceedings*, Valencia, forthcooming.
Dacey, M. 1968. An empirical study of the areal distribution of houses in Puerto Rico *Transactions of the Institute of British Geographers* 45: 51-69.
Dartmouth Medical School, Center for the Evaluative Clinical Sciences. 1998. *Dartmouth Atlas of Health Care 1998*. Chicago: American Hospital Publishing.
David, M. 1977. *Geostatistical Ore Reserve Estimation*. Amsterdam: Elsevier.
de Gruijter, J., and C. ter Braak. 1990. Model-free estimation from spatial samples: A reappraisal of classical sampling theory *Mathematical Geology* 22: 407- 15.
de Jong, P., C. Sprenger, and F. van Veen. 1984. On extreme values of Moran's I and Geary's c *Geographical Analysis* 16: 17-24.
Deutsch, C., and A. Journel. 1992. *GSLIB: Geostatistical Software Library and User's Guide*. New York: Oxford University Press.
Diggle, P. 1993. Point process modelling in environmental epidemiology In *Statistics for the Environment*, edited by V. Barnett and K. Turkman, 89-110. New York: Wiley.
Dingman, S., M. Diana, and R. Reynolds. 1988. Application of kriging to estimating mean annual precipitation in a region of orographic influence *Water Resources Bulletin* 24: 329-39.

Dobson, J. 1996. National Center for Environmental Decision-making Research debuts *GIS World* 9 (no. 1): 38.
Doveton, J. 1973. *Numerical Analysis Relating Location of Hydrocarbon Traps to Structure and Stratigraphy of the Mississippian 'B' of Stafford County, South-Central Kansas* Technical Report, KOX Project. Lawrence: Geological Research Section, Kansas Geological Survey.
Dubin, R. 1992. Spatial autocorrelation and neighborhood quality *Regional Science and Urban Economics* 22: 433-52.
Durrett, R. 1994. Stochastic spatial models *Forefronts* (newsletter of the Cornell Theory Center) 9 (Spring, no. 4): 4-6.
Dutilleul, P. 1993. Modifying the t test for assessing the correlation between two spatial processes *Biometrics* 49: 305-14.
Dutilleul, P., and P. Legendre. 1992. Lack of robustness in two tests of normality against autocorrelation in sample data *J. of Statistical Computation and Simulation* 42: 79-91.
Dutter, R., and C. Reimann. 1998. *Environmental Geochemical Atlas of the Central Barents Region*. Trondheim, Norway: Norges Geologiske Undersokelse.
Earickson, R., and J. Harlin. 1994. *Geographic Measurement and Quantitative Analysis*. New York: Macmillan.
Efron, B. 1982. *The Jackknife, the Bootstrap and Other Resampling Plans*. Monograph. Philadelphia, Pa.: SIAM-NSF.
Emerson, J., and M. Stoto. 1983. Transforming data In *Understanding Robust and Exploratory Data Analysis*, edited by D. Hoaglin, F. Mosteller, and J. Tukey, 97-128. New York: Wiley.
Fairfield Smith, H. 1938. An empirical law describing heterogeneity in the yields of agricultural crops *J. of Agricultural Science* 28: 1-23.
Federer, W., and C. Schlottfildt. 1954. The use of covariance to control gradients in experiments *Biometrics* 10: 282-90.
Freund, J. 1992. *Mathematical Statistics* 5th ed. Englewood Cliffs, N.J.: Prentice Hall.
Gasim, A. 1988. First-order autoregressive models: A method for obtaining eigenvalues for weighting matrices *Journal of Statistical Planning and Inference* 18: 391-98.
Geary, R. 1947. Testing for normality *Biometrika* 34: 209-42.
Getis A. 1990. Screening for spatial dependence in regression analysis *Papers of the Regional Science Association* 69: 69-81.
———. 1995. Spatial filtering in a regression framework: Experiments on regional inequality, government expenditures, and urban crime In *New Directions in Spatial Econometrics*, edited by. L. Anselin and R. Florax, 172-88. Berlin: Springer-Verlag.
Getis A, and J. Ord. 1992. The analysis of spatial association by use of distance statistics. *Geographical Analysis* 24: 189-206.
———. 1998. The use of a local statistic to study the diffusion of AIDS from San Francisco In *Econometric Advances in Spatial Modelling and Methodology: Essays in Honour of Jean Paelinck*, edited by D. Griffith, C. Amrhein, and J.-M. Huriot, 143-58. Dordrecht: Kluwer.
GIS World Inc. 1996. Federal connection *GIS World.* 9 (no. 6): 26-28.
Gleeson, A., and C. McGilchrist. 1980. Bilateral processes on a rectangular lattice *Australian J. of Statistics* 22: 197-206.

Gomez, M., and K. Hazen. 1970. *Evaluating Sulfur and Ash Distribution in Coal Seams by Statistical Response Surface Regression Analysis* Report RI-7377. Washington, D.C.: U.S. Bureau of Mines.
Gonick, L., and W. Smith. 1993. *A Cartoon Guide to Statistics*. New York: Harper-Perennial.
Goodchild, M., L. Anselin, and U. Deichmann. 1993. A framework for the areal interpolation of socioeconomic data *Environment and Planning A* 25: 383-97.
Goodchild, M., R. Haining, and S. Wise. 1992. Integrating GIS and spatial data analysis: Problems and possibilities *International J. of Geographical Information Systems* 6: 407-23.
Goovaerts, P. 1997. *Geostatistics for Natural Resources Evaluation*. New York: Oxford University Press.
Grant, P. 1986. *Ecology and Evolution of Darwin's Finches*. Princeton, N.J.: Princeton University Press.
Grant, P., T. Price, and H. Snell. 1980. The exploration of Isla Daphne Minor *Noticias de Galápagos* 31: 22-27.
Gregoire, T., D. Brillinger, P. Diggle, E. Russek-Cohen, and W. Warren. 1997. *Modelling Longitudinal and Spatially Correlated Data*. New York: Springer-Verlag.
Griffith, D. 1978. A spatially adjusted ANOVA model *Geographical Analysis* 10: 296-301.
———. 1980. Towards a theory of spatial statistics *Geographical Analysis* 20: 176-86.
———. 1981a. Interdependence in space and time: Numerical and interpretative considerations In *Dynamic Spatial Models*, edited by D. Griffith and R. MacKinnon, 258-87. Alphen aan den Rijn, The Netherlands: Sijthoff and Noordhoff.
———. 1981b. Modelling urban population density in a multi-centered city *J. of Urban Economics*. 9: 298-310.
———. 1983. The boundary value problem in spatial statistical analysis *J. of Regional Science* 23: 377-87.
———. 1985. An evaluation of correction techniques for boundary effects in spatial statistical analysis: Contemporary methods *Geographical Analysis* 17: 81-88.
———. 1987a. Toward a theory of spatial statistics: Another step forward *Geographical Analysis* 19: 69-82.
———. 1987b. *Spatial Autocorrelation: A Primer*. Resource Publication. Washington, D.C.: Association of American Geographers.
———. 1988a. *Advanced Spatial Statistics*. Dordrecht: Kluwer.
———. 1988b. Estimating spatial autoregressive model parameters with commercial statistical packages *Geographical Analysis* 20: 176-86.
———. 1989. *Spatial Regression Analysis on the PC: Spatial Statistics Using MINITAB* Discussion Paper no. 1. Ann Arbor, Mich: Institute of Mathematical Geography.
———. 1991. Univariate and Multivariate Modelling of Areal data In *GIS and Spatial Data Analysis: Report on the Sheffield Workshop* Regional Research Laboratory Initiative Discussion Paper no. 11, edited by R. Haining and S. Wise, 22-25. Sheffield, U.K.: University of Sheffield.
———. 1992a. Estimating missing values in spatial urban census data *Operational Geographer* 10 (2): 23-26.
———. 1992b. What is spatial autocorrelation? Reflections on the past 25 years of spatial statistics *L'Espace géographique* 21: 265-80.

———. 1992c. Teaching statistics to geographers using MINITAB *J. of Geography in Higher Education* 16: 45-59.

———. 1992d. A spatially adjusted N-way ANOVA model *Regional Science and Urban Economics* 22: 347-69.

———. 1993a. Advanced spatial statistics for analyzing and visualizing geo-referenced data *International J. of Geographical Information Systems* 7: 107-23.

———. 1993b. *Spatial Regression Analysis on the PC: Spatial Statistics Using SAS.* Resource Publication. Washington, D.C.: Association of American Geographers.

———. 1993c. Which spatial statistics techniques should be converted to GIS functions? In *Geographic Information Systems, Spatial Modelling and Policy Evaluation,* edited by M. Fischer and P. Nijkamp, 103-14. Berlin: Springer-Verlag.

———. 1996a. Spatial autocorrelation and eigenfunctions of the geographic weights matrix accompanying geo-referenced data *Canadian Geographer* 40: 351-67.

———. 1996b. Introduction: The need for spatial statistics In *Practical Handbook of Spatial Statistics,* edited by S. Arlinghaus, 1-15. Boca Raton, Fla.: CRC Press.

———. 1997. Single spatial median problem solutions for maps with hole sin them that have been filled with estimated missing spatial data: Explorations with urban census data *L'Espace Geographique* 26: 173-82.

Griffith, D., ed. 1990. *Spatial Statistics: Past, Present, and Future.* Ann Arbor, Mich.: Institute of Mathematical Geography.

Griffith, D., and C. Amrhein. 1991. *Statistical Analysis for Geographers.* Englewood Cliffs, N.J.: Prentice Hall.

———. 1997. *Multivariate Statistical Analysis for Geographers.* Englewood Cliffs, N.J.: Prentice Hall.

Griffith, D., and A. Can. 1996. Spatial statistical/econometric versions of simple urban population density models In *Practical Handbook of Spatial Statistics,* edited by S. Arlinghaus, 231-49. Boca Raton, Fla.: CRC Press.

Griffith, D., and F. Csillag. 1993. Exploring relationships between semi-variogram and spatial autoregressive models *Papers in Regional Science* 72: 283-96.

Griffith, D., and F. Lagona. 1997. *Specification Errors in Spatial Models: Impacts on Modeling and Estimation* Discussion Paper no. 108. Syracuse, N.Y.: Department of Geography, Syracuse University.

———. 1998. On the quality of likelihood-based estimators in spatial autoregressive models when the data neighborhood structure is misspecified *J. of Statistical Planning and Inference,* 69: 153-174.

Griffith, D., and L. Layne. 1997. Uncovering relationships between geo-statistical and spatial autoregressive models In *Proceedings of the Section on Statistics and the Environment, American Statistical Association,* 91-96. Alexandria, Va.: American Statistical Association.

Griffith, D., and A. Sone. 1995. Trade-offs associated with normalizing constant computational simplifications for estimating spatial statistical models *J. of Statistical Computation and Simulation* 51: 165-83.

Griffith, D., R. Bennett, and R. Haining. 1989. Statistical analysis of spatial data in the presence of missing observations: An application to urban census data *Environment and Planning A* 21: 1511-23.

Griffith, D., P. Doyle, and D. Wheeler. 1997. A GIS and spatial statistical analysis of urban childhood lead pollution exposure: A case study of Syracuse, New York In *Conference Proceedings, First Syracuse Regional Lead Conference,* edited by

A. Hunt, 13-16. Syracuse, N.Y.: Department of Pathology, SUNY Health Science Center at Syracuse.

Griffith, D., R. Haining, and G. Arbia. 1994. Heterogeneity of attribute sampling error in spatial data sets *Geographical Analysis* 26: 300-320.

Griffith, D., L. Layne, and P. Doyle. 1996. Further explorations of relationships between semi-variogram and spatial autoregressive models In *Spatial Accuracy in Natural Resource and Environmental Sciences: Second International Symposium*, edited by H. Mowrer, R. Czaplewski and R. Hamre, 147-54. General Technical Report RM-GTR-277. Ft. Collins, Colo.: Rocky Mountain Forest and Range Experiment Station.

Griffith, D., J. Paelinck, and R. van Gastel. 1998. The Box-Cox transformation: Computational and interpretation features of the parameters In *Advances in Spatial Modelling and Methodology: Essays in Honor of Jean Paelinck*, edited by D. Griffith and C. Amrhein, 45-56. Dordrecht: Kluwer.

Grondona, M., and N. Cressie. 1991. Using spatial considerations in the analysis of experiments *Technometrics* 33: 381-92.

Guyon, X. 1995. *Random Fields on a Network: Modeling, Statistics, and Applications*. New York: Springer-Verlag.

Haan, C. 1977. *Statistical Methods in Hydrology*. Ames: Iowa State University Press.

Hahn, G., and W. Meeker. 1993. Assumptions for statistical inference *American Statistician* 47: 1-11.

Haining, R. 1978. *Specification and Estimation Problems in Models of Spatial Dependence*. Evanston, Ill: Studies in Geography no. 24 Northwestern University.

———. 1981. An approach to the statistical analysis of clustered map pattern In *Dynamic Spatial Models*, edited by D. Griffith and R. MacKinnon, 288-317. Alphen aan den Rijn, The Netherlands: Sijthoff and Noordhoff.

———. 1990. *Spatial Data Analysis in the Social and Environmental Sciences*. Cambridge, U.K.: Cambridge University Press.

———. 1991. Bivariate correlation and spatial data *Geographical Analysis* 23: 210-27.

Haining, R., D. Griffith, and R. Bennett. 1983. Simulating two-dimensional autocorrelated surfaces *Geographical Analysis* 15: 247-55.

———. 1984. A statistical approach to the problem of missing spatial data using a first-order Markov model *Professional Geographer* 36: 338-45.

———. 1989. Maximum likelihood estimation with missing spatial data and with an application to remotely sensed data *Communications in Statistics: Theory and Methods* 18: 1875-94.

Haining, R., J. Ma, and S. Wise. 1996. Design of a software system for interactive spatial statistical analysis linked to a GIS *Computational Statistics* 11: 449-66.

Hair, J., R. Anderson, R. Tatham, and W. Black. 1995. *Multivariate Data Analysis: With Readings* 4th ed. Englewood Cliffs, N.J.: Prentice Hall.

Haith, D. 1976. Land use and water quality in New York rivers *J. of the Environmental Engineering Division* (ASCE) 102: 1-15.

Hamilton, L. 1992. *Regression with Graphics: A Second Course in Statistics*. Pacific Grove, Calif.: Brooks/Cole.

Hamilton, T., I. Rubinoff, R. Barth, and G. Bush. 1963. Species abundance: Natural regulation of insular variation *Science* 142: 1575-77.

Hancock, J., and D. Griffith. 1996. *Understanding Spatial Autocorrelation (USA): A Tutorial for Spatial Statistics*

http://www.geocities.com/CapeCanaveral/4203/USA.html.
Hand, D., F. Daly, A. Lunn, K. McConway, and E. Ostrowski. 1994. *A Handbook of Small Data Sets*. New York: Chapman & Hall.
Harper, W., and J. Furr. 1986. *Geostatistical Analysis of Potentiometric Data in the Wolfcamp Aquifer of the Palo Duro Basin, Texas* Technical Report BMI/ONWI-587. Columbus, Ohio: Battelle Memorial Institute.
Heuvelink, G. 1993. *Error Propagation in Quantitative Spatial Modelling*. Utrecht: Universitiet Utrecht, Faculteit Ruimteliijke Wetenschappen.
Hicks, D., M. McSaveney, and T. Chinn. 1990. Sedimentation in proglacial Ivory Lake, Southern Alps, New Zealand *Arctic and Alpine Research* 22: 26-42.
Hills, M., and the M345 Course Team. 1986. S*tatistical Methods, Unit 3: Examining Straight-line Data*. Milton Keynes, U.K.: The Open University.
Hinkley, D. 1977. On a quick choice of power transformation *Applied Statistics* 26: 67-69.
Hird, J. 1994. *Superfund: The Political Economy of Environmental Risk*. Baltimore, Md: Johns Hopkins University Press.
Hoaglin, D., F. Mosteller, and J. Tukey, eds. 1983. *Understanding Robust and Exploratory Data Analysis*. New York: Wiley.
Hubert, L., R. Golledge, C. Constazo, and N. Gale. 1985. Measuring association between spatially defined variables: An alternative procedure *Geographical Analysis* 17: 36-46.
IBM. 1991. Exploring new worlds with GIS *Directions* (Summer/Fall): 12-19.
Isaaks, E., and R. Srivastava. 1989. *An Introduction to Applied Geostatistics*. Oxford, U.K.: Oxford University Press.
John, P. 1971. *Statistical Design and Analysis of Experiments*. New York: Macmillan.
Johnson, D., K. McDade, and D. Griffith. 1996. Seasonal variation in pediatric blood lead levels in Syracuse, New York *Environmental Geochemistry and Health* 18 (no. 2): 81-88.
Johnson, M., and P. Raven. 1970. Natural regulation of plant species diversity Evolutionary Biology 4: 127-62.
———. 1973. Species number and endemism: The Galapagos revisited *Science* 179: 893-95.
Johnson, R., and D. Wichern. 1992. *Applied Multivariate Statistical Analysis*. Englewood Cliffs, N.J.: Prentice Hall.
Jones, K., and N. Wrigley. 1995. Generalized additive models, graphical diagnostics, and logistic regression *Geographical Analysis* 27: 1-21.
Journel, A., and R. Froidevaux. 1982. Anisotropic hole-effect modeling *International Association for Mathematical Geology J.* 14: 217-39.
Kahn, M. 1997. Particulate pollution trends in the United States *Regional Science and Urban Economics* 27: 87-107.
Kaiser, M., and N. Cressie. 1997. Modeling Poisson variables with positive spatial dependence *Statistics and Probability Letters*, forthcoming.
Kemp, K., and R. Wright. 1997. UCGIS identifies GIScience education priorities Manuscript submitted to *GeoInfoSystems*.
Kennedy, S. 1988. A geographic regression model for medical statistics *Social Science Medicine* 26: 119-29.
King, P., and P. Smith. 1988. Generation of correlated properties in heterogeneous porous media *Mathematical Geology* 20: 863-77.
Kitanidis, P. 1997. *Introduction to GEOSTATISTICS: Applications in Hydrogeology*. New

York: Cambridge University Press.
Kotz, S., N. Johnson, and C. Bread, eds. 1982. *Encyclopedia of Statistical Sciences*. Vol. 1, *A-Circular Probable Error*. New York: Wiley.
Kuensch, H. 1985. Statistical analysis of uniformity trials based on parametric models Manuscript, Fachgruppe fur Statistik, ETH, Zurich.
Larsen, R., and M. Marx. 1986. *An Introduction to Mathematical Statistics and Its Applications*. Englewood Cliffs, N.J.: Prentice-Hall.
Layne, L., A. Sone, and D. Griffith. 1997. Statistical models for spatial data: Some communalities and useful implementation suggestions Manuscript submitted to *American Statistician*.
Legendre, P. 1993. Spatial autocorrelation: Trouble or new paradigm? *Ecology* 74: 1659-73.
Leonte, D., and N. Schofield. 1996. Evaluation of a soil contaminated site and clean-up criteria: A geostatistical approach In *Geostatistics for Environmental and Geotechnical Applications*, edited by R. Srivastava, S. Rouhani, M. Cromer, A. Johnson, and A. Desbarats, 133-45. STP 1283. West Conshohocken, Pa.: American Society for Testing and Materials.
Levine, N. 1996. Spatial statistics and GIS: Software tools to quantify spatial patterns *J. of the American Planning Association* 62: 381-91.
Little, R., and D. Rubin. 1987. *Statistical Analysis with Missing Data*. New York: Wiley.
Lloyd, M. 1967. Mean crowding *J. of Animal Ecology* 36: 1-30.
Lyons, R. 1980. A review of multidimensional scaling Master's thesis, University of Reading.
McBratney, A., and R. Webster. 1981. Detection of ridge and furrow patterns by spectral analysis of crop yield *International Statistical Review* 49: 45-52.
McDade, K. 1994. Factors that relate to children at high risk for elevated blood lead levels in Syracuse, New York Master's thesis, Department of Chemistry, State University of New York College of Environmental Science and Forestry, Syracuse, N.Y.
McKechnie, S., P. Ehrlich, and R. White. 1975. Population genetics of *Euphydryas* butterflies. I. Genetic variation and the neurality hypothesis *Genetics* 81: 571-94.
McLachlan, G. 1997. *The EM Algorithm and Extensions*. New York: Wiley.
Manly, B. 1991. *Randomization and Monte Carlo Methods in Biology*. London: Chapman & Hall.
———. 1994. *Multivariate Statistical Methods: A Primer* 2d ed. New York: Chapman & Hall.
Mantel, N. 1967. The detection of disease clustering and a generalized regression approach *Cancer Research* 27: 209-20.
Mardia, K. 1984. Spatial discrimination and classification maps *Communications in Statistics: Theory and Methods* 13: 2181-97.
———. 1988. Multi-dimensional multivariate Gaussian Markov random fields with applications to image processing *J. of Multivariate Analysis* 24: 265-84.
Mardia, K., and A. Watkins. 1989. On multimodality of the likelihood in the spatial linear model *Biometrika* 76: 289-95.
Martin, R. 1990. The role of spatial statistical process in geographic modelling In *Spatial Statistics: Past, Present, and Future*, edited by D. Griffith, 109-27. Ann Arbor, Mich.: Institute of Mathematical Geography.
Martin, R., and D. Griffith. 1995. Some results on graph spectra, with applications to geographic spatial modelling, Research Report No. 461/95, Department of

Probability and Statistics, University of Sheffield.
Matheron, G., and M. Armstrong, eds. 1987. *Geostatistical Case Studies*. Dordrecht: Reidel.
MathSoft. 1995. *MathCad User's Guide*. Cambridge, Mass.: MathSoft Inc.
Matui, I. 1968. Statistical study of the distribution of scattered villages in two regions of the Tonami Plain, Toyama Prefecture In *Spatial Analysis: A Reader in Statistical Geography*, edited by B. Berry and D. Marble, 149-58. Englewood Cliffs, N.J.: Prentice Hall.
Meade, R. 1967. A mathematical model for the estimation of interplant competition *Biometrics* 23: 189-205.
Mercer, W., and A. Hall. 1911. The experimental error of field trials *J. of Agricultural Science* (Cambridge) 4: 107-32.
Messer, J., R. Linthurst, and W. Overton. 1991. An EPA program for monitoring ecological status and trends *Environmental Monitoring and Assessment* 17: 67-78.
Montgomery, D., and E. Peck. 1982. *Introduction to Linear Regression Analysis*. New York: Wiley.
Moore, D. 1991. *Statistics: Concepts and Controversies* 3d ed. New York: Freeman.
Morrison, A., A. Getis, M. Santiago, J. Rigau-Perez, and P. Reiter. 1998. Exploratory space-time analysis of reported dengue cases during an outbreak in Florida, Puerto Rico, 1991-1992 *American J. of Tropical Medicine and Hygiene*, 58 (3): 287-98.
Mowrer, H., R. Czaplewski, and R. Hamre, eds. 1996. *Spatial Accuracy in Natural Resource and Environmental Sciences: Second International Symposium*. General Technical Report RM-GTR-277. Ft. Collins, Colo.: Rocky Mountain Forest and Range Experiment Station.
Myers, D., and A. Journel. 1990. Variograms with zonal anisotropies and non-invertible kriging systems *Mathematical Geology* 22: 779-85.
National Research Council, Mapping Science Committee; Commission on Physical Sciences, Mathematics, and Resources. 1990a. *Spatial Data Needs: The Future of the National Mapping Program*. Washington, D.C.: National Academy Press.
———, Board on Mathematical Sciences. 1990b. *Renewing U.S. Mathematics: A Plan for the 1990s*. Washington, D.C.: National Academy Press.
———, Panel on Spatial Statistics and Image Processing. 1991. *Spatial Statistics and Digital Image Analysis*. Washington, D.C.: National Academy Press.
———, Committee on Applied and Theoretical Statistics. 1996. *Massive Data Sets: Proceedings of a Workshop*. Washington, D.C.: National Academy Press.
———, Rediscovering Geography Committee. 1997. *Rediscovering Geography: New Relevance of Science and Society*. Washington, D.C.: National Academy Press.
National Science Foundation. 1987. *Solicitation: National Center for Geographic Information and Analysis*. Washington, D.C.: NSF Biological, Behavioral, and Social Sciences Directorate.
Navidi, W. 1997. A graphical illustration of the EM algorithm *American Statistician* 51: 29-31.
Neter, J., Wasserman, W., and M. Kutner. 1996. *Applied Linear Regression Models* 4th ed. Homewood, Ill: Irwin.
Newman, J., ed. 1956. *The World of Mathematics*. New York: Simon and Schuster.
Newsome, T., F. Aranguren, and R. Brinkmann. 1997. Lead contamination adjacent to

roadways in Trujillo, Venezuela *Professional Geographer* 49: 331-41.
Nyman, L. 1997. GIS emerges in public health *GIS World* 10 (no. 9): 86.
Ogata, Y., and M. Tanemura. 1985. Estimation of interaction potentials of marked spatial point patterns through the maximum likelihood method *Biometrics* 41: 421-33.
Okabe, A., Boots, B., and K. Sugihara. 1992. *Spatial Tessellations: Concepts and Applications of Voronoi Diagrams.* New York: Wiley.
Olea, R. 1991. *Geostatistical Glossary and Multilingual Dictionary.* New York: Oxford University Press.
Oliver, M., and R. Webster. 1986. Semi-variograms for modelling the spatial pattern of landform and soil properties *Earth Surface Processes and Landforms* 11: 491-504.
Ord, J. 1975. Estimating methods for models of spatial interaction *J. of the American Statistical Association* 70: 120-26.
Ott, W. 1995. *Environmental Statistics and Data Analysis.* Boca Raton, Fla.: CRC Press.
Owen, D. 1988. The starship *Communications in Statistics: Simulation and Computation* 17: 315-23.
Owen, D., and H. Li. 1988. The starship for point estimates and confidence intervals on a mean and for percentiles *Communications in Statistics: Simulation and Computation* 17: 325-41.
Pace, R., and R. Barry. 1997. Quick computation of spatial autoregressive estimators *Geographical Analysis* 29: 232-47.
Pickle, L., M. Mungiole, G. Jones, and A. White. 1996. *Atlas of United States Mortality.* Hyattsville, Md: National Center for Health Statistics, Centers for Disease Control and Prevention, U.S. Department of Health and Human Services.
Platt, W., G. Evans, and S. Rathbun. 1988. The population dynamics of a long-lived conifer (*Pinus palustris*) *American Naturalist* 131: 491-525.
Press, W., B. Flannery, S. Teukolsky, and W. Vetterling. 1986. *Numerical Recipes: The Art of Scientific Computing.* New York: Cambridge University Press.
Rasmussen, S. 1992. *An Introduction to Statistics with Data Analysis.* Pacific Grove, Calif.: Brooks/Cole.
Rathbun, S. 1995. *Spatial Modeling in Irregularly Shaped Regions: Kriging Estuaries* Technical Report STA 95-1. Athens: Department of Statistics, University of Georgia.
Rayner, A. 1969. *A First Course in Biometry for Agriculture Students.* Pietermaritzburg, South Africa: University of Natal Press.
Rey, S., A. Getis, and A. Bortman. 1994. Estimation of suppressed employment data: An evaluation of spatial and non-spatial approaches, paper presented to the annual meeting of the Regional Science Association International, Niagara Falls, Ontario, Canada, November.
Reynolds, H., and C. Amrhein. 1998. Some effects of spatial aggregation on multivariate regression parameters In *Econometric Advances in Spatial Modelling and Methodology: Essays in Honour of Jean Paelinck,* edited by D. Griffith, C. Amrhein, and J.-M. Huriot, 85-106. Dordrecht: Kluwer.
Richardson, S. 1990. Some remarks on the testing of association between spatial processes In *Spatial Statistics: Past, Present, and Future,* edited by D. Griffith, 277-309. Ann Arbor, Mich.: Institute of Mathematical Geography.
Richardson, S., and D. Hémon. 1981. On the variance of the sample correlation between two independent lattice processes *J. of Applied Probability* 18: 943-48.

Ripley, B. 1981. *Spatial Statistics*. New York: Wiley.
———. 1988. *Statistical Inference for Spatial Processes*. Cambridge, U.K.: Cambridge University Press.
———. 1990. Gibbsian interaction models In *Spatial Statistics: Past, Present, and Future*, edited by D. Griffith, 3-53. Ann Arbor, Mich.: Institute of Mathematical Geography.
Rouhani, S., and D. Myers. 1990. Problems in space-time kriging of hydrogeological data *Mathematical Geology* 22: 611-23.
Russo, D. 1984. Design of an optimal sampling network for estimating the variogram *Soil Science Society of America J.* 48: 708-16.
Schmoyer, R. 1994. Permutation tests for correlation in regression errors *J. of the American Statistical Association* 89: 1507-16.
Sen, A., and M. Srivastava. 1990. *Regression Analysis: Theory, Methods, and Applications*. Berlin: Springer-Verlag.
Shaw, M. 1990. A test of spatial randomness on small scales, combining information from mapped locations within several quadrats *Biometrics* 46: 447-58.
Simpson, D., D. Ruppert, and R. Carroll. 1989. *One-Step GM-Estimates For Regression With Bounded Influence and High Breakdown-Point* Technical Report no. 859. Ithaca, N.Y.: School of Operations Research and Industrial Engineering, Cornell University.
Sokal, R., and F. Rohlf. 1981. *Biometry* 2d ed. San Francisco: Freeman.
Spiegel, M. 1968. *Mathematical Handbook of Formulas and Tables*. New York: McGraw-Hill.
Stehman, S., and W. Overton. 1996. Spatial sampling In *Practical Handbook of Spatial Statistics*, edited by. S. Arlinghaus, 31-63. Boca Raton, Fla.: CRC Press.
Stein, M., and M. Handcock. 1989. Some asymptotic properties of kriging when the covariance function is misspecified *Mathematical Geology* 21: 171-90.
Strauss, D. 1992. The many faces of logistic regression *American Statistician* 46: 321-27.
Streitberg, B. 1979 Multivariate models of dependent spatial data In *Exploratory and Explanatory Statistical Analysis of Spatial Data*, edited by C. Bartels and R. Ketallapper, 139-77. The Hague: Martinus Nijhoff.
Sullivan, J. 1996. Understanding the degrees of freedom concept by computer experiments *American Statistician* 50: 234-37.
Switzer, P., and A. Green. 1984. *MIN/MAX Autocorrelation Factors for Multivariate Spatial Imagery* Technical Report no. 6. Stanford University: Department of Statistics.
Symons, M., R. Grimson, and Y. Yuan. 1983. Clustering of rare events *Biometrics* 39: 193-205.
Thiessen, A. 1911. Precipitation averages for large areas *Monthly Weather Review* 39: 1082-84.
Thomas, E. 1968. Maps of residuals from regression: Their characteristics and uses in geographic research In *Spatial Analysis: A Reader in Statistical Geography*, edited by B. Berry and D. Marble, 326-52. Englewood Cliffs, N.J.: Prentice Hall.
Tiefelsdorf, M., and B. Boots. 1995. The exact distribution of Moran's I *Environment and Planning A* 27: 985-99.
Tietjen, G. 1986. *A Topical Dictionary of Statistics*. New York: Chapman & Hall.
Tobler, W. 1970. A computer movie simulating urban growth in the Detroit region *Economic Geography* 46 (supplement): 234-40.

Tomerlin, C. 1998. From the editor: It's time to focus on imagery data *Business Geographics* 6 (no. 3): 8.
Upton. G. 1990. Information from regional data In *Spatial Statistics: Past, Present, and Future*, edited by D. Griffith, 315-59. Ann Arbor, Mich.: Institute of Mathematical Geography.
Upton, G., and B. Fingleton. 1985. *Spatial Data Analysis by Example: Point Pattern and Quantitative Data* Vol. 1. New York: Wiley.
———. 1989. *Spatial Data Analysis by Example: Categorical and Directional Data* Vol. 2. New York: Wiley.
U.S. Department of Agriculture. 1954-92 (selected census years). *Census of Agriculture: Part 52, Puerto Rico*. Washington, D.C.: U.S. Department of Commerce, Bureau of the Census.
U.S. Environmental Protection Agency. 1987-90. *Toxic Release Inventory*. Washington, D.C.: Office of Toxic Substances, U.S. EPA.
———. 1991. *TRI Location Data Quality Assurance for Geographic Information Systems*. Washington, D.C.: Office of Toxic Substances, U.S. EPA.
———. 1992. *Environmental Equity—Reducing Risk for All Communities*. Washington, D.C.: Office of Toxic Substances, U.S. EPA.
Usmani, R. 1987. *Applied Linear Algebra*. New York: Marcel Dekker.
van Gastel, R., and J. Paelinck. 1995. Computation of Box-Cox transformation parameters: A new method and its application to spatial econometrics In *New Directions in Spatial Econometrics*, edited by L. Anselin and R. Florax, 136-55. Berlin: Springer-Verlag.
Venables, W., and B. Ripley. 1994. *Modern Applied Statistics with S-Plus*. Berlin: Springer-Verlag.
Wackernagel, H. 1995. *Multivariate Geostatistics: An Introduction with Applications*. New York: Springer-Verlag.
Walford, N. 1995. *Geographical Data Analysis*. New York: Wiley.
Waller, L., B. Carlin, H. Xia, and A. Gelfand. 1997. Hierarchical spatio-temporal mapping of disease rates *J. of the American Statistical Association* 92: 607-17.
Walter, S. 1992. The analysis of regional patterns in health data: I. Distributional considerations; II. The power to detect environmental effects *American J. of Epidemiology* 136: 730-41 & 742-59.
Warnecke, L. 1990. GIS in the states: Applications abound *GIS World* 3 (no. 3): 54-58.
———. 1991. *State Geographic Information Activities Compendium*. Lexington, Ky.: Council of State Governments.
Wartenberg, D. 1985. Multivariate spatial correlation: A method for exploratory geographical analysis *Geographical Analysis* 17: 263-83.
Watson, D. 1992. *Contouring: A Guide to the Analysis and Display of Spatial Data*. New York: Pergamon Press.
Welty, E. 1984. *One Writer's Beginnings*. Cambridge, Mass.: Harvard University Press.
White, J., R. Welch, and W. Norvell. 1997. Soil zinc map of the USA using geostatistics and geographic information systems *Soil Science Society of America J.* 61: 185-94.
Whittle, P. 1954. On stationary processes in the plane *Biometrika* 41: 434-49.
Wiebe, G. 1935. Variation and correlation among 1500 wheat nursery plots *J. of Agricultural Research* 50: 331-57.
Wilkinson, G., S. Eckert, T. Hancock, and O. Mayo. 1983. Nearest neighbor (NN) analysis

with field experiments *J. of the Royal Statistical Society* 45B: 151-78.
Wilson, D. 1996. New technologies revitalize the ancient field of geography *Chronicle of Higher Education* 62 (November 29): A23-A24.
Yates, F. 1935. Complex experiments *J. of the Royal Statistical Society* 2 (supplement): 181-247.
Zhang, Z., and D. Griffith. 1997. Developing user-friendly spatial statistical analysis modules for GIS: An example using ArcView *Computers, Environment and Urban Systems* 21: 5-29.
Zimmerman, D., and M. Zimmerman. 1991. A comparison of spatial semivariogram estimators and corresponding ordinary kriging predictors *Technometrics* 33: 77-91.
Zirschky, J., M. ASCE and D. Harris. 1986. Geostatistical analysis of hazardous waste site data *J. of Environmental Engineering* 112: 770-84.

SUBJECT INDEX

AIC (Akaike's information criterion)
37, 137, 470-473

anisotropy
103, 138-142, 153, 238, 307, 309-311, 313-314, 316, 351, 471-472

ANOVA (analysis of variance)
247, 256, 260, 366, 369, 441-444, 456

associations
bivariate
11, 18-19, 21, 24, 36, 74, 80, 94, 140, 211, 266, 308, 312, 338
multivariate
6, 21-24, 69, 84-85, 90, 93, 111, 120, 153, 309, 430, 482
observation indices
30, 88, 89

bootstrap 105, 108-110

cartoons 484-486

computer code
SAS
39, 41-51, 53, 55-60,62-63,67-68,155-158, 165-168, 172-174, 178-192, 431, 433-434, 436, 438-439
SPSS
169-171, 175-177, 193-204

cross-validation
37, 137, 227, 241, 243, 430, 444, 449, 451-456, 466-467, 470, 475, 480

diagnostic statistics
A-D (Anderson-Darling)
72, 76, 77, 104, 208, 215-216, 219, 224-225, 128, 131, 208-209, 211, 216, 220, 225, 237-238, 242, 250, 253, 257, 261, 283, 286, 290, 318, 321-329, 332-333, 335, 340-341, 343-346, 358, 362, 366, 370, 375, 379, 384, 387, 389, 392, 400-401, 407, 412, 415, 435, 438
Bartlett
81, 208-209, 211, 215-216, 219-220, 224, 225, 237-238, 242, 250, 253, 257, 261, 283, 286, 290, 318, 321, 323-329, 332-333, 335, 340-341, 343-346, 358, 362, 366, 370, 375, 379, 384, 387, 389, 392, 400-401, 407, 412, 415, 437, 438, 442-443, 448
GR (Geary Ratio)
15-16, 29, 34, 88, 103-104, 112, 114, 117-118, 130-131, 133, 208-209, 211, 215-216, 219-220, 224-225, 237-238, 243, 250, 253, 257, 261, 283, 286, 290, 318, 321, 323-329, 332-333, 335, 340-341, 343-346, 358, 362, 366, 370, 375, 379, 384, 387, 389, 392, 400-401, 407-408, 412, 415, 435, 437, 442-443, 481
K-S (Kolmogorov-Smirnov)
72, 76-77, 104, 208-209, 211, 215-216, 219-220, 224-225, 128, 131, 237-238, 242, 250, 253, 257, 261, 283, 286, 290, 318, 321-329, 332-333, 335, 340-341, 343-346, 358, 362, 366, 370, 375, 379, 384, 387, 389, 392, 400-401, 407, 412, 415, 435

504 Subject Index

Levene
  81, 208-209, 211, 215-216, 219-220, 224, 225, 237-238, 242, 250, 253, 257, 261, 283, 286, 290, 318, 321, 323-329, 332-333, 335, 340-341, 343-346, 358, 362, 366, 370, 375, 379, 384, 387, 389, 392, 400-401, 407, 412, 415, 437, 438, 442-443, 448
MC (Moran Coefficient)
  11-12, 15-16, 24-25, 29, 31, 61, 88, 100, 103-110, 112-114, 117-118, 121-122, 130, 133, 161, 208-209, 211, 215-216, 219-220, 224-225, 237-238, 243, 250, 253, 257, 261, 283, 286, 290, 318, 321, 323-329, 332-333, 335, 340-341, 343-346, 358, 362, 366, 370, 375, 379, 384, 387, 389, 392, 400-401, 407-408, 412, 415, 435, 437, 442-443, 479, 481
R-J (Ryan-Joiner)
  76-77, 104, 208-209, 211, 215-216, 219-220, 224-225, 128, 131, 237-238, 242, 250, 253, 257, 261, 283, 286, 290, 318, 321-329, 332-333, 335, 340-341, 343-346, 358, 362, 366, 370, 375, 379, 384, 387, 389, 392, 400-401, 407, 412, 415, 435
S-W (Shapiro-Wilk)
  72, 75-77, 82-83, 208-209, 211, 215-216, 219-220, 224-225, 237-238, 242, 250, 253, 257, 261, 283, 286, 290, 318, 321-329, 332-333, 335, 340-341, 343-346, 358, 362, 366, 370, 375, 379, 384, 387, 389, 392, 400-401, 407, 412, 415, 432-433, 436, 442-443, 448

distance clustering
  40, 47-48, 141, 319, 408, 450, 466, 472, 479

eigenfunctions
  25, 118, 130, 144, 153, 475-476
approximations
  130
eigenvalues
  12, 22-23, 25-27, 54, 85, 103, 110, 122-123, 125, 128-130, 134, 141, 143-144, 152-155, 159, 161-162, 297, 299, 304, 307, 419, 421
eigenvectors
  12, 25-26, 28-29, 32, 100, 103-105, 109-110, 117-118, 125, 130-133, 152, 229-230, 233, 293-295, 299, 304, 399-402, 419-421, 424, 476, 478-480, 483

geographic connectivity/contiguity/spatial weights matrix
  5, 29, 54, 110, 118, 143, 149, 155, 159-160, 162, 207, 262, 294, 297, 406, 439, 468, 483

GIS (geographic information system)
  39, 54, 235, 355, 476-477, 479

graphics
  MC scatterplot
    17, 30, 34, 61, 86-87, 104, 131, 209, 212, 216, 221, 224, 231, 247, 262, 267, 274, 286, 294, 298, 313, 317-318, 322, 333, 340, 358, 363, 380, 384, 389, 392, 418, 420, 437
  semiovariogram plot
    34, 36, 92, 95, 96, 98, 99, 101-102, 104, 209, 213, 217, 221, 226, 231, 239-240, 245, 248, 251, 254, 259, 264, 268-269, 272, 275, 277, 282, 284, 287, 289, 296, 301, 309-310, 314-315, 319, 336, 348-349, 360, 365, 368, 372, 377, 382, 385, 388, 390, 393, 403-404, 409, 414, 417, 424, 465

linkages between geostatistics and saptial autoregression
  6-7, 9, 37, 120, 122-124, 133, 145-148, 151, 153, 232-233, 303, 394-395, 425, 469

map 7, 28, 32-33, 94, 115-117, 133, 147-148, 150, 159, 219, 228, 236, 241, 246, 248, 252, 255, 265, 273, 276, 300, 311, 316, 356, 361, 374, 379, 381, 391, 418, 422, 448, 450-451

missing data estimation
  kriging
    9, 13, 36, 42, 89-91, 133-134, 140, 207, 304, 428-430, 438, 440-441, 444-445, 447-454, 456, 461, 463, 470, 473-476, 478, 481-483
  spatial prediction
    7-8, 13, 120, 338, 347, 355, 361-362, 365, 430, 455-456, 459, 461

model estimation
  auto-normal/Gaussian (autocorrelation parameter ρ)
    111, 115-116, 144, 411, 477
  AR (autoregressive response)
    122, 124, 126, 128, 154, 160-161,

164, 208-210, 214-218, 237-238, 253-254, 261, 283, 363, 383, 387, 392, 445-447, 456
CAR (conditional autoregressive)
  8, 122-125, 142, 144, 146-151, 153-155, 160-162, 307-309, 311-312, 314-315, 321-329, 363, 380, 382, 396, 439-440
  due to misspecification
    13-14, 123, 136, 212, 233, 482
  normalizing constant (Jacobian) approximation
    70-71, 122-125, 128-130, 141, 153, 159-162, 209, 214, 297, 304, 307, 400, 423, 473, 475-476
  SAR (simultaneous autoregressive)
    8, 37, 108, 122-124, 126-128, 142, 144-146, 148-149, 151, 153-154, 160-161, 164, 225-226, 237-238, 242, 250, 283, 285-287, 290, 293-294, 299, 307, 317-318, 330-335, 338, 340-347, 358, 362-363, 375-376, 389, 415, 438-441, 444-445, 455, 472
  spatial filtered OLS
    229-230, 383, 400
auto-binomial (logistic)
  111, 130, 132, 232-233, 244, 304, 411, 423
auto-Poisson
  111, 130, 132, 304, 398, 402, 423-425, 476, 478, 480
semivariogram (nugget, range, sill)
  37
  Bessel function
    36-38, 123, 134-136, 139, 145-149, 151, 153, 162, 210, 213, 218, 222, 226-227, 232, 239, 243, 249, 251, 255, 258, 263, 284, 287, 291, 295, 300, 308, 313, 319-320, 337, 347-349, 359, 364, 367-368, 371, 378, 382, 386-387, 390, 393, 403, 405, 408, 414, 417, 422, 444, 447, 449, 452, 466-468
  Cauchy
    462, 464, 466-467
  circular
    36, 134, 462, 464, 466-468
  cubic
    462, 464, 466-467
  De Wijsian
    462-463
  exponential
    36, 134-135, 162, 210, 213, 218, 222, 226-227, 232, 239, 243, 249, 251, 255, 258, 263, 284, 287, 291, 295, 300, 308, 313, 319-320, 337, 347-349, 359, 364, 367-368, 371, 378, 382, 386-387, 390, 393, 403, 405, 408, 414, 417, 422, 444, 447, 449, 452, 466-468, 471-472
  Gaussian
    36, 134-136, 140, 162, 210, 213, 218, 222, 226-227, 232, 239, 243, 249, 251, 255, 258, 263, 284, 287, 291, 295, 308, 313, 319-320, 337, 347-349, 359, 364, 367-368, 371, 378, 382, 386-387, 390, 393, 403, 405, 409, 414, 417, 422, 444, 447, 449, 452, 466-468, 471-472
  linear
    463
  penta-spherical
    462, 464, 466-467
  power
    134-135, 162, 210, 213, 18, 222, 226-227, 232, 239, 243, 249, 251, 255, 258, 263, 284, 287, 291, 295, 309, 313, 319-320, 337, 348-349, 359, 364, 367-368, 371, 378, 382, 386-387, 390, 393, 403, 405, 409, 414, 417, 422, 444, 447, 449, 452, 466-467, 471-472
  rational quadratic
    134, 462, 464, 466-468
  spherical
    36, 134, 210, 213, 18, 222, 226-227, 232, 239, 243, 249, 251, 255, 258, 263, 284, 287, 291, 295, 300, 308, 313, 319-320, 337, 347-349, 359, 364, 367-368, 371, 378, 382, 386-387, 390, 393, 403, 405, 409, 414, 417, 422, 444, 447, 449, 452, 466-467, 471-472
  stable
    462, 464, 466-468
  wave/hole
    134, 136, 162, 210, 213, 18, 222, 226-227, 232, 239, 243, 249, 251, 255, 258, 263, 284, 287, 291, 295, 309, 313, 319-320, 337, 348-349, 359, 364, 367-368, 371, 378, 382, 386-387, 390, 393, 403, 405, 409, 414, 417, 422, 444, 447, 449, 452, 467-468

non-normal distributions
  binomial
    111, 113-114, 117, 131, 229, 244
  Poisson
    90, 111-113, 132-133, 256, 399-401, 406-408, 418-424, 478

OLS (ordinary least squares)
  13, 42, 93, 105, 123, 126, 128, 134, 136, 161, 208, 210, 214-217, 219-222, 224-227, 230-231, 237-240, 242-243, 245, 248, 250-251, 253-255, 257, 259, 261-262, 264, 283, 286-287, 290, 294, 308-310, 313-317, 331, 333-334, 340-342, 344-346, 357, 362, 375, 379, 383, 387, 389, 392, 400-401, 407, 413, 415, 420, 422, 438, 444-447, 449, 452, 472

outlier
  30, 40-41, 77, 79, 82, 88, 91, 93-96, 98-100, 211, 215, 218, 222, 224, 227, 237-239, 246, 253, 287, 290, 293, 302, 357-359, 362-363, 366, 375-376, 379-381, 383-384, 386, 388, 395, 398-402, 404, 406-408, 412-413, 419, 421-424, 431-432, 435, 441, 455, 459-460, 471, 479

spatial
  15, 18, 31, 35, 94-95, 98, 224, 286, 382, 383

simulation
  72, 105-109

spectral density function
  142, 146-151, 153

transformations
  14, 75-77, 80-84, 86, 88-89, 100, 107, 109, 214-216, 218-220, 224, 228-229, 237, 242-244, 246-247, 249, 252, 255-256, 260-261, 285, 289, 293, 296-297, 307, 312, 317, 321, 326, 331-332, 334-335, 338-345, 357, 361-362, 365-366, 369-370, 383, 386, 388, 391, 411, 419, 423, 431-433, 436, 438-443, 448, 455, 460-461, 463, 466, 470, 476-478, 480
  Box-Cox
    70-71, 73-74, 79, 209, 211, 341, 374, 378, 461
  Box-Tidwell
    71, 74, 219, 331, 333-334, 341